VOLUME FIVE HUNDRED AND SIXTY NINE

METHODS IN ENZYMOLOGY

Intermediate Filament Associated Proteins

METHODS IN ENZYMOLOGY

VOLUME FIVE HUNDRED AND SIXTY NINE

METHODS IN ENZYMOLOGY

Intermediate Filament Associated Proteins

Edited by

KATHERINE L. WILSON
Department of Cell Biology,
Johns Hopkins University School of Medicine,
Baltimore, Maryland, USA

ARNOUD SONNENBERG
Division of Cell Biology,
The Netherlands Cancer Institute,
Amsterdam, The Netherlands

AMSTERDAM • BOSTON • HEIDELBERG • LONDON
NEW YORK • OXFORD • PARIS • SAN DIEGO
SAN FRANCISCO • SINGAPORE • SYDNEY • TOKYO
Academic Press is an imprint of Elsevier

Academic Press is an imprint of Elsevier
50 Hampshire Street, 5th Floor, Cambridge, MA 02139, USA
525 B Street, Suite 1800, San Diego, CA 92101–4495, USA
The Boulevard, Langford Lane, Kidlington, Oxford OX5 1GB, UK
125 London Wall, London, EC2Y 5AS, UK

First edition 2016

Notices

Knowledge and best practice in this field are constantly changing. As new research and experience broaden our understanding, changes in research methods, professional practices, or medical treatment may become necessary.

Practitioners and researchers must always rely on their own experience and knowledge in evaluating and using any information, methods, compounds, or experiments described herein. In using such information or methods they should be mindful of their own safety and the safety of others, including parties for whom they have a professional responsibility.

To the fullest extent of the law, neither the Publisher nor the authors, contributors, or editors, assume any liability for any injury and/or damage to persons or property as a matter of products liability, negligence or otherwise, or from any use or operation of any methods, products, instructions, or ideas contained in the material herein.

ISBN: 978-0-12-803469-9
ISSN: 0076-6879

For information on all Academic Press publications visit our website at http://store.elsevier.com/

CONTENTS

Part IV
Functional and Genetic Analysis of Lamin-Associated Proteins

CONTRIBUTORS

Noelia Alonso-García
Instituto de Biología Molecular y Celular del Cáncer, Consejo Superior de Investigaciones Científicas, University of Salamanca, Salamanca, Spain

Ruthellen H. Anderson
Sanford Children's Health Research Center, Sanford Research, Sioux Falls, South Dakota, USA

Christopher Arnette
Department of Pathology, Northwestern University Feinberg School of Medicine, Chicago, Illinois, USA

Peter Askjaer
Andalusian Center for Developmental Biology, CSIC-Junta de Andalucia-Universidad Pablo de Olavide, Carretera de Utrera, Seville, Spain

Petros Batsios
Institute for Biochemistry and Biology, Department of Cell Biology, University of Potsdam, Potsdam, Germany

Nadja Begré
Department of Clinical Research, University of Bern, and Department of Dermatology, Inselspital, Bern University Hospital, Bern, Switzerland

Jason M. Berk
Department of Cell Biology, Johns Hopkins University School of Medicine, Baltimore, Maryland, USA

Veronika Boczonadi
Institute of Genetic Medicine, Newcastle University, Newcastle-upon-Tyne, United Kingdom

Pascale Bomont
Atip-Avenir Team, and INM Inserm U1051, Université de Montpellier, Montpellier, France

Luca Borradori
Department of Clinical Research, University of Bern, and Department of Dermatology, Inselspital, Bern University Hospital, Bern, Switzerland

Jamal-Eddine Bouameur
Department of Clinical Research, University of Bern, and Department of Dermatology, Inselspital, Bern University Hospital, Bern, Switzerland

Benjamin Bourgeois
Institute for Integrative Biology of the Cell, CEA, CNRS, Université Paris-Sud, Gif-sur-Yvette, Paris, France

Brigitte Buendia
Institut de Biologie Fonctionnelle et Adaptative, Université Paris Diderot, Paris, France

Rubén M. Buey
Instituto de Biología Molecular y Celular del Cáncer, Consejo Superior de Investigaciones Científicas, and Metabolic Engineering Group, Department of Microbiology and Genetics, University of Salamanca, Salamanca, Spain

Arturo Carabias
Instituto de Biología Molecular y Celular del Cáncer, Consejo Superior de Investigaciones Científicas, University of Salamanca, Salamanca, Spain

Ana M. Carballido
Instituto de Biología Molecular y Celular del Cáncer, Consejo Superior de Investigaciones Científicas, University of Salamanca, Salamanca, Spain

Hee-Jung Choi
School of Biological Sciences, Seoul National University, Seoul, South Korea

Donald M. Coen
Department of Biological Chemistry and Molecular Pharmacology, Harvard Medical School, Boston, Massachusetts, USA

Victor E. Cruz
Department of Biology, Massachusetts Institute of Technology, Cambridge, Massachusetts, USA

José M. de Pereda
Instituto de Biología Molecular y Celular del Cáncer, Consejo Superior de Investigaciones Científicas, University of Salamanca, Salamanca, Spain

Agnieszka Dobrzynska
Andalusian Center for Developmental Biology, CSIC-Junta de Andalucia-Universidad Pablo de Olavide, Carretera de Utrera, Seville, Spain

F. Esra Demircioglu
Department of Biology, Massachusetts Institute of Technology, Cambridge, Massachusetts, USA

Bertrand Favre
Department of Clinical Research, University of Bern, and Department of Dermatology, Inselspital, Bern University Hospital, Bern, Switzerland

Ricardo A. Figueroa
Department of Neurochemistry, Stockholm University, Stockholm, Sweden

Radia Forteza
Department of Cell Biology, University of Miami Miller School of Medicine, Miami, Florida, USA

Peter Fuchs
Department of Biochemistry and Cell Biology, Max F. Perutz Laboratories, University of Vienna, Vienna Biocenter (VBC), Vienna, Austria

Christelle Gally
IGBMC, Development and Stem Cells Program, CNRS (UMR 7104)/INSERM (U964)/ Université de Strasbourg, Illkirch, France

Spiro Getsios
Department of Dermatology; Department of Cell and Molecular Biology, Northwestern University Feinberg School of Medicine, and Robert H. Lurie Comprehensive Cancer Center, Northwestern University, Chicago, Illinois, USA

María Gómez-Hernández
Instituto de Biología Molecular y Celular del Cáncer, Consejo Superior de Investigaciones Científicas, University of Salamanca, Salamanca, Spain

Dmitry Goryunov
Department of Pathology and Cell Biology, Taub Institute for Research on Alzheimer's Disease and the Aging Brain, Columbia University College of Physicians and Surgeons, New York, USA

Kathleen J. Green
Department of Pathology; Department of Dermatology, Northwestern University Feinberg School of Medicine, and Robert H. Lurie Comprehensive Cancer Center, Northwestern University, Chicago, Illinois, USA

Ralph Gräf
Institute for Biochemistry and Biology, Department of Cell Biology, University of Potsdam, Potsdam, Germany

Yosef Gruenbaum
Department of Genetics, The Alexander Silberman Institute of Life Sciences, The Hebrew University of Jerusalem, Jerusalem, Israel

Ines Hahn
Faculty of Life Sciences, Michael Smith Building, Manchester, United Kingdom

Einar Hallberg
Department of Neurochemistry, Stockholm University, Stockholm, Sweden

Jennifer C. Harr
Department of Biological Chemistry and Center for Epigenetics, Johns Hopkins University, Baltimore, MD, USA

Isaline Herrada
Institute for Integrative Biology of the Cell, CEA, CNRS, Université Paris-Sud, Gif-sur-Yvette, Paris, France

Paul Hoover
Department of Dermatology, Northwestern University Feinberg School of Medicine, Chicago, Illinois, USA

Yu-Shan Huang
Institute of Molecular Medicine, College of Life Sciences, National Tsing Hua University, Hsinchu, Taiwan

Arantza Infante
Stem Cells and Cell Therapy Laboratory, BioCruces Health Research Institute, Cruces University Hospital, Barakaldo, Spain

Mohammed Hakim Jafferali
Department of Neurochemistry, Stockholm University, Stockholm, Sweden

Jeremy P. Kamil
Department of Biological Chemistry and Molecular Pharmacology, Harvard Medical School, Boston, Massachusetts, and Department of Microbiology and Immunology, Louisiana State University Health Sciences Center, Shreveport, Louisiana, USA

Jennifer L. Koetsier
Department of Pathology, Northwestern University Feinberg School of Medicine, Chicago, Illinois, USA

Rashmi Kothary
Regenerative Medicine Program, Ottawa Hospital Research Institute; Department of Cellular and Molecular Medicine; Department of Medicine, University of Ottawa, and University of Ottawa Center for Neuromuscular Disease, Ottawa, Ontario, Canada

Michel Labouesse
IBPS, Paris, France

Ronald K.H. Liem
Department of Pathology and Cell Biology, Taub Institute for Research on Alzheimer's Disease and the Aging Brain, Columbia University College of Physicians and Surgeons, New York, USA

Anisha Lynch-Godrei
Regenerative Medicine Program, Ottawa Hospital Research Institute, and Department of Cellular and Molecular Medicine, University of Ottawa, Ottawa, Ontario, Canada

Arto Määttä
School of Biological and Biomedical Sciences, Durham University, Durham, United Kingdom

Alexandr A. Makarov
Wellcome Trust Centre for Cell Biology, University of Edinburgh, Edinburgh, United Kingdom

José A. Manso
Instituto de Biología Molecular y Celular del Cáncer, Consejo Superior de Investigaciones Científicas, University of Salamanca, Salamanca, Spain

Anastasia Mashukova
Department of Physiology, Nova Southeastern University, Ft. Lauderdale, and Department of Cell Biology, University of Miami Miller School of Medicine, Miami, Florida, USA

Aaron A. Mehus
Sanford Children's Health Research Center, Sanford Research, Sioux Falls, South Dakota, USA

Peter Meinke
Wellcome Trust Centre for Cell Biology, University of Edinburgh, Edinburgh, United Kingdom

Irene Meyer
Institute for Biochemistry and Biology, Department of Cell Biology, University of Potsdam, Potsdam, Germany

Esther Ortega
Instituto de Biología Molecular y Celular del Cáncer, Consejo Superior de Investigaciones Científicas, University of Salamanca, Salamanca, Spain

Ming-Der Perng
Institute of Molecular Medicine, College of Life Sciences, National Tsing Hua University, Hsinchu, Taiwan

Andreas Prokop
Faculty of Life Sciences, Michael Smith Building, Manchester, United Kingdom

Roy A. Quinlan
Biophysical Sciences Institute, University of Durham, Durham, United Kingdom

Karen L. Reddy
Department of Biological Chemistry and Center for Epigenetics, Johns Hopkins University, Baltimore, MD, USA

Günther A. Rezniczek
Department of Obstetrics & Gynecology, Marien Hospital Herne, Ruhr-Universität Bochum, Herne, Germany

Andrea Rizzotto
Wellcome Trust Centre for Cell Biology, University of Edinburgh, Edinburgh, United Kingdom

Clara I. Rodríguez
Stem Cells and Cell Therapy Laboratory, BioCruces Health Research Institute, Cruces University Hospital, Barakaldo, Spain

Matthew Ronshaugen
Faculty of Life Sciences, Michael Smith Building, Manchester, United Kingdom

Kyle J. Roux
Sanford Children's Health Research Center, Sanford Research, and Department of Pediatrics, Sanford School of Medicine, University of South Dakota, Sioux Falls, South Dakota, USA

Inés García Rubio
Centro Universitario de la Defensa, Academia General Militar, Zaragoza, Spain

Pedro J. Salas
Department of Cell Biology, University of Miami Miller School of Medicine, Miami, Florida, USA

Camille Samson
Institute for Integrative Biology of the Cell, CEA, CNRS, Université Paris-Sud, Gif-sur-Yvette, Paris, France

Eric C. Schirmer
Wellcome Trust Centre for Cell Biology, University of Edinburgh, Edinburgh, United Kingdom

Thomas U. Schwartz
Department of Biology, Massachusetts Institute of Technology, Cambridge, Massachusetts, USA

Mayuri Sharma
Department of Biological Chemistry and Molecular Pharmacology, Harvard Medical School, Boston, Massachusetts, USA

Natalia Sánchez-Soriano
Cellular and Molecular Physiology, Institute of Translational Medicine, University of Liverpool, Liverpool, United Kingdom

Sandra Szabo
Department of Biochemistry and Cell Biology, Max F. Perutz Laboratories, University of Vienna, Vienna Biocenter (VBC), Vienna, Austria

Gernot Walko
Centre for Stem Cells & Regenerative Medicine, Faculty of Life Sciences and Medicine, King's College London, London, United Kingdom

William I. Weis
Department of Structural Biology, and Department of Molecular & Cellular Physiology, Stanford University School of Medicine, Stanford, California, USA

Karl L. Wögenstein
Department of Biochemistry and Cell Biology, Max F. Perutz Laboratories, University of Vienna, Vienna Biocenter (VBC), Vienna, Austria

Gerhard Wiche
Department of Biochemistry & Cell Biology, Max F. Perutz Laboratories, University of Vienna, Vienna, Austria

Katherine L. Wilson
Department of Cell Biology, Johns Hopkins University School of Medicine, Baltimore, Maryland, USA

Lilli Winter
Institute of Neuropathology, University Hospital Erlangen, Erlangen, Germany

Howard J. Worman
Department of Medicine, and Department of Pathology and Cell Biology, College of Physicians and Surgeons, Columbia University, New York, USA

Huimin Zhang
Jiangsu Key Laboratory of Infection and Immunity, Institutes of Biology and Medical Sciences, Soochow University, Suzhou, Jiangsu, PR China

Sophie Zinn-Justin
Institute for Integrative Biology of the Cell, CEA, CNRS, Université Paris-Sud, Gif-sur-Yvette, Paris, France

PREFACE

Proteins that bind to intermediate filaments come in hundreds of sizes and "flavors." Some are structural giants that interlink the cytoskeleton. Others reside at the inner nuclear membrane, holding court with lamins and silent genes. Many more remain hidden from view, their potential contributions to cells or disease unknown. Studying such proteins requires a cutting-edge toolkit of methods and expert advice, both featured in this volume of *Methods in Enzymology*.

Roux and colleagues (Chapter 1) detail a powerful and broadly applicable method, biotinylation identification (BioID), to identify proteins located within 10 nm of a BioID-fusion protein in living cells. These "proximal" proteins include direct *in vivo* partners or associated protein complexes, as demonstrated for the nuclear intermediate protein, lamin A. Gräf and colleagues successfully modified the BioID toolkit for use in *Dictyostelium*, now enabling studies of intermediate filament-associated proteins in this evolutionarily diverse organism (Chapter 2). Taking "lamin association" to an entirely new level, Harr and Reddy detail their state-of-the-art new TCIS method to study how specific segments of DNA or chromatin-associated proteins drive intranuclear movement to lamins and the nuclear lamina (Chapter 21).

Proteins that bind lamins are exceedingly diverse. Schirmer and colleagues provide insights into the many biochemical and functional features that should be considered when expressing an emerging spectrum of lamin-binding nuclear membrane proteins as recombinant polypeptides in bacteria (Chapter 5). Successful methods for purification and structural analysis are provided for specific lamin-binding proteins. These include SUN-domain and KASH-domain proteins, which form LINC complexes that transduce force across the nuclear envelope (Schwartz and colleagues, Chapter 4), as well as *L*AP2, *e*merin and *M*AN1, which share the eponymous LEM-domain fold and also have unstructured regions (Zinn-Justin and colleagues, Chapter 3). Beyond proteins, Coen and colleagues describe how to purify entire "nuclear egress" complexes formed by the human cytomegalovirus, which include lamin-associated proteins (Chapter 25).

Advances in the purification and structural analysis of challengingly large cytoplasmic proteins are provided for plectin and BPAG1e (de Pereda and colleagues, Chapter 10) and desmoplakin (Choi and Weis, Chapter 11).

Functional assays are providing significant insight into the biological roles of cytoplasmic intermediate filament binding proteins, including molecular chaperones (Salas and colleagues, Chapter 8; Quinlan and colleagues, Chapter 9) and gigaxonin, an E3 ubiquitin ligase involved in the degradation of intermediate filament proteins (Bomont, Chapter 12). Borradori and colleagues describe a new GFP-based binding assay to study plakin association with intermediate filaments (Chapter 7). Green and colleagues use a 3D culture model to advance the functional interrogation of the intermediate filament-desmosome scaffold during keratinocyte differentiation (Chapter 15). Boczonadi and Määttä discuss general approaches to study the function of periplakin and envoplakin in cell culture models and gene-targeted mice (Chapter 16). The functional diversity of lamin-associated proteins is reflected in protocols to separate functionally distinct populations of lamin-binding proteins (Berk and Wilson, Chapter 6), identify transcription factors relevant to prelamin A pathology (Infante and Rodríguez, Chapter 23), and identify novel protein–protein interactions at the nuclear envelope via chemical cross-linking (Hallberg and colleagues, Chapter 24).

Genetic methods are increasingly important to understand intermediate filament biology and are expanding beyond mice. This volume of *Methods in Enzymology* includes state-of-the-art protocols to study lamin-binding proteins in *Caenorhabditis elegans*, which has one lamin gene (Gruenbaum and colleagues, Chapter 22), and to study the complex functions of the single spectraplakin locus in either *C. elegans* (Labouesse and colleagues, Chapter 20) or *Drosophila* (Prokop and colleagues, Chapter 19). Other chapters describe increasingly sophisticated use of mouse genetics to study the roles of plectin in skin and muscle (Wiche and colleagues, Chapter 13), epiplakin in epithelial cells (Fuchs and colleagues, Chapter 14), BPAG1 in neurons (Lynch-Godrei and Kothary, Chapter 18), and microtubule-actin cross-linking factor 1 in the skin, heart, nervous system, and intestinal epithelia (Goryunov and Liem, Chapter 17). We sincerely hope these protocols will help your research advance.

Arnoud Sonnenberg and Katherine L. Wilson
Division of Cell Biology,
The Netherlands Cancer Institute, Amsterdam, The Netherlands
Department of Biology, The Johns Hopkins
University School of Medicine, Baltimore Maryland, USA

PART I

Identification and Biochemical Analysis of Lamin-Associated Proteins

CHAPTER ONE

BioID Identification of Lamin-Associated Proteins

Aaron A. Mehus*, Ruthellen H. Anderson*, Kyle J. Roux*,†,1
*Sanford Children's Health Research Center, Sanford Research, Sioux Falls, South Dakota, USA
†Department of Pediatrics, Sanford School of Medicine, University of South Dakota, Sioux Falls, South Dakota, USA
[1]Corresponding author: e-mail address: kyle.roux@sanfordhealth.org

Contents

Abstract

A- and B-type lamins support the nuclear envelope, contribute to heterochromatin organization, and regulate a myriad of nuclear processes. The mechanisms by which lamins function in different cell types and the mechanisms by which lamin mutations cause over a dozen human diseases (laminopathies) remain unclear. The identification of proteins associated with lamins is likely to provide fundamental insight into these mechanisms. BioID (proximity-dependent biotin identification) is a unique and powerful method for identifying protein–protein and proximity-based interactions in living cells. BioID utilizes a mutant biotin ligase from bacteria that is fused to a protein of interest (bait). When expressed in living cells and stimulated with excess biotin, this BioID-fusion protein promiscuously biotinylates directly interacting and vicinal endogenous proteins. Following biotin-affinity capture, the biotinylated proteins can be identified using mass spectrometry. BioID thus enables screening for physiologically relevant protein associations that occur over time in living cells. BioID is applicable to insoluble proteins such as

Methods in Enzymology, Volume 569
ISSN 0076-6879
http://dx.doi.org/10.1016/bs.mie.2015.08.008

lamins that are often refractory to study by other methods and can identify weak and/or transient interactions. We discuss the use of BioID to elucidate novel lamin-interacting proteins and its applications in a broad range of biological systems, and provide detailed protocols to guide new applications.

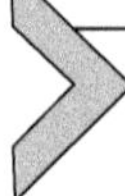

1. INTRODUCTION

1.1 Nuclear Lamina

Eukaryotes have a nuclear envelope (NE) that delineates the nucleus and partitions the DNA from other cellular structures, and includes nuclear pore complexes (NPCs), which mediate transport between the nucleoplasm and cytoplasm (Ptak, Aitchison, & Wozniak, 2014). The NE is continuous with the endoplasmic reticulum (ER), but specialized: the outer membrane (outer nuclear membrane, ONM) is compositionally similar to the ER, whereas the inner nuclear membrane (INM) contains a myriad of integral membrane proteins (NE transmembrane proteins, "NETs"), likely retained by associations with nuclear constituents (Wong, Luperchio, & Reddy, 2014). One special class of NETs, named SUN-domain proteins, associate translumenally with ONM KASH-domain proteins to physically link the nucleoskeleton and cytoskeleton (LINC-complex) (Kim, Birendra, & Roux, 2015; Sosa, Kutay, & Schwartz, 2013).

The metazoan NE is structurally supported by networks of nuclear intermediate filaments, called the nuclear "lamina," near the inner face of the INM. The nuclear lamina is composed of A- and B-type lamins. The *LMNA* gene encodes lamin-A and lamin-C (LaA and LaC), whereas *LMNB1* and *LMNB2* encode lamin-B1 (LaB1) and lamin-B2 (LaB2). Lamins have a central coiled-coil α-helical "rod" domain, a nuclear localization sequence, and a C-terminal immunoglobulin-like fold (Simon & Wilson, 2013). With the exception of LaC, lamins have a C-terminal CaaX motif that is post-translationally prenylated and carboxymethylated. LaA is further processed by the removal of the C-terminal 15 residues from pre-LaA to generate mature LaA. Lamins form stable homodimers that can further assemble via head–tail associations into filaments (Simon & Wilson, 2013). Subsets of lamins localize and function in the nucleoplasm. Early in mitosis, the nuclear lamins are disassembled via specific phosphorylation that occurs in conjunction with NE disassembly (Kochin et al., 2014; Ottaviano & Gerace, 1985). Lamins impact diverse pathways including cell proliferation and senescence, differentiation, transcription, DNA repair, peripheral heterochromatin organization, epigenetic chromatin modifications, NE assembly and disassembly, and the dimensions of the nucleus (Dechat, Adam, Taimen, Shimi, & Goldman, 2010). Even during mitosis, when the NE and nuclear lamina are disassembled, lamins are implicated in the function of the mitotic spindle matrix (Tsai et al., 2006).

Mutations in *LMNA* cause at least 11 rare clinically distinct degenerative diseases including muscular dystrophy, cardiomyopathy, lipodystrophy, and progeria (Chi, Chen, & Jeang, 2009). Duplication of *LMNB1* is associated with adult-onset leukodystrophy and mutations in *LMNB2* lead to partial acquired lipodystrophy (Gao, Li, Fu, & Luo, 2012; Hegele et al., 2006; Padiath et al., 2006). These diverse tissue-specific pathologies are consistent with the tissue-specific phenotypes seen in mice that lack one, two, or all three lamin genes (Chen, Zheng, & Zheng, 2015). Mutations in NE membrane proteins, most of which bind lamins, can also cause diseases ("envelopathies"; Dauer & Worman, 2009; Worman, Ostlund, & Wang, 2010) that phenotypically mimic or overlap with laminopathies, suggesting shared mechanisms. A better understanding of lamin-associated proteins may uncover the fundamental mechanisms by which lamins contribute to normal cellular function and human disease.

1.2 Protein–Protein Interactions

Protein–protein interactions provide critical insights into biological mechanisms, since proteins function collaboratively to drive cellular signaling cascades, regulation, differentiation, and proliferation. Proteins can also become "rogue" if they are mutated or modified. Rogue proteins have the potential to interact aberrantly with other proteins, organelles, or structures and thereby cause disease states.

One hurdle to overcome in studying lamins and associated proteins is the highly insoluble nature of the nuclear lamina itself. Ultrasonification with high concentrations of ionic- or chaotropic detergents to solubilize the lamina can disrupt lamin–protein interactions (Kubben et al., 2010). Various strategies have been used to identify lamin-associated proteins, including immunoprecipitation, cross-linking, fractionation of isolated nuclear membranes coupled with mass spectrometry (MS) analysis, and yeast two-hybrid (Y2H) (Simon & Wilson, 2013). These strategies can be successful but have specific drawbacks. Immunoprecipitation is done using nonphysiological conditions that can be insensitive, lead to loss of weak or transient interactions, or recover nonspecific proteins if wash conditions are insufficiently stringent. Cross-linking prior to harsh lysis is attractive, but interpretation can be complicated by protein aggregation. Fractionation techniques are useful but require large amounts of sample that can be difficult to obtain, and do not reveal specific protein associations. The Y2H method can yield many false-positives. Results can also be affected by posttranslational modifications (PTMs) that may influence protein recovery, or be required for specific partners to interact. A novel technique to screen for potential protein–protein interactions in living cells, named BioID for proximity-dependent biotin identification, avoids most of these issues and offers a powerful new tool for studying lamin-associated proteins (Roux, Kim, Raida, & Burke, 2012).

1.3 BioID

BioID is a recently developed approach that uses a "bait" protein fused to a mutant biotin ligase which promiscuously biotinylates directly interacting or nearby primary amines of "prey" proteins over a period of time. BioID takes advantage of a 35-kDa enzyme, named biotin ligase (BirA), found in prokaryotic cells (*Escherichia coli*). Biotin is a water-soluble molecule found in living cells in small amounts as an enzyme cofactor. Under normal conditions, BirA uses ATP to activate biotin, forming biotinyl-AMP (bioAMP), which then reacts with lysine residues of very specific protein sequences (Chapman-Smith & Cronan, 1999; Lane, Rominger, Young, & Lynen, 1964). The BioID approach utilizes an R118G-mutated biotin ligase (hereafter called BioID) that prematurely releases bioAMP, allowing it to react with any free primary amines nearby (Kwon & Beckett, 2000; Roux et al., 2012). When this promiscuous ligase is fused to a specific protein bait, the resulting BioID-fusion protein can biotinylate directly interacting and/or vicinal proteins, leaving a "trail" of modified proteins that reveal its history of interactions over time.

The BioID method is shown schematically in Fig. 1. BioID can be used to screen potential partners of the bait in virtually any genetically tractable cell type. The bait ORF is recombined in-frame with the BioIDORF in an expression vector plasmid. A stable cell line expressing the BioID protein is then established using common transfection methods or virus-mediated DNA transfer. These stable cells are then incubated with excess biotin (typically 50 μ*M*). Most conventional cell culture media do not contain biotin and the low level of biotin needed by cells is supplied by serum. Under these conditions, there is minimal biotinylation by the BioID-fusion

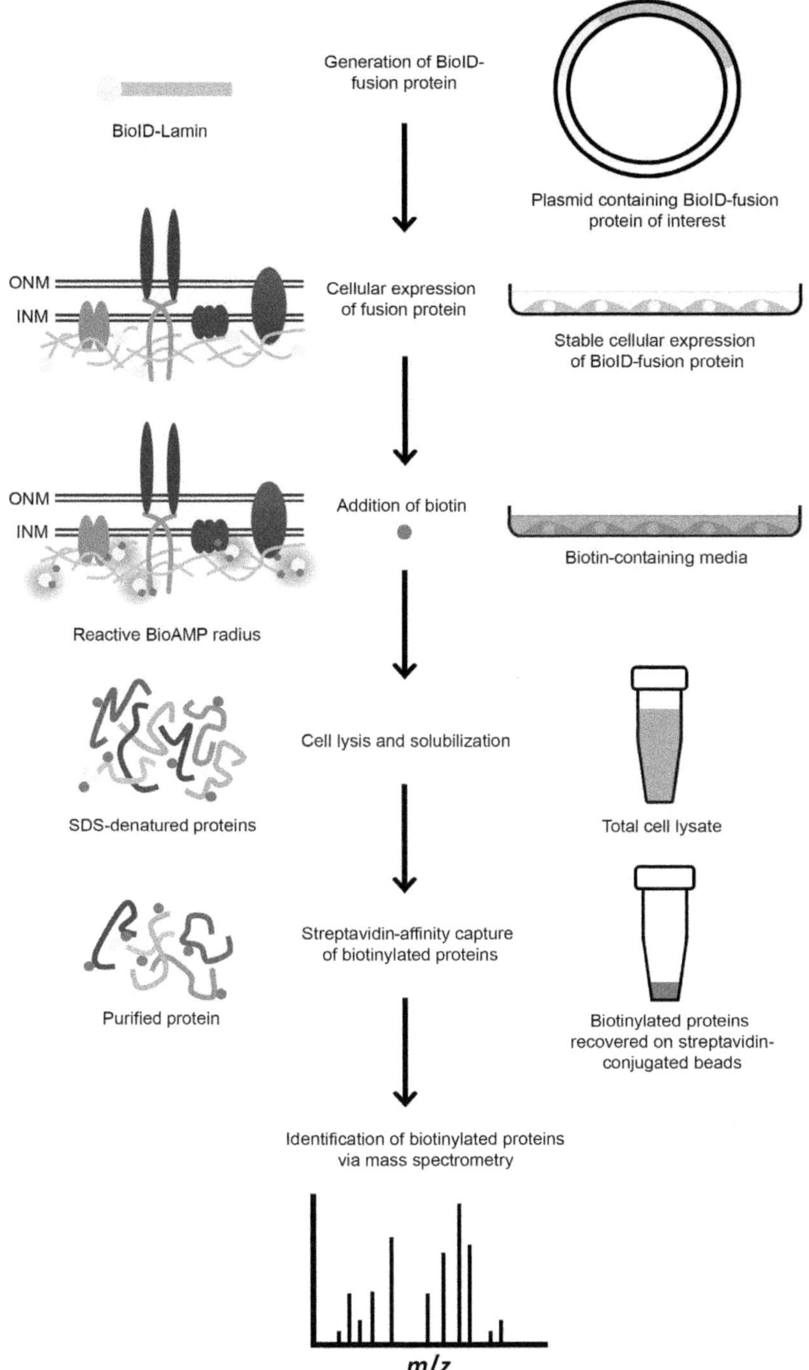

Figure 1 See legend on next page.

protein. The level of biotinylation can be regulated by varying the time of incubation with excess biotin: 15–18 h is typically sufficient to saturate the biotinylation signal (Kim et al., 2014; Roux et al., 2012). The cells are then stringently lysed using SDS/Triton X-100, sonicated, and affinity-captured in a single step using magnetic beads coated with streptavidin, which has a well-known high affinity for biotin. These biotinylated proteins are then analyzed by MS or immunoblotting (IB). Fluorescently labeled streptavidin can be used to visualize biotinylated proteins in fixed and stained cells by immunofluorescence (IF). If antibodies against the bait are unavailable, the BioID-fusion protein can include a MYC or HA tag (Evan, Lewis, Ramsay, & Bishop, 1985; Field et al., 1988). Antibodies specific for BirA are also now available for detection of BioID-fusion proteins by IF and IB (see Section 2.4.1).

BioID is especially powerful for screening protein–protein interactions in living cells. Partners and vicinal proteins are covalently biotinylated in their normal cellular environments and can therefore be recovered using stringent lysis conditions that completely solubilize and denature the proteins. Other major strengths of BioID are that candidate interactors are rapidly and selectively enriched by affinity capture and can be directly identified by MS.

BioID has been successfully applied to abroad range of proteins and cell structures including the nuclear lamina (Roux et al., 2012), NPCs (Kim et al., 2014), chromatin-associated protein complexes (Lambert, Tucholska, Go, Knight, & Gingras, 2014), the trypanosome bilobe (Morriswood et al., 2013), cell junction complexes (Fredriksson et al., 2015; Guo et al., 2014; Steed et al., 2014; Ueda, Blee, Macway, Renner, & Yamada, 2015; Van Itallie et al., 2013, 2014), and centrosomes (Comartin et al., 2013; Firat-Karalar, Rauniyar, Yates, & Stearns, 2014). This method has been used to screen for proteins involved in the Hippo signaling pathway (Couzens et al., 2013). BioID was used to study a novel protein secreted by a bacterial pathogen, *Chlamydia psittaci*, that targets the NE

Figure 1 Application of BioID method to LaA. The N-terminus of lamin-A is fused to a promiscuous biotin ligase that prematurely releases BioAMP. Cells that stably express this fusion protein (BioID-lamin) are incubated in biotin-containing medium, and proteins located within 10 nm of the fusion protein are selectively biotinylated. After cell lysis and protein solubilization, biotinylated proteins are affinity purified using streptavidin-conjugated beads, allowing stringent wash conditions. After qualitative assessment by immunoblotting, candidate proteins are identified by mass spectrometry (MS).

(Mojica et al., 2015), and to identify novel components of the inner membrane complex of *Toxoplasma gondii* (Chen, Kim, et al., 2015), analyze HIV-1 Gag protein interactions (Ritchie, Cylinder, Platt, & Barklis, 2015), and detect c-MYC interacting partners in cultured cells and mice xenograft tumors (Dingar et al., 2014).

2. BioID METHOD

2.1 Construction of a BioID-Fusion Protein

The most important aspect of BioID is the fusion protein itself, especially when aiming to create fusion proteins that function like the endogenous protein. One must choose the most appropriate position within the protein to incorporate the biotin ligase. To guide this decision, investigators should consider all previous fusion constructs for their bait, especially fusions to proteins of similar size (e.g., GFP), and their effects on protein localization and function. Care must be taken not to disrupt PTMs of the N- or C-terminus, such as C-terminal prenylation or N-terminal signal peptides. If the effects of N- versus C-terminal fusions on the bait are unknown, make and test both constructs. Proper subcellular targeting is a good starting point to assess basic function of the fusion protein. If the bait is an enzyme, assay the enzyme activity of the BioID construct. For integral membrane proteins (e.g., INM and ONM proteins), one must also consider the subcellular localization of each domain. Most BioID studies focus on interactions that take place in the cytoplasm or nucleoplasm. BioID can also identify protein–protein interactions in the ER/NE lumen, but is markedly less active (unpublished observations).

2.2 Cell Choice

After generating a BioID expression plasmid and validating the fusion protein, it is important to select an appropriate cell type in which to express the BioID-fusion protein. It is prudent to consider the uniformity and level of expression. For large-scale BioID pull-down and MS analysis, the best results are obtained when all cells in the population express the fusion at similar levels. Thus, transient transfections can be used for preliminary studies of fusion protein functionality and targeting, but not for screening. Stable cell lines expressing the fusion protein can be generated by random integration, using various viral expression systems or by genome editing. In most situations, low-level expression of the BioID fusion is ideal to ensure physiological relevance while providing sufficient candidate proteins. Inducible promoters can be considered if the bait itself is toxic, but are unnecessary to regulation biotinylation which can be controlled by the addition of excess biotin. Overexpression of BioID-fusion proteins, which can lead to aggregation and false association artifacts, should be avoided.

2.3 Functional Assessment of the Fusion Protein

The most important aspect of the fusion protein is its functionality. Ideally, the fusion protein will behave the same as its endogenous counterpart. This can be tested by phenotypic rescue of a knockout/knockdown or mutant system. The minimum level of functionality can be tested by comparing the localizations of the fusion protein and endogenous protein by IF microscopy. Pay special attention to expression levels, since overexpression causes many proteins to mislocalize. In general, expressing the BioID-fusion protein at levels at, or below, that of the endogenous protein is sufficient to identify interacting and/or vicinal proteins.

2.4 Establishing a Stable BioID Cell Line

This protocol describes the creation and characterization of a stable cell line expressing a BioID-fusion protein in mammalian cells. PCR cloning and stable transfection techniques will not be described, as they can vary dramatically. Methods used to assess fusion protein targeting, expression, and activity in cells prior to a large-scale BioID pull-down are described.

2.4.1 Materials

1. Expression plasmid for BioID-fusion protein (see https://www.addgene.org/Kyle_Roux/)
2. cDNA for bait protein
3. Cells of choice for transfection and appropriate medium
4. Biotin stock (1 m*M*; 20 ×; see recipe)
5. Fixative, e.g., paraformaldehyde (PFA)
6. Triton X-100 (TX-100)
7. Streptavidin-Alexa Fluor (Invitrogen, currently Life Technologies)
8. DNA labeling reagent for IF (e.g., Hoechst, DAPI)
9. Phosphate-buffered saline (PBS)
10. SDS-PAGE sample buffer (see recipe)
11. BSA blocking buffer (see recipe)

12. Streptavidin-HRP (High Sensitivity Streptavidin-HRP, Abcam)
13. ABS blocking buffer (see recipe)
14. Enhanced chemiluminescence (ECL) reagent (commercially available, or see recipe)
15. Quenching solution (see recipe)
16. Antibodies to the BioID fusion (e.g., anti-myc/HA) or chicken anti-BirA (Abcam, ab14002)
17. Secondary antibodies (Alexa Fluor form and HRP-conjugated form)
18. 6-well plates
19. Sonicator

2.4.2 Generation and Validation of BioID-Fusion Protein

1. Generate the BioID-fusion protein expression plasmid. This typically requires PCR cloning (Elion, Marina, & Yu, 2007) to generate an expression vector in which the bait protein is fused in-frame with BioID. This requires plasmid DNA of sufficient quality and quantity for cellular transfection. See https://www.addgene.org/Kyle_Roux/ for mammalian BioID expression plasmids. BioID plasmids adapted for use in *Dictyostelium* are also available (see Batsios, Meyer, & Gräf, Proximity-dependent biotin identification (BioID) in *Dictyostelium amoebae*: Identification of lamin protein interactors, this volume).
2. For each experimental condition (e.g., BioID-only control, BioID-fusion protein), plate cells in two wells of a standard 6-well plate, one of which has a single glass coverslip.
3. Express the fusion protein in your chosen cells by transient transfection (Mortensen, Chesnut, Hoeffler, & Kingston, 2003). Lipofectamine 2000 (Life Technologies) reagent works well using the manufacturer's suggested protocol. Process mock-transfected or nontransfected cells in parallel. Apply biotin (50 μ*M* final concentration) to cells either at the time of transfection, or (if you need to replace with fresh medium a short time after transfection) add the biotin with the fresh medium. This excess biotin promotes biotinylation by the BioID-fusion protein, which is usually detectable after 18–24 h by microscopy and blotting.
4. Stain cells on coverslips at ~24 h posttransfection for analysis by IF microscopy. We recommend PFA fixation (3% PFA in PBS) for 10 min, followed by TX-100 permeabilization (0.4% TX-100 in PBS) for 15 min. Methanol fixation enhances background mitochondrial signals in cells stained with fluorescent streptavidin (Streptavidin-Alexa Fluor). Stain cells for biotinylation with Alexa Fluor-conjugated

streptavidin at 1:1000 dilution. Label cells for the BioID-fusion protein using a primary antibody against BirA or the appropriate epitope tag, and the appropriate Alexa Fluor-conjugated secondary antibody at 1:1000 dilution; stain for DNA (e.g., Hoechst or DAPI), then wash, mount, and image (Donaldson, 2001).

5. Process cells for IB analysis at ~24 h after transfection and addition of biotin. To do this, briefly rinse cells in room-temperature PBS to remove serum proteins, then lyse cells in an appropriate volume of SDS-PAGE sample buffer (~200 μl per 1×10^6 cells). Heat samples to 98 °C for 5 min to denature proteins and sonicate to shear DNA.
6. Resolve proteins by SDS-PAGE and transfer to membrane. Block membranes by agitating 20–30 min in BSA blocking buffer at room temperature. BSA (no special grade required) is the best blocking agent, because it eliminates any free biotin that may be present in milk or serum. Free biotin competes with biotinylated proteins for binding to streptavidin-HRP.
7. Agitate membrane in streptavidin-HRP (1:40,000 dilution in BSA blocking buffer) for 40 min at room temperature. Optimize dilution of streptavidin-HRP, if needed.
8. Quickly wash membrane two to three times in PBS to remove unbound streptavidin-HRP.
9. Agitate membrane 5 min in ABS blocking buffer. This step is critical to reduce the background signal on membrane.
10. Quickly wash membrane two to three times in PBS.
11. Agitate membrane 5 min in PBS.
12. Add ECL reagent and observe biotinylated proteins.
13. Following chemiluminescent imaging of the membrane, quench the HRP signal by agitating membrane 20 min in quenching solution. Streptavidin-HRP cannot be removed by standard membrane stripping methods, since biotin binds streptavidin nearly irreversibly. Quenching permanently inactivates HRP, allowing further probing with additional antibodies. Quenching can be confirmed by reapplying the ECL reagents to the membrane, and visualizing.
14. Wash membrane three times with PBS to remove quenching solution. This step removes residual hydrogen peroxide that might disrupt later blotting steps.
15. Probe membrane with antibodies specific to the BioID fusion (e.g., anti-myc/HA/BirA) to confirm its expression and SDS-PAGE migration. Use standard protocols and start by blocking in ABS or equivalent (BSA is not required).

2.4.3 Generate Cells That Stably Express the BioID-Fusion Protein

16. Use your preferred method to generate stable cell lines, e.g., stable transfection or viral transduction. These methods vary and are not described here.
17. If you subclone cells, screen subclones first by IF. Subclones that pass IF screening are then analyzed by IB. For viral infections, screen populations of infected cells using both IF and IB. In all cases, REMEMBER to incubate cells in 50 μ*M* biotin to monitor the biotinylation function of the BioID-fusion protein.
18. Freeze multiple vials of stable cells for future BioID experiments; store in liquid nitrogen.

2.5 Large-Scale BioID Pull-Down to Identify Interacting Proteins

This protocol describes large-scale BioID pull-down experiments from cells that stably express a BioID-fusion protein (along with nonexpressing parental cells or BioID-only control cells). The goal is to isolate enough biotinylated proteins for identification by MS. The number of cells you need will vary based on several factors including the efficiency of biotinylation by your BioID fusion and the number of desired candidate proteins. This protocol describes the analysis of four confluent 10-cm plates (4×10^7 cells) per sample. This protocol ends immediately prior to analysis by MS, since MS expertise is typically provided by a core facility.

2.5.1 Materials

1. Four 10-cm dishes of cells for each experimental condition (BioID constructs and controls)
2. Complete medium
3. Biotin stock (1 m*M*; 20 ×) (see recipe)
4. PBS
5. Lysis buffer (see recipe)
6. 20% TX-100
7. 50 m*M* Tris·Cl, pH 7.4
8. Dynabeads (MyOne Streptavidin C1, Life Technologies)
9. Wash buffer 1 (see recipe)
10. Wash buffer 2 (see recipe)
11. Wash buffer 3 (see recipe)
12. Ammonium bicarbonate 50 m*M* (NH_4HCO_3)
13. SDS-PAGE sample buffer (see recipe)

14. DNase/RNase-free tubes: 15 ml conical and 2 ml microcentrifuge
15. Tube cap opener
16. Sonicator (Branson Sonifier-250 or equivalent)
17. MagneSphere Technology Magnetic Separation Stand (Promega)

2.5.2 Cell Lysis

Unless otherwise specified, perform lysis and wash steps at room temperature to avoid precipitation of SDS and deoxycholic acid. To reduce keratin contamination, use DNase/RNase-free tubes that have not previously been opened, wear gloves, and use a tube cap opener.

1. Begin with four 10-cm dishes for each experimental condition. More cells typically yield more identified candidates (e.g., four 10-cm plates may yield ~150 candidates, whereas two 10-cm plates may yield ~75 candidates), but this can vary with biotin labeling time, expression level, and MS conditions.
2. At ~80% confluency, add fresh complete medium containing 50 μ*M* biotin.
3. Incubate cells with biotin. Incubation time and conditions will depend on your experimental goals (e.g., cell cycle stage-specific labeling). In general, 16–18 h with biotin gives maximal labeling; incubation for 6 h will reduce the level of biotinylated proteins by ~60–70%. Pilot studies and analysis by IF and IB should be done prior to large-scale experiments.
4. Remove medium completely by aspiration and rinse cells twice at room temperature with 3 ml PBS per dish. This step is important to remove residual free-biotin from the medium.
5. Add 540 μl lysis buffer per dish and harvest cells by gentle scraping. Perform this step at room temperature. The goal of this harsh lysis is to completely denature and solubilize all proteins and disrupt their interactions.
6. Transfer lysed cells to a 15-ml conical tube.
7. Add 240 μl of 20% TX-100 (final concentration 2%) and mix by gentle vortexing. Keep tube on ice during subsequent sonication. Adding a fivefold excess of TX-100 dilutes the SDS and prevents its precipitation at 4 °C.
8. Position the sonicator probe tip in the sample, just above bottom of tube.
9. Sonicate in two sessions (30 pulses each) using a Branson Sonifier 250 (or equivalent) at 30% duty cycle and an output level of 4. Let the tube

sit on ice for 2 min between sessions to prevent overheating. If the sample is still viscous and cloudy, sonicate again for 30 pulses. Sonication is important to shear DNA and disrupts protein aggregates.

10. Add 2.4 ml prechilled 50 m*M* Tris·Cl, pH 7.4, and mix well. This dilution improves conditions for affinity capture.
11. Sonicate again (30 pulses at 30% duty cycle and output level 4, using a Branson Sonifier 250 or equivalent). This step helps solubilize proteins that may have precipitated as salt/detergent concentrations were reduced, and helps mix the sample.
12. Aliquot sample evenly into three prechilled 2-ml tubes.
13. Centrifuge 10 min (16,500 × *g*, 4 °C), then carefully place tube on ice until step 17.

2.5.3 Affinity Purification of Biotinylated Proteins

14. While centrifuging in step 13, place 15-ml conical tube(s) in magnetic separation stand; add 1 ml lysis buffer and 1 ml of 50 m*M* Tris·Cl, pH 7.4, both at room temperature, to each tube.
15. Mix the stock of magnetic streptavidin beads well with gentle tapping to resuspend and add 300 μl beads to each tube prepared in step 14. Wait 3 min at room temperature. This step equilibrates the beads in the lysis buffer.
16. After beads accumulate on the tube wall, remove the supernatant gently by pipetting.
17. Carefully transfer the supernatant from step 13 (solubilized proteins; do not disturb the small insoluble pellet on the tube wall) to the tubes prepared in step 16. Work quickly to prevent the beads from drying out.
18. Resuspend the protein samples and beads with gentle pipetting.
19. Rotate tubes at 4 °C overnight.
20. Place the conical tube on the magnetic separation stand and wait 3 min to collect beads at room temperature.
21. Remove the supernatant gently by pipetting. Try not to disturb beads on the tube wall.
22. Add 1 ml wash buffer 1 to the tube, resuspend beads gently by pipetting, and transfer to a new 1.5-ml Eppendorf tube.
23. Rotate tube 8 min at room temperature.
24. Remove the supernatant as described in steps 20 and 21.
25. Repeat steps 22–24 again, using wash buffer 1.
26. Wash once with 1 ml wash buffer 2, as described in steps 22–24.
27. Wash once with 1 ml wash buffer 3, as described in steps 22–24.

28. Add 1 ml of 50 m*M* Tris·Cl, pH 7.4, and resuspend gently with pipetting. This step is important to remove detergents in the sample that may interfere with MS.
29. Save 100 μl (10% of total) of resuspended beads for further analysis by IB.
30. Centrifuge beads (both the 900- and 100-μl samples) for 5 min (2000 × *g*) at room temperature, to collect beads.
31. Remove the supernatant completely. Do not disrupt the pelleted beads.
32. To the tube containing the 900-μl sample (destined for MS analysis), we add 150 μl of 50 m*M* ammonium bicarbonate and resuspend gently with pipetting. This final wash buffer may vary depending your method of MS analysis; please consult your MS experts first. If needed, flash-freeze samples in liquid nitrogen and store at −80 °C until MS analysis.
33. The 100-μl sample (used for qualitative analysis): Remove supernatant (unbound proteins). Add 100 μl SDS-PAGE sample buffer to the beads and pipet gently to resuspend. Heat samples 5 min at 98 °C and store at −20 °C until further analysis by IB (see Basic Protocol 1). We strongly recommend IB analysis (~5 μl per lane) *before* sending samples for MS analysis. A considerable difference in the pattern of biotinylated proteins between the BioID-only and BioID-fusion samples justifies proceeding to MS.

2.6 Reagents and Solutions

Use Milli-Q purified water or equivalent in all recipes.

ABS blocking buffer: 10% (v/v) adult bovine serum and 0.2% (w/v) TX-100, brought up to final volume with 1 × PBS. Store up to 4 weeks at 4 °C.

Biotin, 1 mM (20 × *stock*): Dissolve 12.2 mg biotin (Sigma #B4501) in 50 ml serum-free DMEM (for standard tissue culture medium). Pipetting may be required to dissolve biotin completely. Sterilize by passing through a 0.22-μm syringe-driven filter unit (Millex). Dispense into sterile 50-ml tube; cap tightly. Store up to 8 weeks at 4 °C.

BSA blocking buffer: 1% bovine serum albumin (fraction V), 0.2% (w/v) TX-100, brought up to final volume with 1 × PBS. Store up to 4 weeks at 4 °C.

ECL reagent stock solutions: 250 m*M* luminol (store 500-μl aliquots at −20 °C); 90 m*M* coumaric acid (store 500-μl aliquots at −20 °C); 1 *M* Tris·Cl, pH 8.5 (store at 4 °C); 30% hydrogen peroxide (store at 4 °C).

ECL reagent Solution 1:
1 ml of 1 *M* Tris·Cl, pH 8.5
45 μl coumaric acid
100 μl luminol
8.9 ml H_2O

ECL reagent Solution 2:
1 ml of 1 *M* Tris·Cl, pH 8.5
6 μl of 30% hydrogen peroxide
9 ml H_2O

Store ECL solutions 1 and 2 at 4 °C for up to 1 month. Mix equal volumes of solutions 1 and 2 together and immediately add to membrane for 1 min, then expose to film for ChemiDoc detection.

Lysis buffer: 50 m*M* Tris·Cl pH 7.4, 500 m*M* NaCl, 0.2% SDS (w/v). Store up to 2 weeks at room temperature. Immediately prior to use, add 1 m*M* DTT and 1 × protease inhibitor (Halt Protease Inhibitor Cocktail, EDTA-Free, Thermo Scientific).

Quenching solution: 30% (v/v) hydrogen peroxide.

SDS-PAGE sample buffer: 50 m*M* Tris·Cl pH 6.8, 12% sucrose, 2% SDS, 0.004% bromophenol blue, 20 m*M* DTT.

Wash buffer 1: 2% SDS (w/v). Store at room temperature up to 2 weeks.

Wash buffer 2: 0.1% (w/v) deoxycholic acid, 1% (w/v) TX-100, 1 m*M* EDTA, 500 m*M* NaCl, 50 m*M* HEPES, pH 7.5. Store at room temperature up to 2 weeks. *Note*: Deoxycholic acid stock solution must be protected from light.

Wash buffer 3: 0.5% (w/v) deoxycholic acid, 0.5% (w/v) NP-40, 1 m*M* EDTA, 250 m*M* LiCl, 10 m*M* Tris·Cl, pH 7.4. Store up to 2 weeks at room temperature.

50 mM ammonium bicarbonate (make fresh daily): 198 mg ammonium bicarbonate; add distilled water to 50 ml. Invert several times to completely dissolve.

3. ANTICIPATED RESULTS

A typical BioID-lamin experiment starts with IF and IB analysis of stably expressing cells to localize the fusion protein and assess the number and patterns of biotinylated proteins on blots. Unfused BioID control proteins are diffusely nucleoplasmic and cytoplasmic, whereas BioID-LaA localizes predominantly at the NE with lower signals in the nucleoplasm (Fig. 2A). The migration pattern and intensity of proteins biotinylated by BioID-LaA should be distinct from those biotinylated in the BioID-only or parental (nontransfected) control cells (Fig. 2B). Results such as these would warrant a large-scale BioID pull-down for MS analysis. Typically, 50–250 unique proteins (defined as proteins that are absent or many-fold lower in controls) are detected with 18 h of biotin labeling (Kim et al., 2014; Roux et al., 2012), all of which represent candidate direct partners, or proteins that reside in close proximity (~10 nm) to the BioID-lamin protein.

4. CONCLUSIONS

BioID is not without limitations. The bait itself may be perturbed structurally or functionally when fused to the biotin ligase. Biotinylation, which alters the charge of free amines, might influence proteins by making lysine residues unavailable for normal interactions or posttranslational regulation. Given the breadth of potential candidates, low-abundance partners might be overlooked by MS (false negatives). Finally, like all screening methods, BioID results provide a starting point. Whether an identified protein interacts directly or indirectly, or is simply near the bait, must be further validated.

We are developing a second-generation BioID with a smaller "footprint" that is less likely to affect the function or targeting of the bait protein (unpublished data, Roux lab). To date, most BioID experiments have focused on proteins in cultured cells, but future directions include the development of an animal model. BioID was successfully used in mice tumor xenografts to identify novel protein–protein interactions of the c-MYC on coprotein, identifying >100 high confidence MYC-interacting proteins including >30 known binding partners (Dingar et al., 2014). Animal models that incorporate BioID would significantly improve our ability to study genome biology, cellular pathways, and disease mechanisms in specific tissues.

In summary, BioID is a highly effective strategy for the identification of novel lamin-associated proteins and is also being used to characterize many INM proteins including LBR, emerin, TMPO, MAN1, LAP1, SUN1, and SUN2 (unpublished results, Roux lab). BioID has unique advantages, labeling partners in their natural environments (living cells) over time and enabling new discoveries that may advance our understanding of nuclear structure and laminopathy disease mechanisms.

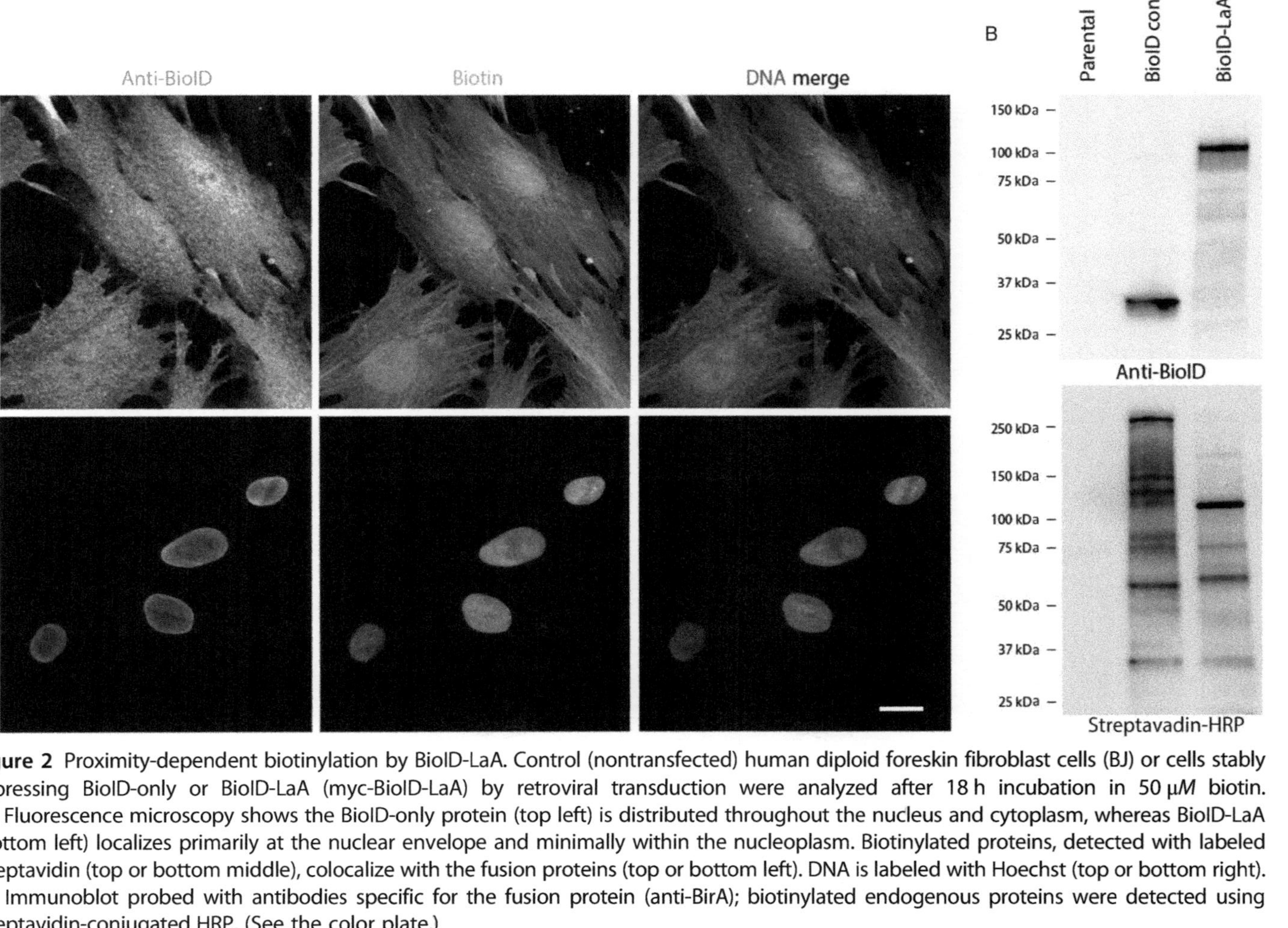

Figure 2 Proximity-dependent biotinylation by BioID-LaA. Control (nontransfected) human diploid foreskin fibroblast cells (BJ) or cells stably expressing BioID-only or BioID-LaA (myc-BioID-LaA) by retroviral transduction were analyzed after 18 h incubation in 50 μ*M* biotin. (A) Fluorescence microscopy shows the BioID-only protein (top left) is distributed throughout the nucleus and cytoplasm, whereas BioID-LaA (bottom left) localizes primarily at the nuclear envelope and minimally within the nucleoplasm. Biotinylated proteins, detected with labeled streptavidin (top or bottom middle), colocalize with the fusion proteins (top or bottom left). DNA is labeled with Hoechst (top or bottom right). (B) Immunoblot probed with antibodies specific for the fusion protein (anti-BirA); biotinylated endogenous proteins were detected using streptavidin-conjugated HRP. (See the color plate.)

ACKNOWLEDGMENT

This work was supported by grants from the National Institutes of Health (RO1GM102203, RO1GM102486, RO1EB014869).

REFERENCES

Chapman-Smith, A., & Cronan, J. E., Jr. (1999). Molecular biology of biotin attachment to proteins. *The Journal of Nutrition, 129*, 477S–484S.

Chen, A. L., Kim, E. W., Toh, J. Y., Vashisht, A. A., Rashoff, A. Q., Van, C., et al. (2015). Novel components of the Toxoplasma inner membrane complex revealed by BioID. *mBio, 6*, e02357.

Chen, H., Zheng, X., & Zheng, Y. (2015). Lamin-B in systemic inflammation, tissue homeostasis, and aging. *Nucleus, 6*, 183–186.

Chi, Y.-H., Chen, Z.-J., & Jeang, K.-T. (2009). The nuclear envelopathies and human diseases. *Journal of Biomedical Science, 16*, 96.

Comartin, D., Gupta, G. D., Fussner, E., Coyaud, E., Hasegan, M., Archinti, M., et al. (2013). CEP120 and SPICE1 cooperate with CPAP in centriole elongation. *Current Biology, 23*, 1360–1366.

Couzens, A. L., Knight, J. D., Kean, M. J., Teo, G., Weiss, A., Dunham, W. H., et al. (2013). Protein interaction network of the mammalian Hippo pathway reveals mechanisms of kinase-phosphatase interactions. *Science Signaling, 6*, rs15.

Dauer, W. T., & Worman, H. J. (2009). The nuclear envelope as a signaling node in development and disease. *Developmental Cell, 17*, 626–638.

Dechat, T., Adam, S. A., Taimen, P., Shimi, T., & Goldman, R. D. (2010). Nuclear lamins. *Cold Spring Harbor Perspectives in Biology, 2*, a000547.

Dingar, D., Kalkat, M., Chan, P. K., Srikumar, T., Bailey, S. D., Tu, W. B., et al. (2014). BioID identifies novel c-MYC interacting partners in cultured cells and xenograft tumors. *Journal of Proteomics, 118*, 95–111.

Donaldson, J. G. (2001). Immunofluorescence staining. *Current Protocols in Cell Biology*. 00:4.3:4.3.1–4.3.6.

Elion, E. A., Marina, P., & Yu, L. (2007). Constructing recombinant DNA molecules by PCR. *Current Protocols in Molecular Biology*. 78:IV:3.17:3.17.1–3.17.12.

Evan, G. I., Lewis, G. K., Ramsay, G., & Bishop, J. M. (1985). Isolation of monoclonal antibodies specific for human c-myc proto-oncogene product. *Molecular and Cellular Biology, 5*, 3610–3616.

Field, J., Nikawa, J., Broek, D., MacDonald, B., Rodgers, L., Wilson, I. A., et al. (1988). Purification of a RAS-responsive adenylyl cyclase complex from Saccharomyces cerevisiae by use of an epitope addition method. *Molecular and Cellular Biology, 8*, 2159–2165.

Firat-Karalar, E. N., Rauniyar, N., Yates, J. R., 3rd, & Stearns, T. (2014). Proximity interactions among centrosome components identify regulators of centriole duplication. *Current Biology, 24*, 664–670.

Fredriksson, K., Van Itallie, C. M., Aponte, A., Gucek, M., Tietgens, A. J., & Anderson, J. M. (2015). Proteomic analysis of proteins surrounding occludin and claudin-4 reveals their proximity to signaling and trafficking networks. *PLoS One, 10*, e0117074.

Gao, J., Li, Y., Fu, X., & Luo, X. (2012). A Chinese patient with acquired partial lipodystrophy caused by a novel mutation with LMNB2 gene. *Journal of Pediatric Endocrinology & Metabolism, 25*, 375–377.

Guo, Z., Neilson, L. J., Zhong, H., Murray, P. S., Zanivan, S., & Zaidel-Bar, R. (2014). E-cadherin interactome complexity and robustness resolved by quantitative proteomics. *Science Signaling*, 7, rs7.

Hegele, R. A., Cao, H., Liu, D. M., Costain, G. A., Charlton-Menys, V., Rodger, N. W., et al. (2006). Sequencing of the reannotated LMNB2 gene reveals novel mutations in patients with acquired partial lipodystrophy. *American Journal of Human Genetics*, *79*, 383–389.

Kim, D. I., Birendra, K. C., & Roux, K. J. (2015). Making the LINC: SUN and KASH protein interactions. *Biological Chemistry*, *396*, 295–310.

Kim, D. I., Birendra, K. C., Zhu, W., Motamedchaboki, K., Doye, V., & Roux, K. J. (2014). Probing nuclear pore complex architecture with proximity-dependent biotinylation. *Proceedings of the National Academy of Sciences of the United States of America*, *111*, E2453–E2461.

Kochin, V., Shimi, T., Torvaldson, E., Adam, S. A., Goldman, A., Pack, C. G., et al. (2014). Interphase phosphorylation of lamin A. *Journal of Cell Science*, *127*, 2683–2696.

Kubben, N., Voncken, J. W., Demmers, J., Calis, C., van Almen, G., Pinto, Y., et al. (2010). Identification of differential protein interactors of lamin A and progerin. *Nucleus*, *1*, 513–525.

Kwon, K., & Beckett, D. (2000). Function of a conserved sequence motif in biotin holoenzyme synthetases. *Protein Science*, *9*, 1530–1539.

Lambert, J. P., Tucholska, M., Go, C., Knight, J. D., & Gingras, A. C. (2014). Proximity biotinylation and affinity purification are complementary approaches for the interactome mapping of chromatin-associated protein complexes. *Journal of Proteomics*, *118*, 81–94.

Lane, M. D., Rominger, K. L., Young, D. L., & Lynen, F. (1964). The enzymatic synthesis of holotranscarboxylase from apotranscarboxylase and (+)-biotin: II. Investigation of the reaction mechanism. *The Journal of Biological Chemistry*, *239*, 2865–2871.

Mojica, S. A., Hovis, K. M., Frieman, M. B., Tran, B., Hsia, R. C., Ravel, J., et al. (2015). SINC, a type III secreted protein of Chlamydia psittaci, targets the inner nuclear membrane of infected cells and uninfected neighbors. *Molecular Biology of the Cell*, *26*, 1918–1934.

Morriswood, B., Havlicek, K., Demmel, L., Yavuz, S., Sealey-Cardona, M., Vidilaseris, K., et al. (2013). Novel bilobe components in Trypanosoma brucei identified using proximity-dependent biotinylation. *Eukaryotic Cell*, *12*, 356–367.

Mortensen, R., Chesnut, J. D., Hoeffler, J. P., & Kingston, R. E. (2003). Selection of transfected mammalian cells. *Current Protocols in Molecular Biology*. 62:I:9.5:9.5.1–9.5.19.

Ottaviano, Y., & Gerace, L. (1985). Phosphorylation of the nuclear lamins during interphase and mitosis. *The Journal of Biological Chemistry*, *260*, 624–632.

Padiath, Q. S., Saigoh, K., Schiffmann, R., Asahara, H., Yamada, T., Koeppen, A., et al. (2006). Lamin B1 duplications cause autosomal dominant leukodystrophy. *Nature Genetics*, *38*, 1114–1123.

Ptak, C., Aitchison, J. D., & Wozniak, R. W. (2014). The multifunctional nuclear pore complex: A platform for controlling gene expression. *Current Opinion in Cell Biology*, *28*, 46–53.

Ritchie, C., Cylinder, I., Platt, E. J., & Barklis, E. (2015). Analysis of HIV-1 Gag protein interactions via biotin ligase tagging. *Journal of Virology*, *89*, 3988–4001.

Roux, K. J., Kim, D. I., Raida, M., & Burke, B. (2012). A promiscuous biotin ligase fusion protein identifies proximal and interacting proteins in mammalian cells. *The Journal of Cell Biology*, *196*, 801–810.

Simon, D. N., & Wilson, K. L. (2013). Partners and post-translational modifications of nuclear lamins. *Chromosoma*, *122*, 13–31.

Sosa, B. A., Kutay, U., & Schwartz, T. U. (2013). Structural insights into LINC complexes. *Current Opinion in Structural Biology*, *23*, 285–291.

Steed, E., Elbediwy, A., Vacca, B., Dupasquier, S., Hemkemeyer, S. A., Suddason, T., et al. (2014). MarvelD3 couples tight junctions to the MEKK1-JNK pathway to regulate cell behavior and survival. *The Journal of Cell Biology*, *204*, 821–838.

Tsai, M. Y., Wang, S., Heidinger, J. M., Shumaker, D. K., Adam, S. A., Goldman, R. D., et al. (2006). A mitotic lamin B matrix induced by RanGTP required for spindle assembly. *Science*, *311*, 1887–1893.

Ueda, S., Blee, A. M., Macway, K. G., Renner, D. J., & Yamada, S. (2015). Force dependent biotinylation of myosin IIA by alpha-catenin tagged with a promiscuous biotin ligase. *PLoS One*, *10*, e0122886.

Van Itallie, C. M., Aponte, A., Tietgens, A. J., Gucek, M., Fredriksson, K., & Anderson, J. M. (2013). The N and C termini of ZO-1 are surrounded by distinct proteins and functional protein networks. *The Journal of Biological Chemistry*, *288*, 13775–13788.

Van Itallie, C. M., Tietgens, A. J., Aponte, A., Fredriksson, K., Fanning, A. S., Gucek, M., et al. (2014). Biotin ligase tagging identifies proteins proximal to E-cadherin, including lipoma preferred partner, a regulator of epithelial cell-cell and cell-substrate adhesion. *Journal of Cell Science*, *127*, 885–895.

Wong, X., Luperchio, T. R., & Reddy, K. L. (2014). NET gains and losses: The role of changing nuclear envelope proteomes in genome regulation. *Current Opinion in Cell Biology*, *28*, 105–120.

Worman, H. J., Ostlund, C., & Wang, Y. (2010). Diseases of the nuclear envelope. *Cold Spring Harbor Perspectives in Biology*, *2*, a000760.

CHAPTER TWO

Proximity-Dependent Biotin Identification (BioID) in *Dictyostelium* Amoebae

Petros Batsios, Irene Meyer, Ralph Gräf[1]

Institute for Biochemistry and Biology, Department of Cell Biology, University of Potsdam, Potsdam, Germany

[1]Corresponding author: e-mail address: rgraef@uni-potsdam.de

Contents

Abstract

The identification of a *bona fide* lamin-like protein in *Dictyostelium* made this lower eukaryote an attractive model organism to study evolutionarily conserved nuclear envelope (NE) proteins important for nuclear organization and human laminopathies. Proximity-dependent biotin identification (BioID), reported by Roux and colleagues, is a powerful discovery tool for lamin-associated proteins. In this method, living cells express a bait protein (e.g., lamin) fused to an R118G-mutated version of BirA, an *Escherichia coli* biotinylase. In the presence of biotin, BirA-R118G biotinylates target proteins in close proximity *in vivo*, which are purified using streptavidin and identified by immunoblotting or mass spectrometry. We adapted the BioID method for use in

Methods in Enzymology, Volume 569
ISSN 0076-6879
http://dx.doi.org/10.1016/bs.mie.2015.09.007

Dictyostelium amoebae. The protocols described here successfully revealed *Dictyostelium* lamin-like protein NE81 proximity to Sun1, a conserved inner nuclear membrane protein.

1. INTRODUCTION

Among the five eukaryotic supergroups including Opisthokonta, Amoebozoa, Archaeplastida, Excavata, and SAR (Stramenopile, Alveolata, Rhizaria), researchers have focused overwhelmingly on opisthokont models (mainly fungi and metazoans; Gräf, Batsios, & Meyer, 2015). *Dictyostelium discoideum*, the best-studied amoebozoan organism (Adl et al., 2012), provides a much-needed alternative model to study general principles in cell biology. *Dictyostelium* cells have proven their value in basic studies (e.g., chemotaxis, phagocytosis, cell differentiation) and as human disease models (Meyer, Kuhnert, & Gräf, 2011). We use *Dictyostelium* cells to study microtubule nucleation, centrosomes, and the nuclear lamina—all deeply interconnected since microtubule organization is a major function of the centrosome, which is connected to the nuclear lamina via nuclear envelope (NE)-spanning LINC complexes (Gräf, 2015). The discovery of NE81, a lamin protein, in *Dictyostelium* makes this organism a crucially diverse model for nuclear lamina structure (Krüger et al., 2012). Lamins were previously considered unique to metazoans, but the existence of NE81 in *Dictyostelium* now moves the evolutionary origin of lamins back to the origin of the Amorphea (synoptic term for Amoebozoa and Opisthokonta), or earlier (Gräf et al., 2015). The identification of NE81-binding partners will be crucial to understand the evolutionarily conserved functions of lamins beyond metazoans. In 2012, Roux and coworkers published proximity-dependent biotin identification (BioID), a powerful new method to screen for protein partners in living cells (Roux, Kim, Raida, & Burke, 2012). This chapter describes the successful adaptation of BioID for use in *Dictyostelium*.

In metazoan cells, type V intermediate filaments formed by A- or B-type lamins are major components of nuclear "lamina" structure. Lamins have a small globular "head" domain, a central ~370-residue coiled coil "rod" domain, and a C-terminal "tail" domain. Other general features include a nuclear localization signal, conserved CDK1 phosphorylation sites, and a C-terminal "CaaX" motif required for posttranslational isoprenylation (Ho & Lammerding, 2012). The "tail" of most lamins includes an Ig-fold domain important for binding to specific partners (Al-Haboubi,

Shumaker, Koser, Wehnert, & Fahrenkrog, 2011). All these features, except the Ig-fold domain, are conserved in *Dictyostelium* NE81 (Krüger et al., 2012). Lamins have diverse roles in nuclear structure, chromatin organization, gene regulation, signaling, and mechanosensing (Fedorchak, Kaminski, & Lammerding, 2014; Simon & Wilson, 2013). Mutations in lamins or lamin-binding proteins cause a spectrum of tissue-specific human diseases ("laminopathies"; Worman & Schirmer, 2015). Disease mechanisms are poorly understood, but may involve perturbed binding to conserved or tissue-specific partners (Worman & Schirmer, 2015). Around 90 lamin-associated proteins have been identified including specific nucleoporins, transcription factors, chromatin modifiers, histones, and numerous inner nuclear membrane (INM) proteins (Dechat, Gesson, & Foisner, 2010; Simon & Wilson, 2013). Certain INM proteins that bind lamins are conserved in other eukaryotic supergroups; these include the LEM-domain proteins (Brachner & Foisner, 2011), Samp1/Ima1 (expressed in mammals, fungi [Buch et al., 2009], and *Dictyostelium* [dictybase.org ID DDB_G0292450]), and SUN-domain proteins (Zhou & Meier, 2013). SUN proteins bind KASH-domain proteins (nesprins) in the outer nuclear membrane, forming "LINC" complexes that link the nucleoskeleton and cytoskeleton (Crisp et al., 2006; Razafsky & Hodzic, 2015).

1.1 Lamins and Their Partners: The Challenge

Studying lamins and their interactions with membrane proteins is challenging. Lamins form insoluble networks under physiological conditions, and conditions required to solubilize lamins might release functionally important partners. Furthermore, the solubilization of native membrane proteins requires detergents and ionic conditions that might disrupt specific partners.

The BioID method, established by Roux and coworkers (Roux, 2013; Roux, Kim, & Burke, 2013; Roux et al., 2012), is a uniquely powerful tool for studying biochemically recalcitrant proteins. Here, the protein of interest ("bait") is fused to an R118G-mutated promiscuous variant of BirA, an *Escherichia coli* biotinylase, and then expressed in cells. Native proteins in close proximity to the BirA-R118G-fused bait can be biotinylated (Figure 1). Biotinylation is a two-step process involving the ATP-dependent formation of a biotinoyl-5′-AMP intermediate, which is conjugated to a lysine residue in the substrate (Kwon & Beckett, 2000). The R118G mutation notably reduces BirA affinity for the highly reactive biotinoyl-5′-AMP intermediate, allowing it to react with any primary amine in its immediate

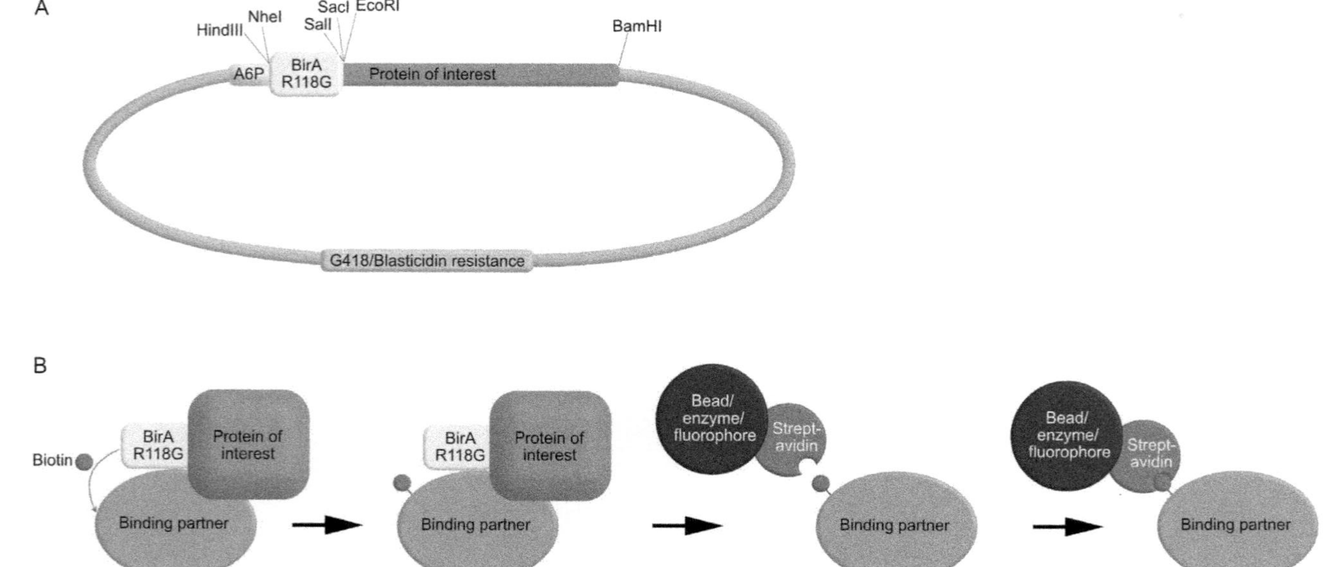

Figure 1 Schematic of BioID-vector design and protein purification. (A) Design of *Dictyostelium* BirA-R118G N-terminal fusion vectors. Useful restriction sites for cloning are indicated. (B) Schematic depicting biotinylation of proximal proteins by the BirA-R118G-fused protein of interest, and subsequent streptavidin-dependent detection or purification.

vicinity *in vitro* (Choi-Rhee, Schulman, & Cronan, 2004). Roux and coworkers exploited this property to identify proximal proteins in living cells (Roux et al., 2012) and showed that proteins must be located within 10 nm of BirA-R118G to become biotinylated (Kim et al., 2014). The primary targets are lysine ε-amino groups, which are frequently surface-exposed on folded proteins. Biotinylated proteins can be detected or purified using streptavidin. Streptavidin-conjugated fluorophores, enzymes (peroxidase, alkaline phosphatase), and beads (agarose, magnetic, latex) are widely available, facilitating both detection (by immunoblotting or fluorescence microscopy) and affinity purification for mass spectrometry (MS) identification. Roux and coworkers proved the feasibility and potential of this method using BirA-R118G fused to lamin A (Roux et al., 2012).

1.2 Advantages and Limitations of BioID

As noted above, BioID is especially powerful for interactions in which one or both partners are insoluble under physiological conditions. This method was applied successfully to identify interactors for the bilobed cytoskeletal complex in *Trypanosoma* (Morriswood et al., 2013), mammalian cell junctional complexes (Van Itallie et al., 2013, 2014), centrosomes (Comartin et al., 2013; Firat-Karalar, Rauniyar, Yates, & Stearns, 2014), nuclear pore complexes (Kim et al., 2014), chromatin (Lambert, Tucholska, Go, Knight, & Gingras, 2014), and INM proteins targeted by a secreted *Chlamydia* protein (Mojica et al., 2015). These reports suggest that BioID strategies are broadly applicable and have rather high specificity.

BioID has a proximity range (10 nm) similar to fluorescence resonance energy transfer (FRET), a well-known fluorescence-based proximity assay (Kim et al., 2014). However, FRET requires expensive equipment and engineered (potentially sterically incompatible) versions of both partners expressed at compatible concentrations. By contrast, BioID involves a single fusion protein and can be used to discover novel partners when combined with MS.

Another BioID advantage is that biotinylation is cumulative in the presence of excess biotin, allowing detection of weak or transient interactions between proteins in their natural subcellular environments under normal posttranslational control. Moreover false positives, a serious problem in yeast two-hybrid screens, are significantly reduced by comparison to BirA-only controls, to identify bait-specific positives (Kim et al., 2014). Biotinylation is a relatively uncommon protein modification in eukaryotes, found mainly on a few carboxylases (Chapman-Smith & Cronan, 1999). Three endogenously biotinylated proteins are present in *Dictyostelium*: namely methylcrotonyl-CoA carboxylase alpha (MccA) (77 kDa), acetyl-CoA carboxylase (AccA) (257 kDa), and propionyl-CoA carboxylase alpha (PccA) (80 kDa) (www.dictybase.org; Basu et al., 2013). Other contaminants in mammalian cells include pyruvate carboxylase, acetyl-CoA carboxylase B, and a few histones, keratins, and ribosomal proteins (Roux et al., 2013). Mitochondria express an endogenously biotinylated protein, named MccA (Sasaki et al., 2003). With these exceptions, in common cell culture media (e.g., DMEM or Dictyostelium HL5c), protein biotinylation is "inducible" because it only occurs when cells are provided with exogenous biotin. Therefore, an inducible promoter is only required if expression of the BirA-R118G fusion protein itself causes a dominant-negative effect.

BioID is also significantly less prone to false-positive identifications because biotin binds streptavidin with $\sim 10^{-14}$ mol/l affinity (among the strongest known noncovalent interactions), enabling harsh wash conditions (e.g., 2% SDS (sodium dodecyl sulfate)) that remove nonspecifically adsorbed proteins.

1.2.1 Limitations of BioID

BioID has four main caveats. First, fusion to BirA-R118G might perturb the subcellular localization or function of the bait protein. Since BirA-R118G (35 kDa) is a globular protein only slightly larger than GFP (27 kDa), this is typically not an issue if the corresponding GFP-fusion protein is known to function normally. One must carefully consider where to place BirA-R118G on the bait protein. For example, fusion to the C-terminus of lamin might disrupt posttranslational processing, whereas fusion to the N-terminus might interfere with polymerization. Proteins with a cleavable N-terminal signal sequence should be tagged on their C-terminus. Second, biotinylation of the target protein might influence its function (biotinylation neutralizes the positive charge of Lys). Third, as with other interaction assays, negative results are uninformative; e.g., *bona fide* partners might lack accessible primary amines. Finally, target proteins must be recovered in sufficient abundance to allow their detection.

1.3 Adaptation of BioID to the *Dictyostelium* System

We modified the BirA-R118G vectors from Roux et al. (2012) for use in *Dictyostelium* cells (Fig. 1B). First, the extremely AT-biased codon usage in *D. discoideum* (GC content is only 27% in exons) required codon optimization of the BirA-R118G coding sequence. Second, expression in *Dictyostelium* required appropriate promoters. We chose the well-characterized *actin6*-promoter to drive expression of the bait protein, since its activity can be controlled by cell culture conditions. When cells are grown on bacterial lawns (*Klebsiella aerogenes* or *E. coli* B/r), *actin6* promoter activity is low; when cells are grown in axenic medium, promoter activity is high (Liu et al., 2002). This flexibility allows cells expressing BirA-R118G-fused baits to be grown under low selective pressure, if overexpression of a specific bait protein is harmful. We created two vectors, one for clone selection with blasticidin S and one for selection with G418. Selection with blasticidin usually leads to single plasmid integrations in the genome, whereas G418 selects for multiple, concatemeric insertions at a single genomic site (Barth, Fraser, & Fisher, 1998). Thus, G418 selection also often yields higher levels of protein expression. To avoid potentially toxic overexpression, it made sense to select clones under conditions that suppress *actin6* promoter activity—namely, selection on lawns of bacteria. To enable this, we removed the usual *actin15*-promoter within the G418 resistance cassette and replaced it in our BioID vectors with the relatively uncommon V18-promoter, which is active when cells are grown on bacteria (Wetterauer et al., 1996).

Our two BirA-R118G vectors for fusion to the N-terminus of the bait protein are based on the GFP-vectors pIS76 (blasticidin) and pIS77 (G418), respectively (Schulz, Baumann, Samereier, Zoglmeier, & Gräf, 2009). The complete BirA-R118G sequence flanked by *Nhe*I and *Sal*I restriction sites was custom synthesized (GenArt, Thermo Fisher Scientific, Waltham, MA, USA), whereby each codon was replaced by the most common codon in the published *Dictyostelium* codon bias table (www.dictybase.org). GFP in pIS76 and pIS77 was then replaced by BirA-R118G using the *Nhe*I and *Sal*I restriction sites, yielding the blasticidin-selectable plasmid pPB86 and G418-selectable plasmid pPB87, respectively. These "BioID-ready" plasmids are available at the Dicty Stock Center (Fey, Dodson, Basu, & Chisholm, 2013).

1.4 Overview of BioID Protocol in *Dictyostelium*

Dictyostelium cells are transformed with plasmids encoding the BirA-R118G-fused bait, or BirA-R118G alone, and grown with or without 50 μ*M* biotin for different times. Successful specific biotinylation of target proteins can be monitored by immunoblotting, or by microscopy as shown in Fig. 2 by fixing cells on coverslips, and staining with fluorescent

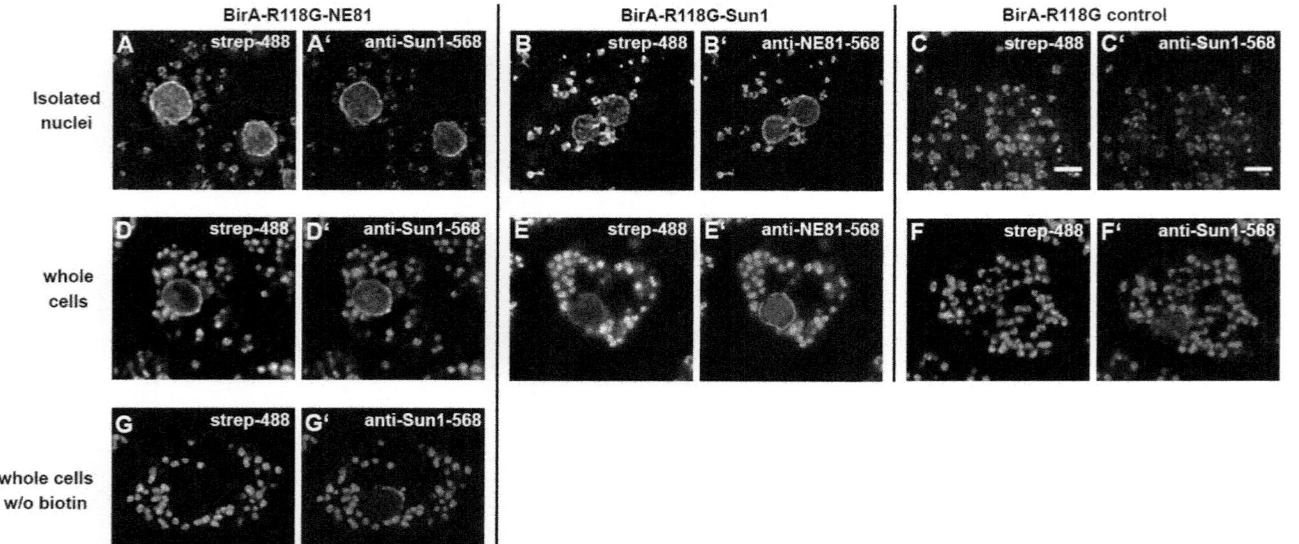

Figure 2 Fluorescence microscopy assay for biotinylated proteins in isolated nuclei (A–C) or whole cells (D–G) from *Dictyostelium* cells that expressed each indicated BioID construct. Grayscale images show biotinylated proteins visualized with streptavidin-AlexaFluor 488. Colored images show merged signals for streptavidin-AlexaFluor 488 (green), DAPI (blue), and specific nuclear envelope proteins (as indicated) stained with AlexaFluor 568 (red). z-Stacks were recorded in 0.25 μm steps. Images were deconvolved using iterative deconvolution (AxioVision 4.8) and a measured PSF (acquired with 200 nm TetraSpeck microspheres, Life Technologies). Maximum intensity projections are shown. Bar, 2 μm. Green speckles represent mitochondria, which stain weakly with DAPI (not visible in merged images). Mitochondrial signals are attributed to the endogenously biotinylated methylcrotonyl-CoA carboxylase, since these signals are seen without added biotin (G, G′). NE proteins were not biotinylated in the absence of biotin (G). Biotinylation at the NE is obvious in isolated nuclei and whole cells expressing BirA-R118G-NE81 or BirA-R118G-Sun1 (A, B, D, E). Biotinylation at the correct location is minimally attributed to autobiotinylation of the fusion protein; further work is needed to assess potential biotinylation of proximal proteins. (See the color plate.)

streptavidin. Background staining of mitochondria, identified by their weak DAPI staining, is expected due to endogenous biotinylation of MccA (Sasaki et al., 2003).

One major issue for protein purification from *Dictyostelium* is that special measures are needed to prevent proteolysis. Proteolysis is a more serious issue in this phagocytic organism than in mammalian cell cultures and is not fully controlled by the usual protease inhibitor cocktails. We use a modified protease inhibitor cocktail, especially enriched in leupeptin (Batsios, Baumann, Gräf, & Meyer, 2013). We strongly discourage detergent lysis methods, which release proteases from lysosomes and microsomes. Instead, as detailed below, we lyse cells mechanically by filtration through polycarbonate filters. When working with nuclear proteins or centrosomes (which are tightly bound to nuclei), we recommend a centrifugation step to separate nuclei from cytoplasm (which contains most proteases). Pilot experiments are fast, and easily evaluated by SDS-PAGE and blotting with conjugated streptavidin and secondary reagents, as shown in Fig. 3A. The BirA-R118G-fused protein itself should be detected, since self-biotinylation should always occur. Biotinylated proteins are easily enriched by affinity purification using magnetic streptavidin microbeads (Miltenyi, Bergisch-Gladbach, Germany) before electrophoresis, as shown in Fig. 3B, and can be scaled up for protein identification by MS.

Two controls are essential. (A) Cells expressing the BirA-R118G-fused bait should also be incubated without added biotin, to confirm that BirA-R118G is functional and inducible. (B) Cells that express BirA-R118G alone should be incubated with biotin, to reveal proteins nonspecifically biotinylated through association with the BirA-R118G-tag itself.

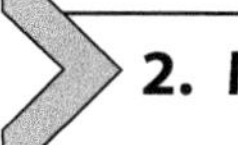

2. MATERIALS

2.1 Materials for Strain Generation

1. HL5c medium (Formedium): 26.55 g powdered medium per liter deionized water; autoclave 20 min at 121 °C.
2. SM agar plates: per liter deionized water add 10 g glucose, 10 g bactopeptone, 1 g yeast extract, 1.9 g KH_2PO_4, 4 ml of 1 M $MgSO_4$, and 20 g agar. Autoclave 20 min.
3. SPB (Sorenson phosphate buffer): 14.6 mM KH_2PO_4, 2 mM Na_2HPO_4, pH 6.0.
4. SPB/G418 plates: 90 mm petri dishes with 1.5% agarose in SPB. Autoclave the SPB agar, wait until solution cools to ~50 °C, and then add

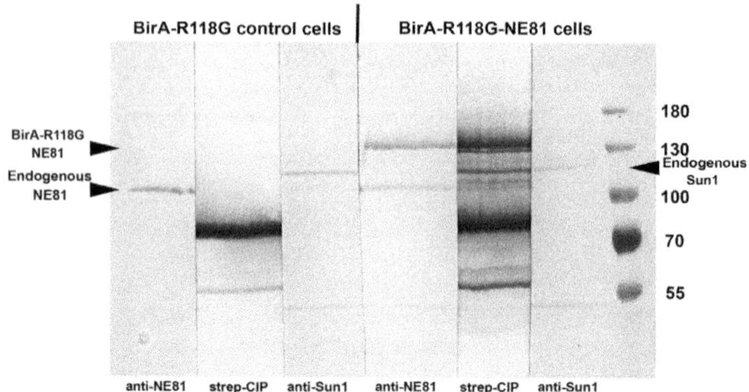

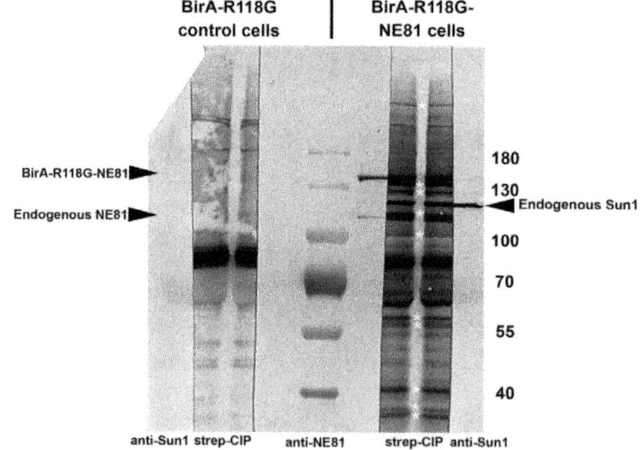

Figure 3 Immunoblots of biotinylated proteins from isolated nuclei (A) or after affinity purification via magnetic streptavidin beads (B), probed with rabbit antibodies specific for NE81 or Sun1, or CIP-conjugated streptavidin. Markers were Prestained PageRuler proteins (Life Technologies). Arrowheads indicate positions of Sun1, NE81, and BirA-fused NE81. (A) Extracts from isolated nuclei from BirA-R118G control cells (left) and BirA-R118G-NE81 cells (right) were resolved using broad SDS-PAGE lanes, cut into strips, and probed as indicated for each lane. (B) Western blot analysis of proteins affinity purified from BirA-R118G controls or BirA-R118G-NE81 cells. Blots were cut through lanes and strips were probed as indicated below each lane to allow direct comparison of strep-CIP and antibody signals. White asterisks indicate candidate proteins for further MS analysis.

G418 to a final concentration of 100 μg/ml. Prepare five plates (20 ml each) per transformation on the day of transformation.

5. Electroporation buffer (EB): 50 m*M* sucrose, 10 m*M* KH_2PO_4 adjusted to pH 6.1 with KOH.
6. 100 × $CaCl_2/MgCl_2$: 100 m*M* of each salt in one stock solution prepared in deionized water.
7. Blasticidin S and G418 antibiotics (Carl Roth, Karlsruhe, Germany).
8. Penstrep (100 ×) stock: 100 U/ml penicillin and 100 μg/ml streptomycin (GE Healthcare).
9. Standard uncoated tissue culture 24-well plates and 175 cm^2 flasks.
10. Conical (50 ml) tubes (BD Falcon 352098) and swinging bucket rotor (e.g. Beckman JS-5.3).

2.2 Materials for Cell Growth and Isolation of Nuclei

All solutions (except HL5c-medium) should be filtered through a 0.45-μm filter. Required centrifuge rotors include a fixed angle rotor for 250 ml centrifuge flasks (e.g., Beckman JLA-16.250) and a swinging bucket rotor (e.g., Beckman JS-5.3) suitable for 15- and 50-ml conical tubes (e.g., BD Falcon 352097 and 352098, respectively).

2.2.1 Cell Preparation

1. HL5c medium (see Section 2.1).
2. SPB (see Section 2.1).
3. Cytochalasin A (Sigma-Aldrich): 5 m*M* stock solution; dissolve in dimethyl sulfoxide and store at −20 °C. Cytochalasin A depolymerizes and solubilizes F-actin and, thus, reduces co-isolation of actomyosin complexes.
4. Biotin stock: 50 m*M* biotin in HL5c medium with 1 × Penstrep. This mixture will form a suspension. Do not autoclave. Resuspend prior to use. Biotin will only dissolve after it is diluted to its working concentration.

2.2.2 Cell Lysis and Isolation of Nuclei

1. 50 × Protease inhibitor cocktail: 50 m*M* Pefabloc SC, 1.25 mg/ml leupeptin, 0.5 mg/ml tosyl-arginin-methylester, 0.5 mg/ml soybean trypsin inhibitor, 0.05 mg/ml aprotinin, 0.05 mg/ml pepstatin, 100 m*M* benzamidine, 50 m*M* Na-ATP, pH 7.0; store at −70 °C. ATP reduces co-isolation of actomyosin complexes by releasing myosin

from the rigor state, thereby helping solubilize F-actin/myosin networks.

2. Dithiothreitol (DTT): 1 *M* stock solution, store at −20 °C.
3. Lysis buffer I: 100 m*M* Na-PIPES, pH 6.9, 2 m*M* EDTA, 10% (w/v) sucrose. Add fresh prior to use: 1 × protease inhibitor cocktail, 2 μ*M* cytochalasin A, 1 m*M* DTT.
4. Sucrose solutions: 30% and 50% (w/v) sucrose in 100 m*M* Na-PIPES, pH 6.9, 2 m*M* EDTA. Add fresh prior to use: 2 m*M* DTT, 1 × protease inhibitor cocktail.
5. 5 μm-Pore polycarbonate filters (diameter 25 mm, Nuclepore, Whatman Inc.) and suitable filter holder for 50-ml syringes (Whatman, Inc.).

2.2.3 Solubilization of Nuclear Material and Isolation of Biotinylated Proteins

1. Solubilization buffer: 300 m*M* NaCl, 1% Triton X-100, 50 m*M* Tris–HCl, pH 8.0; add prior to use: 1 × protease inhibitor cocktail. This buffer worked for BirA-R118G-NE81. Harsher conditions (Mehus, Anderson, & Roux, Chapter "BioID Identification of Lamin-Associated Proteins" in this volume) may also be useful.
2. Urea sample buffer (USB): 150 m*M* Tris–HCl, pH 6.8, 6 *M* urea (never heated), 10% SDS, 100 m*M* DTT, 0.1% bromophenol blue (optional); store at −20 °C.
3. Wash buffer I: 150 m*M* NaCl, 1% NP-40, 0.5% sodium deoxycholate, 0.1% SDS, 50 m*M* Tris–HCl, pH 8.0.
4. Wash buffer II: 20 m*M* Tris–HCl (pH 7.5).
5. Elution buffer for SDS-PAGE: 50 m*M* Tris–HCl (pH 6.8), 50 m*M* DTT, 1% SDS, 1 m*M* EDTA, 0.005% bromophenol blue, 10% glycerol.
6. Streptavidin microbeads (μMACS; Miltenyi Biotec).
7. Magnetic columns (μMACS; Miltenyi Biotec).
8. Magnetic separator (μMACS; Miltenyi Biotec).

2.3 Materials for Light Microscopy

1. Fixing solution: 30 m*M* PIPES, 12.5 m*M* HEPES, 4 m*M* EGTA, 1 m*M* $MgCl_2$, pH 6.9, freshly supplemented with 0.5% glutaraldehyde and 0.5% Triton X-100 (dilute from 25% stock).
2. Borohydride buffer: 0.1% sodium borohydride in SPB. Prepare fresh immediately prior to use. This step reduces unconjugated aldehyde groups, thus preventing unspecific binding of antibodies and streptavidin.

3. PBS (phosphate-buffered saline) 10× stock: 15 m*M* KH_2PO_4, 79 m*M* Na_2HPO_4, 1.38 *M* NaCl, 27 m*M* KCl, pH 7.4.
4. Antibody dilution buffer: 0.1% bovine serum albumin in PBS.
5. Mowiol mounting medium: 120 mg/ml polyvinyl alcohol 4–88, 30% glycerol, 0.2 *M* Tris–HCl, pH 8.8. Dissolve polyvinyl alcohol first in water by stirring and heating to 70 °C and then add glycerol and Tris–HCl (from 1 *M* stock). Centrifuge (10 min, 3000 × *g*), collect supernatant, and store aliquots at −20 °C. Do not freeze-thaw repeatedly.
6. Antibodies as needed. We used primary rabbit polyclonal antibodies against NE81 (Krüger et al., 2012) or Sun1 (Schulz et al., 2009), secondary AlexaFluor 568-conjugated goat antirabbit antibodies, and AlexaFluor 488-conjugated streptavidin (Life Technologies/Molecular Probes).

3. METHODS

3.1 Generate Strains Expressing the BirA-R118G-Fused Protein of Interest

3.1.1 *Cell Growth and Electrotransformation*

1. Per transformation experiment, cultivate *D. discoideum* cells in 25 ml HL5c medium to a density of 2×10^6 cells/ml (exponential growth; 5×10^7 cells total).
2. Chill on ice 20 min.
3. Pellet cells at 400 × *g* for 5 min at 4 °C (1500 rpm, JS-5.3 rotor); wash once with ice-cold SPB and centrifuge again.
4. Wash once with EB and resuspend pellet to 10^8 cells/ml (i.e., add ~700 μl EB).
5. Add 15–30 μg plasmid DNA, gently mix, and then fill into a precooled 4-mm electroporation cuvette.
6. Electroporate with 2 pulses (5 s rest) at 3 μF and 1.1 kV (BioRad Genepulser II).
7. Fill into 1.5-ml tube and incubate 10 min at room temperature. Add 1/100 volume of $CaCl_2$/$MgCl_2$ and rotate 15 min at room temperature.

3.1.2 *Clonal Selection with Blasticidin S*

1. Add cells to 25 ml HL5c medium, distribute over 24-well plate, incubate one day at 21 °C.

2. Replace medium in each well with 1 ml HL5c medium containing 4 μg/ml blasticidin S. Keep at 21 °C. Clones should be visible in about one week, but this can vary a lot for different plasmid constructs. If needed, isolate single clones by replating on bacterial plates as described (Mann et al., 1998).

3.1.3 Clonal Selection with G418

Use this protocol ONLY with plasmids in which G418 resistance is driven by a promoter (e.g., V18 promoter) that is also active when *Dictyostelium* cells are cultured on bacteria.

1. Add cells to 25 ml HL5c medium in a 175 cm^2 T-flask and keep at 21 °C for one day.
2. Plate *K. aerogenes* bacteria on an SM plate and incubate overnight at 37 °C. One plate is sufficient for two *Dictyostelium* transformations.
3. The day after transformation, resuspend bacteria from the plate completely using 5 ml SPB containing 100 μg/ml G418 and 1 × penstrep. Save for Step 5.
4. Tap off *Dictyostelium* cells from the bottom of the T-flask, fill cell suspension into a 50-ml conical tube, and centrifuge 5 min at 400 × *g* at 4 °C (e.g., 1500 rpm in JS-5.3 rotor).
5. Remove supernatant; resuspend cell pellet with 2.5 ml of bacterial suspension from Step 3.
6. Gently plate 0.5 ml cell suspension on SPB plates containing G418; incubate at 21 °C. Small feeding plaques of individual G418-resistant clones should be visible after ~4 days (this strongly varies depending on individual plasmid constructs).
7. Use toothpicks or pipet tips to pick *Dictyostelium* clones from the edges of the feeding plaques, and inoculate HL5c medium containing 10 μg/ml G418 and 1 × penstrep in 24-well plates. Continue cultivation at 21 °C.

3.2 Isolate Nuclei, Prepare Lysates, and Purify Biotinylated Proteins

This protocol is modified from Batsios et al. (2013). Keep all solutions and samples on ice to minimize proteolysis. Wear gloves and, if possible, perform all manipulations in a cold room.

3.2.1 Cell Preparation

1. Cultivate *D. discoideum* cells with biotin (axenic strain AX2). Add biotin (50 μ*M* final) to 400 ml HL5c medium in a 2-L Erlenmeyer flask.

Cultivate cells 24 h at 21 °C on a rotary shaker (150 rpm) to a density of ~4 × 10^6 cells/ml. If mutants do not grow in shaking culture, use adherent culture in T-flasks instead.

2. Sediment cells at 400 × *g* (2000 rpm, JLA-16.250 rotor) for 5 min at 4 °C.
3. Resuspend cell pellets in ~20 ml chilled SPB and transfer into 50-ml conical tube. Centrifuge 5 min at 400 × *g* (1500 rpm, JS-5.3 rotor).
4. Wash once with SPB, and once with SPB containing 2 μ*M* cytochalasin A.

3.2.2 Cell Lysis and Isolation of Nuclei

1. While washing cells (see Section 3.2.1, Step 4), add fresh protease inhibitor cocktail (1 ×) and 2 m*M* DTT to sucrose solutions. Place 1 ml 50% sucrose solution at the bottoms of two 15-ml tubes. Using a pipet, gently overlay each 50% sucrose solution with 2 ml of 30% sucrose solution. Place tubes on ice.
2. Remove the plunger from the syringe and mount syringe to a Nuclepore filter unit with the glossy side of the filter membrane oriented toward the syringe.
3. Resuspend each cell pellet with 20 ml lysis buffer I supplemented with 1 × protease inhibitor cocktail, 2 μ*M* cytochalasin A and 1 m*M* DTT, and fill into the syringe.
4. Filter lysate through the 5 μm-pore polycarbonate filter by thoroughly pushing the plunger. Repeat this step using the same filter. *Dictyostelium* cells are ~6–10 μm diameter and their nuclear diameter is ~2 μm. These shearing forces are sufficient to lyse cells while keeping nuclei intact.
5. Immediately load the filtrate (~10 ml per tube) onto the two sucrose gradients and centrifuge 10 min at 3700 × *g* (4600 rpm, JS-5.3 rotor) at 4 °C.
6. Centrifugation should produce two visible bands, one at the 10%/30% sucrose interface and one at the 30%/50% sucrose interface, and a pellet comprising unlysed cells. Nuclei are located at the 30%/50% sucrose interface.
7. Using a Pasteur pipet, gently remove the supernatant above the 10%/30% zone, then carefully and completely remove the 10%/30% zone without disturbing the nuclei. Finally, collect the nuclei in a total volume of ~1 ml without disturbing the pellet.

3.2.3 Solubilization of Nuclei and Isolation of Biotinylated Proteins

1. Centrifuge isolated nuclei (10 min, 10,000 × *g*, 4 °C). Remove supernatant and immediately add to the pelleted nuclei *either* (a) 1 ml precooled

solubilization buffer (mix well) for subsequent streptavidin affinity purification (Step 2, next) *or* (b) dissolve in 100–200 μl USB for direct analysis by SDS-PAGE and blotting as shown in Fig. 3A.

2. For affinity purification, immediately centrifuge (10 min, 10,000 × *g*, 4 °C) to sediment insoluble material.
3. Transfer supernatant to a fresh 1.5-ml tube and add 50 μl streptavidin beads to the lysate to bind biotinylated proteins. Mix well and incubate on ice 5 min.
4. Place microcolumn in the magnetic field of the μMACS separator. Equilibrate microcolumn by adding 100 μl equilibration buffer for protein application (provided with beads by Miltenyi, Bergisch-Gladbach, Germany). After the equilibration buffer flows through, add 200 μl solubilization buffer to the column and let flow through.
5. Apply the solubilized nuclei from Step 1(a) onto column and let the sample run through. Columns will not run dry.
6. Rinse column four times (200 μl each) with wash buffer I.
7. Rinse column once (100 μl) with wash buffer II.
8. For analysis by SDS-PAGE and blotting: (a) apply 20 μl hot USB buffer (preheated to 70 °C) to the column and incubate 5 min at room temperature; (b) remove column from magnetic separator and place in a fresh 1.5-ml tube; (c) apply 50 μl hot (preheated at 70 °C) USB to the column; and (d) collect eluate for analysis by SDS-PAGE and blotting.
9. For MS identification of gel-resolved proteins, use eluates from $\sim 1 \times 10^8$ cells. Rinse glass plates for SDS-PAGE with acetone or methanol. After electrophoresis, stain protein bands with Roti®-Blue (colloidal Coomassie G250; Carl Roth GmbH & Co. KG, Karlsruhe, Germany) according to manufacturer instructions. After staining, excise candidate bands using sterile scalpels and send to MS core facility. Consult your MS experts before using this protocol. They may prefer more sensitive methods, such as trypsinization of streptavidin-bound proteins.

3.3 Light Microscopy

All steps unless otherwise indicated are performed at room temperature.

3.3.1 Fixation and Fluorescence Microscopy of Whole Cells

1. Insert round 12-mm coverslips into the wells of a 24-well titer plate.
2. Add cell suspension in HL5c medium (4×10^5 cells); let cells settle 15 min.
3. Aspirate medium and gently add fixing solution. Incubate 5 min.

4. Aspirate fixing solution and wash twice (3 min each) with SPB.
5. Wash with freshly prepared borohydride/SPB, incubate 10 min, and wash with SPB.
6. Incubate 1 h with primary antibody diluted in antibody dilution buffer (for most antibodies, 0.5–1 μg/ml is effective). Omit this step if samples will only be stained with conjugated streptavidin.
7. Wash three times, 3 min each, with PBS.
8. Incubate 1 h with secondary antibody and streptavidin conjugate diluted in antibody dilution buffer to a final concentration of 1 μg/ml.
9. Wash three times (5 min each) with PBS. Include 0.1 μg/ml DAPI in the second wash.
10. Mount coverslip on slides using a small droplet of Mowiol.
11. View on a suitable fluorescence microscope. We use a Carl Zeiss CellObserver HS system equipped with a PlanApo 100×/1.4 lens, Axiocam MRm Rev. 3 camera, fiber-coupled HXP mercury halid lamp, high-quality fluorescence filter sets (for DAPI, GFP, and mRFP), ASI piezo stage, and AxioVision 4.8 software (Samereier, Meyer, Koonce, & Gräf, 2010).

3.3.2 Fixation and Fluorescence Microscopy of Isolated Nuclei

Isolated nuclei are fixed and stained as described for whole cells in Section 3.3.1. Instead of cells in Step 2, use nuclei from $\sim 6 \times 10^5$ cells, and sediment nuclei onto coverslips in lysis buffer I by centrifuging the 24-well plate at $2800 \times g$ for 10 min at 4 °C (JS-5.3 rotor with appropriate adaptors for 24-well plates).

3.4 Electrophoresis and Immunoblotting

We use standard procedures for SDS-PAGE, blotting, and immunostaining (e.g., Bollag, Rozycki, & Edelstein, 1996). We stain membranes with Ponceau S to detect protein lanes and cut sections (if needed), prior to blocking and incubating in primary antibody. Blots are incubated with streptavidin-conjugated calf intestine alkaline phosphatase (CIP) (Sigma-Aldrich) or specific antibodies together with secondary CIP-conjugated antibodies, and bands are visualized with nitroblue-tetrazolium chloride and bromo-chloro-indolyl phosphate. Broad-teeth gel combs yield broad lanes that can be cut in multiple strips, probed individually, and realigned seamlessly (Fig 3A). Alternatively, if standard combs are used, cut blots down the middle of individual lanes and probe separately (Fig. 3B).

4. CONCLUSIONS

These protocols have successfully adapted BioID for use in *Dictyostelium*, based on evidence that BirA-R118G-fused NE81 detects proximity to endogenous NE81 and Sun1 (Fig. 3). Both interactions were predicted: NE81 like other lamins is expected to self-assemble into higher order structures and SUN1 is an established lamin-binding protein in mammals and worms (Crisp et al., 2006; Haque et al., 2006). Controls were vital for interpretation: they showed that the strong ~80 kDa band was endogenously biotinylated (likely corresponding to mitochondrial MccA; Fig 3A, "strep-CIP" lanes), and revealed NE81-specific biotinylation of other bands including BirA-R118G-NE81 itself, endogenous NE81, and endogenous Sun1 (Fig. 3A). At least 10 additional prominent bands (asterisks; Fig. 3B) were specifically biotinylated in the presence of BirA-R118G-NE81. These candidate lamin-associated proteins in *Dictyostelium* await future identification by MS. Collectively these results validate these vectors and protocols for BioID applications in *Dictyostelium*.

REFERENCES

Adl, S. M., Simpson, A. G., Lane, C. E., Lukes, J., Bass, D., Bowser, S. S., et al. (2012). The revised classification of eukaryotes. *The Journal of Eukaryotic Microbiology*, *59*, 429–493.

Al-Haboubi, T., Shumaker, D. K., Koser, J., Wehnert, M., & Fahrenkrog, B. (2011). Distinct association of the nuclear pore protein Nup153 with A- and B-type lamins. *Nucleus*, *2*, 500–509.

Barth, C., Fraser, D. J., & Fisher, P. R. (1998). Co-insertional replication is responsible for tandem multimer formation during plasmid integration into the Dictyostelium genome. *Plasmid*, *39*, 141–153.

Basu, S., Fey, P., Pandit, Y., Dodson, R., Kibbe, W. A., & Chisholm, R. L. (2013). DictyBase 2013: Integrating multiple Dictyostelid species. *Nucleic Acids Research*, *41*, D676–D683.

Batsios, P., Baumann, O., Gräf, R., & Meyer, I. (2013). Isolation of Dictyostelium nuclei for light and electron microscopy. *Methods in Molecular Biology*, *983*, 283–294.

Bollag, D. M., Rozycki, M. D., & Edelstein, S. J. (1996). *Protein methods* (2nd ed.). New York: Wiley-Liss, Inc.

Brachner, A., & Foisner, R. (2011). Evolvement of LEM proteins as chromatin tethers at the nuclear periphery. *Biochemical Society Transactions*, *39*, 1735–1741.

Buch, C., Lindberg, R., Figueroa, R., Gudise, S., Onischenko, E., & Hallberg, E. (2009). An integral protein of the inner nuclear membrane localizes to the mitotic spindle in mammalian cells. *Journal of Cell Science*, *122*, 2100–2107.

Chapman-Smith, A., & Cronan, J. E., Jr. (1999). Molecular biology of biotin attachment to proteins. *The Journal of Nutrition*, *129*, 477S–484S.

Choi-Rhee, E., Schulman, H., & Cronan, J. E. (2004). Promiscuous protein biotinylation by Escherichia coli biotin protein ligase. *Protein Science*, *13*, 3043–3050.

Comartin, D., Gupta, G. D., Fussner, E., Coyaud, E., Hasegan, M., Archinti, M., et al. (2013). CEP120 and SPICE1 cooperate with CPAP in centriole elongation. *Current Biology, 23*, 1360–1366.

Crisp, M., Liu, Q., Roux, K., Rattner, J. B., Shanahan, C., Burke, B., et al. (2006). Coupling of the nucleus and cytoplasm: Role of the LINC complex. *The Journal of Cell Biology, 172*, 41–53.

Dechat, T., Gesson, K., & Foisner, R. (2010). Lamina-independent lamins in the nuclear interior serve important functions. *Cold Spring Harbor Symposia on Quantitative Biology, 75*, 533–543.

Fedorchak, G. R., Kaminski, A., & Lammerding, J. (2014). Cellular mechanosensing: Getting to the nucleus of it all. *Progress in Biophysics and Molecular Biology, 115*, 76–92.

Fey, P., Dodson, R. J., Basu, S., & Chisholm, R. L. (2013). One stop shop for everything Dictyostelium: DictyBase and the Dicty Stock Center in 2012. *Methods in Molecular Biology, 983*, 59–92.

Firat-Karalar, E. N., Rauniyar, N., Yates, J. R., 3rd, & Stearns, T. (2014). Proximity interactions among centrosome components identify regulators of centriole duplication. *Current Biology, 24*, 664–670.

Gräf, R. (2015). Microtubule organization in Dictyostelium. In *Encyclopedia of life sciences (ELS)*. Chichester: John Wiley & Sons, Ltd.

Gräf, R., Batsios, P., & Meyer, I. (2015). Evolution of centrosomes and the nuclear lamina: Amoebozoan assets. *European Journal of Cell Biology, 94*, 249–256.

Haque, F., Lloyd, D. J., Smallwood, D. T., Dent, C. L., Shanahan, C. M., Fry, A. M., et al. (2006). SUN1 interacts with nuclear lamin A and cytoplasmic nesprins to provide a physical connection between the nuclear lamina and the cytoskeleton. *Molecular and Cellular Biology, 26*, 3738–3751.

Ho, C. Y., & Lammerding, J. (2012). Lamins at a glance. *Journal of Cell Science, 125*, 2087–2093.

Kim, D. I., Birendra, K. C., Zhu, W., Motamedchaboki, K., Doye, V., & Roux, K. J. (2014). Probing nuclear pore complex architecture with proximity-dependent biotinylation. *Proceedings of the National Academy of Sciences of the United States of America, 111*, E2453–E2461.

Krüger, A., Batsios, P., Baumann, O., Luckert, E., Schwarz, H., Stick, R., et al. (2012). Characterization of NE81, the first lamin-like nucleoskeleton protein in a unicellular organism. *Molecular Biology of the Cell, 23*, 360–370.

Kwon, K., & Beckett, D. (2000). Function of a conserved sequence motif in biotin holoenzyme synthetases. *Protein Science, 9*, 1530–1539.

Lambert, J. P., Tucholska, M., Go, C., Knight, J. D., & Gingras, A. C. (2014). Proximity biotinylation and affinity purification are complementary approaches for the interactome mapping of chromatin-associated protein complexes. *Journal of Proteomics, 118*, 81–94.

Liu, T., Mirschberger, C., Chooback, L., Arana, Q., Dal Sacco, Z., MacWilliams, H., et al. (2002). Altered expression of the 100 kDa subunit of the Dictyostelium vacuolar proton pump impairs enzyme assembly, endocytic function and cytosolic pH regulation. *Journal of Cell Science, 115*, 1907–1918.

Mann, S. K. O., Devreotes, P. N., Eliott, S., Jermyn, K., Kuspa, A., Fechheimer, M., et al. (1998). Cell biological, molecular genetic, and biochemical methods used to examine Dictyostelium. In J. E. Celis (Ed.), *Cell biology: A laboratory handbook* (2nd ed., pp. 431–465). San Diego: Academic Press.

Meyer, I., Kuhnert, O., & Gräf, R. (2011). Functional analyses of lissencephaly-related proteins in Dictyostelium. *Seminars in Cell and Developmental Biology, 22*, 89–96.

Mojica, S. A., Hovis, K. M., Frieman, M. B., Tran, B., Hsia, R. C., Ravel, J., et al. (2015). SINC, a type III secreted protein of *Chlamydia psittaci*, targets the inner nuclear membrane of infected cells and uninfected neighbors. *Molecular Biology of the Cell, 26*, 1918–1934.

Morriswood, B., Havlicek, K., Demmel, L., Yavuz, S., Sealey-Cardona, M., Vidilaseris, K., et al. (2013). Novel bilobe components in *Trypanosoma brucei* identified using proximity-dependent biotinylation. *Eukaryotic Cell, 12*, 356–367.

Razafsky, D., & Hodzic, D. (2015). Nuclear envelope: Positioning nuclei and organizing synapses. *Current Opinion in Cell Biology, 34*, 84–93.

Roux, K. J. (2013). Marked by association: Techniques for proximity-dependent labeling of proteins in eukaryotic cells. *Cellular and Molecular Life Sciences, 70*, 3657–3664.

Roux, K. J., Kim, D. I., & Burke, B. (2013). BioID: A screen for protein–protein interactions. *Current Protocols in Protein Science, 74*, 19.23.1–19.23.14.

Roux, K. J., Kim, D. I., Raida, M., & Burke, B. (2012). A promiscuous biotin ligase fusion protein identifies proximal and interacting proteins in mammalian cells. *The Journal of Cell Biology, 196*, 801–810.

Samereier, M., Meyer, I., Koonce, M. P., & Gräf, R. (2010). Live cell-imaging techniques for analyses of microtubules in Dictyostelium. *Methods in Cell Biology, 97*, 341–357.

Sasaki, E., Okamoto, Y., Yoshida, K., Okamura, H., Shimizu, K., Nasu, F., et al. (2003). Transblot and cytochemical identification of avidin-interacting proteins in mitochondria of cultured cells. *Histochemistry and Cell Biology, 120*, 327–333.

Schulz, I., Baumann, O., Samereier, M., Zoglmeier, C., & Gräf, R. (2009). Dictyostelium Sun1 is a dynamic membrane protein of both nuclear membranes and required for centrosomal association with clustered centromeres. *European Journal of Cell Biology, 88*, 621–638.

Simon, D. N., & Wilson, K. L. (2013). Partners and post-translational modifications of nuclear lamins. *Chromosoma, 122*, 13–31.

Van Itallie, C. M., Aponte, A., Tietgens, A. J., Gucek, M., Fredriksson, K., & Anderson, J. M. (2013). The N and C termini of ZO-1 are surrounded by distinct proteins and functional protein networks. *The Journal of Biological Chemistry, 288*, 13775–13788.

Van Itallie, C. M., Tietgens, A. J., Aponte, A., Fredriksson, K., Fanning, A. S., Gucek, M., et al. (2014). Biotin ligase tagging identifies proteins proximal to E-cadherin, including lipoma preferred partner, a regulator of epithelial cell–cell and cell-substrate adhesion. *Journal of Cell Science, 127*, 885–895.

Wetterauer, B., Morandini, P., Hribar, I., Murgia-Morandini, I., Hamker, U., Singleton, C., et al. (1996). Wild-type strains of Dictyostelium discoideum can be transformed using a novel selection cassette driven by the promoter of the ribosomal V18 gene. *Plasmid, 36*, 169–181.

Worman, H. J., & Schirmer, E. C. (2015). Nuclear membrane diversity: Underlying tissue-specific pathologies in disease. *Current Opinion in Cell Biology, 34*, 101–112.

Zhou, X., & Meier, I. (2013). How plants LINC the SUN to KASH. *Nucleus, 4*, 206–215.

CHAPTER THREE

Purification and Structural Analysis of LEM-Domain Proteins

Isaline Herrada*, Benjamin Bourgeois*,[1], Camille Samson*, Brigitte Buendia†, Howard J. Worman‡,§, Sophie Zinn-Justin*,[2]

*Institute for Integrative Biology of the Cell, CEA, CNRS, Université Paris-Sud, Gif-sur-Yvette, Paris, France
†Institut de Biologie Fonctionnelle et Adaptative, Université Paris Diderot, Paris, France
‡Department of Medicine, College of Physicians and Surgeons, Columbia University, New York, USA
§Department of Pathology and Cell Biology, College of Physicians and Surgeons, Columbia University, New York, USA
[2]Corresponding author: e-mail address: sophie.zinn@cea.fr

Contents

Abstract

LAP2-emerin-MAN1 (LEM)-domain proteins are modular proteins characterized by the presence of a conserved motif of about 50 residues. Most LEM-domain proteins localize at the inner nuclear membrane, but some are also found in the endoplasmic reticulum or nuclear interior. Their architecture has been analyzed by predicting the limits of their globular domains, determining the 3D structure of these domains and in a few cases

[1] Current address: Center for Integrated Protein Science Munich, Department Chemie, Technische Universität München, 85478 Garching, Germany.

Methods in Enzymology, Volume 569
ISSN 0076-6879
http://dx.doi.org/10.1016/bs.mie.2015.07.008

calculating the 3D structure of specific domains bound to biological targets. The LEM domain adopts an α-helical fold also found in SAP and HeH domains of prokaryotes and unicellular eukaryotes. The LEM domain binds to BAF (barrier-to-autointegration factor; *BANF1*), which interacts with DNA and tethers chromatin to the nuclear envelope. LAP2 isoforms also share an N-terminal LEM-like domain, which binds DNA. The structure and function of other globular domains that distinguish LEM-domain proteins from each other have been characterized, including the C-terminal dimerization domain of LAP2α and C-terminal WH and UHM domains of MAN1. LEM-domain proteins also have large intrinsically disordered regions that are involved in intra- and intermolecular interactions and are highly regulated by posttranslational modifications *in vivo*.

1. INTRODUCTION

The nuclear envelope (NE) forms the boundary of the nuclear compartment in eukaryotic cells. The NE includes two nuclear membranes (inner and outer), nuclear pore complexes, and the nuclear lamina. The inner nuclear membrane (INM) has numerous integral membrane proteins and closely contacts nuclear intermediate filament proteins, lamins, to form "nuclear lamina" networks that line the membrane. It is poorly understood how these proteins are organized at the structural level: the nuclear periphery is densely packed with proteins and chromatin, and structural investigation of networks of INM proteins and lamins *in situ* remains challenging (Gruenbaum & Medalia, 2015). Current knowledge of the inner aspect of the NE is therefore based on the identification of direct or indirect interactions between integral INM proteins and their partners including lamins, chromatin, and transcriptional regulators.

The most extensively studied INM proteins and functional complexes involve LAP2-emerin-MAN1 (LEM)-domain proteins. This protein family is characterized by a globular LEM domain, initially identified in *l*amina-associated polypeptide 2 (LAP2), *e*merin, and *M*AN1 (Lin et al., 2000). LEM-domain proteins are often anchored at the INM by binding to lamins. However, INM anchoring is not a feature of all isoforms, and no consensus lamin-binding motif has been identified in LEM-domain proteins. For example, the *LAP2* gene is alternatively spliced to produce six protein isoforms; LAP2β, γ, δ, and ε each have a transmembrane domain, whereas LAP2α and LAP2ξ do not (Berger et al., 1996; Furukawa, Pante, Aebi, & Gerace, 1995). However, the LAP2α isoform-specific region from amino acid 616 to amino acid 693 binds the tail domain of A-type lamins (Dechat et al., 2000). LAP2α is essential for solubilizing a small fraction of lamin A in the nucleoplasm (Naetar et al., 2008). Yeast two-hybrid experiments showed that LAP2β, the largest INM-localized isoform, uses residues 298–373 to interact with lamins B1 and B2 (Furukawa, Fritze, & Gerace, 1998). This lamin-binding region is conserved in most LAP2 isoforms with a transmembrane domain.

The LEM domains of LAP2, emerin, and MAN1 directly bind to barrier-to-autointegration factor (BAF) (Cai et al., 2001; Lee et al., 2001; Mansharamani & Wilson, 2005; Shumaker, Lee, Tanhehco, Craigie, & Wilson, 2001). BAF is highly conserved among multicellular eukaryotes (Cai et al., 1998) and is essential in *C. elegans* (Zheng et al., 2000). BAF colocalizes with chromosomal DNA during both interphase and mitosis in *Drosophila* (Furukawa, 1999), but localizes dynamically at the NE, nucleoplasm, and cytoplasm in *C. elegans* and mammalian cells (Jamin & Wiebe, 2015). Through mechanisms that are not understood, BAF helps tether chromatin to the NE (Asencio et al., 2012; Margalit, Segura-Totten, Gruenbaum, & Wilson, 2005) and functions as an epigenetic regulator (Montes de Oca, Andreassen, & Wilson, 2011).

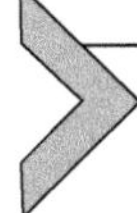

2. "DIVIDE AND CONQUER" APPROACH TO STUDY LEM-DOMAIN PROTEINS

In mammals, three major types of LEM proteins can be distinguished based on their domain organization as shown in Fig. 1 (Brachner & Foisner, 2011). Emerin and LAP2 are INM proteins characterized by one transmembrane segment (Fig. 1A). They have an N-terminal nucleoplasmic LEM domain (shown for LAP2 in Fig. 1A) and a large region predicted as unstructured, followed by the transmembrane segment and short lumenal domain. MAN1 and Lem2 are anchored at the INM by two transmembrane segments, with their N- and C-terminal regions exposed in the nucleoplasm (Fig. 1E). Their LEM domains (Fig. 1F) are each followed by a large region predicted to be unstructured (like emerin and LAP2β), and their C-terminal nucleoplasmic regions include either one (Lem2; not shown) or two globular domains (MAN1; shown in Fig. 1G and H). Ankle1 and Ankle2 (also called Lem3 and Lem4) are characterized by ankyrin repeats, which are common protein–protein interaction motifs mainly found in eukaryotes (Fig. 1). Ankle1 has no transmembrane segment; it is an endonuclease that cleaves DNA and induces a DNA damage response (Brachner et al., 2012). Ankle2 by contrast has an N-terminal transmembrane segment and localizes with the endoplasmic reticulum (Asencio et al., 2012). Ankle2 is essential for

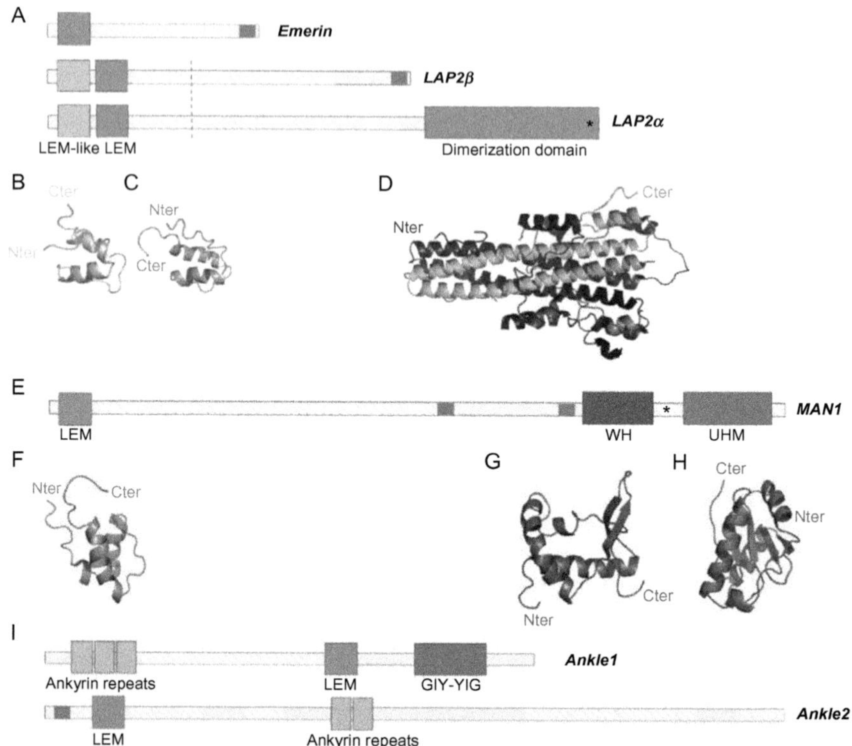

Figure 1 Schematic of LEM-domain protein architecture. Three families of LEM-domain proteins, illustrated by six human proteins. Hatched yellow regions are predicted as intrinsically disordered by Disopred3. Small green boxes indicate transmembrane domains. Blue dotted line indicates that LAP2 isoforms share residues 1–186 but have different C-terminal regions (after the line). (A) LEM-domain proteins with one transmembrane domain (emerin, LAP2β) localize at the INM; alternatively spliced isoform LAP2α has no transmembrane domain. The star in LAP2α indicates the Arg690Cys mutation linked to dilated cardiomyopathy. (B)–(D) 3D structures of the LEM-like domain in light orange (PDB code 1H9E), the LEM domain in dark orange (PDB code 1H9F), and LAP2α-specific dimerization domain in green (PDB code 2V0X). (E) INM-localized LEM-domain proteins with two transmembrane segments are Lem2 (not shown) and MAN1. The star in MAN1 indicates the UHM ligand motif Lys/Arg$_{763}$-X-Trp$_{765}$-Gln$_{766}$-X-X-Ala$_{769}$-Phe$_{770}$. (F)–(H) 3D models of MAN1 globular domains, calculated by homology for the LEM domain in orange, from NMR data for the WH domain in blue (PDB code 2CH0; Caputo et al., 2006), and from NMR chemical shift data and molecular modeling for the UHM domain in red (Konde et al., 2010). (I) LEM-domain proteins with ankyrin repeats: Ankle1 is soluble; Ankle2 is ER-localized. (See the color plate.)

postmitotic NE formation and controls BAF phosphorylation by binding and coordinating the activities of kinase VRK-1 and phosphatase PP2A.

Structural analysis of LEM-domain proteins has been hindered by difficulty expressing and purifying these large modular proteins, especially when they have transmembrane domains. Therefore, a "divide-to-conquer" approach has been chosen by several groups, based on bioinformatics prediction of the limits of the folded regions. Using this approach, globular domains of LEM-domain proteins were systematically produced, purified, and structurally characterized.

2.1 The LEM Domain Common to LAP2, Emerin, and MAN1

Biochemical characterization of the LEM domain started with the chemical synthesis of the LEM domains of human emerin (residues 2–54) and LAP2 (residues 103–159) and the LEM-like domain of LAP2 (residues 2–56) (Laguri et al., 2001; Wolff et al., 2001). The LEM-like domain is a highly divergent version of the LEM domain. Whereas the identity between the LEM domains of emerin and LAP2 is 33%, the LEM and LEM-like domains of LAP2 are only 18% identical. Milligrams of each peptide were produced, purified by HPLC, and dissolved in a sodium phosphate buffer (pH 6.3). They were first analyzed by analytic ultracentrifugation, which showed that all these peptides are monomeric at concentrations less than 50 μM. The three peptides were then analyzed using proton nuclear magnetic resonance (NMR; 500 and 600 MHz spectrometers), at a concentration of about 1 mM at 25 °C, to calculate their three-dimensional structures in solution. Superimposition of the resulting structures revealed that they all adopted the same fold, mainly composed of two large parallel alpha helices stabilized by intramolecular electrostatic interactions and hydrophobic contacts (Fig. 2A).

The N-terminal region of LAP2 containing both the LEM-like and LEM domains was also studied by NMR (Cai et al., 2001). A recombinant His_6-tagged LAP2 fragment (residues 1–168) was produced in *E. coli* as a ^{15}N- and ^{13}C-labeled protein. It was purified by affinity and gel filtration chromatography in phosphate buffer pH 7.2. Multidimensional ^{1}H, ^{15}N, ^{13}C NMR experiments were performed to solve its 3D structure. This fragment contained two structurally independent, noninteracting domains that are connected by a flexible linker and adopt the same α-helical fold. NMR chemical shift mapping demonstrated that the LEM domain binds BAF, whereas the LEM-like domain binds DNA. These interactions involve similar regions of the LEM fold, specifically helix 1, the loop

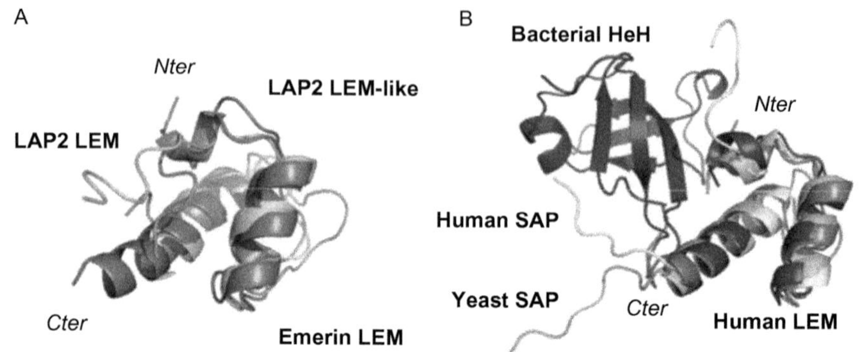

Figure 2 The LEM-domain fold. (A) Superimposition of the LEM structure of emerin (PDB 2ODC in marine blue) with the LEM (PDB 1H9F in cyan; DALI rmsd 2.2 Å) and LEM-like (PDB 1H9E in gray; DALI rmsd 3.0 Å) structures of LAP2. (B) Superimposition of the LEM fold with the bacterial HeH structure from the RNA-binding domain of the transcription termination factor rho (PDB 1A62 in magenta; DALI rmsd 1.7 Å), the yeast SAP structure of the RNA-binding protein Tho1 (PDB 4UZW; DALI rmsd 1.4 Å) and the human SAP structure of the ribonucleoprotein Hcc-1 (PDB 2DO1; DALI rmsd 2.2 Å). (See the color plate.)

connecting the two large helices and the N-terminus of helix 2. The distinct binding properties of the LEM-like and LEM domains are determined by their surface residues in these locations: predominantly positively charged in the case of the LEM-like domain, and mainly hydrophobic for the LEM domain.

There are few structural domains of less than 50 residues in the Protein Data Bank. A DALI search highlighted that the LEM and LEM-like domains are structurally highly related to the SAF-acinus-PIAS (SAP) and the helix-extended loop-helix (HeH) motifs. For example, the emerin LEM structure can be superimposed onto the SAP structures of human nuclear protein Hcc-1 and yeast protein Tho1, and the HeH domain of *E. coli* transcriptional terminator protein Rho (Fig. 2B). The SAP domain is a eukaryotic putative DNA-binding module found in chromatin-associated proteins that likely targets these proteins to specific chromosomal locations (Aravind & Koonin, 2000). HeH domains from bacteria, bacteriophages, and plants are known (or predicted) to bind nucleic acids, suggesting that this was the ancestral function of the HeH fold (Aravind & Koonin, 2001; Aravind, Mazumder, Vasudevan, & Koonin, 2002). The SCOP database gathers all these domains into a category named "LEM/SAP helix–extended loop–helix (HeH) motif." Brachner and Foisner showed in 2011 that orthologs of metazoan LEM-domain proteins in unicellular eukaryotes, which lack BAF, contain a SAP or HeH instead of a LEM motif, suggesting the LEM domain evolved from an ancestral SAP/HeH domain found in chromosome tethering proteins, concomitant with the emergence of BAF (Brachner & Foisner, 2011).

2.2 The C-Terminal Domain Specific to LAP2α

LAP2α contains an N-terminal region from amino acid 1 to amino acid 186 common to all LAP2 isoforms (including LEM-like and LEM domains, see Fig. 1A–C) and a specific C-terminal region from amino acid 187 to amino acid 693 responsible for many of its known functions (see Fig. 1A and D). This last region is specifically involved in LAP2α retention in postmitotic nuclei and essential for A-type lamin binding (Dechat et al., 1998, 2000; Vlcek, Just, Dechat, & Foisner, 1999). It is mutated at position 690 (Arg to Cys) in two brothers with dilated cardiomyopathy, but segregation of the disease-causing allele within the family is unclear (Taylor et al., 2005). LAP2α variant $Arg_{690}Cys$ is reported to be impaired in its ability to bind lamin A. To identify the molecular basis of the LAP2α/lamin A interaction, full-length murine LAP2α was digested by chymotrypsin and fragments were analyzed by mass spectrometry and N-terminal sequencing. A stable fragment was identified that comprised residues 459–693. This fragment was expressed in *E. coli* and purified by metal affinity chromatography followed by size-exclusion chromatography. Its three-dimensional structure was determined by X-ray crystallography (Fig. 1D; Bradley et al., 2007). The C-terminal region of LAP2α is an elongated α-helical dimer. Each monomer exhibits six helices with the fourth and fifth arranged into a four-stranded coiled coil. No complex between this C-terminal dimerization domain and A-type lamin tails could be observed using chromatography, calorimetry, or ultracentrifugation experiments. Only a solid-phase overlay assay suggested transient binding between the two molecules and pointed to six LAP2α residues involved in lamin recognition. The $Arg_{690}Cys$ substitution was introduced into the murine LAP2α fragment, but did not impair lamin binding in the solid-phase overlay assay. This illustrates the difficulty encountered in understanding the molecular basis of lamin recognition by LEM-domain proteins.

2.3 The WH and UHM Domains in MAN1

MAN1 and Lem2 are LEM-domain proteins anchored in the INM by two transmembrane segments. They exhibit two nucleoplasmic regions: an

N-terminal region including the LEM domain and a C-terminal region (see Fig. 1E and F). This last region contains two globular domains in MAN1 (Fig. 1G and H) and one globular domain in Lem2. Both globular domains in the C-terminal region of human MAN1 were analyzed by NMR to determine their three-dimensional structure.

A fragment comprising MAN1 C-terminal residues 655–775, which are shared with Lem2, was produced as a ^{15}N-, ^{13}C-labeled GST fusion protein and purified by affinity chromatography. The GST fragment was cleaved using thrombin and retained on an affinity column. The MAN1 fragment was eluted in the flow-through and then dialyzed to a concentration of about 0.5 m*M* in phosphate buffer (pH 6.0) containing 150 m*M* NaCl. It was analyzed by multidimensional ^{1}H, ^{15}N, and ^{13}C NMR, and the 3D structure of the region between residues 666 and 750 was solved from data acquired at 30 °C on 600 and 900 MHz spectrometers (Caputo et al., 2006). The N-terminal half of this region is mainly α-helical, whereas the C-terminal half is composed of two large β-strands arranged in a twisted antiparallel β-sheet (Fig. 1G). The helices form a three-helix bundle. A three-stranded β-sheet is packed onto the three-helix bundle. The α/β interface is mainly hydrophobic. The DALI server clearly identified the MAN1 region from residue 666 to residue 750 as adopting a winged helix (WH) fold also found in several DNA-binding domains belonging to transcription factors. The helix that contacts DNA in these transcription factors is positively charged in MAN1 and, consistently, the MAN1 fragment binds DNA (Caputo et al., 2006).

The MAN1 fragment from amino acid 775 to amino acid 911 is not found in Lem2. A point mutation in this region abolishes the capacity of MAN1 to bind R-Smads and repress transcription of the R-Smad-mediated signaling pathway (Pan et al., 2005). Consistently, Lem2 is not capable of antagonizing R-Smad-mediated signaling activity (Brachner, Reipert, Foisner, & Gotzmann, 2005). The MAN1 fragment (residues 755–911) was produced as a ^{15}N-, ^{13}C-labeled His_6-tagged protein to characterize its 3D structure by NMR (Konde et al., 2010). It was purified by affinity chromatography and gel filtration in a Tris buffer (pH 6.7) containing 150 m*M* NaCl. NMR analysis provided the secondary structure of the MAN1 fragment. However, conformational exchange precluded the determination of its 3D structure. Sequence alignment strongly suggested that residues 785–880 adopt a U2AF-homology motif (UHM) fold (Kielkopf, Lucke, & Green, 2004). NMR chemical shift data were consistent with this prediction. From NMR and bioinformatics analyses, it was possible to calculate a model for the UHM domain of MAN1 (Fig. 1H). Broad NMR signals suggesting conformational exchange were observed in the linker region between the WH and UHM domains (residues 755–785) as well as around the hydrophobic cavity of the UHM domain. UHM domains interact with peptides exhibiting a UHM ligand motif (ULM). This motif consists of a patch of positively charged amino acids followed by a conserved tryptophan residue (Kielkopf et al., 2004). In the MAN1 linker region, the motif Lys/Arg_{763}-X-Trp_{765}-Gln_{766}-X-X-Ala_{769}-Phe_{770} is highly conserved through metazoans (this motif is represented by a star in Fig. 1E). When Trp_{765} and Gln_{766} are mutated to alanines, the radius of gyration of the MAN1 fragment significantly increases, as shown by small-angle X-ray scattering (SAXS), and the NMR signals of the residues of the UHM hydrophobic cavity are changed. From these data, a model of the MAN1 fragment was proposed in which Trp_{765} anchors the linker onto the UHM domain through a conserved intramolecular interaction. The biological significance of this interaction is currently unknown. However, mutating Trp_{765} and Gln_{766} to alanines significantly weakens the interaction between the MAN1 fragment and the MH2 domain of Smad2.

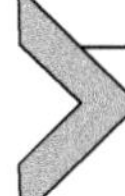

3. ANALYSIS OF PREDICTED UNSTRUCTURED REGIONS IN LEM-DOMAIN PROTEINS

Large intrinsically disordered regions (IDRs) have been identified in LEM-domain proteins, either between two globular domains or between a globular domain and a transmembrane segment (Fig. 1A, E, and I). The functional role of these regions is not yet understood. They might contain MoRFs, small (10–70 residues) fragments that undergo a disorder-to-order transition upon binding to their partners. However, no 3D structure of a LEM-domain protein IDR bound to its partner has been solved. They are largely posttranslationally modified and phosphorylated in a cell-cycle-dependent manner (Ellis, Craxton, Yates, & Kendrick-Jones, 1998; Hirano et al., 2009, 2005; Yip, Cotteret, & Chernoff, 2012). Phosphorylation of the emerin IDR is also triggered by the application of mechanical force on the nucleus (Guilluy et al., 2014).

The IDR located between the WH and UHM domains of MAN1 (residues 750–785) has been studied while analyzing the three-dimensional structure of the whole C-terminal nucleoplasmic region of MAN1. This IDR binds to MAN1 UHM domain, thus regulating its recognition properties (Konde et al., 2010). Mutation of the IDR sequence Trp_{765}-Gln_{766}

weakens both the intramolecular IDR–UHM interaction and the intermolecular MAN1–Smad2 interaction (Fig. 1E). This suggests that the IDR position onto the UHM is critical for Smad2 recognition. Three residues were identified as phosphorylated in this IDR, namely Ser_{777}, Ser_{781}, and Thr_{783} (http://www.phosphosite.org/proteinAction.do?id=7784). Ser_{777} and Thr_{783} are potential targets of Pro-directed kinases such as P38 (mitogen-activated protein kinase); Ser_{777} is also a potential target of glycogen synthase kinase-3β (GSK-3β) (predictions: elm.eu.org). These kinases might regulate the intramolecular interaction between MAN1 IDR and UHM regions, and thus the binding properties of MAN1. Other larger IDRs have been described in emerin and MAN1 N-terminal nucleoplasmic regions. Production and purification of these regions should enable a biophysical study of their structural and binding properties, either before or after posttranslational modification.

3.1 Production and Purification of Predicted Unstructured Regions of Emerin and MAN1

3.1.1 Emerin Nucleoplasmic Region

The emerin nucleoplasmic region contains a LEM domain and a large IDR (Fig. 1A). We attempted to produce in *E. coli* and purify by chromatography the whole nucleoplasmic fragment of emerin from amino acid 1 to amino acid 221 and a smaller fragment from amino acid 1 to amino acid 187 (emerin 1–187), but only the latter was soluble in our conditions. We expressed it as a 8His-tag, followed by a TEV site and the emerin 1–187 sequence. The cDNA coding for this construct was optimized for *E. coli,* synthesized by Genscript and cloned into a pETM13-LIC vector (pETM13 expression vector modified in our laboratory for LIC cloning system). The emerin 1–187 sequence contains four positions with amino acid substitutions or deletions in patients with X-linked Emery-Dreifuss muscular dystrophy (EDMD; positions 54, 95–99, 133, and 183). Studying the structural consequences of these mutations is of particular interest knowing that most cases of X-linked EDMD are caused by mutations leading to loss of emerin expression.

Detailed protocol for emerin 1–187 production and purification

1. *E. coli* cells are grown at 37 °C in LB media (or M9 media for NMR samples). When the OD_{600} is between 1 and 1.5, add 0.5 m*M* IPTG and induce expression overnight at 20 °C.
2. Lyse cells by sonication in lysis buffer (50 m*M* Tris–HCl, pH 7.5; 300 m*M* NaCl; 5% glycerol; and 1% Triton X-100).

3. As the protein is produced in inclusion bodies, the pellet is resuspended in Buffer A8 (50 m*M* Tris–HCl, pH 8; 150 m*M* NaCl; 20 m*M* imidazole; 8 *M* urea) and centrifuged 20 min at 20,000 × *g* at room temperature.
4. Load the supernatant on a Ni-NTA column (GE Healthcare) equilibrated in Buffer A8.
5. Wash the column with Buffer A8 and elute protein using 100% Buffer B8 (50 m*M* Tris–HCl, pH 8; 150 m*M* NaCl; 1 *M* Imidazole; 8 *M* urea).
6. Fractions containing emerin 1–187 are pooled and dialyzed against 20 m*M* Tris–HCl, pH 8; 30 m*M* NaCl at room temperature in three steps (one night and twice for 2 h), and stored at 4 °C.

3.1.2 The N-Terminal Nucleoplasmic Region of MAN1

The MAN1 nucleoplasmic region contains a LEM domain and a large IDR (Fig. 1E). We first attempted to produce recombinant MAN1 (amino acid 1 to amino acid 471; MAN1 1–471) in *E. coli* as a 6His-tagged fusion protein, with a TEV site between the tag and the MAN1 sequence. However, due to weak expression yield and the production of several truncated MAN1 peptides, we optimized the cDNA sequence for expression in *E. coli* and the synthetic gene (ProteoGenix) was cloned into our pETM13-LIC vector.

Detailed protocol for MAN1 1–471 production and purification

1. *E. coli* cells are grown at 37 °C in LB media (or M9 media for NMR samples) and induced 4 h at 37 °C, with 0.5 m*M* IPTG when the OD_{600} is between 0.8 and 1.
2. Lyse cells by sonication: 30 pulses of 1 s each (power from 20% to 70%) separated by intervals of 1 s in lysis buffer (50 m*M* Tris–HCl, pH 7.5; 300 m*M* NaCl; 5% glycerol; 1% Triton X-100).
3. Incubate cell lysates with benzonase for 20 min at room temperature and centrifuge 20 min at 20,000 × *g*.
4. As the protein is soluble, load the supernatant on a Ni-NTA column (GE Healthcare) equilibrated in Buffer A (50 m*M* Tris–HCl, pH 8; 150 m*M* NaCl, 20 m*M* imidazole).
5. Wash the column with Buffer A and elute using a 20 min gradient from 0% to 100% Buffer B (50 m*M* Tris–HCl, pH 8; 150 m*M* NaCl; 1 *M* imidazole).
6. Fractions containing MAN1 1–471 are pooled, concentrated by centrifugation using Vivapsin sample concentrators (GE Healthcare), injected on a gel filtration column (Superdex 200, Hi-load, 120 ml, GE Healthcare) equilibrated in a buffer containing 150 m*M* NaCl and 1 m*M* tris(2-carboxyethyl)phosphine.
7. Store the purified MAN1 1–471 polypeptide at 4 °C.

3.2 Association Between Globular and Predicted Unstructured Regions in MAN1

In the MAN1 C-terminal nucleoplasmic region, a ULM motif was detected in the linker region that interacts with the UHM domain (Konde et al., 2010). Similarly, within MAN1 residues 1–471 a conserved motif Arg_{190}-Arg_{191}-Lys_{192}-Pro_{193}-His_{194}-Ser_{195}-Trp_{196}-Trp_{197}-Gly_{198} could play the role of a ULM motif. We, therefore, tested the entire purified N-terminal nucleoplasmic region of MAN1 for potential binding to the C-terminal nucleoplasmic region using 2D ^{1}H, ^{15}N NMR. We prepared two NMR samples: both contained 200 μl ^{15}N-labeled MAN1 1–471 at 100 μ*M* in phosphate buffer (pH 6.7) with 150 m*M* NaCl; one tube also contained unlabeled MAN1 755–911 at 100 μ*M*. 2D ^{1}H–^{15}N HSQC spectra were acquired on both samples at 20 °C on a 600 MHz spectrometer. Superimposition of these two spectra revealed an interaction between the two MAN1 nucleoplasmic regions (unpublished results). Indeed, several MAN1 1–471 NMR signals decreased in intensity upon addition of MAN1 755–911. These signals were assigned to the MAN1 sequence Trp_{196}-Trp_{197}-Gly_{198} thus revealing that this motif behaves like a new ULM. We suggest that the two MAN1 ULM motifs, Trp_{196}-Trp_{197}-Gly_{198} and Trp_{765}-Gln_{766}, can compete with each other for binding to the UHM domain leading to regulation of MAN1–SMAD2/3 complex formation.

3.3 Self-Association of Emerin Via Predicted Unstructured Region

A role for the emerin IDR in promoting self-association events was suggested by two studies. First, yeast two-hybrid experiments revealed that a fragment comprising emerin residues 1–225 binds to itself, suggesting C-terminally truncated emerin could form homodimers and/or multimers (Sakaki et al., 2001). Second, *in vitro* GST-pull-down experiments using recombinant emerin residues 1–221, and coimmunoprecipitation of full-length emerin from HEK293T cell extracts, also strongly suggested that emerin could oligomerize both *in vitro* and in cells (Berk et al., 2014). To identify which emerin region was responsible for self-association, N-terminally His-tagged emerin fragment was bound to Ni^{2+}-NTA agarose under denaturing conditions and the bead-bound polypeptide was incubated with purified GST-tagged emerin fragments in an appropriate binding buffer. After washing the beads, proteins were eluted using SDS-sample buffer and the binding reactions were resolved on gels. This experiment showed that GST-tagged emerin 170–220 is sufficient to bind emerin 1–221, whereas emerin 1–160 does not bind. Further pull-down assays suggested that emerin 170–220 recognized both fragments 1–132 and 170–220. In Hela cells co-expressing GFP-tagged and Flag-tagged emerin with different deletions, fragment 170–220 was also essential for emerin–emerin coimmunoprecipitation (Berk et al., 2014). Thus, self-association is due to sequences that are predicted to be unstructured within the emerin nucleoplasmic region.

4. LEM-DOMAIN PROTEINS WITH THEIR PARTNERS

LEM-domain proteins interact with lamins and with other INM proteins such as SUN1, SUN2, LAP1, and contribute to NE architecture (Haque et al., 2010; Patel et al., 2014; Shin et al., 2013). These interactions play critical roles in nuclear structure and chromatin organization. However, the difficulty in obtaining pure and functional fragments of LEM-domain proteins in their proper oligomerization and modification states makes it difficult to reconstitute *in vitro* biologically relevant complexes. Thus current knowledge is based on studies that focused on binding between specific functional domains and specific partners.

4.1 3D Structure of the LEM–BAF Complex

The first LEM-domain complex successfully studied consisted of the LEM domain of emerin and the small (10 kDa) protein BAF. BAF is a centrosymmetric homodimer that binds DNA and thereby bridges two DNA molecules, thus contributing to DNA compaction (Cai et al., 1998). Various studies including NMR spectroscopy, X-ray crystallography, and site-directed mutagenesis have shown that BAF is an α-helical protein that recognizes DNA through a pair of helix-hairpin-helix (HhH) motifs (Bradley, Ronning, Ghirlando, Craigie, & Dyda, 2005; Umland, Wei, Craigie, & Davies, 2000). BAF binding to emerin 1–187 can be observed in solution by NMR (unpublished results; Fig. 3A). Clore and colleagues determined the three-dimensional structure of the complex between BAF and the LEM domain of emerin (Fig. 3B; Cai et al., 2007). In this structure, one LEM domain interacts with two BAF monomers, as observed by NMR, light scattering, and analytic centrifugation, with a dissociation constant of 0.6 μM as shown by isothermal titration calorimetry. However, further studies suggested that posttranslational modifications of emerin's IDR also

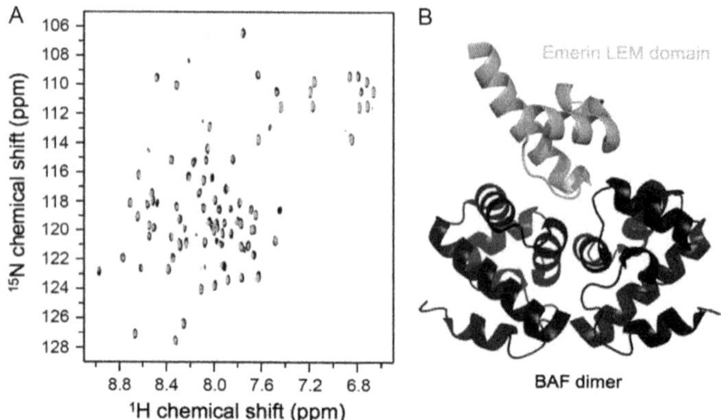

Figure 3 The emerin–BAF interaction. (A) Superimposition of two NMR ^1H–^{15}N HSQC spectra revealing the interaction between emerin 1–187 and BAF: the red and black spectra were recorded on ^{15}N-labeled BAF alone and in complex with emerin, respectively (in 40 m*M* phosphate buffer, pH 6.7, 150 m*M* NaCl, 1 m*M* tris(2-carboxyethyl) phosphine, 1 m*M* EDTA at 20 °C on a 600 MHz spectrometer). Disappearance of ^1H–^{15}N signals indicates binding to the unlabeled partner. (B) 3D structure of the LEM–BAF complex: the emerin LEM-domain is blue, the BAF dimer is red (PDB code 2ODG). (See the color plate.)

regulate BAF recognition (Berk et al., 2013; Tifft, Bradbury, & Wilson, 2009). Information obtained from the structure of the complex showed that BAF-binding sites to the LEM domain and DNA do not overlap. Thus, BAF could bind simultaneously to emerin and DNA. However, additional experiments suggested that, in cells, expression of BAF variant $Gly_{25}Glu$, with Gly_{25} being located at the interface between BAF and DNA, affected emerin localization during telophase (Haraguchi et al., 2001).

4.2 Interaction of the MAN1 WH Domain with DNA

The MAN1 residues 655–775 interact with a 211-base pair linear DNA molecule as observed by electrophoretic mobility shift assay (Caputo et al., 2006). The apparent affinity of the MAN1 fragment for this DNA molecule is 50 n*M*. Mutations of positively charged residues of the helix predicted to contact DNA, abolish DNA binding. Thus, the recognition helix of the MAN1 WH domain is involved in the binding of the carboxyl-terminal region of MAN1 to DNA.

4.3 MAN1 Interactions with SMAD2/3 and PPM1A

Three groups reported that the MAN1 C-terminal region physically interacts with R-Smad proteins to repress the transforming growth factor-β

(TGF-β) signaling pathway (Hellemans et al., 2004; Lin, Morrison, Wu, & Worman, 2005; Pan et al., 2005). Mutations in the gene encoding MAN1 cause sclerosing bone dysplasias, which sometimes have associated skin abnormalities (Hellemans et al., 2004). Fibroblasts from affected individuals are haploinsufficient with respect to MAN1 expression. These perturbations enhanced TGF-β signaling, since downstream genes targeted by this pathway were upregulated (Hellemans et al., 2004).

The MAN1 fragment from amino acid 755 to amino acid 911 recognizes the MH2 domain of Smad2 with micromolar affinity. From biochemical, NMR, and SAXS analyses, it was possible to calculate a model of the complex between this MAN1 fragment and the Smad2 MH2 domain (Bourgeois et al., 2013). As predicted by this model, experiments *in vitro* showed that MAN1 binds to Smad2 alone, and to the activated Smad2–Smad4 complex. However, in cells, MAN1 does not bind Smad4-containing complexes. Overexpression of MAN1 leads to Smad2 dephosphorylation, thus hindering Smad2 binding to Smad4. *In vitro*, MAN1 binds directly to the phosphatase PPM1A, which catalyzes dephosphorylation of Smad2. These results suggest a mechanism through which the MAN1-specific C-terminal region inhibits TGF-β signaling (Bourgeois et al., 2013). They demonstrate that this MAN1 region recognizes different forms of Smad2 (monomers, homo, and heterotrimers) and show that MAN1, by recruiting Smad2 to the NE, facilitates its dephosphorylation by PPM1A and thus its inactivation.

5. CONCLUDING REMARKS

LEM-domain proteins are involved in diverse cellular processes including DNA replication and cell-cycle control, chromatin organization, nuclear assembly, regulation of gene expression, and signaling pathways, as well as retroviral infection (Barton, Soshnev, & Geyer, 2015; Brachner & Foisner, 2011; Li & Craigie, 2006). The INM LEM-domain proteins LAP2, emerin, and MAN1 have been intensively studied using biophysical and cellular approaches. Their LEM domain directly recognizes the DNA-binding protein BAF (Cai et al., 2001). However, these proteins are membrane proteins and their soluble regions are largely predicted as intrinsically disordered. LAP2α dimerizes through its C-terminal α-helical domain (Bradley et al., 2007). Emerin also self-assembles into poorly characterized oligomers (Berk et al., 2014). It is thus difficult to identify their functional states *in vitro* and to reconstitute complexes between these proteins and their biological partners. First attempts to stabilize NE complexes have recently led to the resolution of the X-ray structure of the KASH–SUN complex, located between the outer and INMs (Sosa, Rothballer, Kutay, & Schwartz, 2012). Moreover, development of mass spectrometry and NMR techniques now enables description of posttranslational modification events in LEM-domain proteins, and their consequences on protein structure and binding properties. Integrative approaches, including structural characterization of protein subcomplexes based on a panel of biophysical techniques and development of new tools in biochemistry and imaging techniques for identification of protein–protein interfaces might ultimately provide a better view of inner NE architecture.

ACKNOWLEDGMENTS

This work was supported by AFM grants 17243 (research grant to S. Z. J.) and 18159 (full PhD fellowship to C. S.), by FRM grant FDT20140931008 (1 year PhD fellowship to I. H.), and by the French national infrastructure FRISBI program.

REFERENCES

Aravind, L., & Koonin, E. V. (2000). SAP—A putative DNA-binding motif involved in chromosomal organization. *Trends in Biochemical Sciences*, *25*, 112–114.

Aravind, L., & Koonin, E. V. (2001). Prokaryotic homologs of the eukaryotic DNA-end-binding protein Ku, novel domains in the Ku protein and prediction of a prokaryotic double-strand break repair system. *Genome Research*, *11*, 1365–1374.

Aravind, L., Mazumder, R., Vasudevan, S., & Koonin, E. V. (2002). Trends in protein evolution inferred from sequence and structure analysis. *Current Opinion in Structural Biology*, *12*, 392–399.

Asencio, C., Davidson, I. F., Santarella-Mellwig, R., Ly-Hartig, T. B., Mall, M., Wallenfang, M. R., et al. (2012). Coordination of kinase and phosphatase activities by Lem4 enables nuclear envelope reassembly during mitosis. *Cell*, *150*, 122–135.

Barton, L. J., Soshnev, A. A., & Geyer, P. K. (2015). Networking in the nucleus: A spotlight on LEM-domain proteins. *Current Opinion in Cell Biology*, *34*, 1–8.

Berger, R., Theodor, L., Shoham, J., Gokkel, E., Brok-Simoni, F., Avraham, K. B., et al. (1996). The characterization and localization of the mouse thymopoietin/lamina-associated polypeptide 2 gene and its alternatively spliced products. *Genome Research*, *6*, 361–370.

Berk, J. M., Maitra, S., Dawdy, A. W., Shabanowitz, J., Hunt, D. F., & Wilson, K. L. (2013). O-Linked beta-N-acetylglucosamine (O-GlcNAc) regulates emerin binding to barrier to autointegration factor (BAF) in a chromatin- and lamin B-enriched "niche" *Journal of Biological Chemistry*, *288*, 30192–30209.

Berk, J. M., Simon, D. N., Jenkins-Houk, C. R., Westerbeck, J. W., Gronning-Wang, L. M., Carlson, C. R., et al. (2014). The molecular basis of emerin-emerin and emerin-BAF interactions. *Journal of Cell Science*, *127*, 3956–3969.

Bourgeois, B., Gilquin, B., Tellier-Lebegue, C., Ostlund, C., Wu, W., Perez, J., et al. (2013). Inhibition of TGF-beta signaling at the nuclear envelope: Characterization of interactions between MAN1, Smad2 and Smad3, and PPM1A. *Science Signaling*, *6*, ra49.

Brachner, A., Braun, J., Ghodgaonkar, M., Castor, D., Zlopasa, L., Ehrlich, V., et al. (2012). The endonuclease Ankle1 requires its LEM and GIY-YIG motifs for DNA cleavage *in vivo*. *Journal of Cell Science*, *125*, 1048–1057.

Brachner, A., & Foisner, R. (2011). Evolvement of LEM proteins as chromatin tethers at the nuclear periphery. *Biochemical Society Transactions*, *39*, 1735–1741.

Brachner, A., Reipert, S., Foisner, R., & Gotzmann, J. (2005). LEM2 is a novel MAN1-related inner nuclear membrane protein associated with A-type lamins. *Journal of Cell Science*, *118*, 5797–5810.

Bradley, C. M., Jones, S., Huang, Y., Suzuki, Y., Kvaratskhelia, M., Hickman, A. B., et al. (2007). Structural basis for dimerization of LAP2alpha, a component of the nuclear lamina. *Structure*, *15*, 643–653.

Bradley, C. M., Ronning, D. R., Ghirlando, R., Craigie, R., & Dyda, F. (2005). Structural basis for DNA bridging by barrier-to-autointegration factor. *Nature Structural & Molecular Biology*, *12*, 935–936.

Cai, M., Huang, Y., Ghirlando, R., Wilson, K. L., Craigie, R., & Clore, G. M. (2001). Solution structure of the constant region of nuclear envelope protein LAP2 reveals two LEM-domain structures: One binds BAF and the other binds DNA. *EMBO Journal*, *20*, 4399–4407.

Cai, M., Huang, Y., Suh, J. Y., Louis, J. M., Ghirlando, R., Craigie, R., et al. (2007). Solution NMR structure of the barrier-to-autointegration factor-Emerin complex. *Journal of Biological Chemistry*, *282*, 14525–14535.

Cai, M., Huang, Y., Zheng, R., Wei, S. Q., Ghirlando, R., Lee, M. S., et al. (1998). Solution structure of the cellular factor BAF responsible for protecting retroviral DNA from autointegration. *Nature Structural Biology*, *5*, 903–909.

Caputo, S., Couprie, J., Duband-Goulet, I., Konde, E., Lin, F., Braud, S., et al. (2006). The carboxyl-terminal nucleoplasmic region of MAN1 exhibits a DNA binding winged helix domain. *Journal of Biological Chemistry*, *281*, 18208–18215.

Dechat, T., Gotzmann, J., Stockinger, A., Harris, C. A., Talle, M. A., Siekierka, J. J., et al. (1998). Detergent-salt resistance of LAP2alpha in interphase nuclei and phosphorylation-dependent association with chromosomes early in nuclear assembly implies functions in nuclear structure dynamics. *EMBO Journal*, *17*, 4887–4902.

Dechat, T., Korbei, B., Vaughan, O. A., Vlcek, S., Hutchison, C. J., & Foisner, R. (2000). Lamina-associated polypeptide 2alpha binds intranuclear A-type lamins. *Journal of Cell Science*, *113*, 3473–3484.

Ellis, J. A., Craxton, M., Yates, J. R., & Kendrick-Jones, J. (1998). Aberrant intracellular targeting and cell cycle-dependent phosphorylation of emerin contribute to the Emery-Dreifuss muscular dystrophy phenotype. *Journal of Cell Science*, *111*, 781–792.

Furukawa, K. (1999). LAP2 binding protein 1 (L2BP1/BAF) is a candidate mediator of LAP2-chromatin interaction. *Journal of Cell Science*, *112*, 2485–2492.

Furukawa, K., Fritze, C. E., & Gerace, L. (1998). The major nuclear envelope targeting domain of LAP2 coincides with its lamin binding region but is distinct from its chromatin interaction domain. *Journal of Biological Chemistry*, *273*, 4213–4219.

Furukawa, K., Pante, N., Aebi, U., & Gerace, L. (1995). Cloning of a cDNA for lamina-associated polypeptide 2 (LAP2) and identification of regions that specify targeting to the nuclear envelope. *EMBO Journal*, *14*, 1626–1636.

Gruenbaum, Y., & Medalia, O. (2015). Lamins: The structure and protein complexes. *Current Opinion in Cell Biology*, *32C*, 7–12.

Guilluy, C., Osborne, L. D., Van Landeghem, L., Sharek, L., Superfine, R., Garcia-Mata, R., et al. (2014). Isolated nuclei adapt to force and reveal a mechanotransduction pathway in the nucleus. *Nature Cell Biology*, *16*, 376–381.

Haque, F., Mazzeo, D., Patel, J. T., Smallwood, D. T., Ellis, J. A., Shanahan, C. M., et al. (2010). Mammalian SUN protein interaction networks at the inner nuclear membrane and their role in laminopathy disease processes. *Journal of Biological Chemistry*, *285*, 3487–3498.

Haraguchi, T., Koujin, T., Segura-Totten, M., Lee, K. K., Matsuoka, Y., Yoneda, Y., et al. (2001). BAF is required for emerin assembly into the reforming nuclear envelope. *Journal of Cell Science, 114*, 4575–4585.

Hellemans, J., Preobrazhenska, O., Willaert, A., Debeer, P., Verdonk, P. C., Costa, T., et al. (2004). Loss-of-function mutations in LEMD3 result in osteopoikilosis, Buschke-Ollendorff syndrome and melorheostosis. *Nature Genetics, 36*, 1213–1218.

Hirano, Y., Iwase, Y., Ishii, K., Kumeta, M., Horigome, T., & Takeyasu, K. (2009). Cell cycle-dependent phosphorylation of MAN1. *Biochemistry, 48*, 1636–1643.

Hirano, Y., Segawa, M., Ouchi, F. S., Yamakawa, Y., Furukawa, K., Takeyasu, K., et al. (2005). Dissociation of emerin from barrier-to-autointegration factor is regulated through mitotic phosphorylation of emerin in a Xenopus egg cell-free system. *Journal of Biological Chemistry, 280*, 39925–39933.

Jamin, A., & Wiebe, M. S. (2015). Barrier to autointegration factor (BANF1): Interwoven roles in nuclear structure, genome integrity, innate immunity, stress responses and progeria. *Current Opinion in Cell Biology, 34*, 61–68.

Kielkopf, C. L., Lucke, S., & Green, M. R. (2004). U2AF homology motifs: Protein recognition in the RRM world. *Genes & Development, 18*, 1513–1526.

Konde, E., Bourgeois, B., Tellier-Lebegue, C., Wu, W., Perez, J., Caputo, S., et al. (2010). Structural analysis of the Smad2-MAN1 interaction that regulates transforming growth factor-beta signaling at the inner nuclear membrane. *Biochemistry, 49*, 8020–8032.

Laguri, C., Gilquin, B., Wolff, N., Romi-Lebrun, R., Courchay, K., Callebaut, I., et al. (2001). Structural characterization of the LEM motif common to three human inner nuclear membrane proteins. *Structure, 9*, 503–511.

Lee, K. K., Haraguchi, T., Lee, R. S., Koujin, T., Hiraoka, Y., & Wilson, K. L. (2001). Distinct functional domains in emerin bind lamin A and DNA-bridging protein BAF. *Journal of Cell Science, 114*, 4567–4573.

Li, M., & Craigie, R. (2006). Virology: HIV goes nuclear. *Nature, 441*, 581–582.

Lin, F., Blake, D. L., Callebaut, I., Skerjanc, I. S., Holmer, L., McBurney, M. W., et al. (2000). MAN1, an inner nuclear membrane protein that shares the LEM domain with lamina-associated polypeptide 2 and emerin. *Journal of Biological Chemistry, 275*, 4840–4847.

Lin, F., Morrison, J. M., Wu, W., & Worman, H. J. (2005). MAN1, an integral protein of the inner nuclear membrane, binds Smad2 and Smad3 and antagonizes transforming growth factor-beta signaling. *Human Molecular Genetics, 14*, 437–445.

Mansharamani, M., & Wilson, K. L. (2005). Direct binding of nuclear membrane protein MAN1 to emerin *in vitro* and two modes of binding to barrier-to-autointegration factor. *Journal of Biological Chemistry, 280*, 13863–13870.

Margalit, A., Segura-Totten, M., Gruenbaum, Y., & Wilson, K. L. (2005). Barrier-to-autointegration factor is required to segregate and enclose chromosomes within the nuclear envelope and assemble the nuclear lamina. *Proceedings of the National Academy of Sciences of the United States of America, 102*, 3290–3295.

Montes de Oca, R., Andreassen, P. R., & Wilson, K. L. (2011). Barrier-to-autointegration factor influences specific histone modifications. *Nucleus, 2*, 580–590.

Naetar, N., Korbei, B., Kozlov, S., Kerenyi, M. A., Dorner, D., Kral, R., et al. (2008). Loss of nucleoplasmic LAP2alpha-lamin A complexes causes erythroid and epidermal progenitor hyperproliferation. *Nature Cell Biology, 10*, 1341–1348.

Pan, D., Estevez-Salmeron, L. D., Stroschein, S. L., Zhu, X., He, J., Zhou, S., et al. (2005). The integral inner nuclear membrane protein MAN1 physically interacts with the R-Smad proteins to repress signaling by the transforming growth factor-{beta} superfamily of cytokines. *Journal of Biological Chemistry, 280*, 15992–16001.

Patel, J. T., Bottrill, A., Prosser, S. L., Jayaraman, S., Straatman, K., Fry, A. M., et al. (2014). Mitotic phosphorylation of SUN1 loosens its connection with the nuclear lamina while the LINC complex remains intact. *Nucleus, 5*, 462–473.

Sakaki, M., Koike, H., Takahashi, N., Sasagawa, N., Tomioka, S., Arahata, K., et al. (2001). Interaction between emerin and nuclear lamins. *Journal of Biochemistry*, *129*, 321–327.

Shin, J. Y., Mendez-Lopez, I., Wang, Y., Hays, A. P., Tanji, K., Lefkowitch, J. H., et al. (2013). Lamina-associated polypeptide-1 interacts with the muscular dystrophy protein emerin and is essential for skeletal muscle maintenance. *Developmental Cell*, *26*, 591–603.

Shumaker, D. K., Lee, K. K., Tanhehco, Y. C., Craigie, R., & Wilson, K. L. (2001). LAP2 binds to BAF.DNA complexes: Requirement for the LEM domain and modulation by variable regions. *EMBO Journal*, *20*, 1754–1764.

Sosa, B. A., Rothballer, A., Kutay, U., & Schwartz, T. U. (2012). LINC complexes form by binding of three KASH peptides to domain interfaces of trimeric SUN proteins. *Cell*, *149*, 1035–1047.

Taylor, M. R., Slavov, D., Gajewski, A., Vlcek, S., Ku, L., Fain, P. R., et al. (2005). Thymopoietin (lamina-associated polypeptide 2) gene mutation associated with dilated cardiomyopathy. *Human Mutation*, *26*, 566–574.

Tifft, K. E., Bradbury, K. A., & Wilson, K. L. (2009). Tyrosine phosphorylation of nuclear-membrane protein emerin by Src, Abl and other kinases. *Journal of Cell Science*, *122*, 3780–3790.

Umland, T. C., Wei, S. Q., Craigie, R., & Davies, D. R. (2000). Structural basis of DNA bridging by barrier-to-autointegration factor. *Biochemistry*, *39*, 9130–9138.

Vlcek, S., Just, H., Dechat, T., & Foisner, R. (1999). Functional diversity of LAP2alpha and LAP2beta in postmitotic chromosome association is caused by an alpha-specific nuclear targeting domain. *EMBO Journal*, *18*, 6370–6384.

Wolff, N., Gilquin, B., Courchay, K., Callebaut, I., Worman, H. J., & Zinn-Justin, S. (2001). Structural analysis of emerin, an inner nuclear membrane protein mutated in X-linked Emery-Dreifuss muscular dystrophy. *FEBS Letters*, *501*, 171–176.

Yip, S. C., Cotteret, S., & Chernoff, J. (2012). Sumoylated protein tyrosine phosphatase 1B localizes to the inner nuclear membrane and regulates the tyrosine phosphorylation of emerin. *Journal of Cell Science*, *125*, 310–316.

Zheng, R., Ghirlando, R., Lee, M. S., Mizuuchi, K., Krause, M., & Craigie, R. (2000). Barrier-to-autointegration factor (BAF) bridges DNA in a discrete, higher-order nucleoprotein complex. *Proceedings of the National Academy of Sciences of the United States of America*, *97*, 8997–9002.

CHAPTER FOUR

Purification and Structural Analysis of SUN and KASH Domain Proteins

F. Esra Demircioglu, Victor E. Cruz, Thomas U. Schwartz[1]

Department of Biology, Massachusetts Institute of Technology, Cambridge, Massachusetts, USA
[1]Corresponding author: e-mail address: tus@mit.edu

Contents

Abstract

Molecular tethers span the nuclear envelope to mechanically connect the cytoskeleton and nucleoskeleton. These bridge-like tethers, termed linkers of nucleoskeleton and cytoskeleton (LINC) complexes, consist of SUN proteins at the inner nuclear membrane and KASH proteins at the outer nuclear membrane. LINC complexes are central to a variety of cell activities including nuclear positioning and mechanotransduction, and LINC-related abnormalities are associated with a spectrum of tissue-specific diseases, termed laminopathies or envelopathies. Protocols used to study the biochemical and structural characteristics of core elements of SUN–KASH complexes are described here to facilitate further studies in this new field of cell biology.

Methods in Enzymology, Volume 569
ISSN 0076-6879
http://dx.doi.org/10.1016/bs.mie.2015.08.011

1. INTRODUCTION

The nuclear envelope (NE) physically separates the nucleus from the cytoplasm, generating two distinct compartments. Molecular exchange between the nucleoplasm and cytoplasm is mediated by nuclear pore complexes, which act as selective permeability barriers. Mechanical communication between the nucleus and cytoplasm involves specific tethers, termed linkers of nucleoskeleton and cytoskeleton (LINC) complexes, that span the NE. LINC complexes are formed by a family of KASH (Klarsicht, ANC-1, and Syne Homology) proteins embedded in the outer nuclear membrane (ONM) that interact within the NE lumen with SUN (Sad1 and UNC-84) proteins, which span the inner nuclear membrane (INM). SUN and KASH proteins each project from the NE and directly bind to components of the nucleoskeleton and the cytoskeleton, respectively. These mechanical connections are critically important in a wide range of activities such as nuclear migration and anchorage, meiotic chromosome movements, the centrosome–nucleus connection, signal transduction, and DNA repair (Burke & Roux, 2009; Chang, Worman, & Gundersen, 2015; Luxton & Starr, 2014; Rothballer & Kutay, 2013; Starr & Fridolfsson, 2010).

KASH proteins are tail-anchored, single-span transmembrane proteins, mostly found at the ONM. The C-terminal "KASH motif" comprises the transmembrane helix and the adjacent luminal segment, which consists of 8–30 residues, depending on the specific nesprin gene and species (Starr & Han, 2002). Vertebrate KASH proteins are often called nesprins (NE spectrin repeat proteins), since their cytoplasmic portions typically contain numerous spectrin repeats. The cytoplasmic extensions vary greatly in size due to alternative splicing and transcription initiation of multiple nesprin genes (Zhang et al., 2001). The two longest ("giant"; 0.8–1.0 MDa) nesprin isoforms, Nesprin-1G and Nesprin-2G, each bind to actin filaments via calponin homology (CH) domains at their N terminus. Coupling of actin filaments to LINC complexes has been best visualized in migrating fibroblasts, where SUN2–Nesprin-2G complexes assemble into linear arrays at the NE and form so-called TAN (transmembrane actin-associated nuclear) lines. Formation of TAN lines is instrumental in moving the nucleus rearward by coupling to retrogradely moving actin cables (Luxton, Gomes, Folker, Vintinner, & Gundersen, 2010; Luxton, Gomes, Folker, Worman, & Gundersen, 2011). The much shorter protein Nesprin-3α binds plectin, which in turn binds to cytoplasmic intermediate filaments and/or actin.

Since Nesprin-3α also binds to the CH domains of Nesprin-1G and Nesprin-2G, a nesprin scaffold is proposed to form around the nucleus that might play a role in regulating nuclear size (Lu et al., 2012). Nesprin-4 interacts with microtubules through kinesin-1 and is proposed to function specifically in ear development and hearing (Horn, Brownstein, et al., 2013). Yet another tissue-specific KASH protein, Nesprin-5, binds to microtubules through dynein and functions during meiotic chromosome pairing in germ cells (Horn, Kim, et al., 2013). Finally, a recently recognized sixth KASH protein in zebrafish, lymphoid-restricted membrane protein, is involved in pronuclear congression during fertilization (Lindeman & Pelegri, 2012).

Similarly, most organisms also encode several SUN homologs. Of the five known mammalian SUN proteins, SUN1 and SUN2 are widely expressed (Crisp et al., 2006; Padmakumar et al., 2005), whereas SUN3, SUN4, and SUN5 are expressed during spermatogenesis in testis (Göb, Schmitt, Benavente, & Alsheimer, 2010). SUN proteins have at least one transmembrane helix, which typically anchors them in the INM. The C-terminal ~20-kDa SUN domain, preceded by a predicted coiled-coil segment of variable length, is located in the NE luminal space. The N terminus of SUN proteins extends into the nucleoplasm and binds lamins (nuclear intermediate filament proteins). Mammalian SUN proteins are known to bind to A-type lamins, while interaction with B-type lamins is relatively weak (Crisp et al., 2006; Haque et al., 2006). Although lamina attachment restricts diffusion of the SUNs (Ostlund et al., 2009), lamins do not seem to be the only factors anchoring LINC complexes. Indeed, both SUN1 and SUN2 are properly localized in the absence of A- and B-type lamins (Crisp et al., 2006; Haque et al., 2006; Padmakumar et al., 2005). Lamin-associated proteins such as emerin and SAMP1 likely help anchor the LINC complexes (Borrego-Pinto et al., 2012; Chang, Folker, Worman, & Gundersen, 2013), and the intricate interplay among these proteins is not yet understood. Mutations in A-type lamins, emerin, SUNs, and nesprins can each disrupt nucleocytoskeletal coupling and are also genetically linked to laminopathies such as skeletal and/or cardiac muscular dystrophies, lipodystrophy, dysplasia, or segmental progeroid (accelerated aging) disorders (Worman, 2012). The striated muscle disease EDMD (Emery–Dreifuss muscular dystrophy) and the premature aging syndrome HGPS (Hutchinson–Gilford progeria syndrome) are examples where perturbed functioning of LINC complexes and associated factors contributes to pathology (Bione et al., 1994; Bonne et al., 1999; Chen et al., 2012; Haque et al., 2010; Puckelwartz et al., 2009; Zhang et al., 2007). Further structural and biochemical characterization of LINC complexes is needed to understand the molecular basis of these disorders.

Recent crystallographic studies established the structure of the core element of the LINC complex, namely, the SUN–KASH interaction (Wang et al., 2012; Zhou et al., 2012). Human SUN2 proteins form a triple-stranded coiled-coil stalk to generate a trimeric structure that positions the adjacent, C-terminal β-sandwich-shaped SUN domains to form a globular trefoil. The helical stalk assumes an unusual, right-handed supercoil that positions the SUN domains in a KASH-binding competent state. Three KASH peptides are bound at the three interfaces between adjacent SUN protomers in the trefoil, immediately explaining why monomeric SUN does not bind a KASH peptide. Altogether, the core of the LINC complex is a heterohexameric SUN_3–$KASH_3$ complex. The carboxyl group of the terminal residue in the KASH peptide is specifically recognized by the SUN domain, thereby explaining why KASH peptides are always located at the C terminus of a protein. A cysteine residue at the N terminus of the KASH peptide can form a disulfide bridge with a conserved cysteine on the SUN domain, presumably enhancing the mechanical strength of the complex (Sosa, Rothballer, Kutay, & Schwartz, 2012).

Although these SUN2–KASH1/2 structures revealed crucial information about LINC complex assembly, much more work is needed to fully understand the SUN–KASH interactome. Specific methods for the purification, biochemical analysis, and structure determination of apo-SUN2 and SUN2–KASH complexes are detailed below. In addition to SUN2–KASH1/2 complexes, which have been crystallized, we also describe strategies for purifying SUN2 complexes with KASH3, KASH4, KASH5, or KASH6.

2. PURIFICATION OF SUN PROTEINS AND SUN–KASH COMPLEXES

2.1 Construct Design

We use the pETDuet-1 bacterial expression system (EMD Biosciences) to produce SUN and KASH domain proteins from two different multiple cloning sites (MCS). The cDNA encoding each SUN domain protein is cloned into the first MCS (MCS1), and the cDNA encoding the KASH domain protein is cloned into the second MCS (MCS2), either as single open reading frames for isolation of the apo proteins or in tandem for isolation of SUN–KASH complexes. When cloned together, the dual cassette expression enables SUN–KASH interaction already in the bacterial cell, facilitating isolation of stoichiometric complexes.

We generated N-terminally 6 × His-tagged SUN2 fragments for expression from MCS1. These fragments contain luminal portions of SUN2. Since SUN2 fragments that lack significant portions of the predicted coiled-coil regions might become monomeric in solution, prohibiting KASH binding, short SUN2 fragments are fused to a coiled-coil fragment of engineered triGCN4 (Ciani et al., 2010) at their N termini to restore KASH-binding competence (Fig. 1A). The crystallized part of SUN2 (residues 522–717) is such

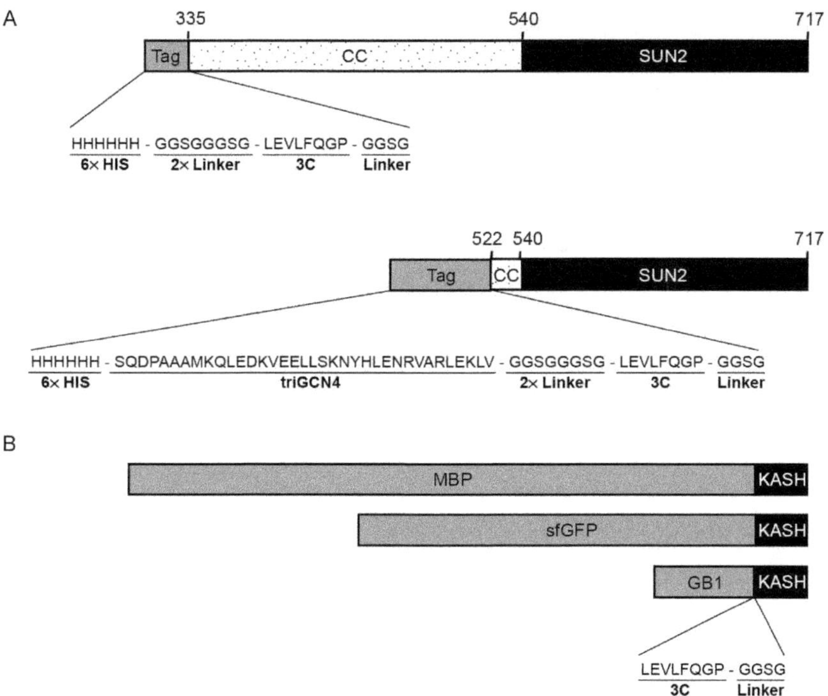

Figure 1 Schematic drawing of the expression constructs for human SUN2 (A) and KASH (B). Each SUN2 fragment includes the SUN domain (residues 540–717) and preceding luminal segments of different length, predicted to form coiled coils (CC). The SUN2 (residue 335–717) fragment is N-terminally fused to a 3C-cleavable 6 × His tag, whereas a 6 × His:triGCN4 tag is used for a shorter SUN2 fragment (residues 522–717). KASH motifs are C-terminally attached to 3C-cleavable MBP, superfolder GFP (sfGFP), or GB1 tags. Short flexible linkers are incorporated around the 3C cleavage sites to enhance removal of the fusion tags.

an example (Sosa et al., 2012). This engineering is not required for longer SUN2 constructs with a native extended coiled-coil stalk, which form stable trimers in solution. A cleavage site for 3C protease is inserted near the N terminus of each SUN2 fragment to facilitate removal of the fusion tag after purification.

For purification of SUN2–KASH complexes, we clone luminal portions of KASH motifs into the MCS2 site of the pETDuet-1 vector (Fig. 1B). KASH1/2/3/4 peptides can be attached to 3C-cleavable maltose-binding protein (MBP) tags (di Guan, Li, Riggs, & Inouye, 1988) at their N termini. MBP tagging helps in various ways. First, it typically yields super-stoichiometric expression of the MBP–KASH fusion protein compared to SUN2, which enables the isolation of stoichiometric SUN2–KASH complexes in large amounts. Second, the MBP moiety provides an orthogonal affinity tag after Ni^{2+} pulldown to isolate stoichiometric complexes in high purity (see Section 2.3).

We faced problems removing the fusion tags during the purification of SUN2–KASH complexes. To increase the efficiency of tag removal, we introduced flexible Gly/Ser-rich linkers on either side of the 3C-cleavage sites (Fig. 1). This linker strategy proved useful for cutting off the 6 × His-triGCN4 tags, but did not improve the removal of MBP tags. As explained in Section 2.3, this problem was solved in some cases by cleaving the MBP tag in multiple steps, followed by chromatography. For example, the SUN2–KASH1/2 crystals were obtained from MBP-tagged complexes (Sosa et al., 2012). However, isolation of the other SUN–KASH complexes (SUN2–KASH3/4/5/6) was more problematic. For these complexes, we tested different tags including superfolder (sf) GFP (Pédelacq, Cabantous, Tran, Terwilliger, & Waldo, 2006), thioredoxin (LaVallie, Lu, Diblasio-Smith, Collins-Racie, & McCoy, 2000), and GB1 (Huth et al., 1997). The GB1 tag significantly enhanced cleavage efficiency and was used to purify SUN2–KASH5/6 complexes (Fig. 1B). Advantages of using different tags during purification will be explained in detail in Section 2.3.

2.2 Purification of Human Apo-SUN2

We purified human apo-SUN2 (residues 335–717) and apo-SUN2 (residues 522–717) in *Escherichia coli* strain LOBSTR-BL21(DE3)-RIL (Kerafast, Inc., Boston, MA) (Andersen, Leksa, & Schwartz, 2013) for increased purity, using the following protocol.

1. Inoculate 3–6 ml Lysogeny Broth (LB) with a single LOBSTR-BL21 (DE3)-RIL colony that was heat-shock transformed with the SUN2-expressing plasmid. Include ampicillin (100 μg/ml) and chloramphenicol (34 μg/ml) to select for the pETDuet-1-derived SUN2-expressing plasmid and the RIL plasmid (Agilent Technologies), respectively. Grow this starter culture overnight at 30 °C.
2. The next morning, inoculate 1 l of LB medium containing 0.4% (w/v) glucose, ampicillin (100 μg/ml), and chloramphenicol (34 μg/ml) with the overnight culture. Grow these bacteria in a 2-l baffled Erlenmeyer flask at 37 °C in a shaker to an OD_{600} of 0.6–0.8, then transfer to 18 °C and incubate 20 min longer. Then add IPTG (0.2 m*M* final) to induce protein expression, and shake overnight at 18 °C.
3. The next morning, record the OD_{600} (usually between 6 and 8) and then harvest. Pellet cells by centrifugation at 6000 rpm for 6 min (e.g., Sorvall SLA-3000 rotor). Resuspend the bacterial pellet (20 ml lysis buffer per 1000 OD_{600}) in ice-cold lysis buffer (50 m*M* potassium phosphate pH 8.0, 400 m*M* NaCl, and 40 m*M* imidazole). Note that lysis and all subsequent steps should be done at 4 °C with prechilled solutions.
4. Resuspend the bacteria homogenously to obtain a clump-free cell suspension, then process using a cell homogenizer (Constant Systems) at 25 kpsi. Mix the collected lysate immediately with 0.1 *M* PMSF (50 μl per 10 ml lysate) and add 250 units of TurboNuclease (Eton Bioscience).
5. Centrifuge the lysate at 9500 rpm for 25 min (e.g., Sorvall SLA-600TC rotor) and recover the supernatant. Mix the supernatant with Ni^{2+} Sepharose 6 Fast Flow (GE Healthcare) slurry equilibrated with lysis buffer. Use approximately 1 ml Ni^{2+} resin per 1000 OD_{600} of cells.
6. Gently stir the mixture for 30 min, collect the Ni^{2+} resin in a 50-ml conical tube via several spins using a tabletop centrifuge, and then batch-wash the Ni^{2+} resin three times with 40 ml lysis buffer. Pour the Ni^{2+}-Sepharose slurry into a disposable Pierce column (Thermo Scientific) and wash with 6 × resin bed volumes of lysis buffer via gravity flow.
7. Once the column is drained, elute proteins using a 6 × resin bed volume of elution buffer (10 m*M* Tris/HCl pH 8.0, 150 m*M* NaCl, and 250 m*M* imidazole).
8. Concentrate that eluted protein to a final volume of ~10 ml using a centrifugal concentrator and then purify by size exclusion chromatography on a HiLoad 26/60 Superdex S200 column (GE Healthcare) in a buffer containing 10 m*M* Tris/HCl pH 8.0 and 150 m*M* NaCl. Pool the peak

corresponding to His_6-tagged SUN2 is pooled, and mix with 3C protease at an enzyme:protein ratio of 1:50 (w/w). Overnight incubation with 3C protease is generally sufficient to remove the fusion tags, but this should be verified by SDS-PAGE analysis of a small aliquot.

9. Concentrate the cleaved SUN2 protein and purify again by size exclusion chromatography on a HiLoad 26/60 Superdex S200 column in 10 m*M* Tris/HCl pH 8.0, and 150 m*M* NaCl. Note that apo-SUN2 (residues 335–717) and apo-SUN2 (residues 522–717) elute differently without their fusion tags. Apo-SUN2 (residues 335–717) includes the entire predicted coiled-coil region and elutes as a homotrimer, as confirmed by analytical ultracentrifugation (Sosa et al., 2012). By contrast, apo-SUN2 (residues 522–717) behaves as a monomer after the His_6-triGCN4 tag is removed, although this behavior is buffer dependent (see Section 3.2). These observations support the notion that the coiled-coil region (residues 335–540) adjacent to the SUN domain (residues 540–717) helps stabilize the SUN homotrimer and, hence, the KASH-binding-competent oligomeric state.

2.3 Purification of Human SUN2–KASH Complexes

Protocols for purifying SUN2 (residues 522–717)–KASH1–6 complexes are described here and shown schematically (Fig. 2).

1. Steps 1–7 described in Section 2.2 remain essentially the same, with one important difference: we now include 1 m*M* KCl in the Ni^{2+} elution buffer (and all subsequent buffers) since a K^+ ion is likely to be coordinated in the cation loop of SUN2–KASH complexes (Sosa et al., 2012).

2. Similar to apo-SUN2, purify SUN2–KASH Ni^{2+} eluates by size exclusion chromatography using a Superdex S200 column equilibrated with 10 m*M* Tris/HCl pH 8.0, 150 m*M* NaCl, and 1 m*M* KCl. This step helps remove aggregates, unbound SUN2, and KASH. Alternatively, the Ni^{2+} eluate can be subjected to a second purification on amylose resin if KASH is MBP-tagged. In this scenario, imidazole is removed by dialyzing the Ni^{2+} eluate against 10 m*M* Tris/HCl pH 8.0, 150 m*M* NaCl, and 1 m*M* KCl prior to binding to amylose, and the SUN2–KASH complexes are then eluted in the presence of 10 m*M* maltose.

3. Mix the resulting SUN2–KASH complexes with 3C protease at an enzyme:substrate ratio of 1:25–1:50 (w/w). Retain an aliquot of uncut sample for SDS-PAGE analysis. After incubating 16 h, resolve a small

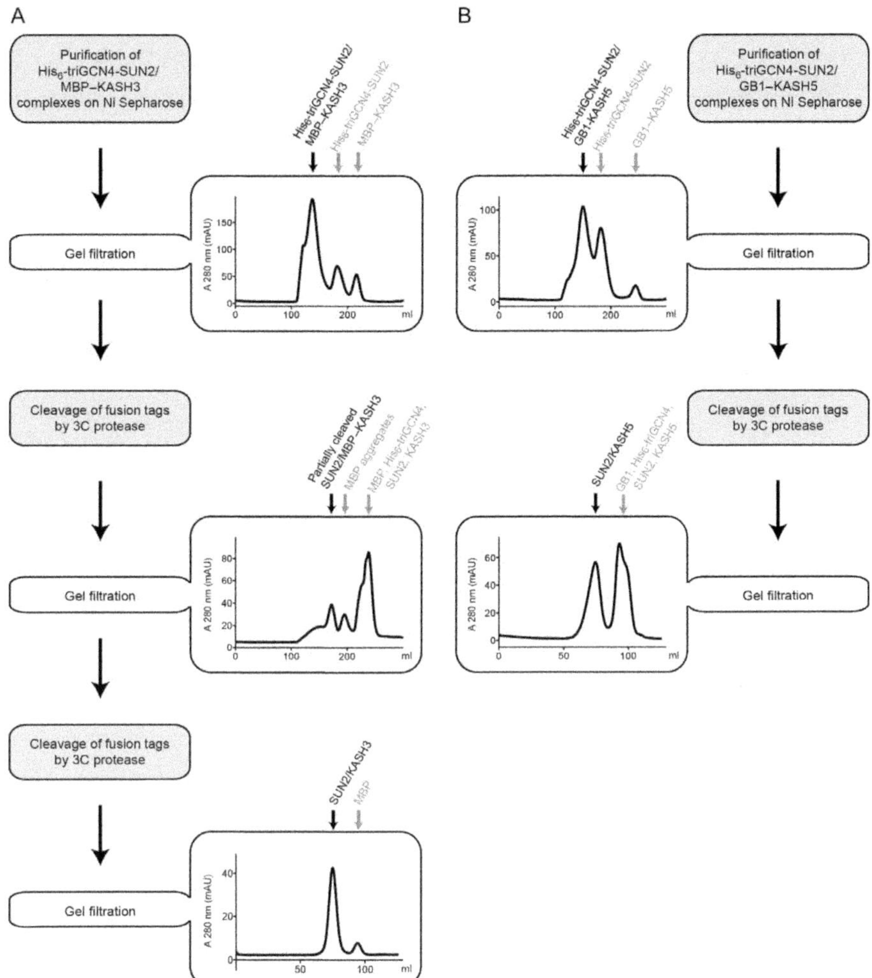

Figure 2 Purification schemes for SUN2–KASH complexes. (A) Purification of the SUN2–KASH3 complex is illustrated to exemplify the strategy used for SUN2/MBP–KASH constructs. A representative gel filtration elution profile at each step is shown, and the proteins eluted under each peak are indicated. (B) Purification of the SUN2–KASH5 complex is illustrated to demonstrate the strategy used for SUN2/GB1–KASH constructs.

aliquot of SUN–KASH complexes by 15% SDS-PAGE to estimate cleavage efficiency. If necessary, add fresh 3C protease and incubate for at least 8 h.

4. (a) The above procedure yields MBP-tagged SUN–KASH complexes that are often incompletely cleaved. Cleaved MBP can also form soluble aggregates of ill-defined size that partially coelute with SUN–KASH

complexes during gel filtration. To overcome both problems, and improve SUN–KASH complex homogeneity, we cleave the MBP tag in two steps (Fig. 2A). First, load the partially cleaved SUN–KASH complex from step 3 of Section 2.3 on a Superdex S200 column (in standard gel filtration buffer, e.g., 10 m*M* Tris/HCl pH 8.0, 150 m*M* NaCl, and 1 m*M* KCl) and purify. To avoid protein precipitation before column loading, the protein concentration should not exceed 1–2 mg/ml at this step. The partially cleaved SUN–KASH complex elutes in a fairly non-homogeneous peak during gel filtration, and is collected with contaminants. Second, pool the SUN–KASH complex-containing fractions and cleave again with 3C protease at an enzyme:substrate ratio of approximately 1:50 (w/w). This second cleavage step should completely remove the fusion tags from SUN2–KASH1/2/3/4 complexes (verify by analytical SDS-PAGE). Reconcentrate these purified complexes and then load onto a HiLoad 16/60 Superdex S200 column and purify using standard gel filtration buffer. SUN–KASH complexes elute in a single peak and are now suitable for crystallization studies or biochemical assays. (b) The above strategies failed to yield pure and homogeneous complexes in the case of SUN2–KASH5/6. We tested alternatives to MBP tags that could sustain super-stoichiometric expression of KASH5/6 peptides, relative to SUN, and could be removed more efficiently by 3C protease in step 3 of Section 2.3. Two candidates, sfGFP and thioredoxin, both reduced KASH-fusion protein expression to substoichiometric levels compared to SUN (data not shown). However, the GB1 tag met both criteria and was completely removed by proteolysis as observed by SDS-PAGE analysis. In a final step, the resulting complexes are concentrated and purified via gel filtration (Fig. 2B). We have not tried the GB1 tag with other SUN–KASH complexes, but suspect it may have consistent advantages over the MBP tag.

3. STRUCTURAL ANALYSIS OF HUMAN SUN2 AND SUN2–KASH1/2 COMPLEXES

3.1 Crystallization and Structure Determination

We deposited three crystal structures in the Protein Data Bank representing human apo-SUN2 (PDB ID: 4DXT) and SUN2–KASH1/2 complexes (PDB ID: 4DXR/4DXS) (Sosa et al., 2012). Independently, apo-SUN2 (PDB ID: 3UNP) and SUN2–KASH2 (PDB ID: 4FI9) structures were solved by the Wang and Zhou labs (Wang et al., 2012; Zhou et al.,

2012). Crystallization in all labs was performed by the hanging-drop vapor diffusion method in distinct experiment drop compositions. Apo-SUN2 crystals have grown in 16% (w/v) polyethylene glycol (PEG) 3350 and 200 m*M* potassium thiocyanate at 18 °C in our lab, whereas the Wang group crystallized it in 100 m*M* imidazole, 1 *M* sodium acetate pH 6.5, and 10 m*M* YCl_3 at 4 °C. The resulting structures are overall very similar, except for a varying conformation of the unstructured KASH "lid" (SUN2 residues 567–587) in the apo-form.

The KASH lid of the SUN domain becomes ordered, and adopts a β-hairpin form, only upon binding to KASH. Reflecting the high similarity between KASH1 and KASH2 in length and amino acid composition, the lid adopts an identical conformation in SUN2–KASH1/2 complexes. Although KASH1/2 peptides are not involved in crystal-packing contacts, SUN2–KASH1 and SUN2–KASH2 complexes crystallize in different conditions. In our lab, the SUN2–KASH1 complex was crystallized in 100 m*M* HEPES pH 7.4, 7% (w/v) PEG 4000, 10% 1,6-hexanediol, and 0.25% *n*-decyl-β-D-maltoside (DM), whereas SUN2–KASH2 complex crystals were grown in 100 m*M* HEPES pH 7.5, 200 m*M* ammonium acetate, 25% 2-propanol, and 0.3% DM. The Zhou lab, on the other hand, crystallized the SUN2–KASH2 complex in 50 m*M* $MgCl_2$, 100 m*M* HEPES pH 7.5, 6% (w/v) PEG monomethyl ether 5000, and 19.5 m*M* methyl-6-*O*-(*N*-heptylcarbamoyl)-α-D-glucopyranoside (HECAMEG).

Interestingly, apo-SUN2 and SUN2–KASH complexes pack in rhombohedral crystals such that a trefoil-to-trefoil assembly occurs between neighboring SUN2 homotrimers, with the coiled-coil stalks pointing away in opposite directions. The major difference between these crystals is that the distance between neighboring apo-SUN2 trimers is much smaller than that of the SUN2–KASH1/2 complexes, due to the conformational change in the KASH lid upon binding to KASH. Therefore, apo-SUN2 crystals have a lower solvent content, which might explain why they tend to diffract to higher resolution. Because the KASH peptide is not directly involved in crystal packing, there is a good chance that other SUN2–KASH complexes can be structurally characterized under similar crystallization conditions.

3.2 *In Vitro* Binding Experiments

Interactions between SUN and KASH proteins have been studied using *in vitro* binding protocols established for apo-SUN2 and KASH2. For these studies, apo-SUN2 (residues 522–717) and KASH2 (residues 6863–6885) are each purified separately. Apo-SUN2 is purified as described in Section 2.2. We purify KASH2 as a fusion to the C terminus of 6× His-sfGFP; this 6× His-sfGFP–KASH2 polypeptide (sfGFP–KASH2) is first purified by Ni^{2+}-affinity as described for apo-SUN constructs. The Ni^{2+-} eluate is further purified by gel filtration using a Superdex S75 column equilibrated into 10 m*M* Tris/HCl pH 8.0 and 150 m*M* NaCl. sfGFP–KASH2 purified in this manner yields two approximately equal-intensity bands on SDS-PAGE, with a small size difference. Only the larger species binds SUN, suggesting the truncation is C-terminal and eliminates (part of) the KASH peptide. Since these two forms of sfGFP–KASH2 are present at a ratio of about 1:1 (by SDS-PAGE), for qualitative binding assays we use a molar stoichiometry of 2:1 (sfGFP–KASH2 to apo-SUN2) to ensure a roughly 1:1 ratio of apo-SUN2 and binding-competent sfGFP–KASH2.

Binding is measured by analytical gel filtration on a Superdex S200 HR10/300 column (Fig. 3). Prior to injecting the sample, apo-SUN2, and apo-SUN2 incubated with sfGFP–KASH2, is dialyzed into the gel filtration buffer overnight. For each gel filtration run, a total volume of 500 μl is loaded onto the column; each sample contains 50 μ*M* SUN2 and 100 μ*M* sfGFP–KASH2 (to achieve a 1:1 ratio of SUN2 to binding-competent undegraded sfGFP–KASH). In the chromatogram, the sfGFP–KASH2 fusion protein is monitored by absorbance at 488 nm, and total protein absorption is measured at 280 nm.

The results of such binding experiments are shown in Fig. 3. When apo-SUN2 (residues 522–717) is run in buffer 1 (20 m*M* HEPES/NaOH pH 8.0, 100 m*M* KCl), it elutes primarily as a trimer ("S3"; Fig. 3A); a small fraction elutes as a monomer ("S1"; Fig. 3A). Buffer 1 was previously used for *in vitro* SUN–KASH-binding assays; however, the effect of monomer–trimer equilibrium on KASH binding was not discussed (Zhou et al., 2012). In contrast, apo-SUN2 elutes as a monomer when using buffer 2 (10 m*M* Tris/HCl pH 8.0, 150 m*M* NaCl, 1 m*M* KCl) (Fig. 3B). SUN trimerization is a prerequisite for KASH binding as shown by gel filtration analysis of apo-SUN2/sfGFP–KASH2 mixtures. Using buffer 1, we obtain a stoichiometric SUN2/sfGFP–KASH2 complex, and excess sfGFP–KASH2 elutes as a separate peak ("K"; Fig. 3C). Using buffer 2, only a fraction of SUN2 elutes as an assembled complex with sfGFP–KASH2, whereas the majority remains unbound (Fig. 3D). Thus, buffer conditions play an important role in SUN–KASH-binding experiments. This buffer dependence is particularly critical for SUN constructs that lack portions of the coiled-coil element.

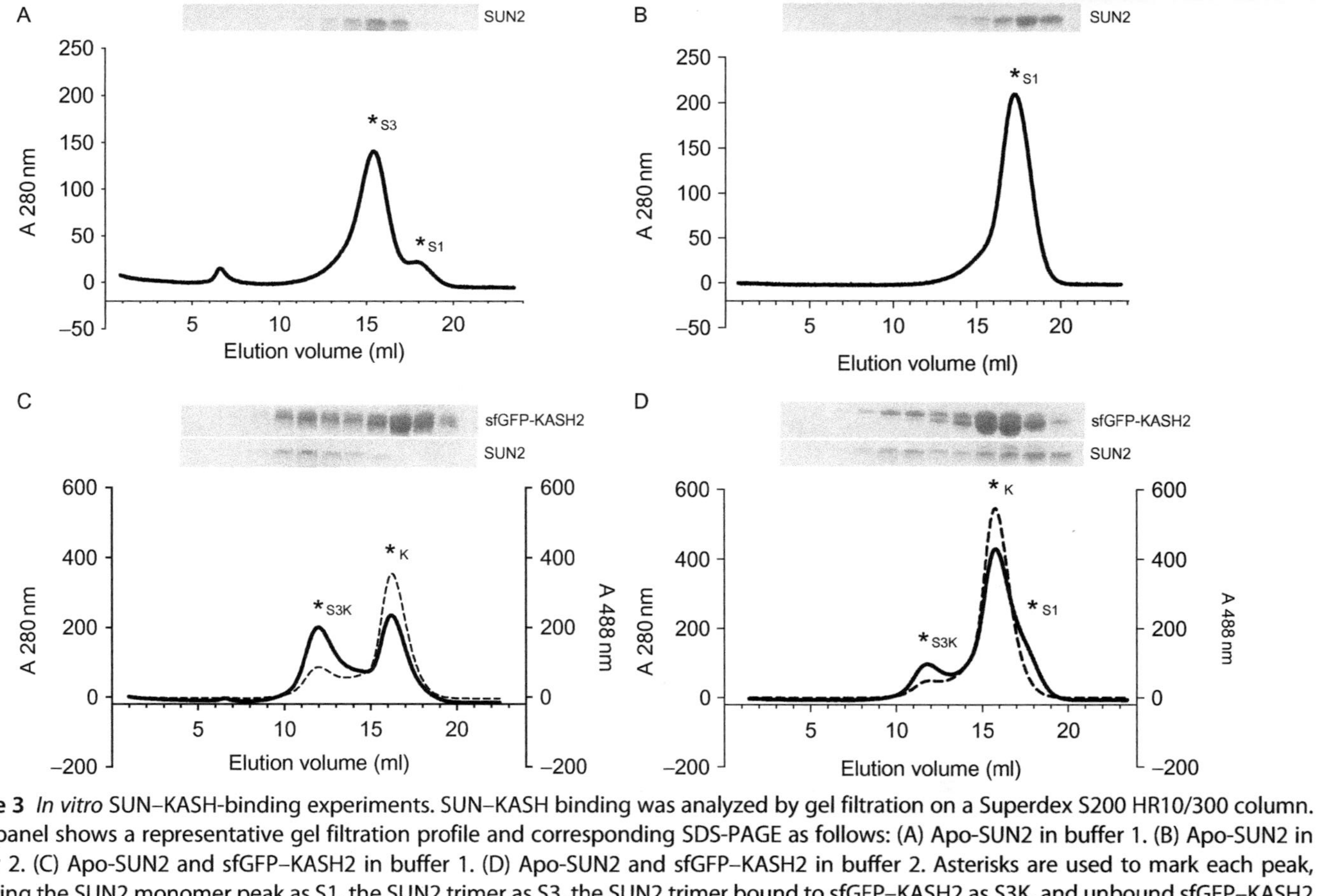

Figure 3 *In vitro* SUN–KASH-binding experiments. SUN–KASH binding was analyzed by gel filtration on a Superdex S200 HR10/300 column. Each panel shows a representative gel filtration profile and corresponding SDS-PAGE as follows: (A) Apo-SUN2 in buffer 1. (B) Apo-SUN2 in buffer 2. (C) Apo-SUN2 and sfGFP–KASH2 in buffer 1. (D) Apo-SUN2 and sfGFP–KASH2 in buffer 2. Asterisks are used to mark each peak, denoting the SUN2 monomer peak as S1, the SUN2 trimer as S3, the SUN2 trimer bound to sfGFP–KASH2 as S3K, and unbound sfGFP–KASH2 as K. Solid lines on each chromatogram represent the 280 nm trace; dashed lines represent the 488 nm trace. Buffer 1 contains 20 m*M* HEPES pH 8.0 and 100 m*M* KCl. Buffer 2 contains 10 m*M* Tris/HCl pH 8.0, 150 m*M* NaCl, and 1 m*M* KCl. Each Coomassie-stained SDS-PAGE gel shows 1-ml fractions covering the 7–18 ml elution segment. SUN oligomerization is buffer dependent. Buffers that enable SUN trimerization are required for SUN–KASH-binding experiments.

4. CONCLUSIONS AND PITFALLS

Our biochemical understanding of SUN–KASH assemblies is still incomplete, and many interesting questions remain open. The buffer dependence of SUN proteins *in vitro* is not yet understood; indeed, the mechanisms of LINC complex assembly and disassembly may be strongly influenced by the unique microenvironment of the NE/ER lumen, which current *in vitro* conditions do not reproduce. The specificity and strength of SUN–KASH interactions are relatively unexamined, yet further analysis of SUN–KASH proteins from different species can potentially help resolve this problem. Finally, apart from the SUN–KASH core, structural knowledge about most other regions of LINC complexes is limited—an open area of research with fascinating implications for understanding the NE and mechanisms of human laminopathy diseases.

ACKNOWLEDGMENTS

We thank Brian A. Sosa for developing the initial protocols for purification and crystallization of human SUN2–KASH1/2 complexes. This work was supported by a grant from the NIH (1RO1-AR065484).

REFERENCES

Andersen, K. R., Leksa, N. C., & Schwartz, T. U. (2013). Optimized E. coli expression strain LOBSTR eliminates common contaminants from His-tag purification. *Proteins*, *81*, 1857–1861.

Bione, S., Maestrini, E., Rivella, S., Mancini, M., Regis, S., Romeo, G., et al. (1994). Identification of a novel X-linked gene responsible for Emery–Dreifuss muscular dystrophy. *Nature Genetics*, *8*, 323–327.

Bonne, G., Di Barletta, M. R., Varnous, S., Becane, H. M., Hammouda, E. H., Merlini, L., et al. (1999). Mutations in the gene encoding lamin A/C cause autosomal dominant Emery-Dreifuss muscular dystrophy. *Nature Genetics*, *21*, 285–288.

Borrego-Pinto, J., Jegou, T., Osorio, D. S., Auradé, F., Gorjánácz, M., Koch, B., et al. (2012). Samp1 is a component of TAN lines and is required for nuclear movement. *Journal of Cell Science*, *125*, 1099–1105.

Burke, B., & Roux, K. J. (2009). Nuclei take a position: Managing nuclear location. *Developmental Cell*, *17*, 587–597.

Chang, W., Folker, E. S., Worman, H. J., & Gundersen, G. G. (2013). Emerin organizes actin flow for nuclear movement and centrosome orientation in migrating fibroblasts. *Molecular Biology of the Cell*, *24*, 3869–3880.

Chang, W., Worman, H. J., & Gundersen, G. G. (2015). Accessorizing and anchoring the LINC complex for multifunctionality. *The Journal of Cell Biology*, *208*, 11–22.

Chen, C.-Y., Chi, Y.-H., Mutalif, R. A., Starost, M. F., Myers, T. G., Anderson, S. A., et al. (2012). Accumulation of the inner nuclear envelope protein Sun1 is pathogenic in progeric and dystrophic laminopathies. *Cell*, *149*, 565–577.

Ciani, B., Bjelic, S., Honnappa, S., Jawhari, H., Jaussi, R., Payapilly, A., et al. (2010). Molecular basis of coiled-coil oligomerization-state specificity. *Proceedings of the National Academy of Sciences of the United States of America, 107*(46), 19850–19855.

Crisp, M., Liu, Q., Roux, K., Rattner, J. B., Shanahan, C., Burke, B., et al. (2006). Coupling of the nucleus and cytoplasm: Role of the LINC complex. *The Journal of Cell Biology, 172*, 41–53.

di Guan, C., Li, P., Riggs, P. D., & Inouye, H. (1988). Vectors that facilitate the expression and purification of foreign peptides in Escherichia coli by fusion to maltose-binding protein. *Gene, 67*, 21–30.

Göb, E., Schmitt, J., Benavente, R., & Alsheimer, M. (2010). Mammalian sperm head formation involves different polarization of two novel LINC complexes. *PLoS One, 5*, e12072.

Haque, F., Lloyd, D. J., Smallwood, D. T., Dent, C. L., Shanahan, C. M., Fry, A. M., et al. (2006). SUN1 interacts with nuclear lamin A and cytoplasmic nesprins to provide a physical connection between the nuclear lamina and the cytoskeleton. *Molecular and Cellular Biology, 26*, 3738–3751.

Haque, F., Mazzeo, D., Patel, J. T., Smallwood, D. T., Ellis, J. A., Shanahan, C. M., et al. (2010). Mammalian SUN protein interaction networks at the inner nuclear membrane and their role in laminopathy disease processes. *The Journal of Biological Chemistry, 285*, 3487–3498.

Horn, H. F., Brownstein, Z., Lenz, D. R., Shivatzki, S., Dror, A. A., Dagan-Rosenfeld, O., et al. (2013). The LINC complex is essential for hearing. *The Journal of Clinical Investigation, 123*, 740–750.

Horn, H. F., Kim, D. I., Wright, G. D., Wong, E. S. M., Stewart, C. L., Burke, B., et al. (2013). A mammalian KASH domain protein coupling meiotic chromosomes to the cytoskeleton. *The Journal of Cell Biology, 202*, 1023–1039.

Huth, J. R., Bewley, C. A., Jackson, B. M., Hinnebusch, A. G., Clore, G. M., & Gronenborn, A. M. (1997). Design of an expression system for detecting folded protein domains and mapping macromolecular interactions by NMR. *Protein Science: A Publication of the Protein Society, 6*, 2359–2364.

LaVallie, E. R., Lu, Z., Diblasio-Smith, E. A., Collins-Racie, L. A., & McCoy, J. M. (2000). Thioredoxin as a fusion partner for production of soluble recombinant proteins in Escherichia coli. *Methods in Enzymology, 326*, 322–340.

Lindeman, R. E., & Pelegri, F. (2012). Localized products of futile cycle/lrmp promote centrosome-nucleus attachment in the zebrafish zygote. *Current Biology, 22*, 843–851.

Lu, W., Schneider, M., Neumann, S., Jaeger, V.-M., Taranum, S., Munck, M., et al. (2012). Nesprin interchain associations control nuclear size. *Cellular and Molecular Life Sciences, 69*, 3493–3509.

Luxton, G. W. G., Gomes, E. R., Folker, E. S., Vintinner, E., & Gundersen, G. G. (2010). Linear arrays of nuclear envelope proteins harness retrograde actin flow for nuclear movement. *Science, 329*, 956–959.

Luxton, G. W. G., Gomes, E. R., Folker, E. S., Worman, H. J., & Gundersen, G. G. (2011). TAN lines: A novel nuclear envelope structure involved in nuclear positioning. *Nucleus, 2*, 173–181.

Luxton, G. W. G., & Starr, D. A. (2014). KASHing up with the nucleus: Novel functional roles of KASH proteins at the cytoplasmic surface of the nucleus. *Current Opinion in Cell Biology, 28*, 69–75.

Ostlund, C., Folker, E. S., Choi, J. C., Gomes, E. R., Gundersen, G. G., & Worman, H. J. (2009). Dynamics and molecular interactions of linker of nucleoskeleton and cytoskeleton (LINC) complex proteins. *Journal of Cell Science, 122*, 4099–4108.

Padmakumar, V. C., Libotte, T., Lu, W., Zaim, H., Abraham, S., Noegel, A. A., et al. (2005). The inner nuclear membrane protein Sun1 mediates the anchorage of Nesprin-2 to the nuclear envelope. *Journal of Cell Science, 118*, 3419–3430.

Pédelacq, J.-D., Cabantous, S., Tran, T., Terwilliger, T. C., & Waldo, G. S. (2006). Engineering and characterization of a superfolder green fluorescent protein. *Nature Biotechnology, 24*, 79–88.

Puckelwartz, M. J., Kessler, E., Zhang, Y., Hodzic, D., Randles, K. N., Morris, G., et al. (2009). Disruption of nesprin-1 produces an Emery-Dreifuss muscular dystrophy-like phenotype in mice. *Human Molecular Genetics, 18*, 607–620.

Rothballer, A., & Kutay, U. (2013). The diverse functional LINCs of the nuclear envelope to the cytoskeleton and chromatin. *Chromosoma, 122*, 415–429.

Sosa, B. A., Rothballer, A., Kutay, U., & Schwartz, T. U. (2012). LINC complexes form by binding of three KASH peptides to domain interfaces of trimeric SUN proteins. *Cell, 149*, 1035–1047.

Starr, D. A., & Fridolfsson, H. N. (2010). Interactions between nuclei and the cytoskeleton are mediated by SUN-KASH nuclear-envelope bridges. *Annual Review of Cell and Developmental Biology, 26*, 421–444.

Starr, D. A., & Han, M. (2002). Role of ANC-1 in tethering nuclei to the actin cytoskeleton. *Science, 298*, 406–409.

Wang, W., Shi, Z., Jiao, S., Chen, C., Wang, H., Liu, G., et al. (2012). Structural insights into SUN-KASH complexes across the nuclear envelope. *Cell Research, 22*, 1440–1452.

Worman, H. J. (2012). Nuclear lamins and laminopathies. *The Journal of Pathology, 226*, 316–325.

Zhang, Q., Bethmann, C., Worth, N. F., Davies, J. D., Wasner, C., Feuer, A., et al. (2007). Nesprin-1 and -2 are involved in the pathogenesis of Emery-Dreifuss muscular dystrophy and are critical for nuclear envelope integrity. *Human Molecular Genetics, 16*, 2816–2833.

Zhang, Q., Skepper, J. N., Yang, F., Davies, J. D., Hegyi, L., Roberts, R. G., et al. (2001). Nesprins: A novel family of spectrin-repeat-containing proteins that localize to the nuclear membrane in multiple tissues. *Journal of Cell Science, 114*, 4485–4498.

Zhou, Z., Du, X., Cai, Z., Song, X., Zhang, H., Mizuno, T., et al. (2012). Structure of Sad1-UNC84 homology (SUN) domain defines features of molecular bridge in nuclear envelope. *The Journal of Biological Chemistry, 287*, 5317–5326.

CHAPTER FIVE

Purification of Lamins and Soluble Fragments of NETs

Alexandr A. Makarov, Andrea Rizzotto, Peter Meinke, Eric C. Schirmer[1]
Wellcome Trust Centre for Cell Biology, University of Edinburgh, Edinburgh, United Kingdom
[1]Corresponding author: e-mail address: e.schirmer@ed.ac.uk

Contents

Abstract

Lamins and associated nuclear envelope transmembrane proteins (NETs) present unique problems for biochemical studies. Lamins form insoluble intermediate filament networks, associate with chromatin, and are also connected via specific NETs to the cytoskeleton, thus further complicating their isolation and purification from mammalian cells. Adding to this complexity, NETs at the inner nuclear membrane function in three distinct environments: (a) their nucleoplasmic domain(s) can bind lamins, chromatin,

Methods in Enzymology, Volume 569
ISSN 0076-6879
http://dx.doi.org/10.1016/bs.mie.2015.09.006

and transcriptional regulators; (b) they possess one or more integral transmembrane domains; and (c) their lumenal domain(s) function in the unique reducing environment of the nuclear envelope/ER lumen. This chapter describes strategic considerations and protocols to facilitate biochemical studies of lamins and NET proteins *in vitro*. Studying these proteins *in vitro* typically involves first expressing specific polypeptide fragments in bacteria and optimizing conditions to purify each fragment. We describe parameters for choosing specific fragments and designing purification strategies and provide detailed purification protocols. Biochemical studies can provide fundamental knowledge including binding strengths and the molecular consequences of disease-causing mutations that will be essential to understand nuclear envelope–genome interactions and nuclear envelope linked disease mechanisms.

1. INTRODUCTION

The nuclear envelope (NE) consists of two lipid bilayers: the outer nuclear membrane (ONM) is continuous with the ER and separated from the inner nuclear membrane (INM) by a lumenal space of ~50 nm (Callan, Randall, & Tomlin, 1949). These membranes join around nuclear pore complexes, which mediate traffic between the nucleus and cytoplasm (Prunuske & Ullman, 2006). Both membranes contain a great many NE transmembrane proteins (NETs) (Korfali et al., 2012; Schirmer, Florens, Guan, Yates, & Gerace, 2003), the vast majority of which have yet to be characterized.

NETs with assigned functions at the ONM tend to be involved in connecting cytoplasmic filaments to the NE. The best-characterized examples are ONM nesprins, which bind in the NE lumen to INM NETs, named SUN proteins, which in turn bind the lamin nucleoskeleton, thus creating LINC complexes (LInker of Nucleoskeleton and Cytoskeleton) (Crisp et al., 2006). Specific LINC complexes are important for the mechanical stability of the nucleus and/or cytoskeleton, nuclear migration within the cell, and cell mobility (Lammerding et al., 2004; Lee et al., 2007; Lombardi et al., 2011; Razafsky & Hodzic, 2015).

Most INM NETs whose functions have been studied interact with lamins and distinct functional regions of chromatin and together contribute to genome organization (Solovei et al., 2013; Zuleger, Robson, & Schirmer, 2011; Zullo et al., 2012), tissue-specific gene regulation, and DNA damage repair (Zuleger, Robson, et al, 2011). NETs can also function in signaling and recruitment of chromatin-modifying enzymes (de Las Heras et al., 2013). Their biological importance is underscored by the wide range of human diseases caused by mutations in these proteins or their lamin partners (de Las Heras et al., 2013).

1.1 Protein–Protein Interactions at the NE

The complexity of protein interactions at the NE is amplified by the extreme differences in the biochemical and biophysical context of the different compartments generated by NE structure and the fact that the constituent proteins can differ dramatically between cell types. For example, LINC complex proteins arise by alternative splicing of multiple nesprin genes, SUN genes, and lamin genes. This complexity extends to single cells, where different LINC complexes may have different roles, with some specialized for mechanotransduction, while others, named TAN-lines, form transiently during nuclear migration (Luxton, Gomes, Folker, Worman, & Gundersen, 2011). Distinct LINC complexes can also associate with specific NETs; e.g., TAN-lines contain NET5/Samp1 (Borrego-Pinto et al., 2012). The relative level of each lamin subtype varies reproducibly between tissues (Lehner, Stick, Eppenberger, & Nigg, 1987; Swift et al., 2013), and lamins A/C, B1, and B2 each appear to form independent networks (Shimi et al., 2008, 2015; von Moeller, Barendziak, Apte, Goldberg, & Stick, 2010), while binding and FRET data suggest that interactions between networks are also likely (Delbarre et al., 2006; Schirmer & Gerace, 2004). Thus, a single INM NET has the potential to bind chromatin, different lamin networks, soluble partners, or other INM proteins on one hand, while its lumenal domain(s) can interact with lumenal partners, potentially including ones that connect through the ONM to the cytoskeleton.

These tissue-specific and dynamic interconnections, as well as posttranslational modifications, can mask the epitopes recognized by specific antibodies. For example in spleen, which clearly expresses the INM protein emerin by Western blotting, six different emerin epitopes recognized by six different immunohistochemistry-competent monoclonal antibodies are masked, suggesting major differences in emerin posttranslational modifications or protein–protein associations (Tunnah, Sewry, Vaux, Schirmer, & Morris, 2005).

Furthermore, NETs are integral membrane proteins with the potential to interact with diverse partners in at least three distinct microenvironments: for INM proteins these environments consist of the nucleoplasm, the lipid bilayer, and the NE/ER lumen. Each environment involves biochemically distinct subdomains of the protein. Transmembrane domains are hydrophobic, interacting with lipids. Nucleoplasmic domains tend to include a positively charged region (possibly allowing interactions with chromatin) and other regions that confer binding to specific partners in the nucleoplasm. By contrast, their lumenal domains function in a strongly reducing environment. Indeed, high-throughput analysis of the predicted membrane topology of 199 NETs identified in liver revealed that predicted nucleoplasmic domains have a strong tendency to have very high isoelectric points, whereas lumenal domains exhibited widely varying characteristics (Zuleger, Kelly, et al., 2011). Similar trends are seen for NETs isolated from other tissues (Fig. 1D). Therefore, solubilizing a native NET is merely the first step. Any further functional or biochemical studies (e.g., binding assays) must be done using polypeptide fragments and conditions appropriate to each distinct functional environment. Ultimately, mutations that affect

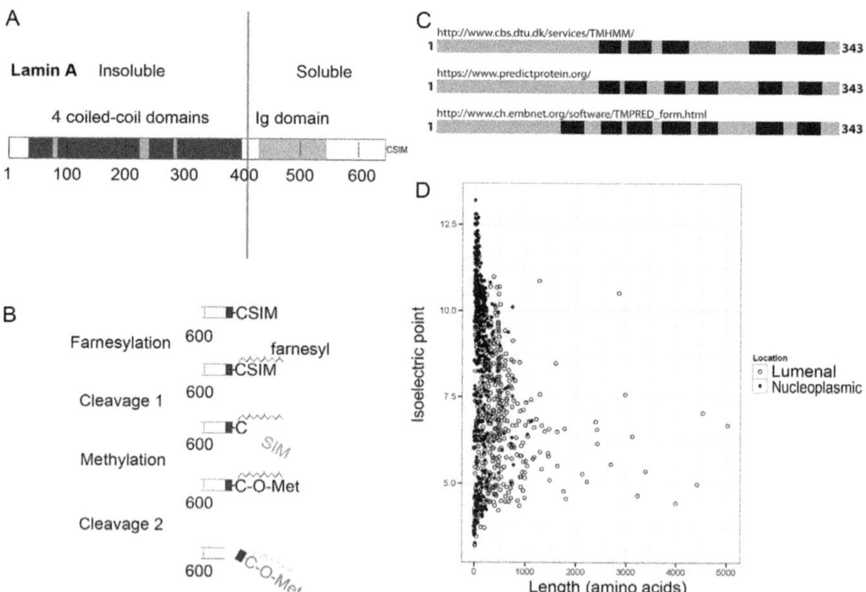

Figure 1 Strategies and considerations. (A) Schematic of lamin A polypeptide showing the locations of soluble versus insoluble subdomains. (B) Sequential posttranslational processing of the C-terminal "CaaX" motif (specifically, CSIM) of prelamin A, then final cleavage to remove the last 16 residues (including the farnesylated Cys) indicated in black. (C) An example in which three different algorithms predict conflicting transmembrane domain organization in one NET (NET29a, also known as TMEM120A); black indicates predicted transmembrane domains. (D) Graph of isoelectric point (p*I*) versus amino acid length shows that the nucleoplasmic domains of many NETs have abnormally high or low isoelectric points, as detailed in text.

specific interactions can then be introduced and studied in the context of the full-length protein *in vivo*. Considerations for choosing purification strategies are presented here, along with protocols for purifying lamin and NET fragments from bacterial expression systems.

2. PURIFICATION PARAMETERS

2.1 Tags on Lamins and Other Considerations

For lamins the purification method, choice of tags and whether or not to generate fragments greatly depends on the experimental question and goal. For example, full-length lamins and their isolated coiled-coil "rod" domain are generally insoluble. By contrast, the large C-terminal "tail" domains of mammalian lamins are soluble when expressed in bacteria (Fig. 1A) and are compatible with most *in vitro* assay systems.

One consideration is that most lamin tail domains are posttranslationally processed. The lamin A precursor (prelamin A), lamin B1, and lamin B2 all have a C-terminal "CaaX" (Cys–aliphatic–aliphatic–any residue) motif; this Cys residue is normally farnesylated and methylated after proteolytic removal of the last three residues. Prelamin A is then further cleaved to remove another 15 residues along with the farnesyl moiety (Fig. 1B) (Camozzi et al., 2014; Sinensky et al., 1994). If modification(s) are potentially relevant to your biological question, consider expressing lamin tails in a eukaryotic (e.g., reticulocyte) lysate system (bearing in mind that the farnesyl moiety is very hydrophobic and will complicate most *in vitro* studies). If modifications are not relevant, lamin tails are readily expressed in bacteria, but you should consider whether to express precursor versus mature polypeptides. Make sure you sequence-verify your construct; not all cDNAs are properly annotated.

Another major posttranslational modification is phosphorylation. Lamins can be phosphorylated during interphase (Kochin et al., 2014; Simon & Wilson, 2013) and become hyperphosphorylated during mitosis to disassemble the lamin polymer; these hyperphosphorylated proteins are soluble (Gerace & Blobel, 1980). One might consider exploiting this solubility to purify native lamins. Extreme hyperphosphorylation will drastically change the parameters for ion exchange purification and make it difficult to cleanly separate lamins from more abundant cytoplasmic intermediate filaments and the many other highly phosphorylated proteins in mitosis. Nonetheless, lamins can be purified from animal tissues by first separating nuclei from cytoplasmic filaments and digesting chromatin to generate NEs, which are then solubilized in urea; lamins are then readily separated by ion exchange chromatography (Florens, Korfali, & Schirmer, 2008; Korfali, Fairley, Swanson, Florens, & Schirmer, 2009; Wilkie & Schirmer, 2008). Tags on lamins can be problematic *in vivo* (e.g., GFP at the N-terminus can interfere with polymerization; Moir, Yoon, Khuon, & Goldman, 2000) and must be characterized.

Different purification strategies reflect a balance between quantity versus quality. Recombinant lamin A purified directly from inclusion bodies is often 80–95% pure, and large quantities (up to 100 mg/L bacterial culture) can be produced. However, further purification steps are less efficient. For example, binding efficiency to nickel beads is relatively poor (~0.7–1 mg protein per mL nickel beads); hence further purification of ~100 mg inclusion body material would require >100 mL of beads, which is not cost effective for most laboratories, and loses roughly 20% of the inclusion body material. However, a much greater percentage of lamin tends to be lost during ion exchange chromatography. Another method, gel extraction (see Section 5), is arduous and time consuming with the lowest yields (0.6 mg recovered from each 1.6 mg resolved on an $18 \times 16 \times 0.15$ cm preparative gel), but is the cleanest and only way to remove degradation products that, deriving largely from the rod domain, are difficult to separate by other methods. The best way to deal with degradation products is to not produce them in the first place. When expressing lamins in bacteria, it is critically important to follow best microbiology practices. These include using fresh transformations or frozen stocks, keeping cells in log-phase growth, avoiding vectors that drive very high-level expression (yields more degradation products), and choosing an appropriate strain of *E. coli* (e.g., taking into account codon preferences etc.). To reduce degradation we use pET28b vectors, which express at lower levels due to a mutation in the origin of replication, in strain BL21 (*fhuA2* [*lon*] *ompT gal* [*dcm*] *ΔhsdS*).

2.2 How to Pick NET Fragments

NETs can have biochemically diverse domains that function in the nucleoplasm or NE lumen, or span the membrane. The first step in choosing fragment(s) for bacterial expression is to determine the membrane topology, if this is not already known. Several prediction algorithms are adequate (e.g., http://www.cbs.dtu.dk/services/TMHMM/, https://www.predictprotein.org/, and http://www.ch.embnet.org/software/TMPRED_form.html). However, different algorithms can yield contradictory results (Fig. 1C).

If all algorithms agree, for example, on whether the N-terminal and C-terminal regions are nucleoplasmic or lumenal, one has more confidence. However, you will still need to experimentally verify protein topology; do this first, if the predictive algorithms disagree. Protein topology can be determined by permeabilizing cells with digitonin versus triton, and staining with specific antibodies (if available) or antibodies against epitope tags located at the N- or C-terminus of the NET (Soderqvist & Hallberg, 1994).

2.3 Tags on NETs

Many NETs have nucleoplasmic domains with extreme isoelectric points (p*I*s ranging from 2.7 to 12.5) as shown in Fig. 1D, which can make the corresponding recombinant fragments extremely insoluble. In such cases, fuse your fragment to a large protein such as GST or MBP with a p*I* close to 6.0, to generate a recombinant protein with better solubility and more predictable purification properties. To improve solubility, use buffers with pH within two units above the p*I* of the recombinant protein, not the same p*I* as the recombinant protein. Overall, take into consideration both the topology and biochemical characteristics of the region to be expressed, when choosing the epitope tag.

3. LAMIN A PURIFICATION

Several protocols for the expression and purification of recombinant lamins are provided here. As noted in Section 2.1, we prefer to use *Escherichia. coli* strain BL21 transformed with lamin cDNAs in the pET28b vector. Starting with isolated inclusion bodies, this protocol typically yields 30–100 mg of lamin A (80–95% pure) per liter of bacterial culture. If cleaner material is required, lamin A can be further (or alternatively) purified using either ion exchange chromatography (Section 3.3; does not require epitope tag) or Ni^{2+} affinity (Section 3.4; His-tagged proteins), or isolated from gels (Section 5). The protocols below are optimized for purification of recombinant lamin A, but with slight optimizing modifications can also be used to purify B-type lamins.

3.1 Buffers and Solutions

LB-Kanamycin (LB-kan) medium: 5 g tryptone, 2.5 g yeast extract, 5 g NaCl, dH_2O to 500 mL, 70 μg/mL kanamycin. Note: kanamycin resistance is conferred by the pET28b vector; if your vector differs, use the appropriate antibiotic.

Lysis buffer: 50 m*M* HEPES, pH 8.0, 3 m*M* $MgCl_2$, 3 m*M* 2-mercaptoethanol, 1 m*M* PMSF (phenylmethylsulfonyl fluoride), 1 μg/mL aprotinin, 1 μ*M* leupeptin, 1 μ*M* pepstatin.

Inclusion bodies buffer (*IBB*): 20 m*M* Tris–HCl, pH 8.0, 8 *M* urea, 2 m*M* dithiothreitol (DTT), 1 m*M* PMSF. Note that this buffer as well as any urea containing buffer should be made 2–3 days in advance and never be heated as that promotes formation of isocyanic acid (a urea derivative) that carbamylates protein (Sun, Zhou, Yang, & Zhang, 2014).

Equilibration buffer: 25 m*M* Tris, pH 8.5, 250 m*M* NaCL, 1 m*M* DTT, 1 m*M* EGTA, 1 m*M* PMSF.

Polymerization buffer: 25 m*M* MES, pH 6.5, 120 m*M* NaCl, 1 m*M* DTT, 1 m*M* EGTA, 1 m*M* PMSF.

3.2 Isolation of Recombinant Lamin A Inclusion Bodies from *E. coli*

1. Transform *E. coli* strain BL21 with pET28b vector encoding lamin A, and allow colonies to grow 12 h at 37 °C. Note: this protocol was optimized for this vector/strain combination; you may need to reoptimize if your conditions differ.
2. Move plates to 4 °C until you initiate protein expression. Note: storing colonies for >1 week can yield strain/vector mutations.
3. Roughly 3 h before ending work for the evening, inoculate a colony into a 4 mL starter culture of LB-kan. Incubate 3 h at 37 °C with shaking.
4. Transfer 1 mL starter culture into a 500 mL flask. Add 100 mL fresh *LB-kan*. Incubate 12 h (overnight) at 37 °C with shaking.
5. Next morning, check the culture OD_{600} against an *LB-kan* blank.
   (A) If the OD_{600} is above 0.6, transfer 1 mL of the overnight culture into 1 L of fresh *LB-kan* in a 4-L flask. This dilution is necessary because the culture was too saturated; saturation induces proteases and many more log-phase cell cycles are required to remove them. Maintain good aeration; smaller media:flask volume ratios are acceptable if your flask is beveled.
   (B) Alternatively, if the OD_{600} is below 0.6, transfer the entire overnight culture into 1-L fresh *LB-kan*.
6. Continue incubating with shaking, measuring every 20–40 min until the OD_{600} is 0.6.

7. Add IPTG (0.3 m*M* final) to induce expression, and incubate 4 h with shaking at 37 °C.
8. Pour culture into 500-mL centrifuge bottles, and pellet bacteria by centrifugation (20 min at 4000 × *g*, 4 °C) using the lowest deceleration settings. If 500-mL bottles are unavailable, use 20 50-mL-capacity centrifuge tubes. Do not pellet faster; inclusion bodies are best preserved when bacteria are pelleted gently into a loose pellet.
9. Carefully discard supernatant. To limit degradation, keep the tubes on ice at all times.

The following steps describe lamin A purification from inclusion bodies:

10. To lyse bacteria, resuspend each 500-ml pellet quickly, but gently in 8 mL *Lysis buffer*. Transfer to a 50-mL centrifuge tube. Wash each 500-ml bottle with 4 mL *Lysis buffer* to recover additional bacteria, and combine for a total of 24 mL/L of starting culture.
11. Add DNaseI (Sigma-Aldrich D4527; 10 U/μL stock) to a final concentration of 1 U/mL. Digest 15 min on ice. Removing bacterial DNA is important, because lamin A binds DNA and chromatin.
12. Lyse cells on ice by sonicating three times (program the sonicator for 120 s per session, with 1 s ON, 1 s OFF pulse intervals, or do 3 min of sonication followed by cooling for ≥5 min) at 30% amplitude (settings for Vibra-Cell VCX 500 Ultrasonic Atomizer, SONICS). Wait 3 min on ice between sessions to avoid overheating the lysate. Note: a 12 mm probe is needed to sonicate 24 mL lysate in a 50-mL conical tube. If only a 6-mm microtip is available, divide lysates among 15-mL tubes or the tip may fracture.
13. Add fresh DNaseI (same amount) and incubate for another 15 min on ice.
14. Repeat step 12, using the same settings. Keep on ice at all times. Save an aliquot of this total cell lysate for later SDS–PAGE analysis.
15. Divide lysate into 12 (2-mL-capacity) microfuge tubes. Note: failure to distribute lysate among smaller tubes yields pellets too large and difficult to resuspend. To separate the inclusion bodies from soluble proteins, centrifuge 10 min at 4000 × *g* (4 °C). This will produce a soft brownish pellet comprising the inclusion bodies and insoluble bacterial material. Save an aliquot of the supernatant for analysis. Note: higher *g*-forces yield a dense pellet that is difficult to disperse in subsequent washing steps.

The following steps use Triton X-100 to remove insoluble bacterial contaminants from the inclusion pellets. For each wash, it is important to disperse the pellet into as many small particles as possible. Successful washing yields pellets that become distinctively whiter than the initial pellet, so adjust the number of washes based on how white/clean the pellet looks.

16. Add 1 mL of 1.5% Triton X-100 in ddH_2O to each tube and use a p1000 pipette tip to disperse each pellet by pipetting until fully homogenized. Then combine the 12 homogenates pairwise into six 2-mL-capacity microfuge tubes to reduce processing volumes and times. Always keep tubes on ice to limit protein degradation.
17. Pellet insoluble lamin A by centrifugation for 10 min at $4000 \times g$ (4 °C). Save an aliquot of the supernatant for analysis.
18. Repeat the wash by dispersing each pellet in 1.5% Triton X-100 and pelleting 10 min at $4000 \times g$ (4 °C) at least once. After each wash, save a small aliquot of the supernatant for analysis.
19. Resuspend pellets in *IBB* for storage at –80 °C.
20. Resolve the collected aliquots by SDS–PAGE to assess purity of the isolated inclusion bodies, as shown in Fig. 2A. Isolating inclusion bodies is an excellent first purification step, whether the recombinant lamins are affinity-tagged or not. This procedure can yield protein clean enough for some applications. Epitope-tagged lamins can be further purified using standard tag-specific protocols.

3.3 Further Purification Without Tags

Untagged lamins can be further purified by ion exchange chromatography. If you wish to purify untagged endogenous lamins from mammalian cells, first isolate NEs as described (Florens et al., 2008; Korfali et al., 2009; Wilkie & Schirmer, 2008), then resuspend in IBB. In either case (inclusion bodies or native NEs), the next step is ion exchange chromatography. This protocol was optimized for lamin polypeptides expressed in bacteria; lamins isolated from eukaryotes may be posttranslationally modified and thus elute slightly differently. Alternatively you may purify untagged lamins by gel extraction, as detailed in Section 5.

Ion exchange chromatography

1. For 6 mg purified inclusion body protein, weigh out 0.17 g P11 cellulose phosphate (Whatman 4071010).
2. Activate by stirring the weighed P11 cellulose phosphate into 25 volumes of 0.5 *M* NaOH and leave for 5 min with gentle agitation.
3. Filter or decant to remove the supernatant, and wash with ddH_2O until the pH is below 11. Then stir the cellulose phosphate into 25 volumes of 0.5 *M* HC, and leave for 5 min with gentle agitation.

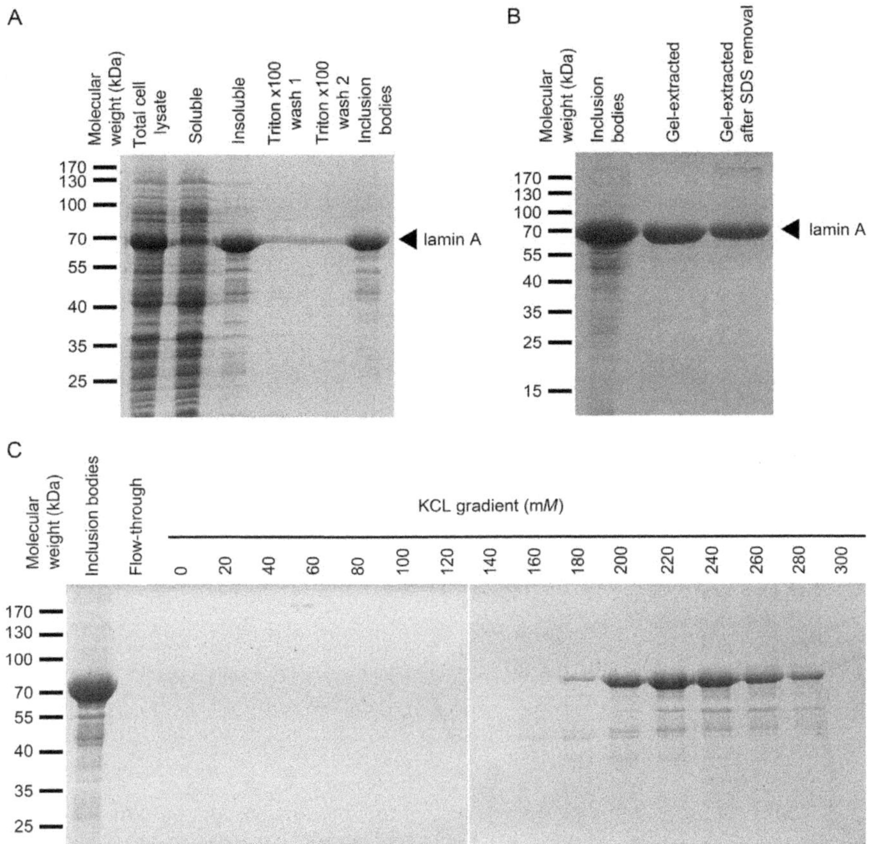

Figure 2 Coomassie-stained gel analysis of fractions obtained during lamin A isolation protocol. (A) Fractions obtained during recombinant lamin A isolation from *E. coli* inclusion bodies. (B) Further purification of lamin A via either (B) gel extraction as described in Section 5; or (C) ion exchange chromatography as described in Section 3.3.

4. Filter or decant off the supernatant and wash with ddH_2O until the pH is above 3.0. Final volume of the swollen P11 media should be around 1 ml. Pack a column of appropriate size; we usually use BioRad 9 cm Poly-Prep chromatography columns.
5. To equilibrate the column, wash with 6 column volumes of *IBB*.
6. Run lamin A protein (resuspended in *IBB*) through the column, twice. Save a small aliquot of each flow-through fraction for analysis and repeated purification, if needed.
7. Wash column with 2 volumes of *IBB* to remove unbound proteins.
8. Elute using a gradient of KCl (from 20 to 400 m*M*) in *IBB*. The gradient can be either continuous (20 column volumes of buffer mixed by a gradient maker) or a step-gradient (20 sequential washes, each with 2

column volumes, increasing the KCl concentration by 20 m*M* each time). In either case, collect 20 separate fractions.

9. Resolve 10-μL aliquots of input material, flow-through and each fraction on SDS–PAGE minigels to assess purification and identify the best elution factions. Lamin A typically elutes between 180 and 260 m*M* KCl, as shown in Fig. 2C. If lamin A continues to elute at higher KCl concentrations, the next time you do this procedure use higher elution volumes between 180 and 260 m*M* KCl.
10. Measure the protein concentration of each lamin A-containing fraction, and resolve 7 μg of each to compare their purity.
11. Combine the cleanest fractions, and dialyze into IBB. Dialyzed fractions can then be frozen and stored indefinitely. Alternatively, if lamins are insufficiently "clean," repeat the cellulose phosphate procedure (steps 1–10), or purify by polymerization and pelleting as described here.
12. Dialyze pooled fractions 3 h at room temperature against 100 volumes of *Equilibration buffer*. Use standard dialysis bags with a molecular weight cutoff below 10 kDa, such as Spectra/Por 6.4 mm, 6–8 kDa dialysis bags from Spectrum Labs.
13. Dialyze 3 h against 100 volumes of *Polymerization buffer*. Full-length lamin A will start polymerizing, forming visible flakes inside the dialysis bags.
14. Shake the bag and carefully pipette the solution and flakes into 1.5-mL centrifuge tubes. Wash the bag with *Polymerization buffer* to collect remaining flakes.
15. Pellet polymerized material at 15,000 × *g* (1 h, 4 °C). Discard the supernatant. Solubilize the lamin A pellet in a buffer of choice containing 8 *M* urea (never heated; see Section 3.5 for buffer considerations).

3.4 Purification of His-Tagged Lamins

His-tagged lamins that include the insoluble rod domain are purified in the presence of *8 M urea*. Urea lowers the binding capacity of nickel beads to approximately 600–700 μg/mL packed resin. The protocol below should saturate the column; some lamin A will be lost, but you should obtain ~1 mg of highly purified lamin A. This procedure must be completed in one day (leaving lamin A on Ni^{2+} beads overnight leads to degradation contaminants).

1. To pack a 1.5-mL column: mix 50% His-Select Nickel Affinity Gel bead slurry (Sigma-Aldrich, P6611) and move 3 mL into the column (e.g., BioRad 9 cm Poly-Prep chromatography column). Let the storage buffer drain without drying the column.
2. Wash column twice with 4.5 mL (3 bead volumes) of ddH_2O to remove residual storage buffer.
3. Equilibrate by washing twice with 4.5 mL (3 bead volumes) of *IBB*.
4. Your starting solutions (purified inclusion body-derived lamin A in 8 *M* urea [IBB]) are quite viscous, due in part to high protein concentrations. To help your columns run faster, dilute lamins in *IBB* to a final protein concentration of 1–2 mg/mL. Save a small aliquot of the flow-through for later analysis by SDS–PAGE.
5. Wash with 4.5 mL *IBB*. Save a small aliquot for analysis.
6. Wash with 4.5 mL *IBB containing freshly added* 5 m*M imidazole* (for 6 × His tags) or 10 m*M* imidazole (for 10 × His tags). Save a small aliquot for analysis.
7. Wash twice with *IBB containing freshly added* 5 m*M imidazole* and 400 m*M* KCl. Save aliquots for analysis.
8. Elute lamin A using 1 mL *IBB containing* 200 m*M imidazole.* Repeat elution four times, collecting four separate elution fractions. Save small aliquots for SDS–PAGE analysis. Store eluted fractions at room temperature in the dark (to prevent imidazole modification of protein) until step 9 is complete.
9. Resolve 10-μL fraction aliquots by 12% SDS–PAGE and stain with Coomassie Blue. Identify and pool lamin A-containing fractions, and dialyze immediately against 1000 volumes of *IBB*.
10. Samples can be stored in IBB at –80 °C indefinitely.

3.5 General Considerations About Buffers for Lamin A

Purified lamin A can be stored in 8 *M* urea buffers (e.g., IBB supplemented with protease inhibitors) at –80 °C almost indefinitely. In these buffers you can also perform binding to nickel resins and some ion exchange resins. However, subsequent experiments may not be compatible with urea. In this case, lamin A may be dialyzed into Tris, HEPES, or NaPi buffers with pH 8.0–8.5 and containing 250 m*M* NaCl and 1 m*M* DTT in which most of the lamin protein should stay soluble at concentrations below 50 μg/mL for several hours. At pH and salt concentrations below this or at higher protein concentrations lamin A will tend to oligomerize. Dialysis into these buffers at higher lamin concentrations has been used to study lamin assembly *in vitro* (Aebi, Cohn, Buhle, & Gerace, 1986; Glass et al., 1993; Heitlinger et al., 1991, 1992). Regardless of your buffers or experimental goals, you should perform all dialysis steps gradually, or the protein is likely to aggregate or assemble filaments.

4. NET FRAGMENT PURIFICATION

4.1 Buffers and Solutions

Lysis buffer 2: 200 m*M* NaPhosphate buffer, pH 7.4, 300 m*M* NaCl, 10% glycerol, 10 U/mL DNaseI (Sigma-Aldrich D4527; 10 U/μL stock), 7 μg/mL RNase A (Sigma-Aldrich R6513, 7 mg/mL stock), 1 mg/mL lysozyme (Sigma-Aldrich L6876), complete EDTA free protease inhibitor (Roche).

5× Loading buffer: 250 m*M* Tris–Cl, pH 6.8, 0.25% bromophenol blue, 300 m*M* DTT, 50% glycerol, 10% SDS.

Washing buffer: 200 m*M* NaPhosphate buffer, pH 7.4, 300 m*M* NaCl, 10 m*M* imidazole.

Elution buffer: 200 m*M* NaPhosphate buffer, pH 7.4, 300 m*M* NaCl, 250 m*M* imidazole.

4.2 Expression Optimization

It is good practice to optimize the expression parameters for each protein construct. Three main variables affect the solubility of recombinant NETs: the expression temperature in bacteria, the IPTG concentration used to induce expression, and the time of induction.

1. Inoculate a single colony of your transfected bacteria (e.g., BL21, or BL21 (DE3) pLys) into a 50 mL LB culture with appropriate antibiotic. Incubate at 30 °C overnight with continuous shaking. Growth at 30 °C (rather than 37 °C) will maintain the culture in logarithmic phase. This prevents the induction of proteases and also avoids inducing B-lactamase, which degrades ampicillin. If you are selecting with ampicillin, and the overnight culture reaches saturation ($OD_{600} > 0.8$), pellet the cells to remove B-lactamase in the culture medium prior to dilution and further culture growth as described in Section 3.1.
2. The next day, dilute the starter culture 1:100 in prewarmed LB. Grow the culture with shaking to a final OD_{600} of 0.6–0.8. Remove 1 mL of

culture and check the OD_{600} at least every hour in a standard spectrophotometer.

3. Split the culture into 16 (50-mL-capacity) conical tubes (3 mL culture each).
4. Add IPTG to each tube; test four different concentrations of IPTG (e.g., 1, 0.5, 0.3, and 0.125 m*M*), in four tubes each.
5. Incubate each set of four cultures (same IPTG concentration) under different conditions:
   37°C for 4 h
   30°C for 4 h
   30°C, overnight
   18°C, overnight
6. At appropriate times, pellet cells 15 min at 4000 × *g* (4 °C), decant the supernatant, let additional liquid settle and remove it by pipetting. The pellet can be (a) directly stored at –80 °C, (b) resuspended in *Lysis buffer 2 without RNAse/DNAse/Lysozyme* and stored at –80 °C, or (c) resuspended in 1 mL *Lysis buffer 2* and processed immediately (step 7).
7. Lyse bacterial pellet 15 s at 20% amplitude (settings for Vibra-Cell VCX 500 Ultrasonic Atomizer, SONICS), keeping samples on ice to limit protein degradation.
8. Pellet cell debris by microfuging 10 min at maximum speed. Collect 80 μL supernatant and mix with 20 μL of *5× Loading buffer* for subsequent SDS–PAGE analysis.
9. Decant remaining supernatant. Resuspend the pellet in 120 μL of *1× Loading buffer.*
10. Heat samples 5 min at 70 °C to allow slow but complete denaturation.
11. Resolve 10 μL of each sample by SDS–PAGE. Stain with Coomassie blue and evaluate results to determine the best conditions for protein expression and the ratio of soluble versus insoluble protein. An example is shown in Fig. 3A–D: expression was induced with different IPTG concentrations (0.125, 0.25, 0.5, and 1 m*M*) for different times and at different temperatures. In this particular case, induction with 0.5 m*M* IPTG at 30 °C overnight yielded the highest levels of soluble NET fragment expression (Fig. 3C, lane 3).

4.3 Purification of His-Tagged NETs from Soluble (or Insoluble) Material

After optimizing expression conditions in Section 4.2, express on a larger scale to obtain enough material for purification by tag-appropriate affinity

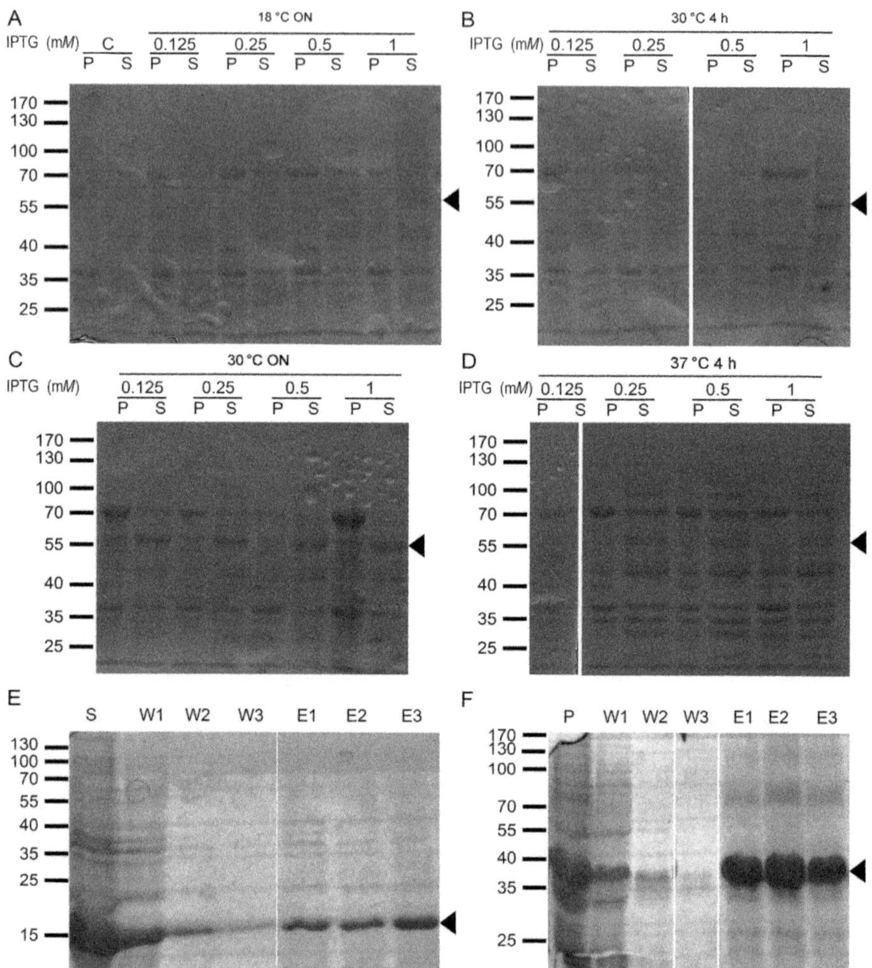

Figure 3 NET purification examples. (A–D) Expression optimization results (Coomassie-stained gels) for a NET named C17Orf62 (also known as NKP9). (A–D) Results for each expression condition described in Section 4.2. Arrowheads indicate the recombinant protein of interest. Note the differing yields for soluble protein fractions (*S*, supernatant) in (B), and the insoluble protein pellet (*P*) fractions in (C). (E) Gel analysis of His-tagged affinity purification of insoluble NET5 residues 59–209. (F) Gel analysis of His-tagged affinity purification of soluble Tmem70 residues 1–110. Arrowheads indicate recombinant protein of interest.

chromatography. The following protocol is for a His-tagged protein in the soluble fraction. If your His-tagged protein is insoluble, the procedure is the same except for step 5, where special instructions are provided.

1. Inoculate and grow a culture from freshly transformed bacterial cells using the best practices described in Section 4.2 to maintain selection,

log-phase growth and reduce protein degradation. However, instead of amplifying the culture for small aliquots as described previously, amplify to 1 L of culture using prewarmed LB.

2. Once the culture reaches OD_{600} 0.6–0.8, induce protein expression using the optimized conditions determined in Section 4.2.
3. Pellet cells in 500-mL centrifuge bottles (4000 × *g*, 21 min, 4 °C), decant the supernatant and remove residual fluids. The pellet now can be stored at −80 °C as described in Section 4.2, or processed immediately by resuspending in 100 mL *Lysis buffer 2* per liter bacterial culture.
4. Lyse bacterial cells as described in Section 3.2 but omitting DNA digestion step.
5. Pellet cell debris (60,000 × *g*, 45 min, 4 °C) and recover the supernatant for step 6. Alternatively, if your NET is insoluble: incubate the sonicated bacterial lysate with 1.5% Triton X-100 for 30 min at 4 °C, centrifuge (4000 × *g*, 30 min, 4 °C), resuspend the pellet in *Lysis buffer 2 containing 2% sarcosyl,* incubate overnight at room temperature, then pellet at 4000 × *g* (30 min, 4 °C), and recover the supernatant.
6. Equilibrate 2 mL Ni-NTA (Qiagen, 30210) agarose with 10 mL *Lysis buffer 2* for 10 min.
7. Incubate the supernatant containing the soluble expressed protein (or resuspended insoluble fraction) with the equilibrated Ni-NTA agarose overnight at 4 °C with gentle rotation.
8. Transfer the column matrix into a gravity column (e.g., BioRad 9 cm Poly-Prep chromatography column) and allow to settle and drain. Do not let the column matrix start drying.
9. Wash the column with 2 column volumes of *Washing buffer*. Collect samples of each flow-through fraction for later SDS–PAGE analysis.
10. Elute the protein sequentially with four 2-mL fractions of *Elution buffer*. Collect 2-mL fractions on ice and store in the dark until step 11 is complete. Reserve a small aliquot of each fraction for SDS–PAGE analysis.
11. Analyse fractions by SDS–PAGE as illustrated in Fig. 3E–F.
12. Dialyse fractions containing pure protein against 1000 volumes of *Elution buffer without imidazole*. Store aliquots of purified protein at −20 °C in *Elution buffer with 20% glycerol.*

5. GEL EXTRACTION OF PROTEINS

Extremely pure protein can be obtained after resolving on Laemmli gels (Fig. 2B). However, this approach is greatly limited by the capacity of protein gels and cannot be easily scaled up. Another disadvantage is that residual SDS interferes with many functional assays, and must be removed. Our typical yield from two 18 × 16 × 0.15 cm preparative gels loaded with 2.5 mg lamin A protein from inclusion bodies and run overnight is 1–1.5 mg pure lamin A in total. Extraction efficiencies for NET fragments vary significantly.

5.1 Buffers and Solutions

Running buffer: 25 m*M* Tris, 200 m*M* glycine, 0.1% SDS.
Zinc stain: 0.1% SDS, 0.2 *M* imidazole
Zinc developer: 0.1% SDS, 0.2 *M* zinc sulfate.
Zinc destain: 2% citric acid in ddH_2O.
ATAc: 86:7:7 acetone:triethylamine:acetic acid mix.
5% ATAc: 5% ATAc in ddH2O.

5.2 Gel Extraction

This protocol uses the BioRad model 422 electro-elutor and fitting gel extraction membrane caps. We use 18 × 16 × 0.15 cm Hoeffer preparative gel slabs for their larger protein resolving capacity. We use reversible Zinc staining (Fernandez-Patron, Castellanos-Serra, & Rodriguez, 1992) to visualize bands without fixing. There are also commercial kits based on the same principle and readily available from, for example, Pierce/Thermo.

1. Resolve 1.5–2 mg protein (lamin or NET) on each preparative 10% Laemmli gel; electrophoresis at 20 mA for 12 h typically places lamin A in the lower third of the gel.
2. An hour before the gel is done, equilibrate gel extraction membrane caps in the *Running buffer* at 65 °C.
3. Wash gel in ddH_2O for 20 s.
4. Incubate gel with the *Zinc stain* for 20 min, wrapped in foil, on a shaker, in the dark as much as possible to avoid potential protein alkylation (Dahl & McKinley-McKee, 1981).
5. Rinse gel 5 s with ddH_2O to remove residual imidazole.
6. Incubate with *Zinc developer*, which precipitates a background of zinc/imidazole chelates (white) where protein is absent from the gel. Within 1–2 min, protein bands will look opaque on this white background. Terminate staining by rinsing in ddH_2O. Bands can be obscured by zinc precipitates at the gel surface; if this happens, wash briefly with *Zinc destain* until bands are obvious.

7. Cut out your protein band, and destain for 15 min in *Running buffer* in the dark.
8. Change to fresh *Running buffer* and wash 15 min in the dark. If gel pieces are not completely transparent, repeat this step.
9. Cut the gel into small (1 mm) cubes and put into the gel extractor tube assembled according to the electro-elutor manual.
10. Gel extractor protocol: Run 2 h in *Running buffer* at 10 mA per tube. Carefully decant top chamber without tilting the extractor tubes. Change buffer in both the top and bottom chambers to 25 m*M* Tris, 192 m*M* glycine (*Running buffer without SDS*), and run for 1 h more.

Final processing to remove SDS is detailed in Section 5.3.

5.3 Considerations for SDS Removal

Gel extraction works best from SDS–PAGE, but must be followed by removal of SDS, which interferes with binding assays and intermediate filament assembly (Steinert, Idler, Cabral, Gottesman, & Goldman, 1981). The following protocol for SDS removal was developed for cytoplasmic intermediate filaments, and we successfully applied it to lamin A. Different NET proteins and fragments will bind SDS with varying efficiency. Inefficient removal of SDS may yield protein solutions that become extra "foamy" after dialyzed out of the running buffer, altered electrophoretic mobility of the purified sample, or loss of function. If so, consider classical methods to remove SDS (using methanol or acetone), or an alternative purification strategy.

1. Remove extracted protein solution from the gel extractor collection chamber (~200–1000 μL) into a 50-mL conical tube.
2. Add 0.2 volumes cold 2 *M* KCl, and observe the formation of crystalline flakes of SDS and lamin A. Incubate 30 min on ice.
3. Centrifuge (4000 × *g*, 15 min, 4 °C). Carefully remove and discard the supernatant using a P200 pipette; the pellet will be loose.
4. Add 20 volumes cold *ATAc*, vortex vigorously until pellet is completely dispersed, and incubate 10 min on ice until insoluble lamin A sediments.
5. Centrifuge (4000 × *g*, 15 min, 4 °C) to pellet the precipitated lamins.
6. Carefully decant the supernatant. Disperse the pellet in 10 volumes of *5% ATAc*. Centrifuge (4000 × *g*, 15 min, 4 °C) to pellet lamins.
7. Wash again, this time using 2 volumes of *5% ATAc*. Pellet by centrifugation and discard the supernatant.
8. *Dry* the pellet for 15 min at room temperature, then resuspend in 1 volume of your buffer of choice containing 8 *M* urea. *Store at −80 °C.*

6. CONCLUSIONS AND POTENTIAL APPLICATIONS

Lamin and NET purification is not trivial even for experienced biochemists. However, the benefits are considerable. Applications can include *in vitro* assembly and partner binding assays to study wild-type versus NE-linked disease-mutated NETs and lamins (Simon & Wilson, 2013). Purified NET fragments are far better antigens than short peptides, enabling the generation of key reagents. We hope these protocols lead to a better understanding of the dynamic functions of the nucleus and its roles in human disease.

ACKNOWLEDGMENTS

We thank Dr. Alastair Kerr for help with bioinformatic analysis. We gratefully acknowledge support from the Wellcome Trust (Senior Research Fellowship 095209 and Centre Grant 092076 to E.C.S.), a Principal's Scholarship from the University of Edinburgh (to A.A.M.) and the Darwin Trust (to A.R.).

REFERENCES

Aebi, U., Cohn, J., Buhle, L., & Gerace, L. (1986). The nuclear lamina is a meshwork of intermediate-type filaments. *Nature, 323*, 560–564.

Borrego-Pinto, J., Jegou, T., Osorio, D. S., Aurade, F., Gorjanacz, M., Koch, B., et al. (2012). Samp1 is a component of TAN lines and is required for nuclear movement. *Journal of Cell Science, 125*, 1099–1105.

Callan, H. G., Randall, J. T., & Tomlin, S. G. (1949). An electron microscope study of the nuclear membrane. *Nature, 163*, 280.

Camozzi, D., Capanni, C., Cenni, V., Mattioli, E., Columbaro, M., Squarzoni, S., et al. (2014). Diverse lamin-dependent mechanisms interact to control chromatin dynamics. Focus on laminopathies. *Nucleus, 5*, 427–440.

Crisp, M., Liu, Q., Roux, K., Rattner, J. B., Shanahan, C., Burke, B., et al. (2006). Coupling of the nucleus and cytoplasm: Role of the LINC complex. *The Journal of Cell Biology, 172*, 41–53.

Dahl, K. H., & McKinley-McKee, J. S. (1981). The imidazole-promoted inactivation of horse-liver alcohol dehydrogenase. *European Journal of Biochemistry/FEBS, 120*, 451–459.

de Las Heras, J. I., Meinke, P., Batrakou, D. G., Srsen, V., Zuleger, N., Kerr, A. R., et al. (2013). Tissue specificity in the nuclear envelope supports its functional complexity. *Nucleus, 4*, 460–477.

Delbarre, E., Tramier, M., Coppey-Moisan, M., Gaillard, C., Courvalin, J. C., & Buendia, B. (2006). The truncated prelamin A in Hutchinson-Gilford progeria syndrome alters segregation of A-type and B-type lamin homopolymers. *Human Molecular Genetics, 15*, 1113–1122.

Fernandez-Patron, C., Castellanos-Serra, L., & Rodriguez, P. (1992). Reverse staining of sodium dodecyl sulfate polyacrylamide gels by imidazole-zinc salts: Sensitive detection of unmodified proteins. *BioTechniques, 12*, 564–573.

Florens, L., Korfali, N., & Schirmer, E. C. (2008). Subcellular fractionation and proteomics of nuclear envelopes. *Methods in Molecular Biology*, *432*, 117–137.

Gerace, L., & Blobel, G. (1980). The nuclear envelope lamina is reversibly depolymerized during mitosis. *Cell*, *19*, 277–287.

Glass, C. A., Glass, J. R., Taniura, H., Hasel, K. W., Blevitt, J. M., & Gerace, L. (1993). The alpha-helical rod domain of human lamins A and C contains a chromatin binding site. *The EMBO Journal*, *12*, 4413–4424.

Heitlinger, E., Peter, M., Haner, M., Lustig, A., Aebi, U., & Nigg, E. A. (1991). Expression of chicken lamin B2 in Escherichia coli: Characterization of its structure, assembly, and molecular interactions. *The Journal of Cell Biology*, *113*, 485–495.

Heitlinger, E., Peter, M., Lustig, A., Villiger, W., Nigg, E. A., & Aebi, U. (1992). The role of the head and tail domain in lamin structure and assembly: Analysis of bacterially expressed chicken lamin A and truncated B2 lamins. *The Journal of Structural Biology*, *108*, 74–89.

Kochin, V., Shimi, T., Torvaldson, E., Adam, S. A., Goldman, A., Pack, C. G., et al. (2014). Interphase phosphorylation of lamin A. *Journl of Cell Science*, *127*, 2683–2696.

Korfali, N., Fairley, E. A., Swanson, S. K., Florens, L., & Schirmer, E. C. (2009). Use of sequential chemical extractions to purify nuclear membrane proteins for proteomics identification. *Method in Molecular Biology*, *528*, 201–225.

Korfali, N., Wilkie, G. S., Swanson, S. K., Srsen, V., de las Heras, J., Batrakou, D. G., et al. (2012). The nuclear envelope proteome differs notably between tissues. *Nucleus*, *3*, 552–564.

Lammerding, J., Schulze, P. C., Takahashi, T., Kozlov, S., Sullivan, T., Kamm, R. D., et al. (2004). Lamin A/C deficiency causes defective nuclear mechanics and mechanotransduction. *The Journal of Clinical Investigation*, *113*, 370–378.

Lee, J. S., Hale, C. M., Panorchan, P., Khatau, S. B., George, J. P., Tseng, Y., et al. (2007). Nuclear lamin A/C deficiency induces defects in cell mechanics, polarization, and migration. *Biophysical Journal*, *93*, 2542–2552.

Lehner, C. F., Stick, R., Eppenberger, H. M., & Nigg, E. A. (1987). Differential expression of nuclear lamin proteins during chicken development. *The Journal of Cell Biology*, *105*, 577–587.

Lombardi, M. L., Jaalouk, D. E., Shanahan, C. M., Burke, B., Roux, K. J., & Lammerding, J. (2011). The interaction between nesprins and sun proteins at the nuclear envelope is critical for force transmission between the nucleus and cytoskeleton. *The Journal of Biological Chemistry*, *286*, 26743–26753.

Luxton, G. W., Gomes, E. R., Folker, E. S., Worman, H. J., & Gundersen, G. G. (2011). TAN lines: A novel nuclear envelope structure involved in nuclear positioning. *Nucleus*, *2*, 173–181.

Moir, R. D., Yoon, M., Khuon, S., & Goldman, R. D. (2000). Nuclear lamins A and B1: Different pathways of assembly during nuclear envelope formation in living cells. *The Journal of Cell Biology*, *151*, 1155–1168.

Prunuske, A. J., & Ullman, K. S. (2006). The nuclear envelope: Form and reformation. *Current Opinion in Cell Biology*, *18*, 108–116.

Razafsky, D., & Hodzic, D. (2015). Nuclear envelope: Positioning nuclei and organizing synapses. *Current Opinion in Cell Biology*, *34*, 84–93.

Schirmer, E. C., Florens, L., Guan, T. L., Yates, J. R., & Gerace, L. (2003). Nuclear membrane proteins with potential disease links found by subtractive proteomics. *Science*, *301*, 1380–1382.

Schirmer, E. C., & Gerace, L. (2004). The stability of the nuclear lamina polymer changes with the composition of lamin subtypes according to their individual binding strengths. *The Journal of Biological Chemistry*, *279*, 42811–42817.

Shimi, T., Kittisopikul, M., Tran, J., Goldman, A. E., Adam, S. A., Zheng, Y., et al. (2015). Structural organization of nuclear lamins A, C, B1 and B2 revealed by super-resolution microscopy. *Molecular Biology of the Cell*. http://dx.doi.org/10.1091/mbc.E15-07-0461. [E-pub ahead of print].

Shimi, T., Pfleghaar, K., Kojima, S., Pack, C. G., Solovei, I., Goldman, A. E., et al. (2008). The A- and B-type nuclear lamin networks: Microdomains involved in chromatin organization and transcription. *Genes and Development*, *22*, 3409–3421.

Simon, D. N., & Wilson, K. L. (2013). Partners and post-translational modifications of nuclear lamins. *Chromosoma*, *122*, 13–31.

Sinensky, M., Fantle, K., Trujillo, M., McLain, T., Kupfer, A., & Dalton, M. (1994). The processing pathway of prelamin A. *Journal of Cell Science*, *107*, 61–67.

Soderqvist, H., & Hallberg, E. (1994). The large C-terminal region of the integral pore membrane protein, POM121, is facing the nuclear pore complex. *European Journal of Cell Biology*, *64*, 186–191.

Solovei, I., Wang, A. S., Thanisch, K., Schmidt, C. S., Krebs, S., Zwerger, M., et al. (2013). LBR and lamin A/C sequentially tether peripheral heterochromatin and inversely regulate differentiation. *Cell*, *152*, 584–598.

Steinert, P. M., Idler, W. W., Cabral, F., Gottesman, M. M., & Goldman, R. D. (1981). In vitro assembly of homopolymer and copolymer filaments from intermediate filament subunits of muscle and fibroblastic cells. *Proceedings of the National Academy of Sciences of the United States of America*, *78*, 3692–3696.

Sun, S., Zhou, J. Y., Yang, W., & Zhang, H. (2014). Inhibition of protein carbamylation in urea solution using ammonium-containing buffers. *Analytical Biochemistry*, *446*, 76–81.

Swift, J., Ivanovska, I. L., Buxboim, A., Harada, T., Dingal, P. C., Pinter, J., et al. (2013). Nuclear lamin-A scales with tissue stiffness and enhances matrix-directed differentiation. *Science*, *341*, 1240104.

Tunnah, D., Sewry, C. A., Vaux, D., Schirmer, E. C., & Morris, G. E. (2005). The apparent absence of lamin B1 and emerin in many tissue nuclei is due to epitope masking. *The Journal of Molecular Histology*, *36*, 337–344.

von Moeller, F., Barendziak, T., Apte, K., Goldberg, M. W., & Stick, R. (2010). Molecular characterization of Xenopus lamin LIV reveals differences in the lamin composition of sperms in amphibians and mammals. *Nucleus*, *1*, 85–95.

Wilkie, G. S., & Schirmer, E. C. (2008). Purification of nuclei and preparation of nuclear envelopes from skeletal muscle. *Methods in Molecular Biology*, *463*, 23–41.

Zuleger, N., Kelly, D. A., Richardson, A. C., Kerr, A. R., Goldberg, M. W., Goryachev, A. B., et al. (2011). System analysis shows distinct mechanisms and common principles of nuclear envelope protein dynamics. *The Journal of Cell Biology*, *193*, 109–123.

Zuleger, N., Robson, M. I., & Schirmer, E. C. (2011). The nuclear envelope as a chromatin organizer. *Nucleus*, *2*, 339–349.

Zullo, J. M., Demarco, I. A., Pique-Regi, R., Gaffney, D. J., Epstein, C. B., Spooner, C. J., et al. (2012). DNA sequence-dependent compartmentalization and silencing of chromatin at the nuclear lamina. *Cell*, *149*, 1474–1487.

CHAPTER SIX

Simple Separation of Functionally Distinct Populations of Lamin-Binding Proteins

Jason M. Berk[1], Katherine L. Wilson[2]
Department of Cell Biology, Johns Hopkins University School of Medicine, Baltimore, Maryland, USA
[2]Corresponding author: e-mail address: klwilson@jhmi.edu

Contents

Abstract

The inner membrane of the nuclear envelope (NE) is home to hundreds of integral membrane proteins (NE transmembrane proteins, "NETs") with conserved or tissue-specific roles in genome organization and nuclear function. Nearly all characterized NETs bind A- or B-type lamins directly. However, hundreds of NETs remain uncharacterized, collectively posing an enormous gap that must be bridged to understand nuclear function and genome biology. We provide technically simple protocols for the separation and recovery of functionally distinct populations of NETs and A-type lamins. This protocol was developed for emerin, an inner nuclear membrane protein that binds lamins and barrier-to-autointegration factor (*BANF1*) as a component of nuclear lamina structure, and has diverse roles in nuclear assembly, signaling, and gene

[1] Current address: Department of Molecular Biophysics and Biochemistry, Yale University, 266 Whitney Ave., New Haven, CT 06511, USA.

Methods in Enzymology, Volume 569
ISSN 0076-6879
http://dx.doi.org/10.1016/bs.mie.2015.09.034

regulation. This protocol separates easily solubilized ("easy") populations of nuclear lamina proteins (emerin, lamin A, BAF) from "sonication-dependent" populations. Depending on cell type, the "easy" and "sonication-dependent" fractions each contain up to about half the available emerin, A-type lamins, and BAF, whereas B-type lamins and histone H3 are predominantly sonication dependent. The two populations of emerin have distinct posttranslational modifications, and only one population associates with BAF. This method may be useful for functional screening or analysis of other lamin-associated proteins, including novel NETs emerging from proteomic studies.

1. INTRODUCTION

Genome organization and function in each tissue are shaped by lamins and nuclear envelope transmembrane (NET) proteins (Worman & Schirmer, 2015), which can localize at the inner or outer nuclear membrane (INM or ONM, respectively) of the nuclear envelope (NE). Characterized NETs such as emerin and other LEM-domain proteins at the INM are widely expressed and bind A- or B-type lamins (or both) directly as conserved components of nuclear lamina structure (Barton, Soshnev, & Geyer, 2015; Berk, Tifft, & Wilson, 2013; Margalit, Brachner, Gotzmann, Foisner, & Gruenbaum, 2007), with roles that include the dynamic "tethering" of silenced chromatin (Demmerle, Koch, & Holaska, 2013; Mattout, Cabianca, & Gasser, 2015; Zheng, Kim, & Zheng, 2015; Zullo et al., 2012) and tissue-specific signaling (Lu, Muchir, Nagy, & Worman, 2011; Muchir, Wu, Sera, Homma, & Worman, 2014; Stubenvoll, Rice, Wietelmann, Wheeler, & Braun, 2015). The NE is customized by the expression of tissue-specific NET proteins that help the genome organize and function appropriately in each cell type (Wong, Luperchio, & Reddy, 2014; Worman & Schirmer, 2015). Dozens of human diseases are linked to mutations in A- or B-type lamins or their INM partners (Bonne, 2014), emphasizing the need to understand the composition, functions, and dynamic regulation of lamin-associated protein complexes.

Loss of emerin, in particular, causes X-linked Emery–Dreifuss muscular dystrophy (EDMD) and related disorders (Berk, Tifft, et al., 2013). Remarkably, EDMD is also caused by mutations in at least six other proteins (lamin A, Luma, Sun1, Sun2, Nesprin-1, and Nesprin-2) that associate with emerin and each other (Haque et al., 2010; Koch & Holaska, 2014; Li, Meinke, Huong le, Wehnert, & Noegel, 2014; Liang et al., 2011; Meinke et al., 2014; Zhang et al., 2007). Hence, EDMD disease likely involves the perturbed assembly, function, or regulation of an emerin-containing "LINC" complex. However, emerin has numerous other direct partners (Berk, Tifft, et al., 2013), notably including BAF (BANF1; barrier-to-autointegration factor). BAF is the "third" component of nuclear lamina structure that links DNA/chromatin to LEM-domain proteins and lamins (Jamin & Wiebe, 2015). Emerin is a component of at least six distinct multiprotein complexes, some of which lack lamins and/or BAF (Holaska & Wilson, 2007). These diverse activities appear to be regulated by differential Ser/Thr or Tyr phosphorylation of emerin (Berk, Tifft, et al., 2013; Tifft, Bradbury, & Wilson, 2009) and *O*-GlcNAc modifications of emerin (Berk, Maitra, et al., 2013). In our attempts to understand this functional complexity, we developed a fractionation protocol that successfully separated two functionally distinct populations of emerin, A-type lamins, and BAF (Berk, Maitra, et al., 2013).

The isolation and biochemical characterization of the nuclear lamina was pioneered nearly 50 years ago (Aaronson & Blobel, 1975; Gerace, Blum, & Blobel, 1978; Zbarsky, Perevoshchikova, Delektorskaya, & Delektorsky, 1969), starting with rat liver cells, and evolving numerous methods for nuclear subfractionation. Current research on the molecular functions of specific INM proteins and their interactions with lamins still faces the same fundamental issues: the insolubility of native lamin filament networks (lamina) and the need for extraction methods stringent enough to solubilize lamina proteins without disrupting biological protein–protein and protein–chromatin interactions. Powerful new methods are available that facilitate the identification of protein–protein interactions *in vivo* (e.g., BioID; Mehus, Anderson, & Roux, 2015; Batsios, Meyer, & Gräf, 2015) or dominantly solubilize lamins *in vivo* (e.g., DARPins; Zwerger et al., 2015). However, these methods require cell engineering or transfection to express the appropriate reagent (e.g., BioID-fused lamin or lamin-specific DARPin, respectively). Other useful methods including chemical crosslinking (Jafferali, Figueroa, & Hallberg, 2015) and conventional coimmunoprecipitation have a different limitation: they rely on specific antibodies that might selectively fail to recover endogenous targets in which the epitope is masked by posttranslational modifications.

We routinely detected interactions of native emerin or GFP-emerin with soluble nucleoplasmic partners, such as BAF and Lmo7 (Berk, Maitra, et al., 2013; Berk et al., 2014; Holaska, Rais-Bahrami, & Wilson, 2006; Tifft et al., 2009), using relatively mild lysis buffers (e.g., 300 m*M* NaCl, 0.3%, v/v, Triton X-100). By contrast, emerin or GFP-emerin interactions with lamins were detectable but puzzlingly unaffected by emerin mutations. Further investigation in HeLa cells showed that these relatively mild conditions solubilized only about half of the endogenous emerin, endogenous A-type lamins, and endogenous BAF. About 80% more emerin, A-type lamins and BAF became soluble after sonication, which thoroughly shears DNA, liberating DNA-bound protein complexes and likely also breaking lamin filaments, and releasing protein complexes that previously pelleted with chromatin. Indeed, endogenous BAF association with GFP-emerin was detected predominantly in sonicated lysates, even though both proteins were present at high levels in both fractions (Berk, Maitra, et al., 2013). Interestingly, emerin in the "easy" nuclear fraction (N_E) was also hyperphosphorylated, compared to that in the sonication-dependent nuclear (N_S) fraction. This simple separation is therefore worth considering as a rapid way to obtain and empirically analyze functionally distinct populations of native endogenous INM proteins (including your protein of interest) and native lamins.

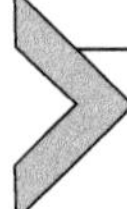

2. CELL CULTURE (WITH OR WITHOUT TRANSFECTION) AND HARVESTING

Our fractionation experiments were done primarily using either HeLa cells (ATCC, #CCL-2, Manassas, VA) or HEK293T cells (ATCC, #CRL-11268) and are applicable to other mammalian cell lines. Cells are cultured at 37 °C in 5% CO_2 and grown in complete medium: high glucose Dulbecco's modified Eagle's medium (DMEM; Life Technologies, #11965-092; Carlsbad, CA) supplemented with 10% (v/v) fetal bovine serum (Life Technologies, #10438-026) and 10 IU/ml Penicillin and 10 μg/ml of Streptomycin (Quality Biological, #A611-K952-06; Gaithersburg, MD). For the assays described here, one confluent 10-cm cell culture plate per condition provided enough emerin from HeLa cells for ~50 lanes in a Western blot, or ~100 lanes from HEK293T cells. Adjust loadings for other proteins of interest as needed. Seed cells at equal densities onto 10 cm plates and grow to confluence (85–95% density).

2.1 Start Here If You Are Transfecting Cells; If Not Transfecting, Begin at Step 2.5

2.1. Plate the required number of 10 cm plates at ~30–40% confluence, 24 h before transfection.

2.2. The day of transfection, bring to room temperature: (a) reduced serum medium Opti-MEM (Life Technologies, #11058021) and (b) transfection reagent *Trans*IT-LT1 (#MIR2300; Mirus Inc., Madison, WI).

2.3. Preincubate transfection mixture for 20 min, as per manufacturer's instructions. We use 1.0 ml Opti-MEM, 5.0 μg DNA (pEGFP-C1 or pEGFP-C1-Emerin), and 25 μl *Trans*IT-LT1 per 10-cm plate.

2.4. Add transfection mixture to cells in a dropwise manner, swirl plate, and incubate 24 h.

2.2 Harvest Cells (Transfected or Untransfected) and Freeze in Preparation for Lysis

2.5. Remove plate from incubator and at room temperature immediately scrape cells off the plate using a Cell Lifter (Corning, #3008; Tewksbury, MA). Tilt the plate and transfer the cells into a 15-ml conical tube. Squirt ~3–5 ml PBS on the plate, scraping any remaining cells, and adding them to the conical tube.

2.6. Centrifuge for 5 min at 400 × *g* (room temperature) in a swinging bucket benchtop centrifuge.

2.7. Decant supernatant and resuspend the cell pellet in 10 ml (room temperature) PBS.

2.8. Centrifuge for 5 min at 400 × *g* (room temperature). Aspirate to remove PBS.

2.9. Resuspend cell pellet in 1 ml PBS and transfer to a 1.5-ml microfuge tube.

2.10. Pellet cells for 5 min at 400 × *g* (~2000 rpm) in a microfuge.

2.11. Aspirate to remove PBS and flash-freeze cell pellet in liquid nitrogen. Store at −80 °C. *Note*: This freezing step is required for cell lysis in Section 3.

3. SEPARATE NUCLEI FROM CYTOPLASM

Nuclei are typically isolated after osmotic shock and hypotonic lysis of the plasma membrane followed by mechanical rupture of the plasma membrane (e.g., dounce homogenizer or small-gauge syringe) and centrifugation to pellet nuclei and other organelle membranes. By contrast, our protocol, adapted from Kazemi, Chang, Haserodt, McKen, and Zachara (2010) and Qing, Yan, and Xiao (2006), bypasses mechanical force and replaces it with a freeze/thaw cycle: frozen cell pellets are thawed in hypotonic lysis buffer. This protocol is a quick, simple, and convenient way to reproducibly isolate a crude fraction containing nuclei and other membrane organelles.

3.1. Remove frozen cell pellets (see Section 2) from the −80 °C freezer and place on ice for 10 min.

3.2. Add 500 μl ice-cold hypotonic lysis buffer (20 m*M* HEPES, pH 7.4, 1 m*M* dithiothreitol [DTT], 100 μ*M* phenylmethylsulfonyl fluoride [PMSF], 1 μg/ml Pepstatin A, and 1 × complete protease inhibitor cocktail [Roche #11697498001, Indianapolis, IN]) on top of each cell pellet. Incubate for 10 min on ice.

3.3. Resuspend the cell pellet to homogeneity by slow, gentle pipetting. Typically, 3–5 plunges are enough to completely resuspend the pellet; use more if needed. Fractionation works best when the pellet is fully resuspended but must be gentle to keep nuclei intact.

3.4. After the pellet is resuspended, incubate on ice for 10 min.

3.5. Centrifuge for 1 min (17,000 × *g* at 4 °C). This supernatant comprises the cytoplasmic fraction, and the pellet contains nuclei and other dense membrane organelles. Carefully collect this cytoplasmic fraction ("C"; Fig. 1) and retain for analysis.

3.6. Wash by slowly pipetting the pelleted nuclei with 500 μl cold hypotonic wash buffer (20 m*M* HEPES, pH 7.4, 1 m*M* DTT) to homogeneity. *Note*: nuclei from different cell types can behave differently: e.g., in contrast to HeLa cell nuclei, nuclei from HEK293T cells are "sticky" and adhere to the inner walls of 1 ml pipette tips. To avoid sample loss with HEK293T or other "problematic" nuclei, wash the

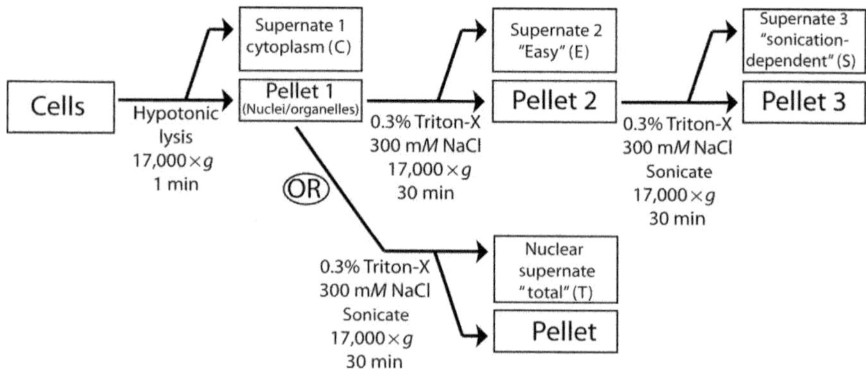

Figure 1 Flowchart of nuclear and subnuclear fractionation steps. Frozen/thawed cells are hypotonically lysed to separate the cytoplasm (Supernant 1; "C") from nuclei and other dense organelles (Pellet 1). This nucleus-enriched pellet can then be processed via the upper pathway, or lower pathway. Upper pathway: The nuclear pellet is lysed and centrifuged, yielding the "easy" nuclear fraction ("E"; Supernant 2); the remaining pellet is then resuspended, sonicated, and centrifuged, yielding the "sonication-dependent" nuclear fraction ("S"; Supernatant 3). Lower pathway: The nuclear pellet is sonicated for one-step recovery of both the "easy" and "sonication-dependent" protein populations ("total" nuclear supernate; "T"). *This research was originally published in the* Journal of Biological Chemistry. *Berk, Maitra, et al., (2013). © The American Society for Biochemistry and Molecular Biology.*

pellet by adding hypotonic lysis buffer and "flicking" the tube, rather than pipetting. Even though the pellet is not resuspended, this is sufficient to wash away remaining cytoplasmic contaminants.

3.7. Pellet the washed nuclei by centrifuging 10 s at 17,000 × *g* (start the timer after centrifuge reaches full speed) at 4 °C. Aspirate to remove the supernatant.

3.8. Repeat washing (Steps 6 and 7) twice, for a total of three washes. The final nuclear pellet should appear much whiter than the starting pellet.

3.9. Resuspend the pelleted nuclei by pipetting in 500 μl nuclear lysis buffer (50 m*M* Tris–HCl pH 7.4, 300 m*M* NaCl, 0.3% Triton-X100, 1 m*M* DTT, 100 μ*M* PMSF, 1 μg/ml Pepstatin A, 1 × Roche complete protease inhibitor cocktail).

Note: Nuclear lysis buffer composition can be manipulated, if you wish. For example, higher concentrations of monovalent salt (≥0.5 *M*) will release additional proteins from the insoluble lamina or DNA (Record, Lohman, & De Haseth, 1976; Senior & Gerace, 1988). You must determine whether these conditions affect key markers (e.g., histones or B-type lamins in sonication-dependent fraction) or your proteins of interest.

3.10. Vortex nuclei at full speed ~10–20 s. Look for nuclei to rupture and form an insoluble mass of fluffy looking chromatin (with experience, this will be obvious). Note that HeLa and HEK293T nuclei differed, again, at this stage. HeLa nuclei could be vortexed thoroughly once and put back on ice, whereas HEK293T nuclei required vortexing to dislodge the nuclear pellet followed by repeated pipetting to lyse the nuclei completely.

From here, you have two choices:

To recover TOTAL nuclear lysate ("easy" and "sonication-dependent" fractions remain together), continue with Step 3.11.

To SEPARATE the "easy" and "sonication-dependent" nuclear fractions, go to Section 4.

3.11. Sonicate the "fluffy" samples using roughly 20 bursts (0.5 s each). Sonicators can differ, so this step may need empirical adjustment based on your specific sonicator and quality of the sonicator tip (old, heavily used sonicator tips require higher power settings than new tips). We use a handheld sonicator (Fisher M-100), at setting number 2 with a 1/8-in. diameter tip. When first attempting this step, start on a very low power setting to avoid sample overheating or sample loss from splashing. Sonication shears genomic DNA and can disrupt detergent/salt-resistant protein complexes. Sonication is complete when the fluffy chromatin pellet is completely dissolved, and the sample appears homogeneous. Avoid over-sonication, which can lead to overheating-induced protein unfolding and aggregation.

3.12. Centrifuge for 30 min at 17,000 × *g* at 4 °C.

3.13. Carefully collect the supernatant; this is your total nuclear fraction (N_T, Fig. 1).

3.14. Resuspend the residual pellet by sonication in 500 μl 1 × SDS-PAGE sample buffer and save for analysis of extraction efficiency (4 × sample buffer stock: 250 m*M* Tris–HCl, pH 6.8; 10%, w/v, SDS; 40%, v/v, glycerol, 4%, v/v, β-mercaptoethanol; bromophenol blue).

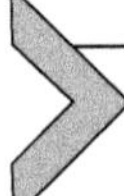

4. FRACTIONATE NUCLEI TO SEPARATE "EASY-" VERSUS "SONICATION-DEPENDENT" PROTEINS

This protocol starts with the crude nuclear pellet from Step 3.10 (see schematic in Fig. 1). Nuclear proteins are solubilized and recovered sequentially before sonication and after sonication, using the same nuclear lysis buffer as Section 3.

4.1. Perform Steps 3.1 through 3.10 to obtain the vortexed "fluffy" nuclear fraction (also collect and save the cytoplasmic fraction).

4.2. Centrifuge the "fluffy" samples for 30 min at 17,000 × *g* at 4 °C, to pellet chromatin and other insoluble material. Collect the supernatant: this is the easily extracted (easy) nuclear fraction (N_E fraction; Fig. 1).

4.3. Add 500 μl nuclear lysis buffer to the pellet, vortex to dislodge the pellet, then quick-spin, and return to ice. The pelleted pieces are now ready for sonication.

4.4. Sonicate using roughly 20 bursts (0.5 s each); see discussion in Step 3.11.

4.5. Centrifuge for 30 min at 17,000 × *g* at 4 °C. Collect this supernatant: this is the sonication-dependent nuclear fraction (N_S fraction; Fig. 1).

4.6. Resuspend the residual pellet by sonication in 500 μl 1 × SDS-PAGE sample buffer and save for Western blot analysis of extraction efficiency.

5. ANALYSIS BY IMMUNOPRECIPITATION AND IMMUNOBLOTTING

5.1 Compare Protein Distributions by Direct Western Blot Analysis

To compare protein distributions between fractions, resolve an equal volume percentage of each fraction by SDS-PAGE and then immunoblot using appropriate primary antibodies. We typically run 2% of each HeLa cell fraction (e.g., 10 μl of the 500 μl lysate stock) or 1% of each HEK293T fraction, as shown in Fig. 2 for HEK293T cells. We recommend gradient gels (e.g., NuPage Novex 4–12% Bis–Tris gradient gels; Life Technologies, #NP0321BOX) to facilitate probing for different-sized proteins of interest in the same blot. See Table 1 for antibodies against useful markers.

5.2 Analysis of Proteins (e.g., GFP-Emerin) Immunoprecipitated from Each Fraction

To detect and quantify binding to partners in each fraction, we immunoprecipitate from an equivalent percentage of each fraction. You will need to

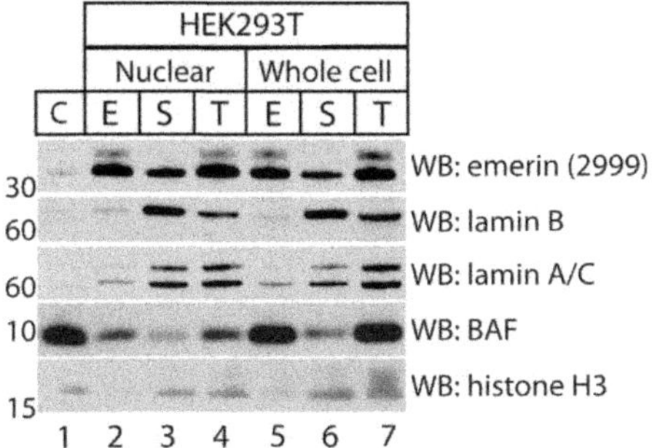

Figure 2 Immunoblot analysis of HEK293T cells after subcellular fractionation. Volume equivalents of the cytoplasm (C), "easy" nuclear fraction (E), sonication-dependent nuclear fraction (S), and total nuclear proteins (T) were resolved using 4–12% Bis–Tris gradient gels and immunoblotted for each indicated protein. Hyperphosphorylated emerin (upper band) is enriched in the E fraction (lanes 2 and 5) compared to the S fraction (lanes 3 and 6). Lamin B extraction requires sonication in HEK293T cells (lanes 3, 4, 6, and 7) and in HeLa cells (not shown). A small fraction of lamin A/C is extracted without sonication from HEK293T cell nuclei (lanes 2 and 5); by contrast, ~50% of lamin A/C is "easy" in HeLa cell nuclei. Histone H3 is a marker for the sonication-dependent nuclear fraction (lanes 3, 4, 6, and 7), although we routinely observed a cytoplasmic signal as well (lane 1). *This research was originally published in the* Journal of Biological Chemistry *as noted in the figure 1 legend.*

Table 1 Commercial Antibodies to Nuclear Marker Proteins.

Antigen	Source	Company	Catalog #	Dilution (v/v)
Actin	Mouse	Millipore, Billerica, MA	MAB1501	1:100,000–500,000
Emerin	Mouse	Novocastra, Newcastle, UK	EMERIN-CE	1:2000–10,000
GFP	Mouse	Santa Cruz Biotechnology, Santa Cruz, CA	sc-9996	1:100,000
Histone H3	Rabbit	Biolegend, San Diego, CA	Poly6019	1:2000
Lamin A/C	Rabbit	Santa Cruz Biotechnology, Santa Cruz, CA	sc-20681	1:2000
BAF	Rabbit[a]	Millipore, Billerica, MA	09-893	1:1000
LAP2 (all isoforms)	Rabbit[b]	Millipore, Billerica, MA	06-1053	1:1000

[a]Pooled sera from rabbits 3273 and 5045 (Haraguchi et al., 2001; Montes de Oca, Lee, & Wilson, 2005).
[b]Rabbit serum 2804 (Fischer, Taysavang, Weber, & Wilson,, 2001).

compare the N_E and N_S fractions from one cell pellet in Section 4, with the N_T fraction from one cell pellet in Step 3.11 that was seeded at the same time at equal density. Before immunoprecipitating, add additional nuclear lysis buffer to increase the total volume from 500 to 1100–1200 μl. The protocol below describes the immunoprecipitation of GFP-fused emerin using the high-affinity alpaca-derived "GFP-Trap" nanobody conjugated to resin (Chromotek #gta-100, Hauppauge, NY). This antibody allows nearly quantitative recovery of GFP-fused proteins (Trinkle-Mulcahy et al., 2008).

5.2.1 Pre-equilibrate enough packed GFP-Trap antibody-conjugated resin (5 μl per reaction) by washing three times (0.5 ml each) in nuclear lysis buffer. After each wash, pellet the resin by centrifuging 4 min at 2000 rpm in a microfuge at room temperature. *Note*: GFP-Trap is excellent for immunoprecipitating GFP-fused proteins of interest, because, in addition to high-affinity recovery, immunoblots are cleaner due to the absence of contaminating light-chain or heavy-chain immunoglobulins, which facilitates detection of coimmunoprecipitating partners.

5.2.2 For each immunoprecipitation, add 1000 μl sample lysate to 5 μl washed resin and rotate overnight at 4 °C. Save an additional 50–100 μl of each sample as your input control.

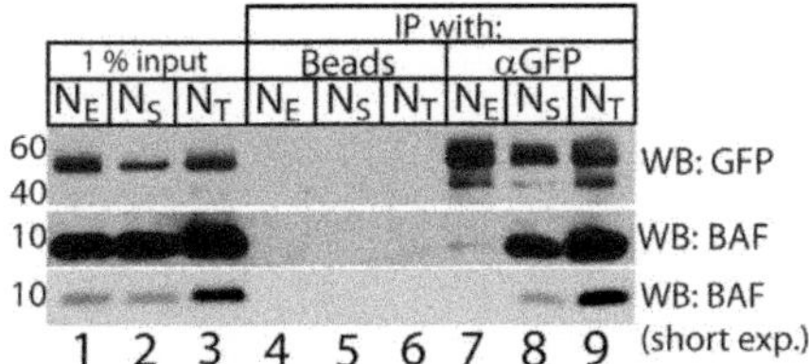

Figure 3 GFP-emerin and endogenous BAF interact robustly in the sonication-dependent (N_S) fraction, not the "easy" (N_E) fraction. HEK293T cells transiently over-expressing GFP-emerin were fractionated to produce equal volumes of "easy" (N_E), "sonication-dependent" (N_S), and "total" (N_T) nuclear fractions. For this experiment, two equivalent frozen cell pellets are required: one is used to generate the N_E and N_S fractions, and the other pellet is the source of N_T. The inputs show that the distribution of GFP-emerin (lanes 1–3) is comparable to that of endogenous emerin in Fig. 2. In the GFP immunoprecipitation (IP with α-GFP) lanes, GFP-emerin interacts with endogenous BAF preferentially in the sonication-dependent fraction (lanes 7 vs. 8). GFP-emerin also associated with BAF in the "total" fraction (lanes 3 and 9). These results suggest that emerin and BAF in each "niche" (N_E and N_S) are differentially regulated and engaged in activities specific to each "niche", including emerin–BAF association in the context of chromatin and B-type lamins (sonication-dependent fraction). *This research was originally published in the* Journal of Biological Chemistry *as noted in the figure 1 legend.*

5.2.3 Wash resin three times (0.5 ml each) with immunoprecipitation wash buffer (50 m*M* Tris–HCl, pH 7.4; 300 m*M* NaCl; 0.3%, v/v, Triton X-100).

5.2.4 Resuspend each final resin pellet in 1 × SDS-PAGE sample buffer, heat for 3 min at 95 °C to recover proteins, and resolve each sample by SDS-PAGE. We typically used 100% of immunoprecipitates from HeLa samples, or 20% from HEK293T samples. Transfer and immunoblot as usual (Fig. 3).

6. CONCLUSIONS

The functions, regulation, and dynamics of lamin-binding proteins are important open frontiers in cell and genome biology. Many new methods are available to study such proteins. Our "partitioning" protocol is valuable because it does not require cell engineering. It can be applied to native cells, or manipulated cells (e.g., transfected cells; stable cell lines), and will allow you to separate and study the functional dynamics (e.g., posttranslational modifications; partners) of specific lamin-associated or INM proteins in three biologically relevant fractions: cytoplasm, the "easy" nuclear fraction, and the "sonication-dependent" (chromatin/lamin B associated) nuclear fraction (Berk, Maitra, et al., 2013). Investigating the functional dynamics of lamin-binding proteins in native cell types may also provide insight into progeroid (accelerated aging) syndromes linked to mutations in lamin A (Gordon, Rothman, Lopez-Otin, & Misteli, 2014) or BAF (Cabanillas et al., 2011; Paquet et al., 2014).

ACKNOWLEDGMENTS

The authors thank Dan Simon for valuable discussions and the NIH (RO1 GM48646) and the Progeria Research Foundation for funding.

REFERENCES

Aaronson, R. P., & Blobel, G. (1975). Isolation of nuclear pore complexes in association with a lamina. *Proceedings of the National Academy of Sciences of the United States of America, 72*, 1007–1011.

Barton, L. J., Soshnev, A. A., & Geyer, P. K. (2015). Networking in the nucleus: A spotlight on LEM-domain proteins. *Current Opinion in Cell Biology, 34*, 1–8.

Batsios, P., Meyer, I., & Gräf, R. (2015). Proximity-dependent biotin identification (BioID) in *Dictyostelium* amoebae. *Methods in Enzymology, 569*. this issue.

Berk, J. M., Maitra, S., Dawdy, A. W., Shabanowitz, J., Hunt, D. F., & Wilson, K. L. (2013). *O*-Linked beta-N-acetylglucosamine (*O*-GlcNAc) regulates emerin binding to barrier to autointegration factor (BAF) in a chromatin- and lamin B-enriched "niche". *Journal of Biological Chemistry, 288*, 30192–30209.

Berk, J. M., Simon, D. N., Jenkins-Houk, C. R., Westerbeck, J. W., Grønning-Wang, L. M., Carlson, C. R., et al. (2014). The molecular basis of emerin-emerin and emerin-BAF interactions. *Journal of Cell Science, 127*, 3956–3969.

Berk, J. M., Tifft, K. E., & Wilson, K. L. (2013). The nuclear envelope LEM-domain protein emerin. *Nucleus, 4*, 3–19.

Bonne, G. (2014). Nuclear envelope proteins in health and diseases. *Seminars in Cell & Developmental Biology, 29*, 93–94.

Cabanillas, R., Cadinanos, J., Villameytide, J. A., Perez, M., Longo, J., Richard, J. M., et al. (2011). Nestor-Guillermo progeria syndrome: A novel premature aging condition with early onset and chronic development caused by BANF1 mutations. *American Journal of Medical Genetics, 155A*, 2617–2625.

Demmerle, J., Koch, A. J., & Holaska, J. M. (2013). Emerin and histone deacetylase 3 (HDAC3) cooperatively regulate expression and nuclear positions of MyoD, Myf5, and Pax7 genes during myogenesis. *Chromosome Research, 21*, 765–779.

Fischer, A. H., Taysavang, P., Weber, C. J., & Wilson, K. L. (2001). Nuclear envelope organization in papillary thyroid carcinoma. *Histology and Histopathology, 16*, 1–14.

Gerace, L., Blum, A., & Blobel, G. (1978). Immunocytochemical localization of the major polypeptides of the nuclear pore complex-lamina fraction. Interphase and mitotic distribution. *Journal of Cell Biology, 79*, 546–566.

Gordon, L. B., Rothman, F. G., Lopez-Otin, C., & Misteli, T. (2014). Progeria: A paradigm for translational medicine. *Cell, 156*, 400–407.

Haque, F., Mazzeo, D., Patel, J. T., Smallwood, D. T., Ellis, J. A., Shanahan, C. M., et al. (2010). Mammalian SUN protein interaction networks at the inner nuclear membrane and their role in laminopathy disease processes. *Journal of Biological Chemistry, 285*, 3487–3498.

Haraguchi, T., Koujin, T., Segura-Totten, M., Lee, K. K., Matsuoka, Y., Yoneda, Y., et al. (2001). BAF is required for emerin assembly into the reforming nuclear envelope. *Journal of Cell Science*, *114*, 4575–4585.

Holaska, J. M., Rais-Bahrami, S., & Wilson, K. L. (2006). Lmo7 is an emerin-binding protein that regulates the transcription of emerin and many other muscle-relevant genes. *Human Molecular Genetics*, *15*, 3459–3472.

Holaska, J. M., & Wilson, K. L. (2007). An emerin "proteome": Purification of distinct emerin-containing complexes from HeLa cells suggests molecular basis for diverse roles including gene regulation, mRNA splicing, signaling, mechanosensing, and nuclear architecture. *Biochemistry*, *46*, 8897–8908.

Jafferali, M. H., Figueroa, R. A., & Hallberg, E. (2015). MCLIP detection of novel protein-protein interactions at the nuclear envelope. *Methods in Enzymology*, *569*. this issue.

Jamin, A., & Wiebe, M. S. (2015). Barrier to autointegration factor (BANF1): Interwoven roles in nuclear structure, genome integrity, innate immunity, stress responses and progeria. *Current Opinion in Cell Biology*, *34*, 61–68.

Kazemi, Z., Chang, H., Haserodt, S., McKen, C., & Zachara, N. E. (2010). *O*-Linked beta-N-acetylglucosamine (*O*-GlcNAc) regulates stress-induced heat shock protein expression in a GSK-3beta-dependent manner. *Journal of Biological Chemistry*, *285*, 39096–39107.

Koch, A. J., & Holaska, J. M. (2014). Emerin in health and disease. *Seminars in Cell and Developmental Biology*, *29C*, 95–106.

Li, P., Meinke, P., Huong le, T. T., Wehnert, M., & Noegel, A. A. (2014). Contribution of SUN1 mutations to the pathomechanism in muscular dystrophies. *Human Mutation*, *35*, 452–461.

Liang, W. C., Mitsuhashi, H., Keduka, E., Nonaka, I., Noguchi, S., Nishino, I., et al. (2011). TMEM43 mutations in Emery-Dreifuss muscular dystrophy-related myopathy. *Annals of Neurology*, *69*, 1005–1013.

Lu, J. T., Muchir, A., Nagy, P. L., & Worman, H. J. (2011). LMNA cardiomyopathy: Cell biology and genetics meet clinical medicine. *Disease Models & Mechanisms*, *4*, 562–568.

Margalit, A., Brachner, A., Gotzmann, J., Foisner, R., & Gruenbaum, Y. (2007). Barrier-to-autointegration factor—A BAFfling little protein. *Trends in Cell Biology*, *17*, 202–208.

Mattout, A., Cabianca, D. S., & Gasser, S. M. (2015). Chromatin states and nuclear organization in development—A view from the nuclear lamina. *Genome Biology*, *16*, 174.

Mehus, A. A., Anderson, R. H., & Roux, K. J. (2015). BioID identification of lamin-associated proteins. *Methods in Enzymology*, *569*. this issue.

Meinke, P., Mattioli, E., Haque, F., Antoku, S., Columbaro, M., Straatman, K. R., et al. (2014). Muscular dystrophy-associated SUN1 and SUN2 variants disrupt nuclear-cytoskeletal connections and myonuclear organization. *PLoS Genetics*, *10*, e1004605.

Montes de Oca, R., Lee, K. K., & Wilson, K. L. (2005). Binding of barrier to autointegration factor (BAF) to histone H3 and selected linker histones including H1.1. *Journal of Biological Chemistry*, *280*, 42252–42262.

Muchir, A., Wu, W., Sera, F., Homma, S., & Worman, H. J. (2014). Mitogen-activated protein kinase kinase 1/2 inhibition and angiotensin II converting inhibition in mice with cardiomyopathy caused by lamin A/C gene mutation. *Biochemical and Biophysical Research Communications*, *452*, 958–961.

Paquet, N., Box, J. K., Ashton, N. W., Suraweera, A., Croft, L. V., Urquhart, A. J., et al. (2014). Nestor-Guillermo progeria syndrome: A biochemical insight into barrier-to-autointegration factor 1, alanine 12 threonine mutation. *BMC Molecular Biology*, *15*, 27.

Qing, G., Yan, P., & Xiao, G. (2006). Hsp90 inhibition results in autophagy-mediated proteasome-independent degradation of IkappaB kinase (IKK). *Cell Research*, *16*, 895–901.

Record, M. T., Jr., Lohman, M. L., & De Haseth, P. (1976). Ion effects on ligand-nucleic acid interactions. *Journal of Molecular Biology*, *107*, 145–158.

Senior, A., & Gerace, L. (1988). Integral membrane proteins specific to the inner nuclear membrane and associated with the nuclear lamina. *Journal of Cell Biology*, *107*, 2029–2036.

Stubenvoll, A., Rice, M., Wietelmann, A., Wheeler, M., & Braun, T. (2015). Attenuation of Wnt/beta-catenin activity reverses enhanced generation of cardiomyocytes and cardiac defects caused by the loss of emerin. *Human Molecular Genetics*, *24*, 802–813.

Tifft, K. E., Bradbury, K. A., & Wilson, K. L. (2009). Tyrosine phosphorylation of nuclear-membrane protein emerin by Src, Abl and other kinases. *Journal of Cell Science*, *122*, 3780–3790.

Trinkle-Mulcahy, L., Boulon, S., Lam, Y. W., Urcia, R., Boisvert, F. M., Vandermoere, F., et al. (2008). Identifying specific protein interaction partners using quantitative mass spectrometry and bead proteomes. *Journal of Cell Biology*, *183*, 223–239.

Wong, X., Luperchio, T. R., & Reddy, K. L. (2014). NET gains and losses: The role of changing nuclear envelope proteomes in genome regulation. *Current Opinion in Cell Biology*, *28*, 105–120.

Worman, H. J., & Schirmer, E. C. (2015). Nuclear membrane diversity: Underlying tissue-specific pathologies in disease? *Current Opinion in Cell Biology*, *34*, 101–112.

Zbarsky, I. B., Perevoshchikova, K. A., Delektorskaya, L. N., & Delektorsky, V. V. (1969). Isolation and biochemical characteristics of the nuclear envelope. *Nature*, *221*, 257–259.

Zhang, Q., Bethmann, C., Worth, N. F., Davies, J. D., Wasner, C., Feuer, A., et al. (2007). Nesprin-1 and -2 are involved in the pathogenesis of Emery–Dreifuss muscular dystrophy and are critical for nuclear envelope integrity. *Human Molecular Genetics*, *16*, 2816–2833.

Zheng, X., Kim, Y., & Zheng, Y. (2015). Identification of lamin B-regulated chromatin regions based on chromatin landscapes. *Molecular Biology of the Cell*, *26*, 2685–2697.

Zullo, J. M., Demarco, I. A., Pique-Regi, R., Gaffney, D. J., Epstein, C. B., Spooner, C. J., et al. (2012). DNA sequence-dependent compartmentalization and silencing of chromatin at the nuclear lamina. *Cell*, *149*, 1474–1487.

Zwerger, M., Roschitzki-Voser, H., Zbinden, R., Denais, C., Herrmann, H., Lammerding, J., et al. (2015). Altering lamina assembly reveals lamina-dependent and -independent functions for A-type lamins. *Journal of Cell Science*, *128*, 3607–3620.

PART II

General Methods for Analysis of Intermediate Filament-Associated Proteins

CHAPTER SEVEN

Development of a Novel Green Fluorescent Protein-Based Binding Assay to Study the Association of Plakins with Intermediate Filament Proteins

Bertrand Favre[*,†,1], **Nadja Begré**[*,†], **Jamal-Eddine Bouameur**[*,†], **Luca Borradori**[*,†]

[*]Department of Clinical Research, University of Bern, Bern, Switzerland
[†]Department of Dermatology, Inselspital, Bern University Hospital, Bern, Switzerland
[1]Corresponding author: e-mail address: bertrand.favre@insel.ch

Contents

Methods in Enzymology, Volume 569
ISSN 0076-6879
http://dx.doi.org/10.1016/bs.mie.2015.06.017

Abstract

Protein–protein interactions are fundamental for most biological processes, such as the formation of cellular structures and enzymatic complexes or in signaling pathways. The identification and characterization of protein–protein interactions are therefore essential for understanding the mechanisms and regulation of biological systems. The organization and dynamics of the cytoskeleton, as well as its anchorage to specific sites in the plasma membrane and organelles, are regulated by the plakins. These structurally related proteins anchor different cytoskeletal networks to each other and/or to other cellular structures. The association of several plakins with intermediate filaments (IFs) is critical for maintenance of the cytoarchitecture. Pathogenic mutations in the genes encoding different plakins can lead to dramatic manifestations, occurring principally in the skin, striated muscle, and/or nervous system, due to cytoskeletal disorganization resulting in abnormal cell fragility. Nevertheless, it is still unclear how plakins bind to IFs, although some general rules are slowly emerging. We here describe in detail a recently developed protein–protein fluorescence binding assay, based on the production of recombinant proteins tagged with green fluorescent protein (GFP) and their use as fluid-phase fluorescent ligands on immobilized IF proteins. Using this method, we have been able to assess the ability of C-terminal regions of GFP-tagged plakin proteins to bind to distinct IF proteins and IF domains. This simple and sensitive technique, which is expected to facilitate further studies in this area, can also be potentially employed for any kind of protein–protein interaction studies.

1. INTRODUCTION

1.1 The Plakins, a Protein Family of Versatile Cytolinkers with Various Structural Domains

The plakin proteins are a family of versatile cytolinkers, which shape the different cytoskeletal networks formed by the microfilaments, intermediate filaments (IFs), and microtubules (MTs). They can connect different cytoskeletal networks with each other and/or tether these networks to distinct plasma membrane sites, the nuclear membrane, and/or organelles in various tissues (Bouameur, Favre, & Borradori, 2014; Sonnenberg & Liem, 2007; Suozzi, Wu, & Fuchs, 2012; Wilhelmsen et al., 2005). In mammals, the plakin family consists of seven members: bullous pemphigoid antigen 1 (BPAG1), desmoplakin, envoplakin, epiplakin, MT-actin cross-linking factor 1 (MACF1), periplakin, and plectin. The plakin family is defined by the presence of a plakin domain in the protein structure of their members, except for epiplakin. The plakin domain consists of several spectrin repeats (SRs) with an atypical src-homology 3 domain, embedded in one of the SRs (Choi & Weis, 2011; Jefferson, Ciatto, Shapiro, & Liem, 2007; Ortega, Buey, Sonnenberg, & de Pereda, 2011; Sonnenberg, Rojas, & de Pereda, 2007). Plakins variably contain an actin-binding domain (Garcia-Alvarez, Bobkov, Sonnenberg, & de Pereda, 2003; Sjoblom, Ylanne, & Djinovic-Carugo, 2008), a coiled-coil rod domain, a rod domain formed by SRs, a plakin-repeat domain (PRD), a growth arrest-specific protein 2 (GAS2)-MT-binding domain, and Gly-Ser-Arg repeats. The atypical plakin epiplakin is uniquely composed of 13 PRDs, linked together by connecting segments (Fujiwara et al., 2001). The isoforms a and b of both BPAG1 and MACF1 are structurally different from the other plakins, as they contain numerous SRs that follow their plakin domain. Consequently, these proteins are often referred to as spectraplakins to emphasize their similarity with proteins of the spectrin family (Huelsmann & Brown, 2014; Roper, Gregory, & Brown, 2002).

The plakins were first identified in epithelial cells, where they link the cytokeratin network to desmosomes and hemidesmosomes. The former are intercellular adhesion complexes found in all epithelia and in specialized cell types, whereas the latter, present in stratified and other complex epithelia, are junctional complexes connecting cells to the extracellular matrix. Later, it was realized that plakins exert other complex functions in a variety of tissues, including the skeletal and cardiac muscle and central nervous system. The modular structure of the various plakins enables them to bind to a variety of proteins in addition to cytoskeletal components. Consequently, plakins are not only cytolinkers but also act as scaffolds and adaptors for signaling proteins that modulate cytoskeletal dynamics, cell migration, differentiation, and stress responses (Bouameur, Favre, & Borradori, 2014; Boyer, Bernstein, & Boudreau-Lariviere, 2010; Sonnenberg & Liem, 2007; Suozzi et al., 2012).

Most plakin members exist as several isoforms due to different transcription initiation sites and alternative splicing. These isoforms exert cell- and tissue-specific functions (Bouameur, Favre, & Borradori, 2014; Sonnenberg & Liem, 2007; Suozzi et al., 2012). For example, the gene encoding BPAG1, called dystonin, is composed of at least 100 exons that give rise to three major isoforms, an epithelial, a neuronal, and a muscular isoform, named BPAG1e (corresponding to the original BPAG1/BP230), BPAG1a, and BPAG1b (Leung, Zheng, Prater, & Liem, 2001). The last two are spectraplakins with structures very similar to MACF1a and b, respectively (Leung, Sun, Zheng, Knowles, & Liem, 1999; Lin, Chen, Leung, Parry, & Liem, 2005).

In the past two decades, the focus of our laboratory has been the study of the functional interplay between several plakins and IF proteins. Specifically, our aim has been to understand the mechanisms by which BPAG1e, desmoplakin, and plectin interact with epidermal and simple epithelial keratins, and type III IF proteins, vimentin, and desmin (Bouameur, Favre, Fontao, et al., 2014; Favre et al., 2011; Fontao et al., 2003; Lapouge et al., 2006).

1.2 The Complexity of the Interaction of Plakins with IFs

In 1988, the group of Wiche first demonstrated that plectin can interact with various IF proteins (Foisner et al., 1988). Subsequently, seminal studies of Wiche's and Green's laboratories showed that the C-terminal region of both plectin and desmoplakin, respectively, contain sequences that are functionally important for binding to IFs (Stappenbeck & Green, 1992; Wiche, Gromov, Donovan, Castanon, & Fuchs, 1993). Fine mapping studies disclosed that the region called the "linker" of plectin, which connects the two last PRDs, PRD-B5 and -C, is essential, but not sufficient, for the interaction of plectin with several IFs (Fig. 1; Bouameur, Favre, Fontao, et al., 2014; Favre et al., 2011; Nikolic, Mac Nulty, Mir, & Wiche, 1996). The

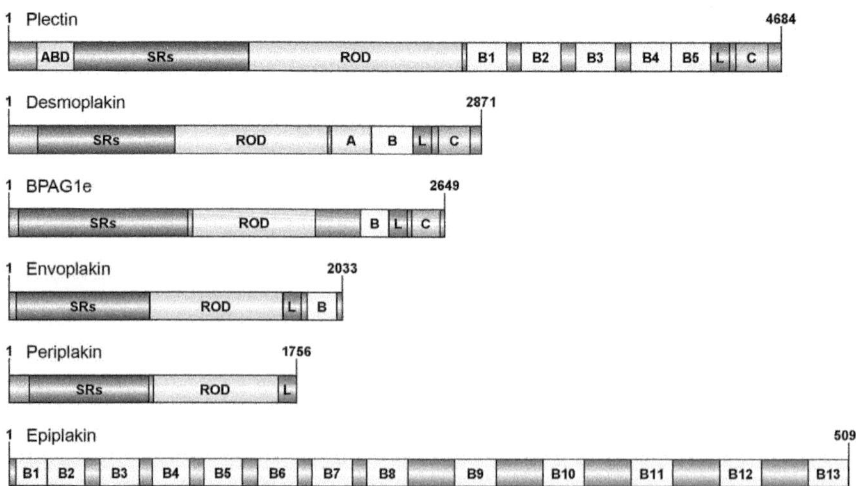

Figure 1 Domain organization in human plakins interacting with IFs. ABD, actin-binding domain; SRs, spectrin repeats; ROD, coiled-coil helix; A, B, and C, plakin-repeat domains; L, linker.

interaction of envoplakin and periplakin with IFs is also dependent on their respective linker regions (Karashima & Watt, 2002). Other studies provided less clear-cut results. They showed that, in addition to the linker and PRDs, the carboxyl extremity of several plakins is also involved in the binding of plakins to IFs (Bouameur, Favre, Fontao, et al., 2014; Fontao et al., 2003; Jang, Kalinin, Takahashi, Marekov, & Steinert, 2005; Karashima et al., 2012; Kazerounian, Uitto, & Aho, 2002; Lapouge et al., 2006; Meng, Bornslaeger, Green, Steinert, & Ip, 1997; Stappenbeck et al., 1993). Moreover, the exact function of the PRDs is still unclear. The PRD modules are made up of 4.5 copies of a 38-residue motif that form a unique globular structure. A conserved basic groove on these domains could constitute an IF-binding site (Choi, Park-Snyder, Pascoe, Green, & Weis, 2002). Nevertheless, the C-terminal PRDs, which are part of the IF-binding domain in some plakins, have a low affinity for IFs (Choi et al., 2002; Favre et al., 2011; Jang et al., 2005; Wang, Sumiyoshi, Yoshioka, & Fujiwara, 2006). Moreover, there is no evidence that the internal PRDs of BPAG1b and MACF1b or the first four PRDs of plectin can bind to IFs (Lin et al., 2005; Nikolic et al., 1996; Steiner-Champliaud et al., 2010). Periplakin, which lacks PRD but has a linker sequence, interacts with IFs (Kalinin, Kalinin, Aho, Uitto, & Aho, 2005; Karashima & Watt, 2002). A number of studies demonstrate that several domains, encompassing different PRDs, the linker, and the C-extremity, are required to have a strong interaction of plakins with IFs (Bouameur, Favre, Fontao, et al., 2014; Fontao et al., 2003; Jang et al., 2005; Karashima et al., 2012; Lapouge et al., 2006; Wang et al., 2006). The association of plakins with IFs is significantly affected by the assembly stage of IF proteins, since dimerization and polymerization of IFs increase the binding of plakins (Bouameur, Favre, Fontao, et al., 2014; Fontao et al., 2003; Jang et al., 2005; Meng et al., 1997), most likely by providing more binding sites (Bouameur, Favre, Fontao, et al., 2014). Although these findings seem logical, based on the fact that IF proteins are obligate dimeric proteins and the intracellular concentration of unpolymerized dimeric IF proteins is very low in normal circumstances (Herrmann & Aebi, 2004; Parry, Strelkov, Burkhard, Aebi, & Herrmann, 2007), the impact of the assembly stage of IFs on plakin binding has not been observed in all studies (Steinbock et al., 2000). This apparent discrepancy is most likely due to the different methodologies used in the various laboratories. Vice versa, plectin has also the ability to affect the assembly of IFs (Steinbock et al., 2000).

Mapping of the binding sites for plakins on IFs has produced conflicting results. Earlier studies have reported that the interaction of epidermal keratins with desmoplakin was dependent on the N-terminal head domain of type II keratins (Kouklis, Hutton, & Fuchs, 1994; Meng et al., 1997), whereas more recent investigations indicate that the central coiled-coil and the carboxyl tail of various types of IFs, such as keratin K5/K14, vimentin, and desmin, are critical for their association with desmoplakin and plectin (Bouameur, Favre, Fontao, et al., 2014; Favre et al., 2011; Fontao et al., 2003; Lapouge et al., 2006). Within the rod of IFs, the coil 1 segment seems to contain essential sequences for binding to plectin (Bouameur, Favre, Fontao, et al., 2014; Favre et al., 2011).

Finally, posttranslational modifications of plakins, mainly phosphorylation, have been shown to modulate their interaction with IFs (Bouameur et al., 2013; Foisner, Malecz, Dressel, Stadler, & Wiche, 1996; Foisner, Traub, & Wiche, 1991; Fontao et al., 2003; Hobbs & Green, 2012; Spurny et al., 2007; Stappenbeck, Lamb, Corcoran, & Green, 1994) or with MTs for spectraplakins (Wu et al., 2011). The effect of the numerous post-translational modifications of IFs (Loschke, Seltmann, Bouameur, & Magin, 2015; Snider & Omary, 2014) on their interaction with plakins has not been thoroughly investigated as yet (Foisner et al., 1991).

Despite this emerging picture about the interaction of plakins with IFs, several questions are still unanswered. It remains unclear how the various C-terminal plakin domains exactly affect the specificity and binding affinity of the different plakins for IFs. Furthermore, it is still not well defined how the structure of IFs modulates their interaction with plakins and how posttranslational modifications of both the plakins and IFs regulate their reciprocal association. Altogether, it is difficult, at the moment, to propose a simple model summarizing the binding mechanisms of plakins to IFs. Based on the recent determination of the three-dimensional structure of the coiled-coil rod of vimentin (Chernyatina, Nicolet, Aebi, Herrmann, & Strelkov, 2012), the next logical step would be to determine the crystal structure of plakin domain(s) bound to the dimeric coil 1 of vimentin.

1.3 Methods to Study Plakin–IF Interactions; Green Fluorescent Protein-Based Binding Assays

Several methods have been used to study the interaction of plakins with IFs and to map their reciprocal binding sites. These include transfection of mammalian cells for co-localization studies by indirect immunofluorescence microscopy (Nikolic et al., 1996; Stappenbeck et al., 1993; Stappenbeck & Green, 1992; Wiche et al., 1993), yeast two/three-hybrid assays (Geerts et al., 1999; Fontao et al., 2003; Meng et al., 1997), overlay assays on purified

IF proteins with purified mammalian plakins (Foisner et al., 1988; Kouklis et al., 1994; Meng et al., 1997), with recombinant plakin domains, fused or not to glutathione *S*-transferase (Favre et al., 2011; Kalinin et al., 2005), or with *in vitro* transcribed-translated, tagged, or radioactively labeled plakin regions (Fontao et al., 2003; Karashima & Watt, 2002), and co-sedimentation assays of plakins or plakin fragments with IFs (Choi et al., 2002; Foisner et al., 1988; Kalinin et al., 2005).

Since detection of fluorescence is sensitive and very convenient, we have recently expressed enhanced green fluorescent protein (EGFP)-tagged plectin domains to analyze their interaction with IFs. Furthermore, we have demonstrated that expression of EGFP-tagged proteins in mammalian cells proved to be useful for mapping the reciprocal binding sites on plakins and IF proteins (Bouameur, Favre, Fontao, et al., 2014). Here, we describe in detail these protein–protein binding assays based on the fluorescence detection of EGFP-tagged proteins.

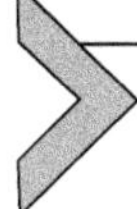

2. RECOMBINANT PROTEINS

2.1 Cloning of cDNA Constructs

Constructs were generated by standard cloning methods, with or without the help of PCR-amplified fragments. Correctness of the DNA sequence of PCR-amplified fragments was verified by sequencing. Human IF cDNAs, encoding K1, K5, K8, K14, K18, desmin, and vimentin, were cloned into pET23 or pET15b (both from Novagen) without modifying the open reading frame (no tag). Human K10 cDNA was cloned into pET15b with an N-terminal H6-tag to facilitate the purification of the protein (see Section 2.3) (Bouameur, Favre, Fontao, et al., 2014). K5 and K14 cDNA fragments corresponding to K5-coil 1 (residue 169–316), K5-coil 2 to tail (res. 317–590), K14-coil 1 (res. 115–262), and K14-coil 2 to tail (res. 261–472) were also cloned into pET15b with an H6-tagged at their N-terminus. Cloning of H6-EGFP into pET28 was performed in two steps. First, the cDNA corresponding to EGFP was excised from pEGFP-C3 (Clontech) and cloned into pET28 to generate a plasmid, which encodes EGFP-H6 (pET28-EGFP-H6). Second, the nucleotide sequence corresponding to an N-terminal H6 tag was excised from pET15b and cloned into pET28-EGFP-H6 to generate the plasmid pET28-H6-EGFP-H6. Mouse plectin 1c cDNA fragments, corresponding to amino acids 3954–4572 (PL-B5-E); aa. 3954–4502

(PL-B5-C); aa. 4289–4506 (PL-C) were cloned into pET28-EGFP-H6, pET28-H6-EGFP-H6, and pEGFP-C vectors. Human desmoplakin cDNA fragments, corresponding to aa. 2194–2871 (DP-B-E), aa. 1946–2871 (DP-A-E), both with the mutation S2849G (Fontao et al., 2003; Stappenbeck et al., 1994), and aa. 1248–2871 (DP-R-E), and human BPAG1e fragment corresponding to aa. 2077–2649 (BP-B-E) were cloned into pET28-H6-EGFP-H6 and pEGFP-C.

2.2 Polyacrylamide Gel Electrophoresis, Transfer, and Western Blotting

Proteins were separated on polyacrylamide gel electrophoresis in the presence of sodium dodecyl (lauryl) sulfate (SDS-PAGE) as previously described (Fontao et al., 2003). Gels were either stained with Coomassie blue without alcohol and acetic acid (Lawrence & Besir, 2009) or electrophoretically transferred to nitrocellulose membrane (Roti-NC, Roth) in a tank (Bio-Rad) filled with PAGE running buffer without SDS and with 20% methanol for 75 min at constant voltage (100 V). We noticed, by staining the proteins on membranes with Ponceau red (Nakamura, Tanaka, Kuwahara, & Takeo, 1985) and in SDS-PAGE with Coomassie blue after the transfer, that keratins of type II do not transfer well in standard blotting buffer. To improve their transfer, 0.05% SDS was added to transfer buffer. However, under these conditions, globular proteins, like bovine serum albumin (BSA), do not bind properly to nitrocellulose membranes. To detect H6-tagged proteins on Western blots, we used a mixture of two mouse anti-H6 monoclonal antibodies (mAb), BMG-His-1, diluted 1/1000 (Roche), and His-1, diluted 1/3000 (Aldrich-Sigma), because binding of these mAbs to the fusion protein is strongly dependent on the position of the H6-tag, N- or C-terminal, and the flanking sequences. Membranes were then incubated with goat anti-mouse secondary antibodies conjugated to IRDye 800cw, followed by scanning with Odyssey infrared imaging system (both from LI-COR). As the structure of EGFP is very stable (Alkaabi, Yafea, & Ashraf, 2005), it can be detected in SDS-PAGE, as long as samples are not heat-denatured. For in-gel detection of EGFP-tagged proteins, we used a Typhoon 9400 imager instrument (GE Healthcare). The relative size of not heat-treated EGFP-fusion proteins can be different from that of completely denatured proteins. After the fluorescence scan, proteins in gels can be either stained with Coomassie brilliant-blue or transferred for Western blotting.

2.3 Expression of Recombinant Proteins in Bacteria

Recombinant IF proteins or their fragments were expressed from pET vectors (Merck Millipore) in the *Escherichia coli* strain Rosetta (DE3) (Novagen) or BL21 (DE3) after induction with 0.2 m*M* isopropyl-beta-D-thiogalactopyranoside (IPTG) (Promega) for 3 h at 37 °C. Except for human K10, the full-length IF proteins, including the human K1, and IF fragments were efficiently expressed. The low level of expression of K10 precluded the purification of this protein to apparent homogeneity in a few steps (our unpublished data). For this reason, we tagged the N-terminus of K10 with H6 (see Section 2.1). Strong expression was also observed for the EGFP-H6 and H6-EGFP fused to various plectin C-terminal domains and the expressed proteins were ≥50% soluble. A small fraction of fluorescent proteins was proteolytically cleaved (Fig. 2A). In contrast to plectin, and despite the successful expression and crystallization of various C-terminal domains of desmoplakin in *E. coli* (Choi et al., 2002), the expression of recombinant desmoplakin proteins H6-EGFP-DP-B-E and H6-EGFP-DP-A-E in *E. coli* was low. Moreover, both proteins in the soluble fraction (and whole extracts, data not shown) were degraded, suggesting an improper folding of the DP domains (Fig. 2A). Expression of H6-EGFP-BP-B-E in *E. coli* also resulted in

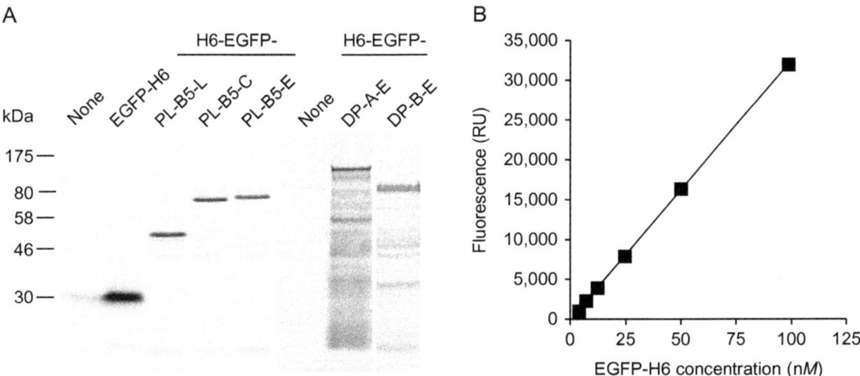

Figure 2 EGFP-fusion proteins can be directly detected in SDS-PAGE and quantified in protein mixtures. (A) EGFP-fluorescence scan of a 10% SDS-PAGE, loaded with 15 μg of the soluble fraction of transformed *E. coli* (strain BL21 (DE3)) expressing the indicated proteins. Bacteria were lysed by sonication in PBS, 2 m*M* DTT, and 1% protease inhibitors in 1/10 volume of the value corresponding to culture volume × OD600 nm and the soluble fraction isolated by centrifugation at 21,000 × *g* for 30 min at 4 °C. (B) Fluorescence of pure EGFP-H6 diluted in PBS, 1% BSA. From this standard curve, the concentration of EGFP-fusion proteins can be estimated in protein mixtures.

abundant proteolytic fragments (data not shown). These inconsistent data among the plakins encouraged us to express EGFP-fusion proteins in mammalian cells to circumvent the folding problems often encountered in bacteria (see Section 4).

2.4 Purification of Recombinant Proteins

All purification chromatography steps were performed with a BioLogic Duo-Flow system (Bio-Rad) at room temperature (RT). After each chromatographic step, samples were immediately put on ice. Purification of untagged recombinant K5, K14, K8, and K18 was performed exactly as previously described (Fontao et al., 2003; Herrmann, Haner, Brettel, Ku, & Aebi, 1999). Vimentin and desmin were purified according to Herrmann et al. (Herrmann, Hofmann, & Franke, 1992; Herrmann, Kreplak, & Aebi, 2004). For the purification of recombinant K1, bacteria pellet from 0.4 L of IPTG-induced culture was resuspended in 10 mL lysis buffer, consisting of 20 m*M* sodium phosphate, pH 7.4, 100 m*M* NaCl, 0.5% Triton X-100, 0.5% deoxycholic acid, 2 m*M* dithiothreitol (DTT), and 1/100 protease inhibitors without EDTA (all from Aldrich-Sigma). Cells were lysed on ice by sonication (8 pulses of 30 s each). Extracts were centrifuged at 21,000 × *g* at 4 °C for 30 min. The pellet was resuspended in lysis buffer without deoxycholic acid and processed as before. This step was repeated a second time. The washed inclusion bodies were dissolved in 25 mL binding buffer: 100 m*M* sodium phosphate buffer, pH 6.2, 6 *M* urea, 2 m*M* DTT, and 0.1 m*M* phenylmethylsulfonyl fluoride (PMSF), on a roller at 4 °C for 3 h to overnight. The suspension was centrifuged for 30 min at 21,000 × *g* at 4 °C. The supernatant was filtrated through a Millex GV filter unit, 0.22 μm (Merck Millipore) and loaded onto the cation-exchanger SP Sepharose FF 1 mL column (GE Healthcare), equilibrated in the binding buffer according to the manufacturer's protocol. The column was washed with 20 mL binding buffer and proteins were eluted with a linear gradient of 25 mL of binding buffer and of binding buffer supplemented with 1 *M* NaCl at a flow rate of 1 mL/min, fraction size: 1 mL. Fractions were analyzed on 10% SDS-PAGE, stained with Coomassie blue, and pooled based on their K1 concentration and contamination level. The pool was concentrated to 1.2 mL with Amicon Ultra-4 10K centrifugal filter device (Merck Millipore) and applied (0.5 mL/run) onto the gel filtration column Superdex 200 10/300 GL (GE Healthcare) equilibrated with 25 m*M* Tris–HCl, pH 7.4, 2 m*M* DTT, 8 *M* urea, 0.3 *M* NaCl, and 0.1 m*M* PMSF (running buffer), at a flow rate of 0.5 mL/min, fraction size: 1 mL. Fractions were analyzed on 10% SDS-PAGE. Only the two first fractions enriched in K1, free of smaller contaminants, were saved and pooled.

For the purification of H6-tagged keratin fragments, which were partially soluble, bacteria were directly lysed by sonication in 20 m*M* sodium phosphate, pH 7.4, 0.5 *M* NaCl, 20 m*M* imidazole, 8 *M* urea, and 0.1 m*M* PMSF (lysis buffer), and centrifuged at 21,000 × *g* for 30 min. For the insoluble H6-K10, washed inclusion bodies (see earlier) were dissolved in lysis buffer. H6-proteins were purified with a His-Trap-FF column according to the manufacturer's protocol (GE Healthcare). They were eluted with lysis buffer supplemented with 0.5 *M* imidazole and dialyzed against phosphate-buffered saline (PBS), 6 *M* urea, 2 m*M* DTT.

Soluble recombinant EGFP-H6 and H6-EGFP-PL-B5-L were affinity-purified as H6 keratins but without urea. Elution of the proteins in the chromatographic fractions was analyzed by fluorescence measurement (see Section 3). For purification of EGFP-H6 to apparent homogeneity, His-Trap fractions were dialyzed against 20 m*M* Tris, pH 8.0, 0.1 m*M* PMSF (dialysis buffer), and further purified on an anion-exchange Hi-Trap-Q-FF column (GE Healthcare), equilibrated in dialysis buffer. EGFP-H6 was eluted with a linear gradient of 20 mL of dialysis buffer and dialysis buffer supplemented with 1 *M* NaCl. Fractions were analyzed on 12% SDS-PAGE. Fractions containing apparently pure EGFP-H6 were dialyzed against PBS, 2 m*M* DTT. For long-term storage, aliquots were frozen at −80 °C. After a few month-storage, the thawed aliquots were as fluorescent as the fresh protein. Addition of 0.1% BSA or 10% glycerol before freezing the samples at −80 °C had no (positive or negative) effect on fluorescence of the thawed protein. However, aliquots containing 50% glycerol and stored at −20 °C were less fluorescent (15%) after a few month-storage than the other samples kept at −80 °C.

Protein concentration was assessed by the Bradford method (Bio-Rad), using BSA as a reference protein. For purified proteins, protein concentration was cross-checked by absorbance measurements at 280 nm. SDS-PAGE analysis of purified proteins is shown in Fig. S1 and S2 in Bouameur, Favre, Fontao, et al. (2014).

3. FLUORESCENCE MEASUREMENT OF SOLUBLE EGFP

Fluorescence intensity of EGFP and EGFP-fused proteins in solution was measured in a final volume of 100 μL, usually consisting of 10 μL sample and 90 μL PBS supplemented with 1% BSA, in wells of a 96-well transparent polystyrol plate with round bottom (Greiner; about twofold more sensitive than a similar plate with flat bottom) with the fluorometer Infinite 200 (Tecan), set the following way: mode, fluorescence top reading; excitation wavelength, 488 nm; excitation bandwidth, 9 nm; emission wavelength, 520 nm; emission bandwidth, 20 nm; gain, 100, manual (for comparative results among different experiments, the auto gain must not be selected); number of flashes, 25; integration time, 20 μs. Our preliminary tests did not reveal any difference with opaque plates, specifically designed for fluorescence/chemiluminescence measurements, but that were less convenient and more expensive. It is important to always use the same type of plates, the same volume in each well of the plate, and the same fluorometer with the same setting for comparative purposes. With pure EGFP-H6 (0.28 mg/mL), it is essential to dilute it in a buffer containing protein (≥0.1% BSA) and optionally detergent (0.1% Triton X-100), otherwise fluorescence rapidly decreases with time (completely "disappears" in a few hours) for unclear reasons (possibly nonspecific adsorption to the plastic well and fluorescence quenching). Determination of the intrinsic fluorescence of pure EGFP (Fig. 2B) enabled us to estimate the concentration of EGFP or EGFP-fusion proteins in cell extracts by using exactly the same conditions for measurements of fluorescence intensity (Cinelli et al., 2000) (see Fig. S1c in Bouameur, Favre, Fontao, et al., 2014).

4. EXPRESSION OF EGFP-FUSION PROTEINS IN MAMMALIAN CELLS

4.1 Culture and Transfection of Human Embryonic Kidney 293T Cells

Human embryonic kidney (HEK) 293T cells were cultured in DMEM (Invitrogen) supplemented with 10% newborn bovine serum, 100 U/mL penicillin, and 100 μg/mL streptomycin (all from Aldrich-Sigma). HEK 293T cells, about 50% confluent, were transfected with pEGFP-C3 (negative control) or pEGFP-C fused to the C-terminal plakin domains, by a standard calcium phosphate method (Jordan & Wurm, 2004). A volume ratio of 9–1 between the medium and the DNA-calcium phosphate/HEPES buffered saline mixture at the time of transfection is advisable to keep the transfected cells in culture for more than 24 h (36–48 h), otherwise the medium becomes too acidic. After 36–48 h, transfected cells were gently washed with PBS at RT and either directly lysed (see later) or frozen at −80 °C. In the latter case, the dish with frozen cells was put on ice before adding the lysis buffer.

4.2 Preparation of Soluble Extracts of Transfected HEK 293T Cells Expressing EGFP-Fusion Proteins

For the mobile phase of the binding assays, fresh or frozen, transfected or nontransfected HEK 293T cells were lysed in ice-cold PBS, 2 m*M* DTT, 0.1% Triton X-100, and 1% protease inhibitor cocktail without EDTA (Aldrich-Sigma). Aliquots of whole extracts were kept behind to assess the solubility of the expressed EGFP-fusion proteins. Whole extracts were centrifuged at 21,000 × *g* for 30 min at 4 °C. Protein concentration in soluble extracts was measured with the Bradford reagent (Bio-Rad, micro method) using BSA as a reference protein. Fluorescence intensity in whole extracts and soluble fractions, diluted 10-fold in lysis buffer without protease inhibitors (addition of BSA is not necessary), was measured in a 96-well plate (0.1 mL/well, round bottom) with a fluorometer (see Section 3). Bubbles at the surface of the liquid must be punctured with a needle before reading. The concentration of the EGFP-fusion proteins was estimated from the standard fluorescence curve of pure recombinant EGFP-H6 in similar conditions (Fig. 2B) (see Fig. S1c in Bouameur, Favre, Fontao, et al., 2014). For comparative purposes, fluorescence levels can be made equal between the various samples by diluting the more concentrated samples with appropriate volumes of soluble cell extracts from nontransfected cells.

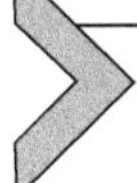

5. ANALYSIS OF THE INTERACTION OF EGFP-FUSION RECOMBINANT PLAKINS WITH IF PROTEINS

5.1 Immobilization of IF Proteins on Nitrocellulose Membrane and Fluorescence Overlay Assay

Keratins 5 and 14 (approximately 1 μg/μL) were polymerized exactly as previously described (Bouameur, Favre, Fontao, et al., 2014; Coulombe & Fuchs, 1990; Ma, Yamada, Wirtz, & Coulombe, 2001). Various amounts of monomeric or polymeric keratins (from 0.1 to 0.4 μg in 2–3 μL; spotting equivalent mole number is even preferable) were spotted on pieces (2.5 cm × 2.5 cm) of dry nitrocellulose membrane. To facilitate the spotting procedure, regularly spaced holes (3 mm diameter) were punctured in 2-mm-thick foam rubber membrane, which was laid on the nitrocellulose membrane. After spotting, the membranes were let dry for 30 min at RT. Quality of protein spotting and immobilization was checked by Ponceau red-staining of randomly picked membrane pieces. Unstained membrane pieces were separately blocked in compartment plastic boxes with PBS supplemented with 5% skimmed milk for 30 min at RT and washed twice with ice-cold PBS. Blocked membranes were then overlaid with purified recombinant H6-tagged EGFP, EGFP-fusion proteins or soluble fractions of transfected HEK 293T cells expressing EGFP or EGFP-fusion proteins (see Section 4.2) (0.1 mL/cm^2) in the cold room for 2 h under constant, gentle agitation, washed three times with an excess of ice-cold PBS for 5 min each, and immediately scanned (wet) with a Typhoon 9400 scanner (GE Healthcare), set for optimal EGFP-fluorescence detection (Fig. 3). Fluorescence was then quantified with ImageQuant software (GE Healthcare). Experiments were performed in triplicate. Among the tested Tris- or phosphate-based buffers with various pH, highest binding of plakins to IF proteins was obtained in PBS. We also found a good correlation between the co-localization degree of recombinant plakins with IFs in transfected

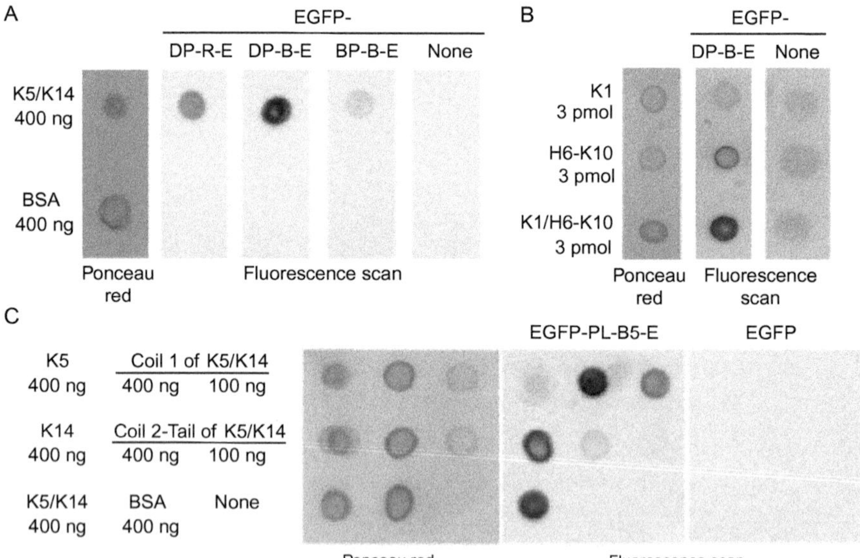

Figure 3 Analysis of the interaction of recombinant plakins with IF proteins by FluoBACE-overlay assay. (A) IF proteins were spotted as indicated. K5/K14 means polymerized IFs. One of the prepared membranes was stained with Ponceau red. Blocked membranes were incubated with the soluble extracts of HEK 293T cells expressing the indicated proteins. Their concentration was estimated according to Fig. 2B, EGFP-DP-R-E, 34 n*M*; EGFP-DP-B-E, 32 n*M*; EGFP-BPAG1e-B-E, 91 n*M*; and EGFP, 99 n*M*. (B) As for (A) with K1/H6-K10, EGFP-DP-B-E, 78 n*M*; EGFP, 246 n*M*. (C) As for (A) PL-B5-E, 27 n*M*; EGFP, 114 n*M*.

mammalian cells and the insolubility of these proteins in cell fractionation assays, when PBS, supplemented with 0.1% Triton X-100, was used to lyse the transfected cells (Bouameur et al., 2013).

5.2 Immobilization of IF Proteins on ELISA Microtiter Plate and Fluorescence Sorbent Assay

For the solid-phase assay, K5 and K14, in 25 m*M* Tris–HCl, 2 m*M* DTT, 6 *M* urea, pH 7.4, approximately 1 μg/μL, were diluted (≥10-fold) in PBS containing 2 m*M* DTT to obtain a 100 n*M* monomeric protein solution. Dimerization/polymerization of K5/K14 was induced by premixing K5 and K14 in stoichiometric ratio and then by directly diluting the mixture in PBS containing 2 m*M* DTT in two steps, to obtain a 100 n*M* K5/K14 solution (considered as dimeric). Noteworthy, these conditions are not controlled for formation of bona fide K5/K14 filaments. BSA (100 n*M*) was used as negative control. All dilutions were performed on ice. Then, 40 μL of diluted proteins (100 n*M*) were added to wells of MaxiSorp 96-well plate (Nunc, Thermo Scientific). The plate was centrifuged at 180 × *g* for 1 min and then incubated overnight under agitation in the cold room to immobilize proteins. The next day, the liquid was aspirated. Wells were blocked with 0.2 mL PBS, 5% skimmed milk for 2 h at RT under agitation and washed twice with ice-cold PBS. Purified recombinant EGFP-H6, H6-EGFP-fusion proteins, or soluble fractions of transfected HEK 293T cells expressing EGFP or EGFP-fusion proteins (see earlier), in a volume of 50 μL, were added to the wells and the plate was incubated under gentle agitation in the cold room for 2 h. The liquid was aspirated (it is also possible to save aliquots to measure the concentration of unbound EGFP and EGFP-fusion proteins versus their total concentration before the incubation, corresponding to concentrations at equilibrium), wells were washed twice with ice-cold 0.2 mL PBS, and fluorescence was measured in a final volume of 0.1 mL PBS/well with a fluorometer (Tecan Infinite 200), set the following way: plate, Corning 96 flat bottom transparent polystyrol; mode, fluorescence top reading; multiple reads per well (square (filled)), 4 × 4; multiple reads per well (border), 500 μm; excitation wavelength, 488 nm; excitation bandwidth, 9 nm; emission wavelength, 520 nm; emission bandwidth, 20 nm; gain, 100, manual; number of flashes, 25; integration time, 20 μs. Experiments were performed in triplicate. Graphics were generated with GraphPad Prism software and apparent binding parameters calculated from the simplified assumption of one binding site (Fig. 4A).

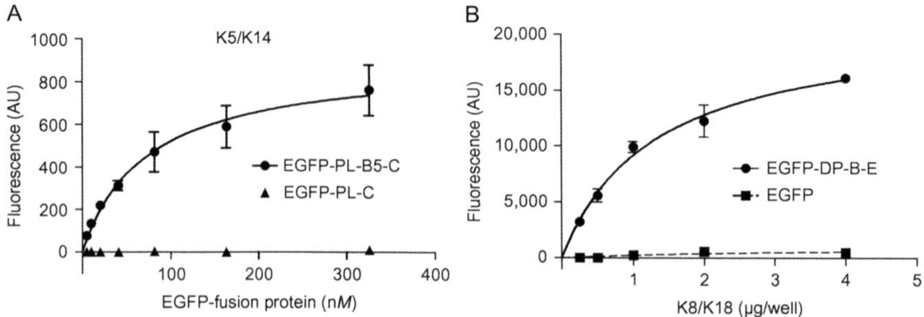

Figure 4 Analysis of the interaction of recombinant plakins with IF proteins by FluoBACE-sorbent assay and -filter plates. (A) Polymerized K5/K14, corresponding to 4 pmol of dimeric proteins, were immobilized in wells of ELISA microtiter plates and incubated with twofold dilutions of the soluble fraction of transfected HEK 293T cells expressing the indicated recombinant EGFP-plectin fusion proteins. (B) Increasing amounts of polymerized K8/K18 were immobilized onto PVDF-filters of a 96-well plate and incubated with the soluble fraction of transfected HEK 293T cells expressing EGFP-DP-B-E (133 n*M*) or EGFP (130 n*M*).

5.3 Immobilization of IF Proteins on Filter Plate and Fluorescence Overlay Assay

It is also possible to perform overlay assays with proteins immobilized on a polyvinyl difluoride (PVDF) membrane in a microtiter plate format, such as Immobilon P microtiter plate (MAIP N45; Merck Millipore). Membranes were activated with 0.1 mL 70% ethanol and the plate centrifuged at 160 × *g* for 2 min. Membranes were wetted with 0.2 mL PBS and the plate centrifuged as before (the plate is designed to use a vacuum manifold device but we have not tested this procedure). Our experience with nitrocellulose membranes on a vacuum manifold device was unsuccessful (as soon as applied solutions contained proteins, homogeneity of the filtration process among the wells was unsatisfactory). Then 50 μL, containing IF proteins or casein as negative control (from 0.25 to 4 μg), were added to the center of the filters and the plate centrifuged as before. Some heterogeneity among the well filters was also observed with these plates; therefore, centrifugation time was increased until all wells were empty but without letting the filters dry. Filters were then blocked with 0.1 mL PBS containing 1% BSA for 30 min and the liquid aspirated. They were washed with 0.2 mL PBS and then incubated with 50 μL containing purified recombinant EGFP-H6, H6-EGFP-fusion proteins, or soluble fractions of transfected HEK 293T cells, expressing EGFP or EGFP-fusion proteins (see Section 4.2). The plate was gently agitated in the cold room for 2 h. The liquid was aspirated (it is also possible to save aliquots to measure the concentration of unbound EGFP and EGFP-fusion proteins versus their total concentration before the incubation, corresponding to concentrations at equilibrium), wells were washed twice with ice-cold 0.2 mL PBS, and GFP fluorescence was measured after adding 50 μL PBS per well with a fluorometer, set for multiple reads/well (see Section 5.2). Experiments were performed in triplicate (Fig. 4B). This last method is more difficult and expensive than the two previous ones.

6. CONCLUDING REMARKS

Protein–protein fluorescence binding assays, based on the fusion of recombinant proteins to EGFP, represents a convenient, fast, simple, and sensitive binding assay, which can be easily used for all kinds of protein–protein interaction studies. Moreover, expression of recombinant proteins in transfected mammalian cells has the advantage of circumventing most of the problems of protein solubility and proper protein folding, which are typically encountered when recombinant proteins are expressed in bacteria. Nonetheless, use of soluble extracts of mammalian cells expressing EGFP-tagged proteins has also its drawbacks, namely (1) other proteins present in the cell extracts may interfere or compete with the binding of overexpressed EGFP-fusion proteins to immobilized IF proteins, (2) proteins expressed in mammalian cells are subject to posttranslational modifications, which could also alter the binding properties of EGFP-fusion proteins (Albrecht et al., 2015; Bouameur et al., 2013; Foisner et al., 1996, 1991; Stappenbeck et al., 1994), and (3) the level of expression of EGFP-tagged proteins in transfected mammalian cells is variable. This fact makes the acquisition and analysis of saturation binding curves difficult or even impossible (Bouameur, Favre, Fontao, et al., 2014) (our unpublished data). Despite these limitations, we found FluoBACE as a very useful qualitative method, which has greatly facilitated our studies aimed at mapping the binding sites on plakins and IF proteins. In conclusion, we can only encourage other laboratories to take advantage of this approach not only in the field of plakin proteins but also in all kinds of investigational area for rapid screening of protein–protein interactions or even fine mapping of binding domains.

ACKNOWLEDGMENTS

We are grateful to H. Herrmann (DKFZ, Heidelberg, Germany) for some IF clones, to S. Yousefi and H.-U. Simon (Institute of Pharmacology, Bern, Switzerland) for free access to their Typhoon 9400 scanner; to E.B. Lane (Institute of Medical Biology, Singapore, Singapore) for keratin cDNA constructs; and to K. Green (Northwestern University, Feinberg School of Medicine, Chicago, IL, USA) and A. Sonnenberg (The Netherlands Cancer Institute, Amsterdam, The Netherlands) for desmoplakin and plectin clones, respectively. This work was supported by grants from the Swiss National Foundation for Research (3100A0-121966 to L. B.) and the Swiss Foundation for Research on Muscle Diseases (to L. B.).

REFERENCES

Albrecht, L. V., Zhang, L., Shabanowitz, J., Purevjav, E., Towbin, J. A., Hunt, D. F., et al. (2015). GSK3- and PRMT-1-dependent modifications of desmoplakin control desmoplakin-cytoskeleton dynamics. *Journal of Cell Biology, 208*, 597–612.

Alkaabi, K. M., Yafea, A., & Ashraf, S. S. (2005). Effect of pH on thermal- and chemical-induced denaturation of GFP. *Applied Biochemistry and Biotechnology, 126*, 149–156.

Bouameur, J. E., Favre, B., & Borradori, L. (2014). Plakins, a versatile family of cytolinkers: Roles in skin integrity and in human diseases. *Journal of Investigative Dermatology, 134*, 885–894.

Bouameur, J. E., Favre, B., Fontao, L., Lingasamy, P., Begre, N., & Borradori, L. (2014). Interaction of plectin with keratins 5 and 14: Dependence on several plectin domains and keratin quaternary structure. *Journal of Investigative Dermatology, 134*, 2776–2783.

Bouameur, J. E., Schneider, Y., Begre, N., Hobbs, R. P., Lingasamy, P., Fontao, L., et al. (2013). Phosphorylation of serine 4642 in the C-terminus of plectin by MNK2 and PKA modulates its interaction with intermediate filaments. *Journal of Cell Science, 126*, 4195–4207.

Boyer, J. G., Bernstein, M. A., & Boudreau-Lariviere, C. (2010). Plakins in striated muscle. *Muscle & Nerve, 41*, 299–308.

Chernyatina, A. A., Nicolet, S., Aebi, U., Herrmann, H., & Strelkov, S. V. (2012). Atomic structure of the vimentin central alpha-helical domain and its implications for intermediate filament assembly. *Proceedings of the National Academy of Sciences of the United States of America, 109*, 13620–13625.

Choi, H. J., Park-Snyder, S., Pascoe, L. T., Green, K. J., & Weis, W. I. (2002). Structures of two intermediate filament-binding fragments of desmoplakin reveal a unique repeat motif structure. *Nature Structural & Molecular Biology, 9*, 612–620.

Choi, H. J., & Weis, W. I. (2011). Crystal structure of a rigid four-spectrin-repeat fragment of the human desmoplakin plakin domain. *Journal of Molecular Biology, 409*, 800–812.

Cinelli, R. A., Ferrari, A., Pellegrini, V., Tyagi, M., Giacca, M., & Beltram, F. (2000). The enhanced green fluorescent protein as a tool for the analysis of protein dynamics and localization: Local fluorescence study at the single-molecule level. *Photochemistry and Photobiology, 71*, 771–776.

Coulombe, P. A., & Fuchs, E. (1990). Elucidating the early stages of keratin filament assembly. *Journal of Cell Biology, 111*, 153–169.

Favre, B., Schneider, Y., Lingasamy, P., Bouameur, J. E., Begre, N., Gontier, Y., et al. (2011). Plectin interacts with the rod domain of type III intermediate filament proteins desmin and vimentin. *European Journal of Cell Biology, 90*, 390–400.

Foisner, R., Leichtfried, F. E., Herrmann, H., Small, J. V., Lawson, D., & Wiche, G. (1988). Cytoskeleton-associated plectin: In situ localization, *in vitro* reconstitution, and binding to immobilized intermediate filament proteins. *Journal of Cell Biology, 106*, 723–733.

Foisner, R., Malecz, N., Dressel, N., Stadler, C., & Wiche, G. (1996). M-phase-specific phosphorylation and structural rearrangement of the cytoplasmic cross-linking protein plectin involve p34cdc2 kinase. *Molecular Biology of the Cell, 7*, 273–288.

Foisner, R., Traub, P., & Wiche, G. (1991). Protein kinase A- and protein kinase C-regulated interaction of plectin with lamin B and vimentin. *Proceedings of the National Academy of Sciences of the United States of America, 88*, 3812–3816.

Fontao, L., Favre, B., Riou, S., Geerts, D., Jaunin, F., Saurat, J. H., et al. (2003). Interaction of the bullous pemphigoid antigen 1 (BP230) and desmoplakin with intermediate filaments is mediated by distinct sequences within their COOH terminus. *Molecular Biology of the Cell, 14*, 1978–1992.

Fujiwara, S., Takeo, N., Otani, Y., Parry, D. A., Kunimatsu, M., Lu, R., et al. (2001). Epiplakin, a novel member of the Plakin family originally identified as a 450-kDa human epidermal autoantigen. Structure and tissue localization. *Journal of Biological Chemistry, 276*, 13340–13347.

Garcia-Alvarez, B., Bobkov, A., Sonnenberg, A., & de Pereda, J. M. (2003). Structural and functional analysis of the actin binding domain of plectin suggests alternative mechanisms for binding to F-actin and integrin beta4. *Structure, 11*, 615–625.

Geerts, D., Fontao, L., Nievers, M. G., Schaapveld, R. Q., Purkis, P. E., Wheeler, G. N., et al. (1999). Binding of integrin alpha6beta4 to plectin prevents plectin association with F-actin but does not interfere with intermediate filament binding. *Journal of Cell Biology, 147*, 417–434.

Herrmann, H., & Aebi, U. (2004). Intermediate filaments: Molecular structure, assembly mechanism, and integration into functionally distinct intracellular scaffolds. *Annual Review of Biochemistry, 73*, 749–789.

Herrmann, H., Haner, M., Brettel, M., Ku, N. O., & Aebi, U. (1999). Characterization of distinct early assembly units of different intermediate filament proteins. *Journal of Molecular Biology, 286*, 1403–1420.

Herrmann, H., Hofmann, I., & Franke, W. W. (1992). Identification of a nonapeptide motif in the vimentin head domain involved in intermediate filament assembly. *Journal of Molecular Biology, 223*, 637–650.

Herrmann, H., Kreplak, L., & Aebi, U. (2004). Isolation, characterization, and *in vitro* assembly of intermediate filaments. *Methods in Cell Biology, 78*, 3–24.

Hobbs, R. P., & Green, K. J. (2012). Desmoplakin regulates desmosome hyperadhesion. *Journal of Investigative Dermatology, 132*, 482–485.

Huelsmann, S., & Brown, N. H. (2014). Spectraplakins. *Current Biology, 24*, R307–R308.

Jang, S. I., Kalinin, A., Takahashi, K., Marekov, L. N., & Steinert, P. M. (2005). Characterization of human epiplakin: RNAi-mediated epiplakin depletion leads to the disruption of keratin and vimentin IF networks. *Journal of Cell Science, 118*, 781–793.

Jefferson, J. J., Ciatto, C., Shapiro, L., & Liem, R. K. (2007). Structural analysis of the plakin domain of bullous pemphigoid antigen1 (BPAG1) suggests that plakins are members of the spectrin superfamily. *Journal of Molecular Biology, 366*, 244–257.

Jordan, M., & Wurm, F. (2004). Transfection of adherent and suspended cells by calcium phosphate. *Methods, 33*, 136–143.

Kalinin, A. E., Kalinin, A. E., Aho, M., Uitto, J., & Aho, S. (2005). Breaking the connection: Caspase 6 disconnects intermediate filament-binding domain of periplakin from its actin-binding N-terminal region. *Journal of Investigative Dermatology, 124*, 46–55.

Karashima, T., Tsuruta, D., Hamada, T., Ishii, N., Ono, F., Hashikawa, K., et al. (2012). Interaction of plectin and intermediate filaments. *Journal of Dermatological Science, 66*, 44–50.

Karashima, T., & Watt, F. M. (2002). Interaction of periplakin and envoplakin with intermediate filaments. *Journal of Cell Science, 115*, 5027–5037.

Kazerounian, S., Uitto, J., & Aho, S. (2002). Unique role for the periplakin tail in intermediate filament association: Specific binding to keratin 8 and vimentin. *Experimental Dermatology, 11*, 428–438.

Kouklis, P. D., Hutton, E., & Fuchs, E. (1994). Making a connection: Direct binding between keratin intermediate filaments and desmosomal proteins. *Journal of Cell Biology, 127*, 1049–1060.

Lapouge, K., Fontao, L., Champliaud, M. F., Jaunin, F., Frias, M. A., Favre, B., et al. (2006). New insights into the molecular basis of desmoplakin- and desmin-related cardiomyopathies. *Journal of Cell Science, 119*, 4974–4985.

Lawrence, A. M., & Besir, H. U. (2009). Staining of proteins in gels with Coomassie G-250 without organic solvent and acetic acid. *Journal of Visualized Experiments, 14*, 351–356, pii: 1350.

Leung, C. L., Sun, D., Zheng, M., Knowles, D. R., & Liem, R. K. (1999). Microtubule actin cross-linking factor (MACF): A hybrid of dystonin and dystrophin that can interact with the actin and microtubule cytoskeletons. *Journal of Cell Biology, 147*, 1275–1286.

Leung, C. L., Zheng, M., Prater, S. M., & Liem, R. K. (2001). The BPAG1 locus: Alternative splicing produces multiple isoforms with distinct cytoskeletal linker domains, including predominant isoforms in neurons and muscles. *Journal of Cell Biology, 154*, 691–697.

Lin, C. M., Chen, H. J., Leung, C. L., Parry, D. A., & Liem, R. K. (2005). Microtubule actin crosslinking factor 1b: A novel plakin that localizes to the Golgi complex. *Journal of Cell Science, 118*, 3727–3738.

Loschke, F., Seltmann, K., Bouameur, J. E., & Magin, T. M. (2015). Regulation of keratin network organization. *Current Opinion in Cell Biology, 32C*, 56–64.

Ma, L., Yamada, S., Wirtz, D., & Coulombe, P. A. (2001). A 'hot-spot' mutation alters the mechanical properties of keratin filament networks. *Nature Cell Biology, 3*, 503–506.

Meng, J. J., Bornslaeger, E. A., Green, K. J., Steinert, P. M., & Ip, W. (1997). Two-hybrid analysis reveals fundamental differences in direct interactions between desmoplakin and cell type-specific intermediate filaments. *Journal of Biological Chemistry, 272*, 21495–21503.

Nakamura, K., Tanaka, T., Kuwahara, A., & Takeo, K. (1985). Microassay for proteins on nitrocellulose filter using protein dye-staining procedure. *Analytical Biochemistry, 148*, 311–319.

Nikolic, B., Mac Nulty, E., Mir, B., & Wiche, G. (1996). Basic amino acid residue cluster within nuclear targeting sequence motif is essential for cytoplasmic plectin-vimentin network junctions. *Journal of Cell Biology, 134*, 1455–1467.

Ortega, E., Buey, R. M., Sonnenberg, A., & de Pereda, J. M. (2011). The structure of the plakin domain of plectin reveals a non-canonical SH3 domain interacting with its fourth spectrin repeat. *Journal of Biological Chemistry, 286*, 12429–12438.

Parry, D. A., Strelkov, S. V., Burkhard, P., Aebi, U., & Herrmann, H. (2007). Towards a molecular description of intermediate filament structure and assembly. *Experimental Cell Research, 313*, 2204–2216.

Roper, K., Gregory, S. L., & Brown, N. H. (2002). The 'spectraplakins': Cytoskeletal giants with characteristics of both spectrin and plakin families. *Journal of Cell Science, 115*, 4215–4225.

Sjoblom, B., Ylanne, J., & Djinovic-Carugo, K. (2008). Novel structural insights into F-actin-binding and novel functions of calponin homology domains. *Current Opinion in Structural Biology, 18*, 702–708.

Snider, N. T., & Omary, M. B. (2014). Post-translational modifications of intermediate filament proteins: Mechanisms and functions. *Nature Reviews Molecular Cell Biology, 15*, 163–177.

Sonnenberg, A., & Liem, R. K. (2007). Plakins in development and disease. *Experimental Cell Research, 313*, 2189–2203.

Sonnenberg, A., Rojas, A. M., & de Pereda, J. M. (2007). The structure of a tandem pair of spectrin repeats of plectin reveals a modular organization of the plakin domain. *Journal of Molecular Biology*, *368*, 1379–1391.

Spurny, R., Abdoulrahman, K., Janda, L., Runzler, D., Kohler, G., Castanon, M. J., et al. (2007). Oxidation and nitrosylation of cysteines proximal to the intermediate filament (IF)-binding site of plectin: Effects on structure and vimentin binding and involvement in IF collapse. *Journal of Biological Chemistry*, *282*, 8175–8187.

Stappenbeck, T. S., Bornslaeger, E. A., Corcoran, C. M., Luu, H. H., Virata, M. L., & Green, K. J. (1993). Functional analysis of desmoplakin domains: Specification of the interaction with keratin versus vimentin intermediate filament networks. *Journal of Cell Biology*, *123*, 691–705.

Stappenbeck, T. S., & Green, K. J. (1992). The desmoplakin carboxyl terminus coaligns with and specifically disrupts intermediate filament networks when expressed in cultured cells. *Journal of Cell Biology*, *116*, 1197–1209.

Stappenbeck, T. S., Lamb, J. A., Corcoran, C. M., & Green, K. J. (1994). Phosphorylation of the desmoplakin COOH terminus negatively regulates its interaction with keratin intermediate filament networks. *Journal of Biological Chemistry*, *269*, 29351–29354.

Steinbock, F. A., Nikolic, B., Coulombe, P. A., Fuchs, E., Traub, P., & Wiche, G. (2000). Dose-dependent linkage, assembly inhibition and disassembly of vimentin and cytokeratin 5/14 filaments through plectin's intermediate filament-binding domain. *Journal of Cell Science*, *113*, 483–491.

Steiner-Champliaud, M. F., Schneider, Y., Favre, B., Paulhe, F., Praetzel-Wunder, S., Faulkner, G., et al. (2010). BPAG1 isoform-b: Complex distribution pattern in striated and heart muscle and association with plectin and alpha-actinin. *Experimental Cell Research*, *316*, 297–313.

Suozzi, K. C., Wu, X., & Fuchs, E. (2012). Spectraplakins: Master orchestrators of cytoskeletal dynamics. *Journal of Cell Biology*, *197*, 465–475.

Wang, W., Sumiyoshi, H., Yoshioka, H., & Fujiwara, S. (2006). Interactions between epiplakin and intermediate filaments. *Journal of Dermatology*, *33*, 518–527.

Wiche, G., Gromov, D., Donovan, A., Castanon, M. J., & Fuchs, E. (1993). Expression of plectin mutant cDNA in cultured cells indicates a role of COOH-terminal domain in intermediate filament association. *Journal of Cell Biology*, *121*, 607–619.

Wilhelmsen, K., Litjens, S. H., Kuikman, I., Tshimbalanga, N., Janssen, H., van den Bout, I., et al. (2005). Nesprin-3, a novel outer nuclear membrane protein, associates with the cytoskeletal linker protein plectin. *Journal of Cell Biology*, *171*, 799–810.

Wu, X., Shen, Q. T., Oristian, D. S., Lu, C. P., Zheng, Q., Wang, H. W., et al. (2011). Skin stem cells orchestrate directional migration by regulating microtubule-ACF7 connections through GSK3beta. *Cell*, *144*, 341–352.

CHAPTER EIGHT

Functional Analysis of Keratin-Associated Proteins in Intestinal Epithelia: Heat-Shock Protein Chaperoning and Kinase Rescue

Anastasia Mashukova*,†, Radia Forteza†, Pedro J. Salas†,1

*Department of Physiology, Nova Southeastern University, Ft. Lauderdale, Florida, USA
†Department of Cell Biology, University of Miami Miller School of Medicine, Miami, Florida, USA
[1]Corresponding author: e-mail address: PSalas@med.miami.edu

Contents

Abstract

A growing body of evidence from several laboratories points at nonmechanical functions of keratin intermediate filaments (IF), such as control of apoptosis, modulation of signaling, or regulation of innate immunity, among others. While these functions are generally assigned to the ability of IF to scaffold other proteins, direct mechanistic causal relationships between filamentous keratins and the observed effects of keratin knockout or mutations are still missing. We have proposed that the scaffolding of chaperones such as Hsp70/40 may be key to understand some IF nonmechanical functions if unique features or specificity of the chaperoning activity in the IF scaffold can be demonstrated. The same criteria of uniqueness could be applied to other biochemical functions of the IF scaffold. Here, we describe a subcellular fractionation technique based on established methods of keratin purification. The resulting keratin-enriched fraction contains several proteins tightly associated with the IF scaffold, including Hsp70/40 chaperones. Being nondenaturing, this fractionation method enables direct testing of chaperoning and other enzymatic activities associated with IF, as well as supplementation experiments

Methods in Enzymology, Volume 569
ISSN 0076-6879
http://dx.doi.org/10.1016/bs.mie.2015.08.019

to determine the need for soluble (cytosolic) proteins. This method also permits to analyze inhibitory activity of cytosolic proteins at independently characterized physiological concentrations. When used as complementary approaches to knockout, knockdown, or site-directed mutagenesis, these techniques are expected to shed light on molecular mechanisms involved in the effects of IF loss of function.

1. INTRODUCTION

1.1 Chaperones Are Associated with the Intermediate Filaments

Intermediate filaments (IF) are components of the cytoskeleton present in most but not all metazoans with remarkable differences with respect to the better-known actin and tubulin cytoskeleton (Koster, Weitz, Goldman, Aebi, & Herrmann, 2015). There are six types of IF proteins. Except for lamins (type V), the rest is expressed in the cytoplasm in a tissue-specific manner (Eriksson et al., 2009). Here, we will focus on types I and II IF, also known as keratins (Krt), which are characteristic of epithelial tissues. Keratins have long been recognized for conferring mechanical properties to epithelia, which are critically important in the skin. However, for several years, many groups recognized that a number of proteins unrelated to IF actually bind to keratin filaments. These observations were common and a consensus was reached accepting that the mechanically strong IF are scaffolds decorated by other proteins.

Chaperones were among the first proteins to be found attached to the IF scaffold. Omary and coworkers (Liao, Lowthert, Ghori, & Omary, 1995) noted that Hsp70 forms a stable complex with keratins early during filament biogenesis. Quinlan, Carte, Sandilands, and Prescott (1996) discovered the association of the small chaperone alpha-crystallin with filensin IF. Mrj, a member of the Hsp40 family, an essential cochaperone of Hsp70, was found to bind the C-terminal region of Krt18 (Izawa et al., 2000).

Bearing in mind that the bulk of the chaperone protein is soluble, i.e., not associated with IF, it is natural to ask what is the function of chaperones attached to the IF scaffold. One possibility is that Hsp70/40 and other chaperones are necessary to maintain the structure and function of the IF themselves. Supporting this view, several investigators showed chaperone functions in keratin turnover (Watson, Geary-Joo, Hughes, & Cross, 2007), folding (Janig, Stumptner, Fuchsbichler, Denk, & Zatloukal, 2005), as well as associated with keratin aggregates in disease (Gu & Coulombe, 2005).

Conversely, a growing body of evidence in keratin knockout mice and knockdown cell lines suggests that the other, nonexcluding, alternative is also possible. Namely, that IF may be necessary for chaperoning non-IF proteins. In this regard, it is important to mention that the Krt8 KO mouse hepatocytes are approximately 100-fold more sensitive to TNFα-induced apoptosis than controls (Caulin, Ware, Magin, & Oshima, 2000). A similar observation was made in Krt18 KO mouse liver (Leifeld et al., 2009). Furthermore, Ser30 O-linked glycosylation of Krt18 protects against apoptosis and is necessary to maintain Akt levels (Ku, Toivola, Strnad, & Omary, 2010) in liver and pancreas. Intriguingly, Coulombe and coworkers (Lessard et al., 2013) uncovered a regulatory function of Krt16 on innate immunity and the role of Krt17 regulating protein synthesis (Kim, Wong, & Coulombe, 2006). Nonmechanical functions of keratin IF have been comprehensively reviewed (Homberg & Magin, 2014; Oriolo, Wald, Canessa, & Salas, 2007; Snider & Omary, 2014). A long list of interactions between keratins and several signaling molecules has been reviewed elsewhere (Pan, Hobbs, & Coulombe, 2013). Overall, however, the mechanistic explanations of how the IF scaffold affects unrelated molecules often physically distant from the IF, remain elusive.

There are different families of chaperones, which not only aid the correct folding or refolding of other proteins (clients) but also facilitate multiprotein arrangements (Quinlan & Ellis, 2013). This review will focus on Hsp70 and Hsp40 chaperones attached to the IF scaffold. Because Hsp or Hsc70 and Hsp40 chaperones work together, control many cellular functions (Clerico, Tilitsky, Meng, & Gierasch, 2015; Saibil, 2013), and show anti-apoptotic properties (Wang, Chen, Zhou, & Zhang, 2014), just like keratins, it is natural to hypothesize that the Hsp70/40 scaffolding on IF may be responsible for some of the nonmechanical functions of keratin IF. Three mechanistic questions come to mind to test this hypothesis:

- Is Hsp70/40 associated with keratin IF active as a chaperone?
- If so, bearing in mind that the bulk of the Hsp70 and 40 chaperones is soluble, and not associated with IF, what fraction of chaperoning occurs on IF as compared with total Hsp70 chaperoning activity?
- If IF-associated Hsp70/40 is active, is there any substrate (client) specificity for the IF-bound chaperones? In other words, is it possible that a subset of Hsp70/40 clients can be folded only on IF, and not in the cytosol?

In the following sections, we will briefly summarize evidence supporting positive answers for all three questions in the case of PKCι in intestinal epithelia. This work will focus on reductionist methods to separate the IF-Hsp70 scaffold from cytosolic Hsp70 aimed to compare the relative chaperoning activity and substrate selectivity in both compartments. Because IF and cytosol are in intimate contact in the cell, it is given that all the proteins in the filaments, including keratins, must be in equilibrium with the cytosol counterparts. Also, because there is no known barrier between filaments and cytosol, one must assume that all the cytosolic components are freely available on the filaments. Therefore, any functional difference, such as substrate selectivity, must be due to the interaction with the keratins or other molecules in the keratin filament.

Along with the methods, we will summarize proof-of-concept evidence supporting the notion of a chaperoning-active, substrate-specific IF scaffold.

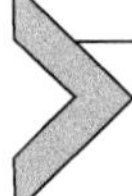

2. ISOLATION OF KERATIN INTERMEDIATE FILAMENTS FOR FUNCTIONAL ASSAYS

The original procedure to isolate an IF fraction was developed by Goldman and coworkers (Steinert, Zackroff, Aynardi-Whitman, & Goldman, 1982). It consists of two parts. In the first one, the cells are extracted in nonionic detergent, and the insoluble fraction is further extracted in 1.5 *M* KCl. The second part involves successive steps of solubilization in urea, followed by dialysis and filament reassembly. When completed, it yields highly purified keratin filaments, devoid of detectable contaminants. We adapted the first part, which yields >90% enriched keratins, with a number of proteins that copurify with keratins as minor components. These proteins can be easily detected in 2D gels stained with silver (e.g., see Fig. 1 in Salas (1999)). The method to obtain keratins along with IF-associated proteins from confluent, differentiated, cultured epithelial cells is as follows (Mashukova, Forteza, Wald, & Salas, 2012; Mashukova et al., 2009):

(a) After extensively washing the cells with saline buffer (e.g., PBS), they are extracted in PBS supplemented with 1% Triton X-100, 2 m*M* EDTA, 1 m*M* ATP, and separate protease and phosphatase inhibitor cocktails at the dilutions recommended by the manufacturer (antiproteases, Sigma P8340, antiphosphatases, Calbiochem 524624 and 524627). Typically, 2×10^6 cells are extracted in 0.5 ml of the extraction buffer. This extraction is performed at room temperature to facilitate solubilization of lipid rafts. Addition of ATP in the extraction buffer is important to dissociate actin–myosin complexes. If ATP is omitted, actin frequently contaminates the final IF fraction (kP). In some functional assays involving dephosphorylation (Section 4), the antiphosphatase cocktail must be omitted.

(b) Immediately sonicate the mix for 15 s on ice (3×5 s intervals with 5 s lapses).

(c) Spin down at 16,000 $\times g$ for 10 min at room temperature. The supernatant of this centrifugation is the soluble fraction (S), and contains cytosol, nucleoplasma, and membrane-associated proteins, including those from organelles.

(d) The resulting pellet is resuspended in 1.5 *M* KCl, sonicated as described in (b), and kept on ice for 10 min. The mix is then spun as described in (c).

(e) The pellet from the centrifugation in (d) is the desired IF fraction (kP), and the supernatant is enriched in actin and actin-binding proteins.

To process these three fractions for SDS-electrophoresis and ensuing immunoblot, the S fraction and the KCl supernatant from (d) are acetone precipitated. The latter, originally from the pellet in (c) is, therefore, named aP (actin pellet, after the acetone step). Then, the resulting pellets, as well as kP are washed three times in distilled water. For the 2×10^6 cells, the protein pellet from S is solubilized in 0.5 ml, and aP or kP in 0.25 ml of 1% SDS in loading buffer without bromophenol blue. Samples of these solutions are used to determine protein by Lowry assay. Then, the extracts are diluted in complete SDS sample buffer up to 1 mg/ml and run in SDS-electrophoresis (usually 40 μg/lane, depending on gel size). A typical blot from these fractions is shown in Fig. 1, using Ponceau S staining for total protein (Fig. 1A) and immunoblot for chaperones, actin, and Krt8 (Fig. 1B). It is important to note that, although the same amount of protein was loaded in the three lanes, to our knowledge there is no valid loading control possible. Accordingly, we use keratins, actin, and fully soluble proteins (e.g., GAPDH) to validate the extraction conditions and to compare loading among the same fractions from different preparations or from different experimental conditions.

The presence of various chaperones, atypical protein kinase C (aPKC), actin-binding proteins, and cytoskeletal proteins in these three fractions, as determined by immunoblot in previous publications, is shown in Table 1. It is important to highlight that all cellular proteins are included in one or more fractions, as the procedure does not discard any fraction. Also, it is important to consider that the S fraction represents the bulk of the cellular protein (75%), while the kP fraction represents less than 10% of the total cellular

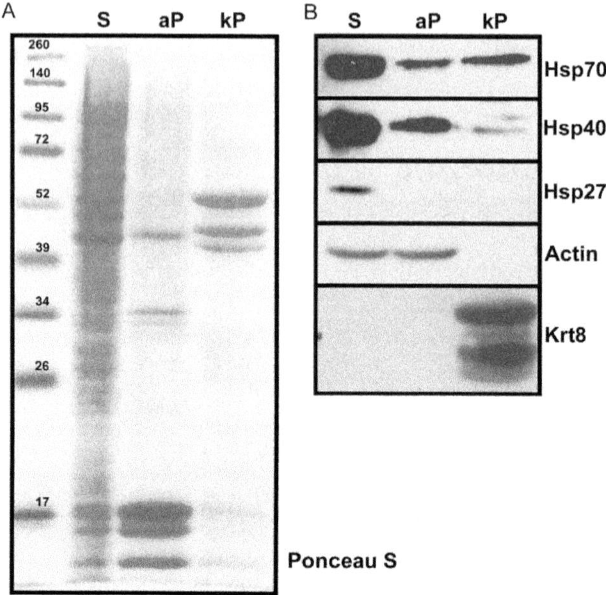

Figure 1 (A) Blot of S, aP, and kP fractions obtained from a confluent differentiated culture of Caco-2 cells (40 μg/lane) stained with Ponceau S. First lane on the left-hand side are MW standards, values represent Mrx 10^3. (B) Immunoblots from the same (or parallel) membrane for the antibodies indicated on the left were developed by chemiluminescence.

protein. Therefore, functions associated with IF, such as chaperoning, must display some different or unique feature as compared to the cytosol to be quantitatively relevant at the cellular level.

The results in Table 1 apply specifically for Caco-2 (human intestinal epithelial) cells. In other cell types or for different IF proteins, the results may vary. For example, Hsp27 preferentially binds to glial fibrillary acidic protein and not keratin (Perng et al., 1999). Accordingly, we have found it in the S fraction in intestinal cells (Fig. 1).

For functional assays in the following sections, processing of the fractions should be different. To maintain nondenaturing conditions and change the buffers to those used in specific biochemical reactions, S and aP, the extract from the KCl extraction, need to be desalted. We use centrifugal ultrafiltration devices with 3000 kD cutoff (e.g., Centricon Ultracel YM-3). We replace the extraction buffer by the buffer desired for a functional assay by 2-volume washes (two consecutive protein concentrations followed by resuspension in the buffer for a functional assay) of the retenates from S and aP. This step is used also for protein concentration or dilution to adjust

Table 1 Proteins Copurifying with Keratin (kP), Actin (aP), and Soluble (S) Fractions in Caco-2 Intestinal Epithelial Cells

Protein	S	aP	kP	References
Keratins 8/18	–	–	+++	Mashukova et al. (2009)
Actin	++	++	–	Fig. 1
Tubulin	+++	+	–	Mashukova et al. (2012)
PKCι	+	+	+	Mashukova et al. (2009)
pT555–PKCι	+++	++	+	Mashukova et al. (2009)
Hsc/Hsp70	+++	++	++	Mashukova, Wald, and Salas (2011)
Hsp 40	+++	++	+	Fig. 1
Hsp 27	++	–	–	Fig. 1
14-3-3	+	–	–	Mashukova et al. (2009)
Bag-1M	+++	+	+	Mashukova et al. (2014)
Bag-1S	+++	+	–	Mashukova et al. (2014)
ZO-1	++	+	–	Mashukova et al. (2009)
Par6	+++	+	–	Mashukova et al. (2009)
Par3	+	–	–	Mashukova et al. (2009)
Pals1	++	+	–	Mashukova et al. (2009)
PDK1	+	+	–	Mashukova et al. (2012)
Fraction of cellular protein (%)	75	18	7	Mashukova et al. (2009)

The + signs represent relative magnitudes of the bands as compared among the three fractions. For example, +++ means the band is stronger than ++. The – sign indicates that no band could be detected, even in overexposed immunoblots. Note that 14-3-3 has been shown to bind to IF (Ku, Michie, Resurreccion, Broome, & Omary, 2002). Therefore, it is an example of IF-associated proteins that do not copurify in the kP fraction.

identical protein concentrations in all the fractions. In addition, this step enables to remove ATP from S fraction which is necessary for negative controls. The kP fraction which is a pellet is washed twice in the desired buffer and finally resuspended in the same by a 2-s sonication step.

There are two caveats that need to be considered. First, the extraction procedures yielding the kP are stringent and involve at least 1 h of manipulation of the keratins. Therefore, it is conceivable that keratin-associated proteins with a weak interaction or a relatively fast dissociation may be solubilized in the process and appear either in S or aP. Longer manipulations of the keratins must be avoided as the keratin monomers slowly dissociate from the pellets, resulting in a slow loss of filamentous material if several changes of solution are performed over a period of hours. Second, conversely, this is a copurification of highly insoluble protein aggregates. We have not determined the presence of lamins or insoluble chromatin scaffold proteins in kP, but it is conceivable that contaminants from other insoluble subcellular pools of protein may be present in kP as well. Accordingly, copurification must be validated independently by fluorescence or EM colocalization, to ensure that copurifying proteins are actually from the keratin IF. In the case of Hsp70, colocalization with keratin IF has been reported before (Liao et al., 1995). Likewise, Hsp40/keratin colocalization was reported as well (Yamazaki, Uchiumi, & Katagata, 2012).

3. DETECTION OF CHAPERONING ACTIVITY ON KERATIN INTERMEDIATE FILAMENTS, RELATIVE TO CYTOSOLIC ACTIVITY

For a direct measurement of Hsp70 chaperoning activity, the luciferase refolding assay is commonly used (Lu & Cyr, 1998). The general principle is to measure luminescence from luciferase before and after chemical denaturation in guanidinium. Then, upon removal of the denaturing agent, luciferase refolds in the presence of a complete Hsp70 chaperoning complex and ATP. In doing so, it regains most (typically 60–70%) of the original luminescence within 4–5 h.

(a) The fractions are resuspended by ultrafiltration (S, aP) or sonication (kP) in refolding buffer (25 m*M* HEPES, pH 7.4, 50 m*M* KCl, 5 m*M* $MgCl_2$, 1 m*M* ATP), as described in the previous section.

(b) Luciferase is diluted to 0.3 mg/ml in denaturation buffer (25 m*M* HEPES, pH 7.4, 50 m*M* KCl, 5 m*M* $MgCl_2$, 6 *M* guanidinum–HCl, and 5 m*M* DTT). Denaturation is carried out at 25 °C for 1 h. Also, 1 μl samples in triplicate of luciferase are equally diluted in 125 μl PBS in parallel, to be used as a measure of 100% luciferase activity. These samples are further diluted as described in the next paragraph.

(c) Chemically denatured luciferase (1 μl) is diluted in 125 μl refolding buffer to dilute guanidinium. Then, 15 μg protein from each fraction (S, aP, and kP) are supplemented with 1 μl of diluted luciferase solution and incubated at 25 °C. Another set of samples without ATP is prepared as negative controls. Additional samples are supplemented with an identical amount of PBS or albumin to measure residual luminescence after guanidinium denaturation. Ideally, all samples should be prepared in triplicates.

(d) Aliquots of 1 μl are removed from the folding reactions at various times (e.g., 1, 2, and 4 h refolding) and luminescence is measured with a Glo™ Luciferase Assay System (Promega) in a luminometer.

S and kP display similar levels of ATP-dependent luciferase refolding chaperone activity for similar protein concentrations. This is not surprising because of the high concentration of soluble Hsp70/40 chaperones in S, but considering the enrichment of IF in kP, it is remarkable that chaperoning activity actually enriches in kP as much as keratins, which represent the bulk of the protein in that fraction. On the other hand, there is no specificity for luciferase refolding in any of the fractions. Soluble or keratin-associated chaperones seem to be equally effective (Mashukova et al., 2014).

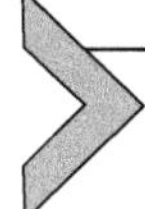

4. REFOLDING OF aPKC BY HSP70/40 IN KERATIN INTERMEDIATE FILAMENTS

PKCs are AGC kinases which comprise, in humans, 10 isoforms divided in three categories: conventional (PKCα, βI, βII, and γ), novel (PKCδ, ε, θ, and η), and atypical (PKCι/λ and ζ). We have focused our study on aPKCs, but some of the concepts that follow can be also extended to conventional or novel PKCs. All isoforms share a highly conserved catalytic domain. They are activated by PDK1-mediated phosphorylation of the activation loop and autophosphorylation of the turn motif (T555 in human PKCι) (Newton, 2003). Importantly, kinase activity leads to a faster dephosphorylation of these phosphosites (Gould & Newton, 2008), by exposing them to phosphatases (Gould et al., 2011). Dephosphorylated PKC molecules are subjected to rapid ubiquitination and degradation (Lu et al., 1998). An Hsp70-mediated rescue mechanism enables refolding of PKC (Gao & Newton, 2006), immediately followed by the same sequence of PDK1-dependent phosphorylation (Mashukova et al., 2012) and autophosphorylation.

Because there is no direct readout of aPKC folding, the autophosphorylation of the turn motif may be used as an indirect surrogate. The basis of the assay is to enable endogenous PKC activity in various cell fractions (S, aP, and kP) with ATP and a substrate peptide until PKC becomes fully dephosphorylated in about 5 h. This process is dependent on the endogenous phosphatases. We have not determined which phosphatases are present, but it is known that PP2A is in a complex with aPKC in epithelial cells (Nunbhakdi-Craig et al., 2002). Ubiquitin is partially lost in the ultra-filtration (~80% retention expected in each cycle for a 8.5-kDa protein, http://kirschner.med.harvard.edu/files/protocols/Millipore_Centricons.pdf) and then diluted. For this reason, we speculate that aPKC protein is not degraded during the assay (Mashukova et al., 2009). Upon removal of the substrate peptide by ultrafiltration, aPKC is allowed to undergo Hsp70-dependent refolding and PDK1 phosphorylation. The readouts are immunoblots for PKCι protein and pT555–PKCι and ζ (the phosphoepitope antibody recognizes the phosphorylated turn domain of both kinases).

(a) Dephosphorylation step: Cell fractions (S, aP, and kP) are supplemented with 150 μ*M* PKC substrate peptide (ERMRPRKRQGSVRRRV) and 1 m*M* ATP. The peptide phosphorylation is allowed to proceed at 30 °C for 5 h.

(b) Peptide is removed by ultrafiltration and replaced by the rephosphorylation buffer as described in Section 2.

(c) Refolding/rephosphorylation: 50 μg of protein from S, or 20 μg of kP protein are incubated separately or together with 1 m*M* ATP (or negative controls without ATP) at 30 °C for 4 h. Samples from the fractions before and after step (a) (original phosphorylation level and fully dephosphorylated level, respectively), and after the incubation in step (c), are analyzed by immunoblot using antibodies against pT555 turn domain and total PKCι protein.

We have shown that, separately, each fraction is unable to rephosphorylate PKCι or ζ. However, combining S and kP enables rephosphorylation, while combinations of either one with aP do not (Mashukova et al., 2009). Bearing in mind that all three fractions contain Hsp70 and 40, the result suggested the specific need of IF for the rephosphorylation. It was later shown that supplementing kP with recombinant purified PDK1 alone was sufficient to enable rephosphorylation of aPKC. Also, immunodepletion of endogenous PDK1 from S prevented the S+kP mixture from supporting aPKC rephosphorylation (Mashukova et al., 2012). The IF fraction, therefore, is sufficient to refold endogenous aPKC. The only soluble component necessary to complete the readout of the reaction is PDK1, which is present in S.

Our laboratory has started to use these techniques to test the loss of function of keratin mutants when and if such mutants are able to form filaments. For this purpose, keratin-deficient cells such as SW-13/cl 2 (Schweitzer et al., 2001) or keratin-deficient keratinocytes (Seltmann, Cheng, Wiche,

Eriksson, & Magin, 2015) represent the ideal background to express mutant keratin along with the corresponding normal keratin partner.

We have also used the same fractionation assay to measure PKC kinase activity using a PKC Kinase Non-Radioactive Assay kit (AssayDesigns, Stressgen, Ann Arbor, MI). Because there is very little isoform specificity in this phosphorylation assay, the three cellular fractions can also be obtained from cells depleted in specific isoforms with RNAi treatments to assess isoform specificity of the activity. Alternatively, individual fractions can be treated with PKC inhibitors, including pseudosubstrate peptides. Using these techniques, we showed that aPKC is responsible for most of PKC activity in the kP fraction (Mashukova et al., 2009). It is conceivable that other enzymes may be enriched in the IF scaffold as well.

5. SUPPLEMENTATION ASSAYS USING CYTOSOL (S) AND ISOLATED KERATIN INTERMEDIATE FILAMENTS (kP)

Using separate fractions not only allows specific immunodepletion, but also to supplement the fractions with purified proteins. For example, supplementing the S fraction with 15 μg (for every 50 μg of S protein) fully purified filamentous keratins without detectable contaminants (either recombinant or after cycles of urea denaturation followed by dialysis), is sufficient to enable aPKC rephosphorylation in the assay described above. Conversely, single recombinant keratins (5 μg Krt8 or 18/50 μg of S protein), which do not form filaments, do not allow rephosphorylation in the S fraction (Mashukova et al., 2009). Finally, the S fraction immunodepleted in Hsp70 fails to supplement highly purified (with no Hsp70) IF to enable aPKC rephosphorylation. Accordingly, this rephosphorylation can be rescued with recombinant Hsp70. Those results show specificity of the Hsp70/40 attached to IF to refold aPKC.

It has been extensively shown that individual keratin molecules assemble into IF directly from the cytosol and likewise are shed back to the cytosol as nonfilamentous subunits (Kolsch, Windoffer, Wurflinger, Aach, & Leube, 2010). Although there are areas of the cytoplasm where incorporation or disassembly is favored, there is consensus that filamentous and soluble nonfilamentous subunits are in continuous exchange. Likewise, supplementation of highly purified filamentous keratins with Hsp70 described above supports the notion of equilibrium between soluble chaperones and chaperones attached to IF. While we and others have measured physiological changes in total cellular Hsp/Hsc70 concentrations (Mashukova et al.,

2009), those changes are quantitatively small (e.g., up to threefold) to allow meaningful changes in the keratin-bound chaperones. One of the small chaperones, however, shows greater variability.

The small chaperone Bag-1 binds Hsp70 via a conserved N-terminal Bag domain. There are four isoforms of Bag-1, products of alternative start codons on the same transcript: Bag-1L (large) which contains a nuclear localization signal, Bag-1M (medium), Bag-1, and Bag-1S (small) (Alberti, Esser, & Hohfeld, 2003). The Bag-1 cellular concentrations are variable over a 20-fold range (Maki et al., 2007). We found that Bag-1M total cellular content varies six- to sevenfold in mouse small intestine villus enterocytes under proinflammatory stimulation and three- to fourfold in Caco-2 human intestinal epithelial cells under TNFα stimulation. Importantly, only Bag-1M and not Bag-1S was found attached to the IF fraction (Fig. 1B). Furthermore, the levels of IF-bound Bag-1M were dependent on the total cellular Bag-1M (Mashukova et al., 2014). These data suggest that attached Bag-1M is in equilibrium with soluble Bag-1M. To confirm the functional implications of Bag-1M—IF attachment supplementation assays can be used. However, to supplement kP fractions, it is critically important to know the cytosolic concentration of the ligand (Bag-1M in this case). This can be achieved by determining the total cellular concentration and then measuring the fraction of the total cellular volume represented by the cytosol.

(a) Total cellular Bag-1M can be determined by semiquantitative immunoblot at different protein loads per lane to determine Bag-1M/total cellular protein. Then, the same are compared with known amounts of recombinant Bag-1M in the same blots.

(b) To determine a concentration, it is necessary to estimate cell volumes. Total protein per cell volume is estimated at 200 mg/ml (Milo, 2013). Using this constant, we estimated the basal levels of Bag-1M in nonstimulated cells at 0.8–1 μ*M*. Because Bag-1M is excluded from the nucleus or organelles and mostly cytosolic, an estimation of the volume of the cytosol will provide a better estimation of the Bag-1M concentration around IF. These measurements were performed for different cell lines using fluorophores (Tan et al., 2012). For Caco-2 cells, in particular, the cytosol represents 50% of the cell volume.

Using this technique, we supplemented the fractions for luciferase refolding assays (as in Section 3) with recombinant Bag-1M at the predetermined concentrations. We found that Bag-1M is necessary for Hsp70 activity at basal concentrations (around 1 μ*M*) but inhibits chaperoning activity in kP at concentrations usually found in epithelial cells under inflammatory stimulation (around 6 μ*M*). Importantly, the inhibitory concentration is higher for soluble, cytosolic Hsp70 which showed another aspect of specificity for chaperones attached to the keratin scaffold (Mashukova et al., 2014). Moreover, these results suggested that the Hsp70 keratin scaffold may not be functional in cells expressing high levels of Bag-1M, such as some carcinomas (Maki et al., 2007).

6. SUMMARY

The reductionist methods described here need to be complementary of cell- or animal-based evidence indicating that IF are necessary for a certain function. These approaches are designed to provide mechanistic insight on whether or not a given biochemical activity, in the cases described here protein chaperoning, can be carried out specifically by keratin filaments. For example, we demonstrated that Krt8 KO mice which lack IF in the small intestine villus enterocytes, also display decreased steady-state levels of aPKC. Conversely, mice overexpressing keratins display increased amounts of aPKC in the same cells. Finally, while we have only tested the role of Hsp70 in aPKC refolding, we also found steep decreases in the steady-state levels of protein or turn-motif phosphorylation in PKCβII, ε, γ, and PKBα (Akt) in keratin-deficient cells. On the other hand, other known Hsp70 clients such as Chk-1 were unaffected by keratin knockdown (Mashukova et al., 2009). These results suggest that only a subset of Hsp70 clients is dependent on IF. Finding that IF knockout or knockdown negatively impacts on the steady-state levels of a protein, without affecting transcription/translation, seems to be a prerequisite to hypothesize that IF-based chaperoning is necessary for that protein.

"Scaffolding" of a certain protein on IF has been shown several times. Considering that in many cases the same protein is present in the cytosol or other compartments in larger amounts than in IF, it is difficult to understand how passive binding of a small population of the molecules may have a biologically significant impact at all. Conversely, if the isolated IF scaffold performs a unique function different from the cytosol or other cellular compartments then it seems valid to conclude that association to IF is essential.

Although progress has been made to unravel the molecular structure of filamentous keratins (Lee, Kim, Chung, Leahy, & Coulombe, 2012), the organization of the protein machinery around IF is still poorly understood. The molecular reasons why Hsp70/40 refold dephosphorylated aPKC specifically in the IF scaffold are unknown. Likewise, the binding mechanism of Bag-1M to IF is also unknown. Binding affinity may explain why Bag-1M is more efficient inhibiting Hsp70 on the filaments, but direct mechanistic evidence supporting that is still missing. Finally, to our knowledge, the approach described here has not been used in other types of IF. The techniques shown here may be helpful to understand the functional need of IF for other "downstream" mechanisms and functions identified by keratin loss-of-function mutations or keratin knockouts.

ACKNOWLEDGMENTS

This work was supported by a grant to P.J.S. from the National Institute of Diabetes, Digestive and Kidney Diseases (R01-DK076652) and a NIH Ruth L. Kirschstein National Research Service Award Fellowship to R.F. (F32-DK095503).

REFERENCES

Alberti, S., Esser, C., & Hohfeld, J. (2003). BAG-1—A nucleotide exchange factor of Hsc70 with multiple cellular functions. *Cell Stress & Chaperones*, *8*(3), 225–231.

Caulin, C., Ware, C. F., Magin, T. M., & Oshima, R. G. (2000). Keratin-dependent, epithelial resistance to tumor necrosis factor-induced apoptosis. *The Journal of Cell Biology*, *149*(1), 17–22.

Clerico, E. M., Tilitsky, J. M., Meng, W., & Gierasch, L. M. (2015). How Hsp70 molecular machines interact with their substrates to mediate diverse physiological functions. *Journal of Molecular Biology*, *427*(7), 1575–1588.

Eriksson, J. E., Dechat, T., Grin, B., Helfand, B., Mendez, M., Pallari, H. M., et al. (2009). Introducing intermediate filaments: From discovery to disease. *The Journal of Clinical Investigation*, *119*(7), 1763–1771.

Gao, T., & Newton, A. C. (2006). Invariant Leu preceding turn motif phosphorylation site controls the interaction of protein kinase C with Hsp70. *The Journal of Biological Chemistry*, *281*(43), 32461–32468.

Gould, C. M., Antal, C. E., Reyes, G., Kunkel, M. T., Adams, R. A., Ziyar, A., et al. (2011). Active site inhibitors protect protein kinase C from dephosphorylation and stabilize its mature form. *The Journal of Biological Chemistry*, *286*(33), 28922–28930.

Gould, C. M., & Newton, A. C. (2008). The life and death of protein kinase C. *Current Drug Targets*, *9*(8), 614–625.

Gu, L. H., & Coulombe, P. A. (2005). Defining the properties of the nonhelical tail domain in type II keratin 5: Insight from a bullous disease-causing mutation. *Molecular Biology of the Cell*, *16*(3), 1427–1438.

Homberg, M., & Magin, T. M. (2014). Beyond expectations: Novel insights into epidermal keratin function and regulation. *International Review of Cell and Molecular Biology*, *311*, 265–306.

Izawa, I., Nishizawa, M., Ohtakara, K., Ohtsuka, K., Inada, H., & Inagaki, M. (2000). Identification of Mrj, a DnaJ/Hsp40 family protein, as a keratin 8/18 filament regulatory protein. *The Journal of Biological Chemistry*, *275*(44), 34521–34527.

Janig, E., Stumptner, C., Fuchsbichler, A., Denk, H., & Zatloukal, K. (2005). Interaction of stress proteins with misfolded keratins. *European Journal of Cell Biology*, *84*(2-3), 329–339.

Kim, S., Wong, P., & Coulombe, P. A. (2006). A keratin cytoskeletal protein regulates protein synthesis and epithelial cell growth. *Nature*, *441*(7091), 362–365.

Kolsch, A., Windoffer, R., Wurflinger, T., Aach, T., & Leube, R. E. (2010). The keratin-filament cycle of assembly and disassembly. *Journal of Cell Science, 123*(Pt. 13), 2266–2272.

Koster, S., Weitz, D. A., Goldman, R. D., Aebi, U., & Herrmann, H. (2015). Intermediate filament mechanics in vitro and in the cell: From coiled coils to filaments, fibers and networks. *Current Opinion in Cell Biology, 32c*, 82–91.

Ku, N. O., Michie, S., Resurreccion, E. Z., Broome, R. L., & Omary, M. B. (2002). Keratin binding to 14-3-3 proteins modulates keratin filaments and hepatocyte mitotic progression. *Proceedings of the National Academy of Sciences of the United States of America, 99*(7), 4373–4378.

Ku, N. O., Toivola, D. M., Strnad, P., & Omary, M. B. (2010). Cytoskeletal keratin glycosylation protects epithelial tissue from injury. *Nature Cell Biology, 12*(9), 876–885.

Lee, C. H., Kim, M. S., Chung, B. M., Leahy, D. J., & Coulombe, P. A. (2012). Structural basis for heteromeric assembly and perinuclear organization of keratin filaments. *Nature Structural & Molecular Biology, 19*(7), 707–715.

Leifeld, L., Kothe, S., Sohl, G., Hesse, M., Sauerbruch, T., Magin, T. M., et al. (2009). Keratin 18 provides resistance to Fas-mediated liver failure in mice. *European Journal of Clinical Investigation, 39*(6), 481–488.

Lessard, J. C., Pina-Paz, S., Rotty, J. D., Hickerson, R. P., Kaspar, R. L., Balmain, A., et al. (2013). Keratin 16 regulates innate immunity in response to epidermal barrier breach. *Proceedings of the National Academy of Sciences of the United States of America, 110*(48), 19537–19542.

Liao, J., Lowthert, L. A., Ghori, N., & Omary, M. B. (1995). The 70-kDa heat shock proteins associate with glandular intermediate filaments in an ATP-dependent manner. *The Journal of Biological Chemistry, 270*(2), 915–922.

Lu, Z., & Cyr, D. M. (1998). Protein folding activity of Hsp70 is modified differentially by the hsp40 co-chaperones Sis1 and Ydj1. *The Journal of Biological Chemistry, 273*(43), 27824–27830.

Lu, Z., Liu, D., Hornia, A., Devonish, W., Pagano, M., & Foster, D. A. (1998). Activation of protein kinase C triggers its ubiquitination and degradation. *Molecular and Cellular Biology, 18*(2), 839–845.

Maki, H. E., Saramaki, O. R., Shatkina, L., Martikainen, P. M., Tammela, T. L., van Weerden, W. M., et al. (2007). Overexpression and gene amplification of BAG-1L in hormone-refractory prostate cancer. *The Journal of Pathology, 212*(4), 395–401.

Mashukova, A., Forteza, R., Wald, F. A., & Salas, P. J. (2012). PDK1 in apical signaling endosomes participates in the rescue of the polarity complex atypical PKC by intermediate filaments in intestinal epithelia. *Molecular Biology of the Cell, 23*(9), 1664–1674.

Mashukova, A., Kozhekbaeva, Z., Forteza, R., Dulam, V., Figueroa, Y., Warren, R., et al. (2014). The BAG-1 isoform BAG-1M regulates keratin-associated Hsp70 chaperoning of aPKC in intestinal cells during activation of inflammatory signaling. *Journal of Cell Science, 127*(Pt. 16), 3568–3577.

Mashukova, A., Oriolo, A. S., Wald, F. A., Casanova, M. L., Kroger, C., Magin, T. M., et al. (2009). Rescue of atypical protein kinase C in epithelia by the cytoskeleton and Hsp70 family chaperones. *Journal of Cell Science, 122*(Pt. 14), 2491–2503.

Mashukova, A., Wald, F. A., & Salas, P. J. (2011). Tumor necrosis factor alpha and inflammation disrupt the polarity complex in intestinal epithelial cells by a posttranslational mechanism. *Molecular and Cellular Biology, 31*(4), 756–765.

Milo, R. (2013). What is the total number of protein molecules per cell volume? A call to rethink some published values. *Bioessays, 35*(12), 1050–1055.

Newton, A. C. (2003). Regulation of the ABC kinases by phosphorylation: Protein kinase C as a paradigm. *The Biochemical Journal, 370*(Pt. 2), 361–371.

Nunbhakdi-Craig, V., Machleidt, T., Ogris, E., Bellotto, D., White, C. L., 3rd., & Sontag, E. (2002). Protein phosphatase 2A associates with and regulates atypical PKC and the epithelial tight junction complex. *The Journal of Cell Biology, 158*(5), 967–978.

Oriolo, A. S., Wald, F. A., Canessa, G., & Salas, P. J. (2007). GCP6 binds to intermediate filaments: A novel function of keratins in the organization of microtubules in epithelial cells. *Molecular Biology of the Cell, 18*(3), 781–794.

Pan, X., Hobbs, R. P., & Coulombe, P. A. (2013). The expanding significance of keratin intermediate filaments in normal and diseased epithelia. *Current Opinion in Cell Biology, 25*(1), 47–56.

Perng, M. D., Cairns, L., van den IJssel, P., Prescott, A., Hutcheson, A. M., & Quinlan, R. A. (1999). Intermediate filament interactions can be altered by HSP27 and alphaB-crystallin. *Journal of Cell Science, 112*(Pt. 13), 2099–2112.

Quinlan, R. A., Carte, J. M., Sandilands, A., & Prescott, A. R. (1996). The beaded filament of the eye lens: An unexpected key to intermediate filament structure and function. *Trends in Cell Biology, 6*(4), 123–126.

Quinlan, R. A., & Ellis, R. J. (2013). Chaperones: Needed for both the good times and the bad times. *Philosophical Transactions of the Royal Society of London, Series B: Biological Sciences, 368*, 20130091.

Saibil, H. (2013). Chaperone machines for protein folding, unfolding and disaggregation. *Nature Reviews Molecular Cell Biology, 14*(10), 630–642.

Salas, P. J. (1999). Insoluble gamma-tubulin-containing structures are anchored to the apical network of intermediate filaments in polarized CACO-2 epithelial cells. *The Journal of Cell Biology, 146*(3), 645–658.

Schweitzer, S. C., Klymkowsky, M. W., Bellin, R. M., Robson, R. M., Capetanaki, Y., & Evans, R. M. (2001). Paranemin and the organization of desmin filament networks. *Journal of Cell Science, 114*(Pt. 6), 1079–1089.

Seltmann, K., Cheng, F., Wiche, G., Eriksson, J. E., & Magin, T. M. (2015). Keratins stabilize hemidesmosomes through regulation of beta4-integrin turnover. *The Journal of Investigative Dermatology, 135*, 1609–1620. http://dx.doi.org/10.1038/jid.2015.46.

Snider, N. T., & Omary, M. B. (2014). Post-translational modifications of intermediate filament proteins: Mechanisms and functions. *Nature Reviews Molecular Cell Biology, 15*(3), 163–177.

Steinert, P., Zackroff, R., Aynardi-Whitman, M., & Goldman, R. D. (1982). Isolation and characterization of intermediate filaments. *Methods in Cell Biology, 24*, 399–419.

Tan, C. W., Gardiner, B. S., Hirokawa, Y., Layton, M. J., Smith, D. W., & Burgess, A. W. (2012). Wnt signalling pathway parameters for mammalian cells. *PLoS One*, 7(2), e31882.

Wang, X., Chen, M., Zhou, J., & Zhang, X. (2014). HSP27, 70 and 90, anti-apoptotic proteins, in clinical cancer therapy (Review). *International Journal of Oncology, 45*(1), 18–30.

Watson, E. D., Geary-Joo, C., Hughes, M., & Cross, J. C. (2007). The Mrj co-chaperone mediates keratin turnover and prevents the formation of toxic inclusion bodies in trophoblast cells of the placenta. *Development, 134*(9), 1809–1817.

Yamazaki, S., Uchiumi, A., & Katagata, Y. (2012). Hsp40 regulates the amount of keratin proteins via ubiquitin-proteasome pathway in cultured human cells. *International Journal of Molecular Medicine, 29*(2), 165–168.

CHAPTER NINE

Purification of Protein Chaperones and Their Functional Assays with Intermediate Filaments

Ming-Der Perng*,[1], Yu-Shan Huang*, Roy A. Quinlan†

*Institute of Molecular Medicine, College of Life Sciences, National Tsing Hua University, Hsinchu, Taiwan
†Biophysical Sciences Institute, University of Durham, Durham, United Kingdom
[1]Corresponding author: e-mail address: mdperng@life.nthu.edu.tw

Contents

Abstract

Intermediate filament (IF) scaffolds facilitate small heat shock protein (sHSP) function, while IF function is sHSP dependent. sHSPs interact with IFs and the importance of this interaction is to maintain the individuality of the IFs and to modulate interfilament interactions both in networks and in assembly intermediates. Mutations in both sHSPs and their interacting IF proteins phenocopy each other in the human diseases they cause. This establishes a key functional relationship between these two very distinct protein families, and it also evidences the role of this cytoskeleton–chaperone complex in the cellular stress response. In this chapter, we describe the detailed experimental protocols for the preparation of purified IF proteins and sHSPs to facilitate the study *in vitro* of their functional interactions. In addition, we describe the detailed biochemical procedures to assess the effect of sHSP on the assembly of IFs, the binding to IFs, and the prevention of noncovalent filament–filament interactions using *in vitro* cosedimentation, electron microscopy, and viscosity assays. These assays are valuable research tools to study and manipulate sHSP–IF complexes *in vitro* and therefore to

Methods in Enzymology, Volume 569
ISSN 0076-6879
http://dx.doi.org/10.1016/bs.mie.2015.07.025

determine the structure–function detail of this complex, and how it contributes to cellular, tissue, and organismal homeostasis and the *in vivo* stress response.

ABBREVIATIONS

μl microliter
μ*M* micromolar
A260 absorbance at 260 nm
A280 absorbance at 280 nm
AKAP A-kinase-anchoring protein
CCD charge-coupled device
cm centimeter
DRM desmin-related myopathy
DTT dithiothreitol
EDTA 2-({2-[bis(carboxymethyl)amino]ethyl}(carboxymethyl)amino)acetic acid
EGTA ethyleneglycol-bis(2-aminoethylether)-*N*,*N*,*N*′,*N*′-tetraacetic acid
GFAP glial fibrillary acidic protein
IF intermediate filament
IFAP IF-associated protein
IPTG isopropyl-1-thio-β-D-galactopyranoside
KCl potassium chloride
kDa kiloDalton
K_m dissociation constant
kV kilovolt
LB Luria–Bertani
M molar
MFM myofibrillar myopathy
ml/min milliliters per minute
m*M* millimolar
NaCl sodium chloride
OD600 optical density at 600 nm
PMSF phenylmethylsulfonyl fluoride
RPM revolutions per minute
SDS sodium dodecyl sulfate
SDS-PAGE sodium dodecyl sulfate polyacrylamide gel electrophoresis
sHSP small heat shock protein
TEM transmission electron microscope
TEN buffer Tris, EDTA, sodium chloride buffer
w/v weight per 1 volume

1. INTRODUCTION

1.1 Intermediate Filaments

Intermediate filaments (IFs) are a highly dynamic cytoskeletal component found in most eukaryotic cells. They form extensive networks that maintain mechanical strength and shape of the cell and provide dynamic platforms for the organization of the cytoplasm on a structural and functional level (Goldman, Grin, Mendez, & Kuczmarski, 2008; Kim & Coulombe, 2007). In human, at least 70 different genes in the human genome have been identified as members of the IF protein family (Hesse, Magin, & Weber, 2001), and there is a complex expression pattern specific for every cell type. Although highly heterogeneous in size and amino acid composition, the structural feature of IF proteins is highly conserved with the defining feature being a centrally located, α-helical domain containing heptad repeats of hydrophobic and apolar residues. The central rod domain that mediates coiled-coil dimer formation represents the major driving force sustaining filament assembly (Herrmann & Aebi, 2004). The rod is flanked, at both ends, by non-α-helical N-terminal and C-terminal domains that differ in length, sequence, and substructure (Parry, 2005). Variations in these so-called head and tail domains account for the marked heterogeneity in IF protein size and other attributes.

IFs were initially identified with mechanical roles, but now the regulation of cellular architecture (Herrmann, Bar, Kreplak, Strelkov, & Aebi, 2007; Herrmann, Strelkov, Burkhard, & Aebi, 2009), cell growth (Kim & Coulombe, 2007; Kim, Wong, & Coulombe, 2006) and metabolism (Alam et al., 2013), cell migration (Leduc & Etienne-Manneville, 2015; Nieminen et al., 2006; Seltmann, Fritsch, Kas, & Magin, 2013), and signaling cascades (Pallari & Eriksson, 2006; Toivola, Tao, Habtezion, Liao, & Omary, 2005). These processes are regulated by the incorporation of coassembly partners (Eliasson et al., 1999; Schweitzer et al., 2001), posttranslational modifications (Sihag, Inagaki, Yamaguchi, Shea, & Pant, 2007; Snider & Omary, 2014), the associations with IF-associated proteins (Green, Bohringer, Gocken, & Jones, 2005; Herrmann & Aebi, 2000; Li et al., 2006; Satoh, Yamamura, & Arima, 2004), and the expression of splice variant (Perng et al., 2008; Roelofs et al., 2005). Although most of IF proteins are very capable of forming 10-nm filaments *in vitro* (Herrmann & Aebi, 2004) and in cells (Goldman et al., 2008), the dynamics of IF assembly as well as the formation and maintenance of a functional IF network involve the small heat shock proteins (sHSPs) as their symbiotic partners with IFs in the cellular stress response (Landsbury, Perng, Pohl, & Quinlan, 2010; Perng & Quinlan, 2015; Quinlan, 2002).

1.2 Small Heat Shock Proteins

sHSPs form a large protein family linked to the cellular stress response. As most have masses less than 30 kDa, the term "small" was assigned within the context of HSP60, 70, and 90 chaperone classes. sHSPs are molecular chaperones, capable of blocking protein aggregation and protecting cells against stresses (Cox, Carver, & Ecroyd, 2014). Of the 10 currently known human sHSPs (Kappe et al., 2003), some are widely expressed, whereas others are restricted to only one or two tissues (Morrow & Tanguay, 2012). This suggests general (Quinlan & Ellis, 2013) as well as specialized functions in cells from cell survival (Hu et al., 2012; Liu et al., 2004) to apoptosis (Acunzo, Katsogiannou, & Rocchi, 2012), from proteolysis (Boelens, Croes, & de Jong, 2001; den Engelsman, Keijsers, de Jong, & Boelens, 2003) to proteostasis (Treweek, Meehan, Ecroyd, & Carver, 2015; van Oosten-Hawle & Morimoto, 2014), and from the cytoskeleton (Landsbury et al., 2010; Wettstein, Bellaye, Micheau, & Bonniaud, 2012) to metabolic enzymes (Brewer, Mustafi, Murray, Rajasekaran, & Benjamin, 2013; Quinlan & Ellis, 2013). Structurally, all sHSP share an 80–90 residue domain known as the α-crystallin domain. This signature feature is a structural scaffold onto which flexible arms capable of preventing irreversible protein aggregation, as well as facilitating the assembly of protein complexes are attached (Basha, O'Neill, & Vierling, 2012). Indeed, one of the key functions of the sHSPs is to organize cytoskeletal networks (Landsbury et al., 2010; Wettstein et al., 2012) and specifically to chaperone individual IFs (Perng et al., 1999).

1.3 sHSP-IF Interaction in Physiology and Pathology

Initial studies on the lens cytoskeleton led to the discovery that the lenticular sHSPs, α-crystallin, interact with soluble and filamentous vimentin (Nicholl & Quinlan, 1994). *In vitro* assembly studies also showed that αB-crystallin can inhibit IF assembly (Perng et al., 1999) and in the case of GFAP, inhibition is independent of phosphorylation (Nicholl & Quinlan, 1994). Subsequently, the αB-crystallin binding to IFs was shown biochemically to be cation, pH, and temperature dependent (Elliott et al., 2013). The temperature-dependent property also allowed the dissociation constant (K_m) to be measured for αB-crystallin binding to desmin (Perng, Wen, van den, Prescott, & Quinlan, 2004). Associations of sHSP with IF networks have also been visualized in a range of cell lines with different IF compositions using immunofluorescence microscopy (Perng et al., 1999; Wisniewski & Goldman, 1998). While these data all point to a direct physical association of sHSP with IFs, the most compelling evidence in support of the functional relationship of this interaction came from the identification of the R120G mutation in αB-crystallin that causes IF aggregation in desmin-related myopathy, a myofibrillar myopathy (MFM), an inherited disease characterized by skeletal muscle weakness and heart failure (Vicart et al., 1998). The characteristic histopathology was filament-containing aggregates in the muscles of the affected individuals. Interestingly, desmin mutations phenocopy MFM caused by αB-crystallin mutations (Landsbury et al., 2010). Protein aggregates containing both IFs and αB-crystallin are a prominent histopathological feature in human diseases caused by mutant cytoplasmic IF proteins that form aggregates (Bjorkdahl et al., 2008; Dabir, Trojanowski, Richter-Landsberg, Lee, & Forman, 2004; Quinlan & Van Den Ijssel, 1999; Selcen, 2011). The phenocopying of mutant sHSPs and IFs in human disease demonstrates the functional symbiosis of sHSPs and IFs (Landsbury et al., 2010; Perng & Quinlan, 2015).

Here, we describe the detailed experimental procedures to the preparation and purification of IF proteins and sHSPs, which allow their functional interaction to be studied *in vitro* using cosedimentation assay, solution viscosity, and electron microscopy. These assays have proved of great value not only in helping to uncover the functional aspects of sHSP–IF interaction *in vitro* but also to understand how sHSPs manage IF organization to prevent unintended filament–filament interactions. The detailed biochemical procedures described here provide a convenient reference for the IF research community for reliable preparation and isolation of both IF proteins and sHSPs to study their functional interactions *in vitro*. The procedures can be performed in as little as 1 day, or may take longer, depending on the precise functional assays used.

2. PURIFICATION OF sHSPs

Recombinant sHSPs have properties similar to those of native sHSPs isolated from tissues (Perng, Sandilands, et al., 2004), and they can be induced to be expressed at relatively high levels in bacteria. In addition, recombinant sHSPs are more amenable to genetic manipulations by site-directed mutagenesis. These manipulations provide a clear advantage to study the effect of specific amino acids on the structure and function of sHSPs, as well as interactions with their client proteins, including IFs.

Using αB-crystallin as an example, we describe the detailed experimental protocols for the preparation of purified sHSPs in a two-step procedure, using DEAE column and gel filtration columns (Fig. 1A). All purification steps were performed at 4 °C unless otherwise stated.

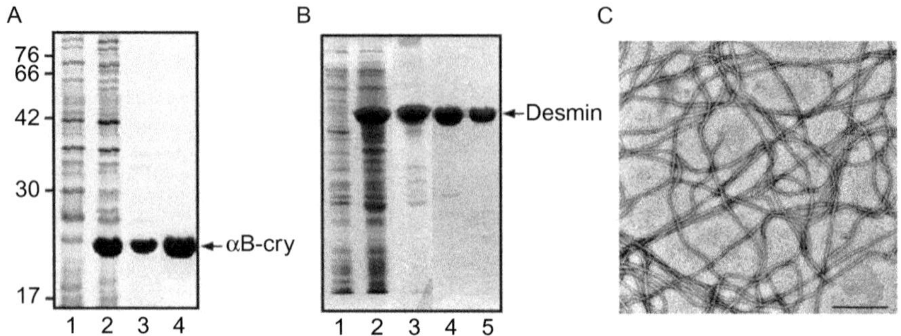

Figure 1 Electrophoretic analysis of major fractions obtained during purification of small heat shock protein and intermediate filament proteins. (A) The sHSP αB-crystallin was expressed in *Escherichia coli* BL21 pLysS strain using a pET-based vector system. Total protein profiles of extracts from uninduced (lane 1) and IPTG-induced (lane 2) bacteria were shown. The αB-crystallin was purified to homogeneity by column chromatography using DEAE (lane 3) and gel filtration (lane 4) columns. Samples were separated by SDS-PAGE, and proteins were visualized by Coomassie Blue staining. The relative electrophoretic mobility of molecular weight markers is indicated on the left side of lane 1. (B) Recombinant human desmin was produced in bacteria with a pET-based expression system. Total protein profiles from uninduced (lane 1) and IPTG-induced (lane 2) bacteria are shown. Overexpressed desmin formed inclusion bodies, which were prepared (lane 3) and further purified by DEAE (lane 4) and CM (lane 5) columns. Position of αB-crystallin and desmin is indicated by arrows. In panels A and B, lanes were run on the same gel but were noncontiguous. (C) Purified recombinant desmin was assembled *in vitro* and visualized by electron microscope. Bar, 200 nm.

1. Plasmids containing either wild-type or mutant αB-crystallin in the pET23b vector (Novagen, Nottingham, United Kingdom) were transformed into *Escherichia coli* BL21 (DE3) pLysS strain, which drive recombinant sHSP expression under the control of T7 promoter (Tabor and Richardson, 1985).
2. A saturated overnight culture of transformed bacteria was inoculated at 1:100 (v/v) into a 2 l Luria–Bertani (LB) medium containing 100 μg/ml ampicillin and 34 μg/ml chloroamphenicol. Bacteria were grown at 37 °C with vigorous shaking at 225 rpm.
3. Once the bacterial culture had reached an OD_{600} of 0.6, protein expression was induced using 0.5 m*M* isopropyl-1-thio-β-D-galactopyranoside (IPTG) for 3 h.
4. Bacteria were harvested by centrifugation at 5000 rpm for 30 min at 4 °C in a JA-10 rotor (Beckman, United Kingdom).
5. Bacterial pellets were resuspended in 50 ml TEN buffer (50 m*M* Tris–HCl, pH 8.0, 1 m*M* EDTA, 300 m*M* NaCl) supplemented with

0.2 m*M* phenylmethylsulfonyl fluoride (PMSF) and protease inhibitor cocktails (Roche Diagnostics, Indianapolis, IN) and homogenized using a Dounce homogenizer.

6. Benzonase nuclease (Novagen) was added to the bacterial lysates to a final concentration of 10 U/ml. After incubation at room temperature for 30 min, lysates were centrifuged at 15,000 rpm in a JA-20 rotor (Beckman Coulter, High Wycombe, United Kingdom) for 30 min at 4 °C.
7. The supernatant was collected and filtered through a 0.2-μm filter (Whatman, Maidstone, United Kingdom). Polyethyleneimine (Sigma Chemical, Poole, Dorset, United Kingdom) was then added to the filtrate to a final concentration of 0.06% (v/v). After incubation on ice for 5 min, the mixture was centrifuged at 15,000 rpm for 10 min to pellet the DNA.
8. The clarified supernatant was dialyzed against buffer A (20 m*M* Tris–HCl, pH 7.4, 20 m*M* NaCl, 1 m*M* $MgCl_2$, 1 m*M* EDTA, 1 m*M* dithiothreitol (DTT), 0.2 m*M* PMSF) and then loaded onto a DEAE–Sepharose column (GE Healthcare (Buckinghamshire, UK)) preequilibrated in the same buffer.
9. The column was washed with six bed volumes of buffer A and then eluted with a linear gradient of 20–200 m*M* NaCl in the same buffer.
10. Column fractions enriched in αB-crystallin were concentrated using Ultrafree-15 concentrator (Millipore, Watford, United Kingdom) with a 100-kDa molecular weight cutoff membrane, dialyzed into buffer B (20 m*M* Tris–HCl, pH 7.4, 100 m*M* NaCl), and further purified using a gel filtration column (60 × 1.6 cm; Fractogel EMD BioSEC Superformance) equilibrated in the same buffer.
11. The column was developed at a flow rate of 0.5 ml/min in buffer B. Fractions were analyzed by SDS-PAGE and those containing purified protein were pooled, concentrated, and stored at −80 °C.
12. The molecular size was measured by gel filtration using the reference molecular weight size markers, thyroglobulin (669 kDa), apoferritin (440 kDa), alpha-amylase (200 kDa), bovine serum albumin (67 kDa), and carbonic anhydrase (29 kDa). Dextran blue (2000 kDa) was used to determine the column void volume. The molecular size of wild-type and mutant αB-crystallin was then calculated from the elution volume.

Pitfalls

1. Nucleotides that bind the recombinant sHSPs preparations could be a major source of contamination, which will not show on the SDS gels.

Therefore, absorption spectra of the preparation should be taken and the ratio of A280–A260 should be determined. A clean preparation of recombinant sHSPs should have an A_{280}/A_{260} ratio of >1.5.

2. Excess polyethyleneimine (>0.12% (v/v)) also precipitates proteins.
3. Protein degradation can occur requiring fresh protease inhibitors to be added at each step of the purification procedure.
4. Incomplete bacterial lysis reduces yield. Several freeze and thaw cycles are recommended to facilitate efficient lysis. We do not recommend using lysozyme as this can contaminant the protein of interest, requiring additional purification steps for its removal.

3. PURIFICATION OF IF PROTEINS

The IFs in muscle cells function as mechanical integrators of the cellular space (Lazarides, 1980). They provide cytoskeletal integrity and strength, as well as the organization necessary to support muscle contraction (Balogh, Li, Paulin, & Arner, 2005; Boriek et al., 2001; Carlsson et al., 2000; Li, Andersson-Lendahl, Sejersen, & Arner, 2013; Shah, Love, O'Neill, Lovering, & Bloch, 2012). Desmin is the major IF protein in skeletal, cardiac, and smooth muscles. By interacting with other muscle IF proteins, such as paranemin (Schweitzer et al., 2001), synemin (Chourbagi et al., 2011; Garcia-Pelagio et al., 2015; Li et al., 2014; Lund et al., 2012; Xue et al., 2004), syncoilin (Kemp et al., 2009; McCullagh et al., 2008; Poon, Howman, Newey, & Davies, 2002; Zhang et al., 2008), and the cytolinker protein plectin (Favre et al., 2011; Schroder et al., 1999; Winter & Wiche, 2013), the phosphoinositide phosphatase myotubularin 1 that affects both desmin filament and mitochondrial organization in skeletal muscle (Hnia et al., 2011; Joubert et al., 2013) and the AKAP, myospryin (Tsoupri & Capetanaki, 2013), desmin forms a transcytoplasmic IF network around the Z-discs and helps to thus maintain the lateral alignment of striated myofibrils and integrity of the cytoskeletal lattice of the muscle (Capetanaki, Papathanasiou, Diokmetzidou, Vatsellas, & Tsikitis, 2015). This key role for desmin in muscle function is confirmed by the discovery of myopathy-causing mutation in desmin (Goldfarb et al., 1998; Munoz-Marmol et al., 1998) and by the muscle phenotypes observed in the desmin knockout mice (Capetanaki et al., 2015; Milner, Weitzer, Tran, Bradley, & Capetanaki, 1996).

Previous *in vitro* assembly studies have demonstrated that purified desmin was very capable of forming 10 nm filaments that are morphologically similar to native muscle IFs, confirming that desmin is the major structural component of the muscle IFs (Huiatt, Robson, Arakawa, & Stromer, 1980). Here, we describe the detailed biochemical procedure for purification of recombinant human desmin and its subsequent biochemical characterization.

Recombinant IFPs have properties similar to those of native IFPs isolated from a wide range of tissues. The ability to genetically manipulated recombinant IFPs by site-directed mutagenesis allows the effect of specific amino acid substitutions or disease-causing mutations on its assembly properties to be studied *in vitro*. It is simple to perform and easy to scale up as two significant practical advantages. We used recombinant human desmin as an example to detail the experimental procedures for the expression and purification of recombinant IF proteins in bacteria. Recombinant human desmin was expressed in bacteria using a pET-based prokaryotic expression system. Overexpressed desmin formed inclusion bodies, which were prepared and further purified by ion exchange chromatography (Fig. 1B).

Procedures

1. *E. coli* strain BL21(DE3) pLysS (Novagen) were transformed with pET 23b vector containing cDNA of human desmin (Elliott et al., 2013).
2. After transformation, bacterial cultures were grown in LB medium supplemented with 100 μg/ml ampicillin and 34 μg/ml chloroamphenicol to OD_{600} of 0.5–0.6.
3. Protein expression was induced by the addition of 0.5 m*M* IPTG at 37 °C for 3 h. After induction, bacteria were harvested by centrifugation at 6000 rpm for 30 min at 4 °C.
4. Harvested bacterial pellets were resuspended in TEN buffer and lysed by homogenization in a Dounce homogenizer. Bacterial lysis can be facilitated by several freeze and thaw cycles.
5. Bacterial lysates were centrifuged at 18,000 rpm for 30 min at 4 °C, and the resulting pellet containing inclusion bodies was first extracted in detergent buffer (20 m*M* Tris–HCl, pH 7.5, 1% (v/v) Triton X-100, 150 m*M* NaCl, 5 m*M* EDTA, 1 m*M* EGTA, 0.5 m*M* DTT), and then in the high salt buffer (20 m*M* Tris–HCl, pH 7.5, 0.5% (v/v) Triton X-100, 1.5 *M* KCl, 5 m*M* EDTA, 1 m*M* EGTA, 0.5 m*M* DTT).
6. After centrifugation, the final pellet was washed with low salt buffer (10 m*M* Tris–HCl, pH 7.5, 150 m*M* NaCl, 5 m*M* EDTA, 1 m*M* EGTA, 0.5 m*M* DTT), followed by extracting in urea buffer (8 *M* urea, 20 m*M* Tris–HCl [pH 7.4], 5 m*M* EDTA, 1 m*M* EGTA, 1 m*M* DTT, and 1 m*M* PMSF) by stirring for 3 h.

7. Urea extracts were clarified by centrifugation at 100,000 × *g* in a bench-top Optima MAX Ultracentrifuge with use of an MLA-80 rotor (Beckman Coulter).
8. The supernatant was dialyzed into buffer A (6 *M* urea, 10 m*M* Tris–HCl [pH 8.0], 5 m*M* EDTA, 1 m*M* EGTA, 1 m*M* DTT, and 1 m*M* PMSF) and then loaded onto a Fractogel EMD TMAE 650S column (Merck) preequilibrated in the same buffer.
9. Proteins were eluted from the column with a linear gradient of 0–0.5 *M* NaCl in buffer A over 1 h at a flow rate of 1 ml/min.
10. Desmin-enriched fractions were pooled, dialyzed against buffer B (6 *M* urea, 20 m*M* sodium formate (pH 4.0), 5 m*M* EDTA, 1 m*M* EGTA, 1 m*M* DTT, and 1 m*M* PMSF), and then applied to a Fractogel EMD COO-650S column (Merck) preequilibrated with buffer B.
11. After being washed with buffer B, proteins were eluted from the column with a linear gradient of 0–0.5 *M* NaCl in the same buffer.
12. Column fractions were analyzed by SDS-PAGE, and those containing purified desmin were collected and stored at −80 °C.

Pitfalls

1. A detailed protocol to evaluate protein expression in bacteria can be found in the pET system manual (11th edition, Novagen, http://www.emdbiosciences.com/docs/docs/PROT/TB055.pdf). The authors have used this system to express a number of IFPs, including GFAP, desmin, vimentin, and keratins and found that all of them were expressed very well using this expression system.
2. Extraction and column chromatography of the IF proteins were carried out in buffers containing 6–8 *M* urea. Solutions of high concentration of urea tend to be hydrolyzed into ammonium and cyanate ions, which can carbamylate IF proteins, potentially compromising IF assembly. Therefore, all urea solutions are treated with mixed-bed ion exchanger beads (Amberlite, Merck-BDH, United Kingdom). Urea solutions were prepared fresh as needed and stored at 4 °C to minimize cyanate formation.
3. IF proteins are very susceptible to proteolysis during preparation and storage. As a result, protease inhibitors should be included in all buffers. Whenever possible, all purification steps should be conducted at 4 °C.

4. ANALYSIS OF sHSP AND IF PROTEINS INTERACTION

By taking advantage of the self-assembly properties of desmin, it is possible to determine IF protein interaction with sHSP such as αB-crystallin. Here, we focus on *in vitro* assembly and cosedimentation assays (Fig. 2) using purified desmin and αB-crystallin as described above.

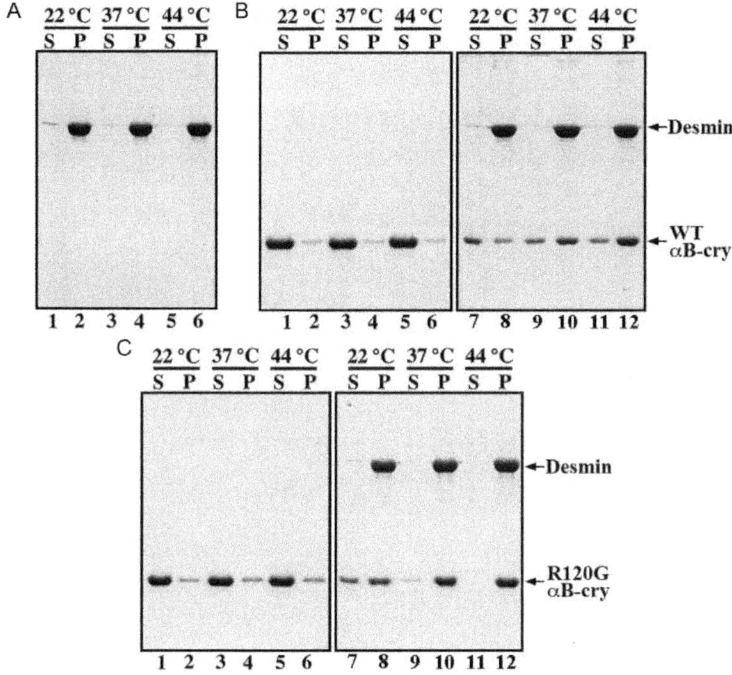

Figure 2 Cosedimentation of wild-type and R120G αB-crystallin with desmin filaments *in vitro*. Desmin was assembled either alone (A) or with additions of either wild-type (B) or R120G αB-crystallin (C) at a molar ratio of 1:1 at the indicated temperatures. The supernatant (S) and pellet (P) fractions were analyzed by SDS-PAGE followed by Coomassie Blue staining. Representative gels from the cosedimentation assay were shown. The positions of desmin, wild-type, and R120G αB-crystallin are indicated. Under these assay conditions, desmin alone assembled efficiently at all selected temperatures and more than 95% of the proteins were found in the pellet fractions (A; lanes 2, 4, and 6, labeled P). In the absence of desmin, both the wild-type and R120G αB-crystallin remained mostly in the supernatant (B and C; lanes 1, 3, and 5, labeled S) with only about 10% being sedimented into the pellet fractions (B and C; lanes 2, 4, and 6, labeled P). When desmin was included in the assay, both the wild-type (B; lanes 8, 10, and 12) and R120G αB-crystallin (C; lanes 8, 10, and 12) cosedimented with desmin filaments in a temperature-dependent manner because increasing proportions of αB-crystallin were found in the pellet fractions at elevated temperatures. The R120G αB-crystallin apparently cosedimented more efficiently than wild-type protein at all selected temperatures (B and C; lanes 7–12), suggesting an increased binding of R120G αB-crystallin to desmin.

4.1 *In Vitro* Assembly and Cosedimentation Assay

Procedure

1. Desmin was diluted to 0.3 mg/ml in 6 *M* urea in low-ionic strength buffer (20 m*M* Tris–HCl, pH 8.0, 1 m*M* EDTA, 1 m*M* EGTA, 1 m*M* DTT, and 0.2 m*M* PMSF).

2. Protein samples were then dialyzed stepwise against 3 *M* urea in low-ionic strength buffer for 4–6 h and then against the same buffer without urea at 4 °C overnight.
3. Filament assembly was completed by dialyzing against assembly buffer (20 m*M* imidazole–HCl [pH 6.8], 100 m*M* NaCl, and 1 m*M* DTT) for 12–16 h at room temperature.
4. The filament forming efficiency was assessed by high-speed sedimentation assay. The assembly mixtures layered onto a cushion of 0.85 *M* sucrose made in the assembly buffer were centrifuged at 80,000 × *g* for 30 min at 20 °C in a TL-100 Bench Top centrifuge (Beckman Coulter) using a TLS 55 rotor (Beckman Coulter).
5. In some experiments, the extent of filament–filament interactions in the whole filament population was assessed by a low-speed centrifugation at 3000 × *g* for 5 min at room temperature in a benchtop centrifuge (5417R; Eppendorf, Hamburg, Germany).
6. The supernatant and pellet fractions were separated by SDS-PAGE and visualized by Coomassie Blue staining. The amount of protein in the supernatant and pellet fractions was analyzed by a luminescent image analyser (LAS-1000plus; Fuji Film, Tokyo, Japan) and quantified using the Image Gauge software (version 4.0; Fuji Film).
7. For cosedimentation assays, αB-crystallin was mixed with desmin in low-ionic strength buffer at different molar ratios as indicated (Fig. 3).
8. Desmin was assembled in the presence or absence of αB-crystallin. Assembly was initiated by the addition of a 20-fold concentrated binding buffer to give a final concentration of 100 m*M* imidazole–HCl, pH 6.8, 1 m*M* DTT, and 0.2 m*M* PMSF.
9. After incubation for 1 h at 37 °C, samples were centrifuged at 80,000 × *g* for 30 min at 20 °C. The supernatant and pellet fractions were compared by SDS-PAGE as described above.
10. The binding curve of αB-crystallin to desmin can be determined by *in vitro* cosedimentation assay using a fixed concentration of desmin and varying the concentration range of αB-crystallin. Scatchard analysis of these data can be used to estimate the dissociation constant (K_d) for αB-crystallin binding to desmin.

4.2 Functional Analyses of sHSP–IF Interaction

sHSPs are important modulators of IF assemblies. They can prevent filament–filament interactions from occurring on the basis of an *in vitro*-based

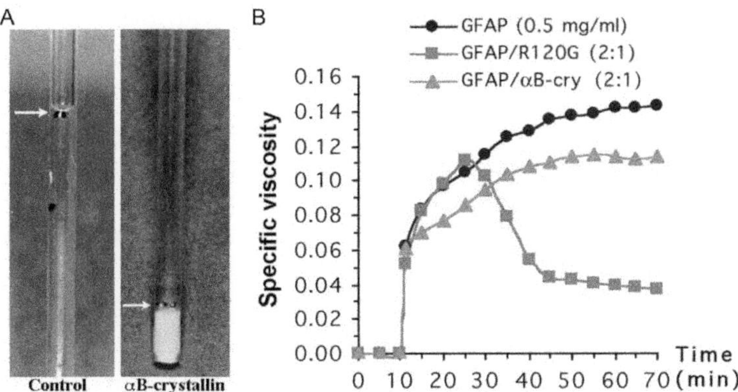

Figure 3 Viscometric analyses of GFAP assembly in the absence or presence of αB-crystallin. (A) Falling ball viscosity assay was performed by assembling of GFAP (0.5 mg/ml) in a glass tube at 37 °C. Prior to this assay, a measurement of the solution viscosity was made. After assembly of GFAP alone in the control tube, a gel had formed preventing the ball from falling. The arrow indicates the ball supported at the top of the filament solution in the control tube (Control, arrow). In contrast, in the presence of αB-crystallin the ball was able to drop to the bottom of the tube (αB-crystallin, arrow), suggesting that αB-crystallin prevents GFAP filaments from forming a gel. (B) Ostwald-type viscometer assay. GFAP (0.5 mg/ml) was assembled either alone or coassembled with either WT or R120G αB-crystallin in a molar ratio of 1:2 at 37 °C. Assembly was initiated at the 10th minute time point by the addition of a 20-fold concentrated binding buffer. Specific viscosity was measured 1 min after assembly and then at 5 min intervals for 60 min. The profile obtained for GFAP reflects the normal increase of specific viscosity after initiation of filament assembly. In the presence of WT αB-crystallin, GFAP exhibited a slightly decreased specific viscosity over a period of 40–45 min before reaching a plateau. In contrast, for GFAP assembled in the presence of R120G αB-crystallin, the specific viscosity increased normally only during the first 25 min. Afterward it decreased dramatically, indicative of GFAP filament aggregation. (See the color plate.)

viscosity assay. The viscosity of IF after assembly was measured at 37 °C using a falling ball viscometer (Fig. 3A), a well-established technique in the actin field for investigating actin-binding protein activity (MacLean-Fletcher & Pollard, 1980).

4.2.1 Falling Ball Viscosity Assay

Procedure

1. IF protein was assembled either alone or in the presence of sHSP by the addition of 20 × binding buffer as described in the cosedimentation assay.
2. Immediate after assembly, 100 μl of protein solution was loaded into a glass tube (100-μl micropipette, Corning, United Kingdom) with 1 mm i.d. and 12.5 cm in length by capillary action.

3. The capillary tubes were sealed at one end with plasticine and then immersed in a water bath to allow GFAP or desmin assembly in the presence of sHSPs at 37 °C for 1 h.
4. Before measurement, the capillary tubes were set up at an angle of 80°. The viscosity was determined at this angle by measuring the time required for a stainless steel ball to fall through a 6-cm path along the capillary tube.

4.2.2 Ostwald Viscometry

Ostwald-type viscometer is a rapid and simple method to measure IF assembly (Fig. 3B). The viscometer is a U-shaped glass tube that includes a section with a narrow diameter through which the sample flows driven by hydrostatic pressure (Hofmann, 1998).

Procedure

1. Viscometers with a flow time of 30 s and a sample volume of 1 ml were used.
2. The viscometer is filled with assembly buffer, but without IF protein as a control.
3. During assembly, solution levels in the viscometer were adjusted by applying a negative pressure so that measurements could be made.
4. The time required for the solution meniscus to pass between two marks on the glass capillary was measured every 5 min. Filament assembly increased viscosity and increases the time needed for the meniscus to pass the two marks.
5. Specific viscosity is calculated by the equation $V_{sp} = (T_s - T_b)/T_b$, where T_s is the flow time of the sample and T_b is the flow time of the buffer.

Note

1. Air bubbles should not be introduced into the viscometer capillary during measurement as this disrupts accurate measurement.
2. After each measurement, the viscometer was cleaned with 10% nitric acid followed by extensive washing with distilled water.

4.3 Negative Staining and Electron Microscopy

This section describes the application of negative staining and electron microscopy to study the *in vitro* assembly of IFs (Fig. 4). Negative staining is a widely used routine technique to prepare biological samples for imaging under the transmission electron microscope (TEM). Compared to other electron microscopic techniques, negative staining is relatively simple and

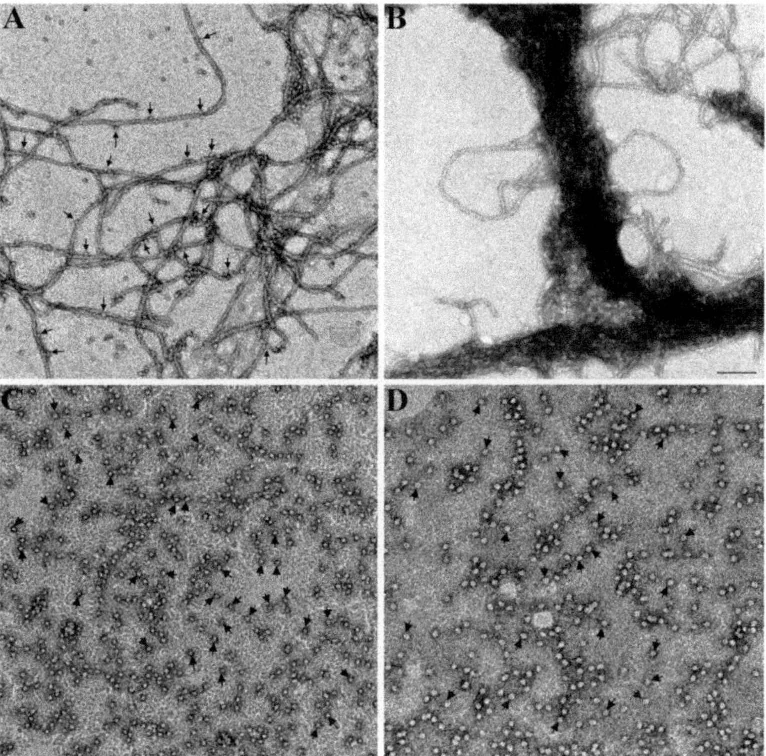

Figure 4 Visualization of wild-type and R120G αB-crystallin binding to desmin filaments *in vitro* by electron microscopy. Desmin was assembled in the presence of either wild-type (A) or R120G αB-crystallin (B) at 37 °C for 1 h. After assembly, protein samples were negatively stained with 1% (w/v) uranyl acetate and examined by electron microscopy. In the presence of wild-type αB-crystallin, desmin filaments were long and had some αB-crystallin particles attached (A, arrows). In contrast, filaments coassembled in the presence of R120G αB-crystallin (B) were clumped. Note that the increased binding of R120G αB-crystallin leading to a very extensive coating of desmin filaments was clearly seen. In the absence of desmin, both wild-type (C) and R120G αB-crystallin (D) formed discrete particles (arrows). The average diameter is 11–13 nm for wild-type αB-crystallin and 14–16 nm for R120G αB-crystallin. All micrographs are at the same magnification. Bar, 200 nm.

can quickly provide reliable structural information to very high resolutions (Harris, 1999).

1. Protein samples were adsorbed for 2 min onto Formvar/carbon-coated grids (Ted Pella Inc.) that were glow discharged before the application of 10 μL sample at a concentration of 0.1 mg/ml.

2. Excess protein solution was removed by wicking the edge of the grid on a piece of blotting paper.
3. Samples were negatively stained for 1 min using 20 μl 1% (w/v) uranyl acetate and examined with an Hitachi H-7600 TEM (Hitachi High-Technologies, Tokyo, Japan), using an accelerating voltage of 100 kV.
4. Images were acquired using a CCD camera (Advanced Microscopy Techniques, Danvers, MA) before being further processed in Adobe Photoshop CS (Adobe Systems, San Jose, CA).
5. The diameters of the assembled filaments and the sHSP particles were measured on enlarged electron micrographs using ImageJ software (National Institutes of Health).

   Note: Glow discharging the grids for 1–2 min in a vacuum evaporator before adding the protein sample improves sample analysis.

ACKNOWLEDGMENTS

The financial support of the Wellcome Trust (R.A.Q., M.D.P., Grant no. 46747), the COST action BM1002 (Travel fellowship to J.L.E.), the Leverhulme Trust (Research Leave Fellowship and Project Grant to R.A.Q.), Fight for Sight (Project Grant 1763 to R.A.Q.), and the National Science Council (M.D.P., Grant no. 98-2311-B-007-007-MY3) is gratefully acknowledged. We thank previous laboratory members Paul van den Ijssel, Aileen M. Hutcheson, Terry Gibbons, Andrew Landsbury, and Antal Tapodi as well as previous collaborators for their contributions to this chapter. E. Peter M. Candido (Department of Biochemistry and Molecular Biology, University of British Columbia, Vancouver, Canada) is thanked for providing the HSP16.2 cDNA and the polyclonal antibody to this sHSP. We thank Prof. Dr. H. Herrmann (German Cancer Research Center, Heidelberg, Germany) for desmin cDNA and advice on the viscosity assays.

REFERENCES

Acunzo, J., Katsogiannou, M., & Rocchi, P. (2012). Small heat shock proteins HSP27 (HspB1), alphaB-crystallin (HspB5) and HSP22 (HspB8) as regulators of cell death. *The International Journal of Biochemistry & Cell Biology*, *44*, 1622–1631.

Alam, C. M., Silvander, J. S., Daniel, E. N., Tao, G. Z., Kvarnstrom, S. M., Alam, P., et al. (2013). Keratin 8 modulates beta-cell stress responses and normoglycaemia. *Journal of Cell Science*, *126*, 5635–5644.

Balogh, J., Li, Z., Paulin, D., & Arner, A. (2005). Desmin filaments influence myofilament spacing and lateral compliance of slow skeletal muscle fibers. *Biophysical Journal*, *88*, 1156–1165.

Basha, E., O'Neill, H., & Vierling, E. (2012). Small heat shock proteins and alpha-crystallins: Dynamic proteins with flexible functions. *Trends in Biochemical Sciences*, *37*, 106–117.

Bjorkdahl, C., Sjogren, M. J., Zhou, X., Concha, H., Avila, J., Winblad, B., et al. (2008). Small heat shock proteins Hsp27 or alphaB-crystallin and the protein components of neurofibrillary tangles: Tau and neurofilaments. *Journal of Neuroscience Research*, *86*, 1343–1352.

Boelens, W. C., Croes, Y., & de Jong, W. W. (2001). Interaction between alphaB-crystallin and the human 20S proteasomal subunit C8/alpha7. *Biochimica et Biophysica Acta*, *1544*, 311–319.

Boriek, A. M., Capetanaki, Y., Hwang, W., Officer, T., Badshah, M., Rodarte, J., et al. (2001). Desmin integrates the three-dimensional mechanical properties of muscles. *American Journal of Physiology. Cell Physiology*, *280*, C46–C52.

Brewer, A. C., Mustafi, S. B., Murray, T. V., Rajasekaran, N. S., & Benjamin, I. J. (2013). Reductive stress linked to small HSPs, G6PD, and Nrf2 pathways in heart disease. *Antioxidants & Redox Signaling*, *18*, 1114–1127.

Capetanaki, Y., Papathanasiou, S., Diokmetzidou, A., Vatsellas, G., & Tsikitis, M. (2015). Desmin related disease: A matter of cell survival failure. *Current Opinion in Cell Biology*, *32*, 113–120.

Carlsson, L., Li, Z. L., Paulin, D., Price, M. G., Breckler, J., Robson, R. M., et al. (2000). Differences in the distribution of synemin, paranemin, and plectin in skeletal muscles of wild-type and desmin knock-out mice. *Histochemistry and Cell Biology*, *114*, 39–47.

Chourbagi, O., Bruston, F., Carinci, M., Xue, Z., Vicart, P., Paulin, D., et al. (2011). Desmin mutations in the terminal consensus motif prevent synemin-desmin heteropolymer filament assembly. *Experimental Cell Research*, *317*, 886–897.

Cox, D., Carver, J. A., & Ecroyd, H. (2014). Preventing alpha-synuclein aggregation: The role of the small heat-shock molecular chaperone proteins. *Biochimica et Biophysica Acta*, *1842*, 1830–1843.

Dabir, D. V., Trojanowski, J. Q., Richter-Landsberg, C., Lee, V. M., & Forman, M. S. (2004). Expression of the small heat-shock protein alphaB-crystallin in tauopathies with glial pathology. *The American Journal of Pathology*, *164*, 155–166.

den Engelsman, J., Keijsers, V., de Jong, W. W., & Boelens, W. C. (2003). The small heat-shock protein alpha B-crystallin promotes FBX4-dependent ubiquitination. *The Journal of Biological Chemistry*, *278*, 4699–4704.

Eliasson, C., Sahlgren, C., Berthold, C. H., Stakeberg, J., Celis, J. E., Betsholtz, C., et al. (1999). Intermediate filament protein partnership in astrocytes. *The Journal of Biological Chemistry*, *274*, 23996–24006.

Elliott, J. L., Der Perng, M., Prescott, A. R., Jansen, K. A., Koenderink, G. H., & Quinlan, R. A. (2013). The specificity of the interaction between alphaB-crystallin and desmin filaments and its impact on filament aggregation and cell viability. *Philosophical Transactions of the Royal Society of London. Series B, Biological Sciences*, *368*, 20120375.

Favre, B., Schneider, Y., Lingasamy, P., Bouameur, J. E., Begre, N., Gontier, Y., et al. (2011). Plectin interacts with the rod domain of type III intermediate filament proteins desmin and vimentin. *European Journal of Cell Biology*, *90*, 390–400.

Garcia-Pelagio, K. P., Muriel, J., O'Neill, A., Desmond, P. F., Lovering, R. M., Lund, L., et al. (2015). Myopathic changes in murine skeletal muscle lacking synemin. *American Journal of Physiology. Cell Physiology*, *308*, C448–C462.

Goldfarb, L. G., Park, K. Y., Cervenakova, L., Gorokhova, S., Lee, H. S., Vasconcelos, O., et al. (1998). Missense mutations in desmin associated with familial cardiac and skeletal myopathy. *Nature Genetics*, *19*, 402–403.

Goldman, R. D., Grin, B., Mendez, M. G., & Kuczmarski, E. R. (2008). Intermediate filaments: Versatile building blocks of cell structure. *Current Opinion in Cell Biology*, *20*, 28–34.

Green, K. J., Bohringer, M., Gocken, T., & Jones, J. C. (2005). Intermediate filament associated proteins. *Advances in Protein Chemistry*, *70*, 143–202.

Harris, J. R. (1999). Negative staining of thinly spread biological particulates. *Methods in Molecular Biology*, *117*, 13–30.

Herrmann, H., & Aebi, U. (2000). Intermediate filaments and their associates: Multi-talented structural elements specifying cytoarchitecture and cytodynamics. *Current Opinion in Cell Biology*, *12*, 79–90.

Herrmann, H., & Aebi, U. (2004). Intermediate filaments: Molecular structure, assembly mechanism, and integration into functionally distinct intracellular scaffolds. *Annual Review of Biochemistry*, *73*, 749–789.

Herrmann, H., Bar, H., Kreplak, L., Strelkov, S. V., & Aebi, U. (2007). Intermediate filaments: From cell architecture to nanomechanics. *Nature Reviews. Molecular Cell Biology*, *8*, 562–573.

Herrmann, H., Strelkov, S. V., Burkhard, P., & Aebi, U. (2009). Intermediate filaments: Primary determinants of cell architecture and plasticity. *The Journal of Clinical Investigation*, *119*, 1772–1783.

Hesse, M., Magin, T. M., & Weber, K. (2001). Genes for intermediate filament proteins and the draft sequence of the human genome: Novel keratin genes and a surprisingly high number of pseudogenes related to keratin genes 8 and 18. *Journal of Cell Science*, *114*, 2569–2575.

Hnia, K., Tronchere, H., Tomczak, K. K., Amoasii, L., Schultz, P., Beggs, A. H., et al. (2011). Myotubularin controls desmin intermediate filament architecture and mitochondrial dynamics in human and mouse skeletal muscle. *The Journal of Clinical Investigation*, *121*, 70–85.

Hofmann, I. (1998). Measuring the assembly kinetics and binding properties of intermediate filament proteins. *Sub-Cellular Biochemistry*, *31*, 363–380.

Hu, W. F., Gong, L., Cao, Z., Ma, H., Ji, W., Deng, M., et al. (2012). AlphaA- and alphaB-crystallins interact with caspase-3 and Bax to guard mouse lens development. *Current Molecular Medicine*, *12*, 177–187.

Huiatt, T. W., Robson, R. M., Arakawa, N., & Stromer, M. H. (1980). Desmin from avian smooth muscle. Purification and partial characterization. *The Journal of Biological Chemistry*, *255*, 6981–6989.

Joubert, R., Vignaud, A., Le, M., Moal, C., Messaddeq, N., & Buj-Bello, A. (2013). Site-specific Mtm1 mutagenesis by an AAV-Cre vector reveals that myotubularin is essential in adult muscle. *Human Molecular Genetics*, *22*, 1856–1866.

Kappe, G., Franck, E., Verschuure, P., Boelens, W. C., Leunissen, J. A., & de Jong, W. W. (2003). The human genome encodes 10 alpha-crystallin-related small heat shock proteins: HspB1-10. *Cell Stress & Chaperones*, *8*, 53–61.

Kemp, M. W., Edwards, B., Burgess, M., Clarke, W. T., Nicholson, G., Parry, D. A., et al. (2009). Syncoilin isoform organization and differential expression in murine striated muscle. *Journal of Structural Biology*, *165*, 196–203.

Kim, S., & Coulombe, P. A. (2007). Intermediate filament scaffolds fulfill mechanical, organizational, and signaling functions in the cytoplasm. *Genes & Development*, *21*, 1581–1597.

Kim, S., Wong, P., & Coulombe, P. A. (2006). A keratin cytoskeletal protein regulates protein synthesis and epithelial cell growth. *Nature*, *441*, 362–365.

Landsbury, A., Perng, M. D., Pohl, E., & Quinlan, R. A. (2010). Functional symbiosis between the intermediate filament cytoskeleton and small heat shock proteins. In S. Simon & A. P. Arigo (Eds.), *Small stress proteins and human diseases* (pp. 55–87). Hauppauge, New York: Nova Science.

Lazarides, E. (1980). Intermediate filaments as mechanical integrators of cellular space. *Nature*, *283*, 249–256.

Leduc, C., & Etienne-Manneville, S. (2015). Intermediate filaments in cell migration and invasion: The unusual suspects. *Current Opinion in Cell Biology*, *32C*, 102–112.

Li, M., Andersson-Lendahl, M., Sejersen, T., & Arner, A. (2013). Knockdown of desmin in zebrafish larvae affects interfilament spacing and mechanical properties of skeletal muscle. *The Journal of General Physiology*, *141*, 335–345.

Li, H., Guo, Y., Teng, J., Ding, M., Yu, A. C., & Chen, J. (2006). 14-3-3gamma affects dynamics and integrity of glial filaments by binding to phosphorylated GFAP. *Journal of Cell Science*, *119*, 4452–4461.

Li, Z., Parlakian, A., Coletti, D., Alonso-Martin, S., Hourde, C., Joanne, P., et al. (2014). Synemin acts as a regulator of signalling molecules during skeletal muscle hypertrophy. *Journal of Cell Science*, *127*, 4589–4601.

Liu, J. P., Schlosser, R., Ma, W. Y., Dong, Z., Feng, H., Liu, L., et al. (2004). Human alphaA- and alphaB-crystallins prevent UVA-induced apoptosis through regulation of PKCalpha, RAF/MEK/ERK and AKT signaling pathways. *Experimental Eye Research, 79*, 393–403.

Lund, L. M., Kerr, J. P., Lupinetti, J., Zhang, Y., Russell, M. A., Bloch, R. J., et al. (2012). Synemin isoforms differentially organize cell junctions and desmin filaments in neonatal cardiomyocytes. *The FASEB Journal, 26*, 137–148.

MacLean-Fletcher, S., & Pollard, T. D. (1980). Identification of a factor in conventional muscle actin preparations which inhibits actin filament self-association. *Biochemical and Biophysical Research Communications, 96*, 18–27.

McCullagh, K. J., Edwards, B., Kemp, M. W., Giles, L. C., Burgess, M., & Davies, K. E. (2008). Analysis of skeletal muscle function in the C57BL6/SV129 syncoilin knockout mouse. *Mammalian Genome: Official Journal of the International Mammalian Genome Society, 19*, 339–351.

Milner, D. J., Weitzer, G., Tran, D., Bradley, A., & Capetanaki, Y. (1996). Disruption of muscle architecture and myocardial degeneration in mice lacking desmin. *The Journal of Cell Biology, 134*, 1255–1270.

Morrow, G., & Tanguay, R. M. (2012). Small heat shock protein expression and functions during development. *The International Journal of Biochemistry & Cell Biology, 44*, 1613–1621.

Munoz-Marmol, A. M., Strasser, G., Isamat, M., Coulombe, P. A., Yang, Y., Roca, X., et al. (1998). A dysfunctional desmin mutation in a patient with severe generalized myopathy. *Proceedings of the National Academy of Sciences of the United States of America, 95*, 11312–11317.

Nicholl, I. D., & Quinlan, R. A. (1994). Chaperone activity of alpha-crystallins modulates intermediate filament assembly. *The EMBO Journal, 13*, 945–953.

Nieminen, M., Henttinen, T., Merinen, M., Marttila-Ichihara, F., Eriksson, J. E., & Jalkanen, S. (2006). Vimentin function in lymphocyte adhesion and transcellular migration. *Nature Cell Biology, 8*, 156–162.

Pallari, H. M., & Eriksson, J. E. (2006). Intermediate filaments as signaling platforms. *Science's STKE, 2006*, 53.

Parry, D. A. (2005). Microdissection of the sequence and structure of intermediate filament chains. *Advances in Protein Chemistry, 70*, 113–142.

Perng, M. D., Cairns, L., van den, I. P., Prescott, A., Hutcheson, A. M., & Quinlan, R. A. (1999). Intermediate filament interactions can be altered by HSP27 and alphaB-crystallin. *Journal of Cell Science, 112*(Pt. 13), 2099–2112.

Perng, M. D., & Quinlan, R. A. (2015). The dynamic duo of small heat proteins and Ifs maintain cell homeostasis, resist cellular stress and enable evolution in cells and tissues. In R. M. Tanguay & L. E. Hightower (Eds.), *The big book of small heat shock proteins* (pp. 401–434). Springer International Publishing AG.

Perng, M. D., Sandilands, A., Kuszak, J., Dahm, R., Wegener, A., Prescott, A. R., et al. (2004). The intermediate filament systems in the eye lens. *Methods in Cell Biology, 78*, 597–624.

Perng, M. D., Wen, S. F., Gibbon, T., Middeldorp, J., Sluijs, J., Hol, E. M., et al. (2008). Glial fibrillary acidic protein filaments can tolerate the incorporation of assembly-compromised GFAP-delta, but with consequences for filament organization and alphaB-crystallin association. *Molecular Biology of the Cell, 19*, 4521–4533.

Perng, M. D., Wen, S. F., van den, I. P., Prescott, A. R., & Quinlan, R. A. (2004). Desmin aggregate formation by R120G alphaB-crystallin is caused by altered filament interactions and is dependent upon network status in cells. *Molecular Biology of the Cell, 15*, 2335–2346.

Poon, E., Howman, E. V., Newey, S. E., & Davies, K. E. (2002). Association of syncoilin and desmin: Linking intermediate filament proteins to the dystrophin-associated protein complex. *The Journal of Biological Chemistry, 277*, 3433–3439.

Quinlan, R. (2002). Cytoskeletal competence requires protein chaperones. *Progress in Molecular and Subcellular Biology*, *28*, 219–233.

Quinlan, R. A., & Ellis, R. J. (2013). Chaperones: Needed for both the good times and the bad times. *Philosophical Transactions of the Royal Society of London. Series B, Biological Sciences*, *368*, 20130091.

Quinlan, R., & Van Den Ijssel, P. (1999). Fatal attraction: When chaperone turns harlot. *Nature Medicine*, *5*, 25–26.

Roelofs, R. F., Fischer, D. F., Houtman, S. H., Sluijs, J. A., Van Haren, W., Van Leeuwen, F. W., et al. (2005). Adult human subventricular, subgranular, and subpial zones contain astrocytes with a specialized intermediate filament cytoskeleton. *Glia*, *52*, 289–300.

Satoh, J., Yamamura, T., & Arima, K. (2004). The 14-3-3 protein epsilon isoform expressed in reactive astrocytes in demyelinating lesions of multiple sclerosis binds to vimentin and glial fibrillary acidic protein in cultured human astrocytes. *The American Journal of Pathology*, *165*, 577–592.

Schroder, R., Warlo, I., Herrmann, H., van der Ven, P. F., Klasen, C., Blumcke, I., et al. (1999). Immunogold EM reveals a close association of plectin and the desmin cytoskeleton in human skeletal muscle. *European Journal of Cell Biology*, *78*, 288–295.

Schweitzer, S. C., Klymkowsky, M. W., Bellin, R. M., Robson, R. M., Capetanaki, Y., & Evans, R. M. (2001). Paranemin and the organization of desmin filament networks. *Journal of Cell Science*, *114*, 1079–1089.

Selcen, D. (2011). Myofibrillar myopathies. *Neuromuscular Disorders*, *21*, 161–171.

Seltmann, K., Fritsch, A. W., Kas, J. A., & Magin, T. M. (2013). Keratins significantly contribute to cell stiffness and impact invasive behavior. *Proceedings of the National Academy of Sciences of the United States of America*, *110*, 18507–18512.

Shah, S. B., Love, J. M., O'Neill, A., Lovering, R. M., & Bloch, R. J. (2012). Influences of desmin and keratin 19 on passive biomechanical properties of mouse skeletal muscle. *Journal of Biomedicine & Biotechnology*, *2012*, 704061.

Sihag, R. K., Inagaki, M., Yamaguchi, T., Shea, T. B., & Pant, H. C. (2007). Role of phosphorylation on the structural dynamics and function of types III and IV intermediate filaments. *Experimental Cell Research*, *313*, 2098–2109.

Snider, N. T., & Omary, M. B. (2014). Post-translational modifications of intermediate filament proteins: Mechanisms and functions. *Nature Reviews. Molecular Cell Biology*, *15*, 163–177.

Tabor, S., & Richardson, C. C. (1985). A bacteriophage T7 RNA polymerase/promoter system for controlled exclusive expression of specific genes. *Proceedings of the National Academy of Sciences of the United States of America*, *82*, 1074–1078.

Toivola, D. M., Tao, G. Z., Habtezion, A., Liao, J., & Omary, M. B. (2005). Cellular integrity plus: Organelle-related and protein-targeting functions of intermediate filaments. *Trends in Cell Biology*, *15*, 608–617.

Treweek, T. M., Meehan, S., Ecroyd, H., & Carver, J. A. (2015). Small heat-shock proteins: Important players in regulating cellular proteostasis. *Cellular and Molecular Life Sciences*, *72*, 429–451.

Tsoupri, E., & Capetanaki, Y. (2013). Myospryn: A multifunctional desmin-associated protein. *Histochemistry and Cell Biology*, *140*, 55–63.

van Oosten-Hawle, P., & Morimoto, R. I. (2014). Organismal proteostasis: Role of cell-nonautonomous regulation and transcellular chaperone signaling. *Genes & Development*, *28*, 1533–1543.

Vicart, P., Caron, A., Guicheney, P., Li, Z., Prevost, M. C., Faure, A., et al. (1998). A missense mutation in the alphaB-crystallin chaperone gene causes a desmin-related myopathy. *Nature Genetics*, *20*, 92–95.

Wettstein, G., Bellaye, P. S., Micheau, O., & Bonniaud, P. (2012). Small heat shock proteins and the cytoskeleton: An essential interplay for cell integrity? *The International Journal of Biochemistry & Cell Biology, 44*, 1680–1686.

Winter, L., & Wiche, G. (2013). The many faces of plectin and plectinopathies: Pathology and mechanisms. *Acta Neuropathologica, 125*, 77–93.

Wisniewski, T., & Goldman, J. E. (1998). Alpha B-crystallin is associated with intermediate filaments in astrocytoma cells. *Neurochemical Research, 23*, 385–392.

Xue, Z. G., Cheraud, Y., Brocheriou, V., Izmiryan, A., Titeux, M., Paulin, D., et al. (2004). The mouse synemin gene encodes three intermediate filament proteins generated by alternative exon usage and different open reading frames. *Experimental Cell Research, 298*, 431–444.

Zhang, J., Bang, M. L., Gokhin, D. S., Lu, Y., Cui, L., Li, X., et al. (2008). Syncoilin is required for generating maximum isometric stress in skeletal muscle but dispensable for muscle cytoarchitecture. *American Journal of Physiology. Cell Physiology, 294*, C1175–C1182.

CHAPTER TEN

Purification and Structural Analysis of Plectin and BPAG1e

José A. Manso*, Inés García Rubio[†], María Gómez-Hernández*, Esther Ortega*, Rubén M. Buey*,[‡], Ana M. Carballido*, Arturo Carabias*, Noelia Alonso-García*, José M. de Pereda*,[1]

*Instituto de Biología Molecular y Celular del Cáncer, Consejo Superior de Investigaciones Científicas, University of Salamanca, Salamanca, Spain
[†]Centro Universitario de la Defensa, Academia General Militar, Zaragoza, Spain
[‡]Metabolic Engineering Group, Department of Microbiology and Genetics, University of Salamanca, Salamanca, Spain
[1]Corresponding author: e-mail address: pereda@usal.es

Contents

Abstract

Plectin and BPAG1e belong to the plakin family of high-molecular-weight proteins that interconnect the cytoskeletal systems and anchor them to junctional complexes. Plectin and BPAG1e are prototypical plakins with a similar tripartite modular structure. The N- and C-terminal regions are built of multiple discrete structural domains, while the central rod domain mediates dimerization by coiled-coil interactions. Owing to the mosaic organization of plakins, the structure of their constituent individual domains or small multi-domain segments can be analyzed isolated. Yet, understanding the integrated function of large regions, oligomers, and heterocomplexes of plakins is difficult due to the large and segmented structure. Here, we describe methods for the production of plectin and BPAG1e samples suitable for structural and biophysical

Methods in Enzymology, Volume 569
ISSN 0076-6879
http://dx.doi.org/10.1016/bs.mie.2015.05.002

analysis. In addition, we discuss the combination of hybrid methods that yield information at several resolution levels to study the complex, multi-domain, and flexible structure of plakins.

1. INTRODUCTION

Plectin and the epithelial bullous pemphigoid antigen 1 (BPAG1e or BP230) belong to the plakin family of cytolinkers that interconnect the intermediate filaments, microtubules, and actin microfilaments and tether them to junctional complexes (Bouameur, Favre, & Borradori, 2014). In (pseudo-) stratified epithelia, plectin and BPAG1e form part of hemidesmosomes where they link the integrin α6β4 to the cytokeratins. Mammalian plakins also include desmoplakin, microtubule actin crosslinking factor 1 (MACF1/ACF7), envoplakin, periplakin, and epiplakin.

Plakins are large proteins that have a mosaic structure composed by multiple structural domains. Plectin, BPAG1e, desmoplakin, envoplakin, and periplakin have a three-segment structure in which the N- and C-terminal regions are separated by a central fibrous rod domain that mediates dimerization through coiled-coil interactions (Fig. 1).

The N-terminal segment corresponds mostly to the plakin domain, which is conserved among most plakins. The plakin domain contains up to nine spectrin repeats (SR1–SR9). This was initially illustrated by the structures of the SR3–SR4 of BPAG1 (Jefferson, Ciatto, Shapiro, & Liem, 2007) and the SR1–SR2 of plectin (Sonnenberg, Rojas, & de Pereda, 2007). The SR fold consists of three α-helices (A–B–C) that form a left-handed bundle. In tandem arrays of SRs, such as the plakin domain, helix C of a repeat is fused to helix A of the downstream domain. As a consequence, the plakin domain has an elongated shape. The plakin domain also contains a SH3 (Src homology-3) domain inserted in between helices B and C of the SR5 (Choi & Weis, 2011; Ortega, Buey, Sonnenberg, & de Pereda, 2011). Upstream of the plakin domain, plectin, MACF1, and the non-epithelial variants of BPAG1 contain an F-actin-binding domain (ABD) formed by two calponin homology domains (CH1 and CH2), which is similar to ABDs found in the spectrin superfamily. BPAG1e lacks the ABD and SR1 domains.

The C-terminal regions of plectin and BPAG1e contain six and two plakin repeat domains (PRDs), respectively, which mediate the interaction with intermediate filaments. The PRD fold consists of 4.5 copies of a

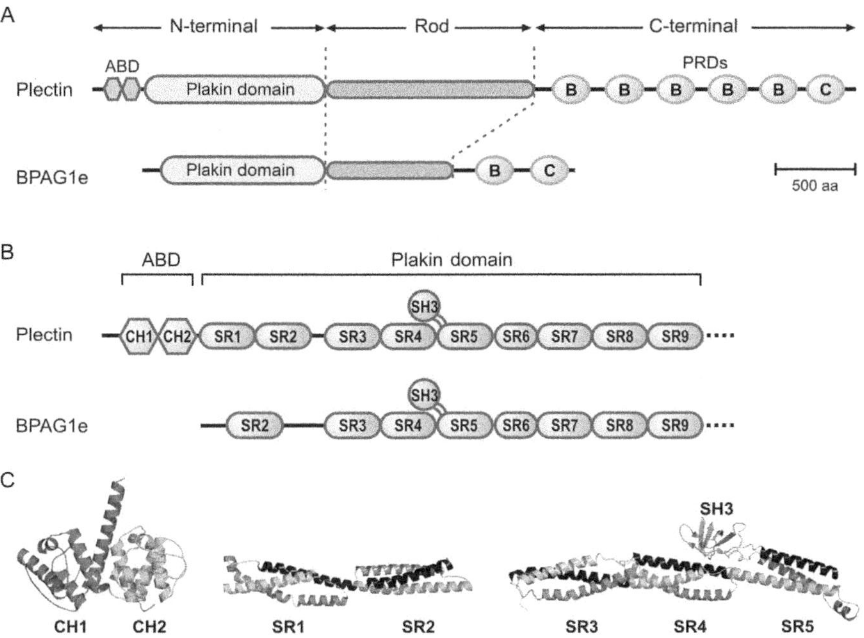

Figure 1 Structure of plectin and BPAG1e. (A) Overall domain organization. (B) Detailed domain structure of the N-terminal regions of plectin and BPAG1e. (C) Structures of the ABD (PDB entry 1MB8) and SR1–SR2 (2ODU) of plectin, and composite 3D model of the SR3–SR5 region constructed using the crystal structures of the SR3–SR4 (3PDY) and SR4–SR5–SH3 (3PE0). (See the color plate.)

38-amino acid repeat motif, each forming a β-hairpin and two antiparallel helices (Choi, Park-Snyder, Pascoe, Green, & Weis, 2002).

Since the last decade, there have been significant advances to understand the structural basis of the contribution of plakins to the cellular architecture. Achievements include detailed structural characterization of individual domains and of protein–protein complexes. This chapter summarizes some methods for the structural analysis of plectin and BPAG1e, focusing in specific techniques useful to analyze these large multi-domain proteins.

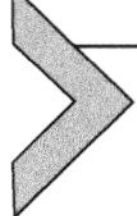

2. PURIFICATION OF FRAGMENTS OF PLECTIN AND BPAG1e FOR STRUCTURAL ANALYSIS

2.1 Design of Fragments

The modular organization of plakins allows the structural and functional analysis of individual domains or small multi-domain segments. Whenever possible, it is advisable to produce fragments that do not contain long additional sequences at the N- or C-termini of the structural domains of interest, provided that such tails do not have a functional role. Inclusion of incomplete regions of adjacent domains might reduce the stability and solubility. In addition, disordered tails might interfere with crystallization.

Special attention should be taken in the design of fragments of the plakin domain. There are well-defined transitions between repeats in the pairs SR1–SR2 and SR4–SR5. On the other hand, the helical linkers in the pairs SR3–SR4 and SR5–SR6 are shared by the anterior and posterior repeats; hence, the limits of the two domains partially overlap. The boundaries of the domains in the N-terminal regions of plectin and BPAG1e are listed in Table 1 as a guide.

2.2 Protein Overexpression in Bacteria and Purification

Fragments of the N-terminal region of plectin and BPAG1e are amiable for expression in *E. coli*. Large amounts of these proteins can be expressed soluble employing the T7 promoter-based pET system. We have extensively used a derivative of the pET15b vector, which codes for an octa-His N-terminal tag followed by a sequence recognized by the tobacco etch virus

Table 1 Domain Boundaries of the N-Terminal Region of Human Plectin and BPAG1e

	hPlectin[a]	**hBPAG1e**[b]
ABD	68–293	n/a
SR1	303–419	n/a
SR2	420–520	59–159
SR3	545–650	273–380
SR4	644–749	374–479
SR5	750–921	480–651
SH3	820–886	550–616
SR6	919–1001	649–731
SR7	1004–1117	734–847
SR8	1118–1233	848–963
SR9	1234–1372	964–1103

[a]Uniprot Q15149-2.
[b]Uniprot Q03001-3.

(TEV) protease, and includes *Nde*I, *Xho*I, and *Bam*HI sites for cloning (Alonso-García, Ingles-Prieto, Sonnenberg, & De Pereda, 2009).

Materials and methods for the expression of proteins in *E. coli*

- Terrific Broth (TB) bacterial growth medium (Tartoff & Hobbs, 1987).
- Binding buffer: 20 m*M* Tris–HCl, 5 m*M* imidazole, 500 m*M* NaCl, pH 7.9.

1. Grow *E. coli* BL21(DE3) carrying the desired plasmid in TB medium with 100 μg/ml ampicillin. Typically, 1.5 l of culture yield multi-milligram amounts of purified protein. Incubate at 37 °C until the OD_{600} reaches 0.6–0.8. Cultures should not grow to saturation at any time (e.g., do not use overnight precultures).
2. Induce the expression of the target protein by adding 0.2 m*M* isopropyl β-D-1-thiogalactopyranoside (IPTG) and incubate 3 h at 37 °C. Alternatively, chill the cultures to 15 °C, add 0.2 m*M* IPTG, and incubate shaking at 15 °C overnight. Expression at low temperature frequently increases the presence of target protein in the soluble fraction.
3. Harvest the bacteria by centrifugation at 12,000 × *g* at 4 °C for 20 min; discard the medium.
4. Resuspend the pellets in 60 ml binding buffer per liter of bacterial culture. The bacterial slurry can be stored at −80 °C.

Materials and methods for the purification of His-tagged proteins

- Elution buffer: 20 m*M* Tris–HCl, 500 m*M* imidazol, 500 m*M* NaCl, pH 7.9.
- Buffer A: 20 m*M* Tris–HCl, 150 m*M* NaCl, pH 7.5.
- HisTrap FF chelating column 5 ml (GE Healthcare cat no. 17-5255-01).
- Chromatography system (e.g., ÄKTA Prime, GE Healthcare).
- Dialysis tubing.
- Recombinant His-tagged TEV protease (His-rTEV). A plasmid for expression of a His-rTEV in *E. coli* (pRK793) can be obtained from AddGene (http://www.addgene.org/). The purification of His-rTEV is described elsewhere (Tropea, Cherry, & Waugh, 2009).

1. Add 1 m*M* phenylmethylsulfonyl fluoride to the bacterial slurry.
2. Lyse the bacteria by sonication on ice. For example, employ a Vibra-Cell 500 W ultrasonic processor (Sonics & Materials, Inc.) at 25% amplitude. Use pulses of 50 s for 6 min.
3. Remove cell debris by centrifugation at 40,000 × *g* at 4 °C for 30 min.
4. Collect the supernatant (Sn) and sonicate it on ice for 1 min to fragment remaining DNA.

5. Immobilized-metal ion affinity chromatography (IMAC). Apply the Sn to the HisTrap column equilibrated in binding buffer. This can be done using a peristaltic pump.
6. Connect the column to the chromatography system and run the following method collecting fractions:
   - Wash with 8 column volumes (CV) of 100% binding buffer.
   - Wash with 5 CV of 91% binding buffer, 9% elution buffer.
   - Elute the target protein with a 13 CV gradient from 9% to 100% elution buffer; and then 4 CV of elution buffer.
7. Analyze the fractions by sodium dodecyl sulfate polyacrylamide gel electrophoresis (SDS-PAGE). Pool the positive fractions that contain the target protein and take a sample for further SDS-PAGE analysis.
8. Add 0.1–0.5 mg of His-rTEV and transfer the sample to a dialysis tube. Dialyze against 1 l of buffer A supplemented with 0.1 m*M* dithiothreitol (DTT) and 0.1 m*M* EDTA, for 4–5 h at room temperature. Proteolytic digestion occurs during the buffer exchange. Analyze the samples before and after digestion by SDS-PAGE to confirm the cleavage.
9. Dialyze overnight against 1 l of buffer A at 4 °C.
10. If the sample shows precipitation, clarify by centrifugation at 40,000 × *g* at 4 °C for 30 min.
11. Remove the His-rTEV and traces of undigested protein with a second IMAC. Apply the sample to the HisTrap column equilibrated in binding buffer. Collect the flow-through, which may contain the target protein. Wash the column with 5 CV of binding buffer, 5 CV of 91% binding, 9% elution buffer, and 5 CV of elution buffer. Collect fractions of the washing steps. Frequently, target proteins bind weakly to the column and elute during the first wash. Analyze the flow-through and the fractions by SDS-PAGE. Pool the positive fractions.
12. If necessary, further purify the sample by size-exclusion chromatography (SEC).
    - Concentrate the sample to ≤5 ml and clarify by centrifugation at 16,000 × *g* at 4 °C for 30 min.
    - Load in a HiPrep Sephacryl S300 26/60 column (GE Healthcare cat no. 17-1196-01) equilibrated in the desired buffer. Run at 0.27 cm/min linear flow. Collect 4 ml fractions.
    - Analyze the fractions by SDS-PAGE and combine those containing the target protein.
13. Dialyze against the desired buffer. This is optional after SEC.

14. Concentrate the protein by ultrafiltration using Amicon stirring cells or centrifugal filter units (Merck Millipore).
15. Remove potential aggregates by centrifugation at 100,000 × *g* 4 °C for 30 min using an Optima TLX ultracentrifuge (Beckman Coulter).
16. For crystallization experiments, it is advisable to use the sample immediately. Alternatively, make ~100 μl aliquots, flash freeze them in liquid nitrogen, and store at −80 °C.

3. STRUCTURAL ANALYSIS OF PLECTIN AND BPAG1e

Owing to the large size, complex domain composition, and potential segmental flexibility of plectin and BPAG1e, their global structures are difficult to characterize by individual conventional methods. Here, we focus on the application of X-ray crystallography, small-angle X-ray scattering (SAXS), and double electron–electron resonance spectroscopy (DEER), and their synergistic combination to analyze the large multi-domain structure of plakins.

3.1 Macromolecular X-ray Crystallography

Macromolecular X-ray crystallography (MX) has proven suitable to elucidate the structure of isolated domains (e.g., ABD) or small multi-domain segments (e.g., fragments of the plakin domain) of plakins. A comprehensive list of the structures of plectin and BPAG1e fragments solved by MX to date is shown in Table 2. In all cases, crystallization was done using conventional vapor diffusion methods.

3.1.1 Phasing of Plectin and BPAG1e Structures

The initial structures of the ABD of plectin were solved by molecular replacement (MR) using the structures of ABDs of spectrins (Garcia-Alvarez et al., 2003; Sevcik et al., 2004). The structures of the SR3–SR4 of BPAG1e, the SR1–SR2 of plectin, and the SR3–SR6 of desmoplakin were experimentally phased using Se-Met derivatives (Choi & Weis, 2011; Jefferson et al., 2007; Sonnenberg et al., 2007). The structure of the SR4–SR5–SH3 of plectin was solved by single isomorphous replacement with anomalous scattering (SIRAS) using a mercurial derivative (Ortega et al., 2011).

The use of experimental phasing methods to solve those structures of the plakin domain illustrates the difficulties in the application of MR using the structures of SRs of closely related proteins, such as spectrins. Nonetheless,

Table 2 Crystal Structures of Plectin and BPAG1e: Crystallization Conditions and Phasing Methods

Protein	Region	PDB	Resolution	Phasing	Crystallization			References
					Prot. conc.	Temp.	Crystallization Solution	
hPlectin	ABD	1MB8	2.15 Å	MR	20 mg/ml	RT	0.1 *M* Tris–HCl (pH 7.0), 15% (v/v) 2-propanol	Garcia-Alvarez, Bobkov, Sonnenberg, and de Pereda (2003)
mPlectin	ABD	1SH5	2.00 Å	MR	20 mg/ml	RT	0.1 *M* Tris–HCl (pH 8.5), 10% (w/v) PEG 8000, 2% (v/v) dioxane	Sevcik, Urbanikova, Kost'an, Janda, and Wiche (2004)
mPlectin	ABD	1SH6	2.00 Å	MR	10 mg/ml	RT	0.1 *M* cacodylate (pH 6.5), 0.2 *M* Ca acetate, 8% (w/v) PEG 8000, 2% (v/v) dioxane	Sevcik et al. (2004)
mPlectin	ABD	4Q59	2.30 Å	MR	n/a	22 °C	0.05 *M* K phosphate (pH 7.5), 20% (w/v) PEG 8000	Song et al. (2015)
hPlectin	SR2–SR3	2ODU 2ODV	2.30 Å 2.05 Å	Se-MAD	7 mg/ml	RT	0.1 *M* citrate–phosphate (pH 4.6), 10% (v/v) 1,2-propanediol, 5% (w/v) PEG 3000, 4% glycerol	Sonnenberg et al. (2007)
hPlectin	SR3–SR4	3PDY	2.22 Å	MR	31 mg/ml	4 °C	0.1 *M* HEPES (pH 7.5), 17% (w/v) PEG 4000, 8% (v/v) 2-propanol	Ortega et al. (2011)
hPlectin	SR4–SR5–SH3	3PE0	2.95 Å	Hg-SIRAS	15 mg/ml	4 °C	0.1 *M* imidazole (pH 7.2), 0.2 *M* Ca acetate, 9% (w/v) PEG 8000, 2 m*M* DTT	Ortega et al. (2011)

hPlectin	Rod (fragment)	4GDO	1.70 Å	Library-MR	25 mg/ml	RT	0.1 *M* Na acetate (pH 4.5), 0.2 *M* Li_2SO_4, 2.3 *M* NaCl.	Sammito et al. (2013)
hPlectin-integrin β4	ABD bound to β4 FnIII-1,2	3F7P	2.75 Å	MR	30 mg/ml	RT	0.1 *M* Na acetate (pH 5.8), 0.2 *M* $CaCl_2$, 10% (w/v) PEG 6000.	de Pereda, Lillo, and Sonnenberg (2009)
mPlectin-integrin β4	ABD bound to β4 FnIII-1,2	4Q58	4.00 Å	MR	12 mg/ml	22 °C	20 m*M* HEPES–NaOH (pH 6.5), 150 m*M* Na formate, 7.5% (w/v) PEG 550 MME, 3% (w/v) sucrose	Song et al. (2015)
mPlectin-calmodulin	ABD bound to calmodulin	4Q57	1.80 Å	MR	11 mg/ml	4 °C	0.1 *M* Bis-Tris (pH 6.5), 0.2 *M* $MgCl_2$, 13% (w/v) PEG 8000	Song et al. (2015)
mBPAG1e	SR3–SR4	2IAK	3.00 Å	Se-SAD	n/a	20 °C	0.05 *M* 3-(cyclohexylamino)-2-hydroxy-1-propanesulfonic acid (pH 9.2), 1.6 *M* $(NH_4)_2SO_4$, 5 m*M* DTT	Jefferson et al. (2007)

the structure of individual SRs and the organization of adjacent repeats are highly conserved between equivalent regions of plakins. Given the increasing repertoire of representative structures of plakins, which currently cover the ABD and the SR1–SR6, it is reasonable that future structures that include these regions could be solved by MR. In fact, MR has been used to solve the structure of the SR3–SR4 of plectin (Ortega et al., 2011), two structures of the ABD of plectin in complex with the integrin β4 (de Pereda et al., 2009; Song et al., 2015), and the ABD of plectin bound to calmodulin (Song et al., 2015).

Elucidation of the structures of other regions of plectin and BPAG1e will probably require the application of experimental phasing methods. *De novo* phasing using Se-Met-labeled proteins and single (SAD) or multiple wavelength anomalous diffraction (MAD) experiments are well-established techniques (Dauter, Dauter, & Dodson, 2002; Hendrickson, 1991) that will not be described here. Methods for the expression of Se-Met-labeled proteins, including the use of nonauxotrophic *E. coli* strains, are described elsewhere (Doublie, 2007).

Alternatively, classical derivatization of crystals with heavy atoms, such as mercury, is very effective for phasing using SIRAS or SAD. We have frequently obtained Hg-derivatives that are isomorphous with the native crystals, by incubating the crystals in the crystallization solution supplemented with 1 m*M* ethylmercurithiosalicylate (EMTS) for a few hours. If soaking degrades the crystal, the concentration of mercurial and the incubation time should be optimized. The excess of EMTS is removed by shortly washing the crystals in an Hg-free solution before data collection. Heavy-atom derivatization does not require metabolic labeling, purification, and crystallization of the labeled proteins. Incorporation of Hg atoms frequently results in large isomorphous differences between native and Hg-derivative datasets. Moreover, Hg has strong anomalous scattering signal at the wavelength of the Cu *Kα* radiation (the f'' term for Hg is ~7.76 e^- at 1.5418 Å), allowing for straightforward phasing using data collected in in-house X-ray generators.

Figure 2 illustrates the phasing of the SR4–SR5–SH3 of plectin (PDB entry 3PE0) using an Hg-derivative. The structure was initially phased by SIRAS (Ortega et al., 2011), but it can be equally solved by SAD using only the anomalous scattering information of the EMTS-derivative dataset. In both cases, phasing was done with the programs SHELXC/D/E (Sheldrick, 2010) (http://shelx.uni-ac.gwdg.de/SHELX/). This example illustrates that when Hg-soaked crystals are nonisomorphous with the native crystals, Hg-SAD is frequently sufficient to phase the structure.

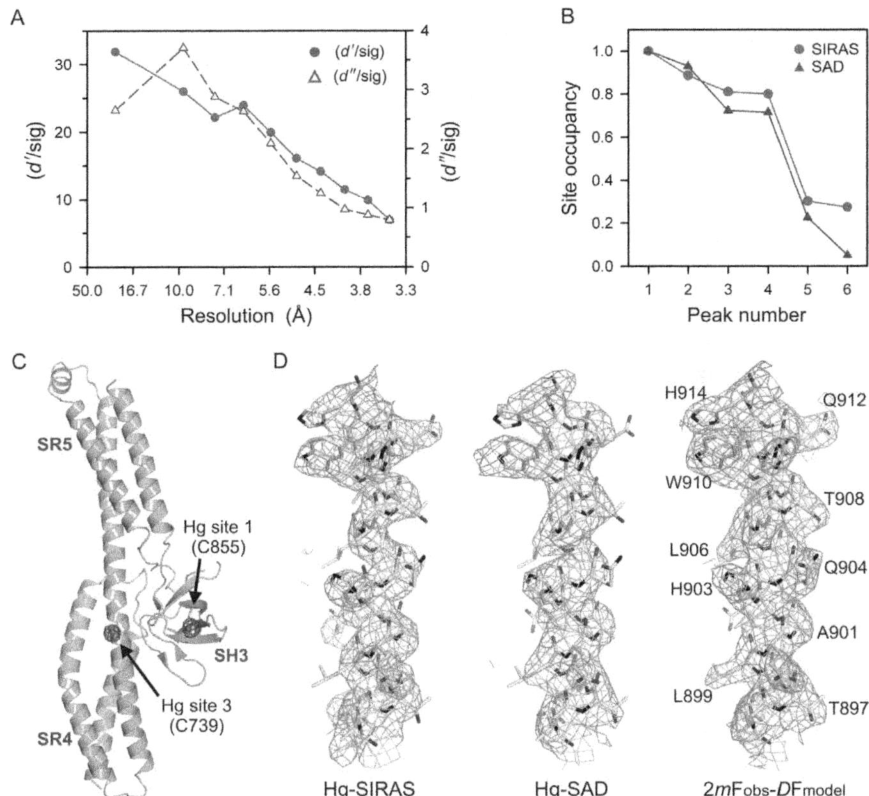

Figure 2 Phasing of the SR4–SR5–SH3 of plectin by Hg-SIRAS and Hg-SAD. (A) Statistics of the isomorphous differences between the EMTS and native datasets (circles) and anomalous signal of the EMTS dataset (triangles) calculated with SHELXC. (B) Occupancies of the Hg sites identified by SIRAS or SAD with SHELXD. The Hg-substructure was solved including data to a maximum resolution of 3.7 Å and 3.9 Å for SIRAS and SAD, respectively. (C) Structure of a plectin molecule (shown as ribbon) and position of the two Hg atoms (sites 1 and 3 in A) attached to cysteines C855 and C739, respectively. The asymmetric unit contains a second plectin molecule with two Hg attached to equivalent positions. A difference anomalous density map calculated using the EMTS-derivative dataset is shown contoured at 5σ. (D) Representative region of the electron density maps calculated after density modification with SHELXE using SIRAS (left) or SAD (center). A $2mF_{obs}$-DF_{model} map calculated using phases of the refined structure is show for comparison (right). Maps are contoured at 1σ. The refined structure is superimposed in all maps. (See the color plate.)

Phasing of fragments of the rod domain should be in principle possible by MR; yet, coiled-coil structures are frequently difficult to tackle by conventional MR. We solved the structure of a 43-amino acid segment of the rod domain of human plectin with the program BORGES that performs MR using libraries of tertiary structure fragments extracted from the PDB (Sammito et al., 2013) (http://chango.ibmb.csic.es). BORGES MR is likely to be useful for phasing other regions of the rod domain. Alternatively, a method for phasing coiled-coil structures based on *de novo* structure prediction might also be applied to solve structures of the rod domains of plakins (Ramisch, Lizatovic, & Andre, 2015).

3.2 Small-Angle X-ray Scattering

SAXS is a powerful method for analyzing the structure of proteins in solution (Petoukhov & Svergun, 2013). SAXS has been applied to elucidate the shape of large multi-domain segments of plakins, which have been elusive for crystallographic analysis so far. SAXS also allows the identification and analysis of segmental flexibility, and the characterization of plectin and BPAG1e oligomers and protein complexes.

3.2.1 Sample Preparation for SAXS

In order to extract useful information from SAXS experiments, it is essential to analyze monodisperse samples. High-molecular-weight aggregates are particularly harmful for the quality of SAXS data due to their large contribution to the scattering. Sample homogeneity can be analyzed by SEC, multi-angle light scattering, dynamic light scattering, or analytical ultracentrifugation.

SAXS is a contrast technique that requires precise subtraction of the contribution of the solvent to the scattering. A practical method to match the solvent is to measure the exactly same buffer used to prepare the protein samples. We routinely equilibrate plectin and BPAG1e proteins for SAXS experiments in 20 m*M* sodium phosphate (pH 7.5), 150 m*M* NaCl, 5% (v/v) glycerol, 3 m*M* DTT by SEC as a final purification step, which also removes possible aggregates. The central fractions of the elution peak are combined and concentrated by ultrafiltration. Glycerol and DTT minimize radiation damage during data collection.

3.2.2 SAXS Data Collection

Methods for SAXS data collection and analysis are described in detail elsewhere (Skou, Gillilan, & Ando, 2014; Svergun, Koch, Timmins, & May, 2013). To accurately restore the size and shape, the data should extend to a minimum value of the scattering vector q ($q=(4\pi\sin\theta)/\lambda$, where 2θ is the scattering angle and λ is the wavelength) that is related to the inverse of the maximum dimension (D_{max}) of the particle,

$q_{min} < \pi / D_{max}$ (Svergun et al., 2013). Since fragments of the plakin domain have an elongated rod-like shape with large D_{max}, the experimental setup should be specifically configured to measure scattering data to very low angle.

3.2.3 Analysis of Segmental Flexibility

SAXS analysis allows the identification of flexibility within multi-domain segments of plakins by inspection of the dimensionless Kratky (Durand et al., 2010) and Porod–Debye (Rambo & Tainer, 2011) plots of the scattering data. Alternatively, flexibility can be analyzed with ensemble fitting to the scattering profiles using methods such as EOM (Tria, Mertens, Kachala, & Svergun, 2015) and MES (Pelikan, Hura, & Hammel, 2009), among others. For a review of the use of SAXS to analyze protein flexibility, see Hammel (2012).

3.2.4 Ab Initio *Shape Reconstruction*

If the sample is monodisperse, low-resolution envelopes can be modeled from the scattering data using *ab initio* methods. Programs for shape reconstruction include DALAI_GA (Chacon, Diaz, Moran, & Andreu, 2000) (http://chaconlab.org/methods/modeling/dalai-ga), DAMMIN (Svergun, 1999), DAMMIF (Franke & Svergun, 2009), and GASBOR (Svergun, Petoukhov, & Koch, 2001) (http://www.embl-hamburg.de/biosaxs/software.html). Details about the fundaments and use of these methods are described in the corresponding publications and Web sites.

3.3 Site-Directed Spin Labeling and Double Electron–Electron Resonance Spectroscopy

DEER (also known as PELDOR) is a pulse electron paramagnetic resonance (EPR) technique sensitive to distances between unpaired spins. Combined with site-directed spin labeling (SDSL), DEER is a powerful method to measure distance distributions between the paramagnetic labels in the range ~10–80 Å in biological macromolecules (Jeschke, 2012). The long-range measurements determined by DEER complement other structural methods and are useful to study protein structure and dynamics, local and large-scale conformational changes, and the structure of homo- and heterocomplexes. Given the modular structure of plakins, DEER can help elucidate the interdomain arrangement and conformational variability.

3.3.1 SDSL and Sample Preparation for DEER

Genetically engineered Cys at selected positions can be labeled with the paramagnetic spin probe (1-Oxyl-2,2,5,5-tetramethyl-Δ3-pyrroline-3-methyl) methanethiosulfonate (MTSL) that carries a thiol-reactive group. As plakins are naturally diamagnetic, the only interaction among paramagnetic centers will be necessarily between artificially introduced probes. Therefore, and depending on whether inter- or intramolecular distances want to be measured, each protein unit should contain one or two Cys amenable to labeling. While reactive Cys in the wild-type sequence can be used for spin labeling, frequently they represent unwanted labeling positions and they have to be substituted by site-directed mutagenesis, see Bachman (2013) for method details.

Spin labels should be introduced at positions where neither the Cys nor the probe distorts the structure of the protein. Conservative mutations on the surface of the protein offer the advantage that their side chains can be frequently replaced without altering the overall native structure and they normally result in high labeling-yields of the solvent-exposed thiol groups. The conformational flexibility of the MTSL reduces the probability of steric hindrance when attached to Cys residues (Jeschke, 2013). Modeling the possible conformations of the probe attached to the protein (see below) helps finding favorable sites. Sites for which only a small number, or strained rotamers can be modeled should be avoided, as labeling could not take place or it could distort the structure. Also, labeling residues in well-ordered secondary structure elements is preferred to sites in loops, where the larger flexibility of the backbone would broaden the inter-spin distance distributions and therefore the uncertainty in the Cys position. Of course, these sites can be targeted if information about flexibility or disorder of such structural element is required.

The relative arrangement of two rigid domains is determined by six parameters, three for the position and three for the orientation. Normally, with the measurement of 12 distance constraints, double the number strictly necessary, the system is considered sufficiently overdetermined. This is necessary to compensate for the fact that the experimental results are error-bound distance distributions of probe conformers that are usually unknown and not exact distances between fixed points. The labeling sites should be spread throughout the protein and the distance measurements should involve a good number of different residues to counterbalance possible errors introduced by the modeling of probe conformers.

Materials and methods for the labeling of plectin and BPAG1e with MTSL

- MTSL, 100 m*M* stock in dimethyl sulfoxide.
- Buffer B: 20 m*M* sodium phosphate (pH 7.5), 150 m*M* NaCl.

1. Prepare a protein solution at a known concentration between 100 and 200 μ*M* in buffer B.
2. Add a 10-fold molar excess of MTSL.
3. Incubate with gentle rotation 1 h at room temperature and then overnight at 4 °C.
4. Remove excess of MTSL by extensive dialysis against buffer B at 4 °C.
5. Centrifuge at 16,000 × *g* at 4 °C for 30 min to remove possible aggregates.
6. Measure the protein concentration.
7. Determine the degree of labeling by measuring the number of free thiols per protein molecule before and after MTSL labeling. Titrate the thiol groups under denaturing conditions using 5,5′-dithio-bis(2-nitrobenzoic acid) as described by Garcia-Alvarez et al. (2003). Effective MTSL labeling should block all the sulfhydryls. The spin-labeling efficiency can also be determined measuring the concentration of spin probe with CW-EPR.
8. Mix the protein solution in a 1:1 ratio with fully deuterated glycerol. The glycerol will provide a good glassy matrix upon freezing and deuteration will increase the relaxation times of the paramagnetic probe.
9. MTSL-labeled proteins can be flash frozen in liquid nitrogen and stored at −80 °C until DEER analysis.

3.3.2 DEER Data Collection and Analysis

The most popular DEER pulse sequence has four pulses and uses a $\pi/2$–π–π train of pulses of microwave (mw_1) to generate a refocused echo that detects a fraction of the spins (observed spins). Between the π pulses, a π pulse of mw_2 is intercalated whose effect is to invert a different fraction of the spins (pumped spins). If a spin interacting with an observed spin is inverted, its resonance frequency will change. This results in an oscillation of the experimental trace when the intensity of the refocused echo is detected as a function of the pump-pulse position. The frequency of this oscillation is inversely proportional to the cube of the inter-spin distance.

To perform DEER experiments, a pulse EPR spectrometer equipped with a second mw source is required. The resonator should have enough bandwidth to cover the difference in the two microwave values (from 80 to 100 MHz) and should allow enough mw power in the sample to have a π pump-pulse as short as possible (typically 12 ns). This will increase the inversion efficiency and the intensity of the oscillation. DEER measurements at Q-band frequencies are gaining popularity due to their higher sensitivity (Polyhach et al., 2012).

The DEER trace should be long enough to allow a reliable separation of the oscillation due to spin pairs from the background decay due to intermolecular interactions (which, of course, depends on sample concentration). To optimize the relaxation times, which limit the length of the trace, the measurements are performed typically at 50 K.

DEER traces are phase corrected, background fitted and the form factor is extracted. For soluble proteins, the background decay is modeled by an exponential. The conversion of the oscillating form factor into a distance distribution is then done via Tikhonov regularization. All this processing is implemented in a free toolbox called DeerAnalysis (http://www.epr.ethz.ch/software/index) (Jeschke et al., 2006).

Spin pairs with broadened distance distributions could indicate a structural alteration induced by the mutation or labeling process and should be discarded. To check for the structural integrity of the labeled protein one can also measure SAXS, which will immediately identify large/middle-range distortions.

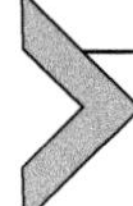

4. ANALYSIS OF MULTI-DOMAIN STRUCTURES COMBINING MX, DEER, AND SAXS

A procedure to combine multi-resolution information is summarized in Fig. 3. The starting point of the analysis is the high-resolution structures of the individual domains obtained by MX, which are assumed to be conserved in the whole. On these structures, the conformations of the MTSL probe attached to the selected sites are modeled. The approach is to select only a set of conformers that do not collide with the structure, calculate its partition function, related to how comfortably the probe sits on the site, and obtain the probability of every individual conformer. The open-source package MMM (http://www.epr.ethz.ch/software/index) has implemented this approach and incorporates a precalculated library of rotamers using molecular dynamics methods (Polyhach, Bordignon, & Jeschke, 2011). With the probabilities for individual rotamers given by MMM, the average position of the spin probe is calculated for every labeled site.

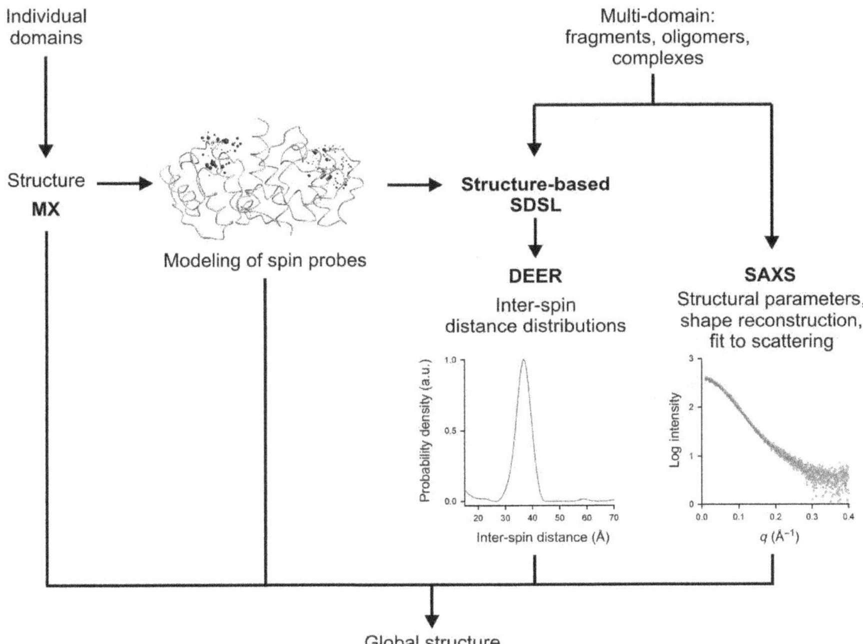

Figure 3 Combination of MX, SDSL–DEER, and SAXS. The crystallographic structures of individual domains are used to predict the positions of the paramagnetic centers of the probe (shown as spheres) and to design the SDSL. Multiple distance distributions between labeling pairs are determined by DEER and are combined with SAXS information, atomic structures, and MTSL modeling to elucidate the multi-domain global structure.

Then, the relative arrangement of two domains is scanned systematically in a 6D space. Normally, a grid of 2 Å for translation and 10° for rotation parameters is adequate. For every relative arrangement, the distance between the measured spin pairs is calculated ($\delta_{i,\mathrm{MODEL}}$) and compared with the experimental value ($\delta_{i,\mathrm{DEER}}$). The rmsd value (σ_{DEER}) is calculated taking into account all experimental distances:

$$\sigma_{\mathrm{DEER}} = \sqrt{\frac{1}{N}\sum_{i=1}^{N}\left(\delta_{i,\mathrm{DEER}} - \delta_{i,\mathrm{MODEL}}\right)^2}$$

Only the structures with $\sigma_{\mathrm{DEER}} < 3$ Å are preselected. This threshold has been found adequate to take into account the deviation between experimental and modeled conformation distributions of the probes (Alonso-Garcia et al., 2015). Then, the SAXS profiles of the structures compatible with DEER constraints are calculated using the program CRYSOL (Svergun, Barberato, & Koch, 1995) and ranged according to the Chi error-estimating parameter (χ_{SAXS}). The set of structures fitting both methods best can be identified in a χ_{SAXS} versus σ_{DEER} plot.

If there are more than two domains, the methodology would be the same with an increased number of degrees of freedom and, consequently, of necessary experimental distance constraints.

5. CONCLUSION

The structural characterization of plakins poses several challenges due to their large and mosaic domain organization. MX is very effective to elucidate the accurate structure of individual domains and small multi-domain segments of the plakin domain, and to reveal in detail the interactions of plakins with other proteins. SAXS allows to study large regions in solution and to detect and model segmental flexibility. Yet, even when crystallographic structures are available, precise modeling of multi-domain structures is frequently hindered because numerous different orientations cannot be discriminated at the resolution of the SAXS data. This limitation may be overcome by incorporating long-range distances determined by SDSL–DEER combined with modeling of the restricted conformations of the attached spin probes based on the 3D crystal structure. In summary, the synergistic combination of MX, DEER, and SAXS is a powerful hybrid approach to understand the complex structure of plectin and BPAG1e, and their contribution to the cellular architecture.

ACKNOWLEDGMENTS

This work was supported by the Spanish Ministry of Economy and Competitiveness and the European Regional Development Fund (grant BFU2012-32847 to J. M. d. P.).

REFERENCES

Alonso-Garcia, N., Garcia-Rubio, I., Manso, J. A., Buey, R. M., Urien, H., Sonnenberg, A., et al. (2015). Combination of X-ray crystallography, SAXS and DEER to obtain the structure of the FnIII-3,4 domains of integrin [alpha]6[beta]4. *Acta Crystallographica Section D: Biological Crystallography*, *71*(4), 969–985.

Alonso-García, N., Ingles-Prieto, A., Sonnenberg, A., & De Pereda, J. M. (2009). Structure of the Calx-beta domain of the integrin beta4 subunit: Insights into function and cation-independent stability. *Acta Crystallographica Section D: Biological Crystallography*, *65*(8), 858–871.

Bachman, J. (2013). Site-directed mutagenesis. *Methods in Enzymology*, *529*, 241–248.

Bouameur, J. E., Favre, B., & Borradori, L. (2014). Plakins, a versatile family of cytolinkers: Roles in skin integrity and in human diseases. *The Journal of Investigative Dermatology*, *134*(4), 885–894.

Chacon, P., Diaz, J. F., Moran, F., & Andreu, J. M. (2000). Reconstruction of protein form with X-ray solution scattering and a genetic algorithm. *Journal of Molecular Biology*, *299*(5), 1289–1302.

Choi, H. J., Park-Snyder, S., Pascoe, L. T., Green, K. J., & Weis, W. I. (2002). Structures of two intermediate filament-binding fragments of desmoplakin reveal a unique repeat motif structure. *Nature Structural Biology*, *9*(8), 612–620.

Choi, H. J., & Weis, W. I. (2011). Crystal structure of a rigid four-spectrin-repeat fragment of the human desmoplakin plakin domain. *Journal of Molecular Biology*, *409*(5), 800–812.

Dauter, Z., Dauter, M., & Dodson, E. (2002). Jolly SAD. *Acta Crystallographica Section D: Biological Crystallography*, *58*(Pt. 3), 494–506.

de Pereda, J. M., Lillo, M. P., & Sonnenberg, A. (2009). Structural basis of the interaction between integrin alpha6beta4 and plectin at the hemidesmosomes. *The EMBO Journal*, *28*(8), 1180–1190.

Doublie, S. (2007). Production of selenomethionyl proteins in prokaryotic and eukaryotic expression systems. *Methods in Molecular Biology*, *363*, 91–108.

Durand, D., Vives, C., Cannella, D., Perez, J., Pebay-Peyroula, E., Vachette, P., et al. (2010). NADPH oxidase activator p67(phox) behaves in solution as a multidomain protein with semi-flexible linkers. *Journal of Structural Biology*, *169*(1), 45–53.

Franke, D., & Svergun, D. I. (2009). DAMMIF, a program for rapid ab-initio shape determination in small-angle scattering. *Journal of Applied Crystallography*, *42*(2), 342–346.

Garcia-Alvarez, B., Bobkov, A., Sonnenberg, A., & de Pereda, J. M. (2003). Structural and functional analysis of the actin binding domain of plectin suggests alternative mechanisms for binding to F-actin and to integrin a6b4. *Structure*, *11*(6), 615–625.

Hammel, M. (2012). Validation of macromolecular flexibility in solution by small-angle X-ray scattering (SAXS). *European Biophysics Journal*, *41*(10), 789–799.

Hendrickson, W. A. (1991). Determination of macromolecular structures from anomalous diffraction of synchrotron radiation. *Science*, *254*(5028), 51–58.

Jefferson, J. J., Ciatto, C., Shapiro, L., & Liem, R. K. (2007). Structural analysis of the plakin domain of bullous pemphigoid antigen1 (BPAG1) suggests that plakins are members of the spectrin superfamily. *Journal of Molecular Biology*, *366*(1), 244–257.

Jeschke, G. (2012). DEER distance measurements on proteins. *Annual Review of Physical Chemistry*, *63*, 419–446.

Jeschke, G. (2013). Conformational dynamics and distribution of nitroxide spin labels. *Progress in Nuclear Magnetic Resonance Spectroscopy*, *72*, 42–60.

Jeschke, G., Chechik, V., Ionita, P., Godt, A., Zimmermann, H., Banham, J., et al. (2006). DeerAnalysis2006—A comprehensive software package for analyzing pulsed ELDOR data. *Applied Magnetic Resonance*, *30*(3–4), 473–498.

Ortega, E., Buey, R. M., Sonnenberg, A., & de Pereda, J. M. (2011). The structure of the plakin domain of plectin reveals a non-canonical SH3 domain interacting with its fourth spectrin repeat. *The Journal of Biological Chemistry*, *286*(14), 12429–12438.

Pelikan, M., Hura, G. L., & Hammel, M. (2009). Structure and flexibility within proteins as identified through small angle X-ray scattering. *General Physiology and Biophysics*, *28*(2), 174–189.

Petoukhov, M. V., & Svergun, D. I. (2013). Applications of small-angle X-ray scattering to biomacromolecular solutions. *The International Journal of Biochemistry & Cell Biology*, *45*(2), 429–437.

Polyhach, Y., Bordignon, E., & Jeschke, G. (2011). Rotamer libraries of spin labelled cysteines for protein studies. *Physical Chemistry Chemical Physics*, *13*(6), 2356–2366.

Polyhach, Y., Bordignon, E., Tschaggelar, R., Gandra, S., Godt, A., & Jeschke, G. (2012). High sensitivity and versatility of the DEER experiment on nitroxide radical pairs at Q-band frequencies. *Physical Chemistry Chemical Physics, 14*(30), 10762–10773.

Rambo, R. P., & Tainer, J. A. (2011). Characterizing flexible and intrinsically unstructured biological macromolecules by SAS using the Porod-Debye law. *Biopolymers, 95*(8), 559–571.

Ramisch, S., Lizatovic, R., & Andre, I. (2015). Automated de novo phasing and model building of coiled-coil proteins. *Acta Crystallographica Section D: Biological Crystallography, 71*(Pt. 3), 606–614.

Sammito, M., Millan, C., Rodriguez, D. D., de Ilarduya, I. M., Meindl, K., De Marino, I., et al. (2013). Exploiting tertiary structure through local folds for crystallographic phasing. *Nature Methods, 10*(11), 1099–1101.

Sevcik, J., Urbanikova, L., Kost'an, J., Janda, L., & Wiche, G. (2004). Actin-binding domain of mouse plectin. Crystal structure and binding to vimentin. *European Journal of Biochemistry, 271*(10), 1873–1884.

Sheldrick, G. M. (2010). Experimental phasing with SHELXC/D/E: Combining chain tracing with density modification. *Acta Crystallographica Section D: Biological Crystallography, 66*(Pt. 4), 479–485.

Skou, S., Gillilan, R. E., & Ando, N. (2014). Synchrotron-based small-angle X-ray scattering of proteins in solution. *Nature Protocols, 9*(7), 1727–1739.

Song, J. G., Kostan, J., Drepper, F., Knapp, B., de Almeida Ribeiro, E., Jr., Konarev, P. V., et al. (2015). Structural insights into Ca(2+)-calmodulin regulation of plectin 1a-integrin beta4 interaction in hemidesmosomes. *Structure, 23*(3), 558–570.

Sonnenberg, A., Rojas, A. M., & de Pereda, J. M. (2007). The structure of a tandem pair of spectrin repeats of plectin reveals a modular organization of the plakin domain. *Journal of Molecular Biology, 368*(5), 1379–1391.

Svergun, D. I. (1999). Restoring low resolution structure of biological macromolecules from solution scattering using simulated annealing. *Biophysical Journal, 76*(6), 2879–2886.

Svergun, D., Barberato, C., & Koch, M. H. J. (1995). CRYSOL—A program to evaluate X-ray solution scattering of biological macromolecules from atomic coordinates. *Journal of Applied Crystallography, 28*(6), 768–773.

Svergun, D. I., Koch, M. H. J., Timmins, P. A., & May, R. P. (2013). *Small angle X-ray and neutron scattering from solutions of biological macromolecules.* Oxford: Oxford University Press.

Svergun, D. I., Petoukhov, M. V., & Koch, M. H. (2001). Determination of domain structure of proteins from X-ray solution scattering. *Biophysical Journal, 80*(6), 2946–2953.

Tartoff, K. D., & Hobbs, C. A. (1987). Improved media for growing plasmids and cosmid clones. *Bethesda Research Laboratories Focus, 9*, 12.

Tria, G., Mertens, H. D. T., Kachala, M., & Svergun, D. I. (2015). Advanced ensemble modelling of flexible macromolecules using X-ray solution scattering. *International Union of Crystallography, 2*(2), 207–217.

Tropea, J. E., Cherry, S., & Waugh, D. S. (2009). Expression and purification of soluble His(6)-tagged TEV protease. *Methods in Molecular Biology, 498*, 297–307.

CHAPTER ELEVEN

Purification and Structural Analysis of Desmoplakin

Hee-Jung Choi*, William I. Weis†,‡,1

*School of Biological Sciences, Seoul National University, Seoul, South Korea
†Department of Structural Biology, Stanford University School of Medicine, Stanford, California, USA
‡Department of Molecular & Cellular Physiology, Stanford University School of Medicine, Stanford, California, USA
[1]Corresponding author: e-mail address: bill.weis@stanford.edu

Contents

Abstract

Desmoplakin (DP) is an obligate component of desmosomes, where it links the desmosomal cadherin/plakoglobin/plakophilin assembly to intermediate filaments. DP contains a large amino-terminal domain (DPNT) that binds to the cadherin/plakoglobin/plakophilin complex, a central coiled-coil domain that dimerizes the molecule, and a C-terminal domain (DPCT) that binds to intermediate filaments. DPNT contains a plakin domain, comprising a set of spectrin-like repeats. DPCT contains three plakin repeat domains, each formed by 4.5 repeats of a sequence motif known as a plakin repeat that bind to intermediate filaments. Here, we review purification, biochemical characterization, and structural analysis of the DPNT plakin domain and the DPCT plakin repeat domains.

Methods in Enzymology, Volume 569
ISSN 0076-6879
http://dx.doi.org/10.1016/bs.mie.2015.05.006

1. INTRODUCTION

Desmosomes are cell–cell adhesion structures that link cadherin cell adhesion molecules to the intermediate filament (IF) cytoskeleton (Kowalczyk & Green, 2013). As IFs are principal load-bearing elements of the cytoskeleton (Koster, Weitz, Goldman, Aebi, & Herrmann, 2015; Parry & Steinert, 1999), desmosomes impart mechanical strength to tissues and are especially prominent in tissues subject to mechanical stress such as the skin and heart. Their importance to tissue homeostasis is emphasized by a number of cardiovascular and skin diseases associated with genetic mutation or autoimmune reactions to desmosomal components (Al-Jassar, Bikker, Overduin, & Chidgey, 2013).

In the desmosome, the desmosomal cadherins, desmogleins (Dsg), and desmocollins (Dsc) are linked to the IF system through several proteins (Kowalczyk & Green, 2013). Portions of the Dsg and Dsc cytoplasmic domains are homologous to those of classical cadherins, and both bind to plakoglobin (Choi, Gross, Pokutta, & Weis, 2009; Huber & Weis, 2001; Mathur, Goodwin, & Cowin, 1994; Wahl et al., 1996; Witcher et al., 1996). These assemblies also contain plakophilins (PKPs), members of the p120 subfamily of armadillo proteins that appear to have multiple roles in desmosome assembly and signaling (Bass-Zubek, Godsel, Delmar, & Green, 2009). Desmoplakin (DP), a member of the plakin family of cytolinkers (Sonnenberg & Liem, 2007), is found in all desmosomes, where it links the cadherin/plakoglobin/PKP complex to IFs (Kowalczyk & Green, 2013; Sonnenberg & Liem, 2007). Given its role in desmosomes, it is not surprising that mutations of DP are associated with a skin and cardiovascular diseases, which have as their common feature loss of mechanical integrity (Al-Jassar, Bikker, et al., 2013).

The primary structure of desmoplakin I (DPI) comprises a large N-terminal domain (DPNT), a central coiled-coil rod that dimerizes the molecule, and a C-terminal region that binds to IFs (DPCT) (Fig. 1). A splice variant eliminates part of the central rod domain to produce DPII (Green et al., 1990). DPNT consists of an amino-terminal basic region (residues 1–170) that contains the binding site for plakoglobin (Kowalczyk et al., 1997; Yang et al., 2006) followed by a plakin domain, which binds Dsg1 and PKP1 (Al-Jassar et al., 2011; Kami, Chidgey, Dafforn, & Overduin, 2009; Kowalczyk et al., 1999). The plakin domains of plakin family members, which include DP, bullous pemphigoid antigen 1 (BPAG1), plectin,

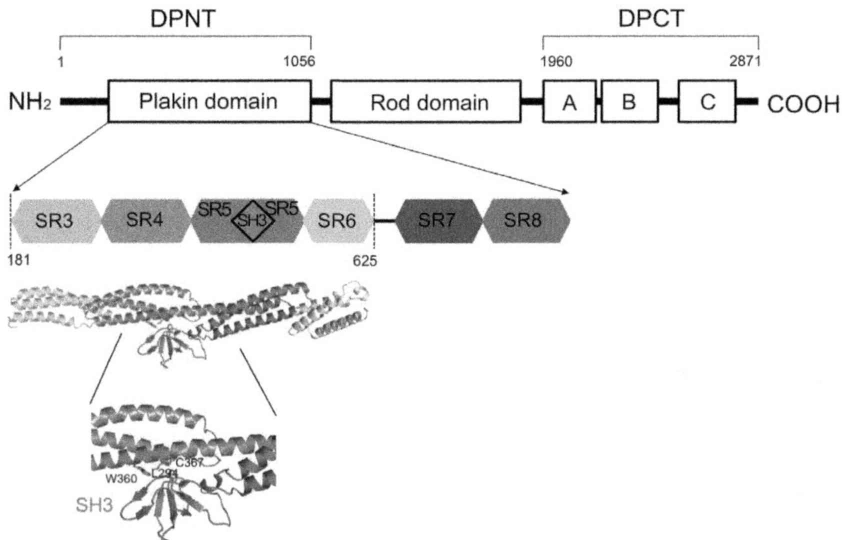

Figure 1 Structure of human desmoplakin I. *Top*, primary structure. The amino acid boundaries of DPNT and DPCT are indicated. Small rectangles labeled as A, B, and C in DPCT represent PRD-A, PRD-B, and PRD-C, respectively. The plakin domain in DPNT, containing six spectrin repeats (SRs) is shown as a close-up view. *Bottom*, crystal structure of DP(175–630). A close-up view of the SH3 domain and the interacting hydrophobic residues from SR4 is also shown. Top: *Adapted with permission from Choi and Weis (2011).* (See the color plate.)

envoplakin, and periplakin, contain variable numbers of spectrin-like repeats, as first shown by sequence analysis enabled by the crystal structures of di-repeat fragments of plectin and BPAG1 (Jefferson, Ciatto, Shapiro, & Liem, 2007; Sonnenberg, Rojas, & de Pereda, 2007). The plakin domain of DPNT consists of six spectrin repeats (SRs), labeled as SR3–SR8 based on sequence alignment to other plakin family members (Sonnenberg et al., 2007; Fig. 1). Amino acid sequence analysis and protease sensitivity (see later) suggest that there is a flexible linker between DP SR6 and SR7, and a single SH3 domain is inserted in the middle of SR5 (Al-Jassar, Bernado, Chidgey, & Overduin, 2013; Al-Jassar et al., 2011; Choi & Weis, 2011; Jefferson et al., 2007; Sonnenberg et al., 2007).

The C-terminal region of DP (DPCT) binds to IFs, including epidermal keratins, simple epithelial keratins, desmin, and vimentin (Bornslaeger, Corcoran, Stappenbeck, & Green, 1996; Choi, Park-Snyder, Pascoe, Green, & Weis, 2002; Kouklis, Hutton, & Fuchs, 1994; Meng, Bornslaeger, Green, Steinert, & Ip, 1997; Stappenbeck & Green, 1992).

The primary structure of DPCT consists of three plakin repeat domains (PRDs), each comprised of 4.5 copies of a 38 amino acid sequence motif known as a plakin repeat (PR) (Green et al., 1990; Fig. 1). The three domains, designated PRD-A, PRD-B, and PRD-C, share ~30% sequence similarity with one another. PRD-B and PRD-C are separated by a ~150 amino acid linker region. The three PRDs are followed by a C-terminal Ser/Arg-rich domain that is subject to posttranslational modifications that modulate association with IFs and with desmosomes (Albrecht et al., 2015).

Here, we review the protocols for purification of recombinantly expressed domains of DP for biochemical and structural studies. We also summarize three-dimensional structural analyses of these proteins using X-ray crystallography and small angle X-ray scattering (SAXS).

2. PURIFICATION AND STRUCTURAL ANALYSIS OF DPNT

2.1 Recombinant Expression and Purification of DPNT

Biochemical and structural studies of DPNT have used full-length DPNT (residues 1–1056, human sequence) or fragments, all expressed in *Escherichia coli* (Al-Jassar, Bernado, et al., 2013; Al-Jassar, Bikker, et al., 2013; Choi & Weis, 2011). Here, we describe purification protocols used in our laboratories; similar protocols have been described by Al-Jassar et al. (2011). DNA encoding the full-length DPNT (residues 1–1056) is cloned in the expression vector pGEX-TEV, which is a modified version of the pGEX-KG vector. This vector produces DPNT N-terminally tagged with glutathione-*S*-transferase (GST), with the GST tag linked to DPNT by a tobacco etch virus (TEV) protease-cleavable sequence. pGEX-TEV-DPNT is transformed into the Rosetta strain of *E. coli*. Bacteria are grown at 37 °C in Luria broth (LB) supplemented with 100 μg ml^{-1} ampicillin and 35 μg ml^{-1} chloramphenicol to reach an OD_{600} of about 0.6, and protein expression is then induced by the addition of isopropyl-β-D-thiogalactopyranoside (IPTG) to a final concentration of 0.5 m*M*. To improve solubility of the expressed protein, the temperature of the culture is lowered to 28 °C after induction and incubated for additional 4 h.

A pellet from a 1 l culture is resuspended with 20 ml of ice-cold PBST buffer (PBS buffer with 1 *M* NaCl and 0.05% Tween 20), supplemented with 50 U DNAse I and protease inhibitor cocktail, and the cells lysed using an Emulsiflex high-pressure cell homogenizer (Avestin, Ontario, Canada). The lysates are centrifuged at 12,000 × *g* for 30 min at 4 °C. Supernatants are recovered and incubated with ||glutathione-agarose beads, which are preequilibrated with PBST buffer for 1 h at 4 °C. Beads are washed with 10 column volumes of PBSTR buffer (PBST with 5 m*M* DTT) and equilibrated with TEV cleavage buffer (25 m*M* Tris–Cl, pH 8.0, 0.1 *M* NaCl, and 2 m*M* DTT). After overnight incubation of TEV protease with the beads (substrate:TEV = 20:1, w/w) at 4 °C, cleaved DPNT is recovered from the flow-through, while the GST remains bound to the beads. Further purification of DPNT is accomplished by anion exchange chromatography and size exclusion chromatography. DPNT is eluted with a linear gradient of NaCl from a MonoQ column (GE Healthcare) in Q buffer (25 m*M* Tris–Cl, pH 8.5, 0.5 m*M* EDTA, and 2 m*M* DTT). After equilibrating a Superdex S200 column (GE Healthcare) with Q buffer containing 100 m*M* NaCl, the protein is applied to the column, where it elutes at the size expected of a monomer (~120 kDa). The protein can be concentrated to 5 mg ml^{-1} using an Amicon spin concentrator (Millipore).

Limited proteolysis of purified proteins can be used to define stable subdomains for structural studies and to discover potential regions of flexibility. This approach was used to characterize DPNT flexibility and to obtain fragments for structural analysis (Al-Jassar, Bernado, et al., 2013; Al-Jassar et al., 2011; Choi & Weis, 2011). Limited proteolysis of DPNT was carried out with trypsin, chymotrypsin, or V8 protease. Separate proteolytic reactions of 10 μ*M* DPNT with 0.01 mg ml^{-1} sequencing grade protease (Roche) in 20 m*M* Tris–Cl (pH 8.0), 100 m*M* NaCl and 2 m*M* DTT were carried out for 10, 30, and 60 min at room temperature and stopped by the addition of SDS sample buffer and boiling at each time point. The reaction mixtures were run on SDS-PAGE and visualized by Coomassie Blue staining. Treatment with trypsin or chymotrypsin gave a ~55 kDa stable fragment on SDS-PAGE. After blotting onto a PVDF membrane, this band was cut out and analyzed by N-terminal (Edman) sequencing. This analysis shows that trypsin and chymotrypsin cleaved DPNT after Lys167 and Tyr172, respectively, suggesting that this region is a flexible, protease accessible linker.

Based on the N-terminal sequencing data and the estimated molecular weight of the stable fragment resulting from limited proteolysis, a new construct of the plakin domain, DP(175–630) was designed for structural analysis. A DNA fragment encoding DP(175–630) was subcloned into the pGEX-TEV vector, and the protein was expressed and purified similarly to the full DPNT. The only modification was to use Tris–Cl, pH 8.0 buffer instead of Tris–Cl, pH 8.5 for anion exchange and gel filtration chromatography. DP(175–630) is present as a monomer in solution, verified by size exclusion chromatography and the overall yield is about 8 mg l^{-1} of culture.

2.2 Structural Analysis of DPNT

2.2.1 Crystal Structure of DP(175–630)

Purified DP(175–630) at 20 mg ml^{-1} was crystallized by vapor diffusion methods, and the structure was solved by single-wavelength anomalous scattering phasing (Choi & Weis, 2011), using selenomethionine-substituted (SeMet) DP(175–630) produced by metabolic inhibition of methionine synthesis (Yu, Jahn, & Brünger, 1999). The structure was refined at 2.95 Å resolution (Choi & Weis, 2011).

DP(175–630) forms an elongated structure with approximate dimensions 180 × 40 × 40 Å produced by a linear arrangement of four SRs (Fig. 1). Each SR consists of three α-helices, designated as A, B, and C, which form an antiparallel three-helix bundle. The SRs are structurally similar, with root-mean-square deviations of 0.8–1.2 Å, except for SR6, which has much shorter helices A and B. The structural variation of SR6 correlates with its sequence: SR3, SR4, and SR5 have 23–27% sequence similarities to one another, whereas SR6 is about 13% similar to the others. A key difference is the replacement of a conserved aromatic residue in helix A of SRs 3, 4, and 5 with aspartic acid in SR6 (Asp556). This change leads to differences in hydrophobic core packing in the N-terminal region of the SR, which result in a large structural deviation of helix A from the canonical SR structure. Accordingly, a DALI search reveals that the SR6 structure is more similar to other three-helix bundle proteins, such as the syntaxin Habc domain and Target of Myb1, than to canonical SRs. Interestingly, sequence analysis indicates that other plakin family proteins, such as envoplakin and periplakin, do not contain SR6.

DPNT and other plakins feature an SH3 domain inserted into a loop between helices B and C of SR5. Classical SH3 domains mediate intracellular protein–protein interactions by recognizing proline-rich sequences (PXXP motif). Although the SH3 domain in DP has the basic SH3 fold, consisting of two orthogonal β sheets and three large loops, called RT, n-Src, and distal loops, it has several distinctive features. Proline-rich ligands bind to a site in classical SH3 domains formed by conserved aromatic residues located on strands β3 and β4, and in the RT and n-Src loops. In contrast, nonproline hydrophobic residues of SR4 occupy the equivalent site of the DP SH3 domain (Fig. 1). In particular, Trp360 located on helix C of SR4 sits in the shallow groove of SH3 domain, interacting with the conserved aromatic residue Tyr469. In addition, several hydrophobic residues from SR4, including Ile364 and Cys367 in helix C and Leu294 in helix A, sit in this site. To accommodate the presence of these large hydrophobic side chains, several of the conserved aromatic residues that form the ligand-binding site in classical SH3 domains are replaced with nonaromatic residues, such as Gly509 and Lys492. For example, replacement of a tyrosine with glycine at residue 509 avoids a potential steric clash with Trp360.

The close contacts between SR4 and the SH3 domain observed in the crystal structure of DP(175–630) indicate that the SH3 would have limited, if any, mobility with respect to the SRs of DP. Indeed, 23% of the solvent accessible surface area of SH3 domain, including the putative ligand-binding site, is buried in the interface with SR4. No ligand for the SH3 domain of DPNT has been reported. A deletion mutant of DPNT, whose SH3 domain is substituted with a six amino acid Gly–Ser linker, proves to be very unstable and prone to aggregation (Choi & Weis, 2011). Moreover, NMR analysis of the isolated DP SH3 domain indicates that it is unstructured, and a fragment encompassing SR3–4 could not be obtained as a soluble protein (Al-Jassar et al., 2011), presumably due to the exposure of hydrophobic residues of SR4 that normally interact with the SH3 domain in SR5. These observations suggest that the SH3 domain of DP functions as an important structural element that stabilizes the structure of tandem SRs, rather than a signaling molecule.

The four SRs of DP(175–630) are connected by continuous helices. The linker helices display different degrees of bending, suggesting that they provide interrepeat flexibility. Superposition of the two crystallographically independent copies of DP(175–630) showed that the degree of linker bending varies from 0.3° to 15.8° (Choi & Weis, 2011). Interestingly, the SR4–5 di-repeat containing the SH3 domain shows the least flexibility, whereas the SR5–6 di-repeat shows the most variation.

2.2.2 SAXS Analysis of DPNT

No structural information has been reported for full-length DPNT, but the molecular architecture of the full DPNT plakin domain comprising SRs 3–8 and the linker to the rod domain (residues 180–1022) has been determined by SAXS analysis (Al-Jassar et al., 2011). Molecular envelopes of two SR di-repeats, SR5–6 and SR7–8, were also determined in this study.

The SAXS-derived molecular envelope of DP(180–1022) reveals an a L-shaped molecule whose radius of gyration (R_g) is 94 Å and whose longest dimension is 335 Å (Al-Jassar et al., 2011; Fig. 2). Although the crystal structure of SR3–6 is known (Choi & Weis, 2011), there are no high-resolution structural data for SR7–8. Thus, the authors produced a SAXS envelope of SR7–8, and generated a three-dimensional model based on the structure of

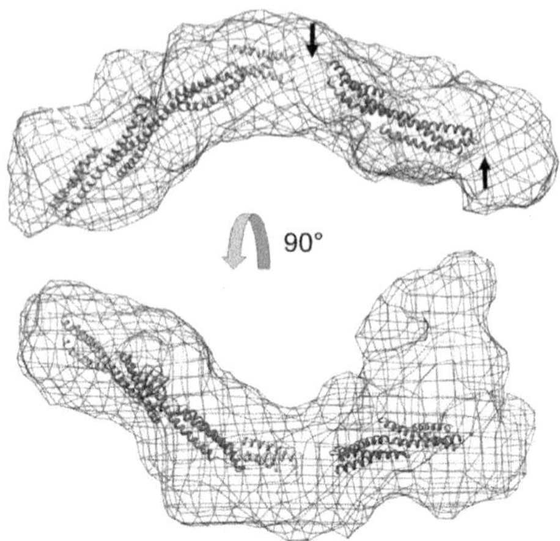

Figure 2 Molecular envelope of DP(180–1022) obtained by SAXS data analysis. The best fit for the structure of SR3-6 and the SR7-8 model is shown. Each SR is colored as Fig. 1. The flexible hinge region between SR6 and SR7 is not modeled and is indicated by a black arrow. The unmodeled envelope density at the right (black arrow) corresponds to non-SR sequence that links to the rod domain. *Adapted with permission from Al-Jassar et al. (2011).* (See the color plate.)

erythroid spectrin SR8–9, which has 14% amino acid sequence identity with DPNT SR7–8. This structure, which has a linear arrangement of the two repeats, fits the SAXS data well. The SR7–8 model and the crystal structure of SR3–6 fit the short and long arms of the DP(180–1022) SAXS envelope (Fig. 2). Limited proteolysis of DPNT produces stable fragments corresponding to SR3–6 and SR7–8 (Al-Jassar et al., 2011; Choi & Weis, 2011), implying a flexible linker between SR3–6 and SR7–8 corresponding to the unmodeled "elbow" in the SAXS envelope.

Recently, the ensemble optimization method of Bernardo, Mylonas, Petoukhov, Blackledge, & Svergun (2007) was used to investigate the flexibility of the SR6–SR7 hinge (Al-Jassar, Bernado, et al., 2013). In this analysis, a random 10,000 conformations are generated by the program RANCH and an ensemble of 50 conformations that fit the SAXS data is chosen by the program Genetic Algorithm Judging Optimization of Ensembles (GAJOE). The best ensemble selected by GAJOE has two different values of R_g, suggesting two different conformations in solution; one corresponds to an extended linear conformation and the other to a U-shaped conformation. In this study, the SR3–8 region of two other plakin family proteins, envoplakin and periplakin, were studied to characterize the nature of flexible hinge between SR3–5 and SR7–8 (these two family members lack SR6) (Al-Jassar, Bernado, et al., 2013). In contrast to DPNT, the envoplakin structures display a much narrower R_g distribution, implying that it has less mobility; the flexibility of periplakin was intermediate between DP and envoplakin. The different levels of hinge motion within plakin domains could be related with the different length of the linker, which is longer in DPNT than in envoplakin and periplakin. Although more structural and biophysical characterization of plakin domains is needed, these results suggest that different flexibility of plakin domains might have functional implications in junction assembly and maintenance of tissue integrity as well as elasticity in response to tension.

As noted above, the SR5–6 linker shows some flexibility, with a ~15° difference in the relative positions of SRs 5 and 6 in the two crystallographically independent copies (Choi & Weis, 2011). SAXS analysis of the isolated DP SR5–6 di-repeat revealed a distinct bend that is not consistent with the crystal structure of DP(175–630); to model the data, SR5 and SR6 coordinates were separately placed into the envelope (Al-Jassar et al., 2011). Thus, it is not unreasonable that there may be significant flexibility between SR5 and SR6. On the other hand, the SR5–6 junction is not susceptible to digestion in limited proteolysis analysis (Al-Jassar, Bernado, et al., 2013; Al-Jassar et al., 2011; Choi & Weis, 2011). It is conceivable that the severe bend observed by SAXS is due to the lack of SR4; since the SR5 SH3 domain packs against SR4, perhaps there are propagated changes that make the SR5–6 linker more flexible. Indeed, the SAXS envelope of the full plakin regions DP(180–1022) (SRs3–8) indicates an arrangement of SR3–6 similar to that of the crystal structure.

3. PURIFICATION AND STRUCTURAL ANALYSIS OF DPCT

3.1 Recombinant Expression and Purification of DPCT and PRDs

Full-length DPCT (residues 1946–2871, human) with an N-terminal His_6 affinity tag is expressed in our laboratories in *E. coli* strain JM109. Cells are grown at 37 °C, induced with 2 m*M* IPTG and grown for 4 h at 37 °C, then harvested and lysed with a French Press in a buffer containing 100 m*M* HEPES, pH 7.8, 0.3 *M* KCl, 2 m*M* β-mercaptoethanol, 10 μg ml^{-1} leupeptin, 2 μg ml^{-1} pepstatin, and 1 m*M* PMSF. DPCT is purified by Ni^{2+}-NTA agarose chromatography, followed by gel filtration (Superdex S200) in a buffer containing 20 m*M* HEPES, pH 7.8, 0.15 *M* KCl, 2 m*M* β-mercaptoethanol, and 1 m*M* EDTA. Purified DPCT shows degradation products by SDS-PAGE suggesting flexibility that would interfere with crystallization. To define the physical boundaries of PRD-A, PRD-B, and PRD-C, limited proteolysis of full-length DPCT was performed using chymotrypsin. Purified DPCT at 2.5 mg ml^{-1} in 100 m*M* Tris–HCl, pH 8.0, and 10 m*M* $CaCl_2$ was mixed with 0.2 mg ml^{-1} chymotrypsin in 50 m*M* Tris–HCl, pH 8.0, and 10 m*M* $CaCl_2$ in a 6:1 (v/v) ratio (i.e., 75:1 (w/w) DPCT:chymotrypsin). The digestion mixture was further purified by MonoS and MonoQ columns and three purified stable fragments with molecular weight of 20–27 kDa were analyzed by N-terminal amino acid sequencing (Choi et al., 2002).

Based on the proteolysis data, three constructs were designed. PCR products encoding PRD-A (residues 1960–2208), PRD-B (residues 2209–2456), and PRD-C (residues 2609–2822) were subcloned into a pPROEX-HTc vector, which has TEV-cleavable N-terminal His_6 tag. Each construct is transformed into *E. coli* strain BL21, and the cells induced with 0.5 m*M* IPTG for protein expression. After a 4 h incubation at 30 °C, cells are harvested and stored at −80 °C until use. A pellet from a 1 l culture is resuspended with ice-cold Ni^{2+}-binding buffer (50 m*M* Tris–Cl, pH 8.0, 0.1 *M* NaCl, 5 m*M* imidazole, and 2 m*M* β-mercaptoethanol) and lysed by French Press or Emulsiflex. Cell lysates are cleared by centrifugation at 12,000 × *g* for 30 min and the recovered supernatant is incubated with Ni-NTA beads, which are preequilibrated with Ni^{2+}-binding buffer. After 1 h incubation at 4 °C, beads are washed with two column volumes of high-salt buffer (Ni^{2+}-binding buffer with 1 *M* NaCl, instead of 0.1 *M* NaCl) and two column volumes of washing buffer (Ni^{2+}-binding buffer with 20 m*M* imidazole instead of 5 m*M* imidazole). Each protein is eluted with 250 m*M* imidazole. Recombinant TEV protease is added to the elution fractions (substrate:TEV = 20:1, w/w) to cleave off the His_6 tag. After overnight incubation at 4 °C, the cleavage reaction mixture is dialyzed against Ni^{2+}-binding buffer for 3 h at 4 °C, and reloaded onto a Ni^{2+}-NTA agarose column to separate cleaved protein; cleaved protein is recovered from the flow-through, while uncleaved protein and TEV protease remain bound to Ni^{2+}-NTA column. PRD-A is further purified by cation exchange chromatography, with a MonoS column preequilibrated with S buffer (20 m*M* HEPES, pH 7.2, 20 m*M* NaCl, and 2 m*M* DTT) and eluted with linear gradient of NaCl. In the case of PRD-B and PRD-C, anion exchange on MonoQ with Q buffer (20 m*M* Tris–Cl, pH 8.0, 20 m*M* NaCl, and 2 m*M* DTT) is used. The pooled fractions from the ion exchange column are loaded onto a Superdex S200 gel filtration column. Each protein elutes as a monomer, and the overall yield is about 6–8 mg protein per 1 l culture.

3.2 Crystal Structure of Subdomains B and C of DPCT

The purified PRD-B and PRD-C proteins were concentrated to 30 mg ml^{-1} for crystallization. Along with native proteins, SeMet-substituted proteins were overexpressed in the *E. coli* methionine auxotroph strain DL41, purified similarly to the native proteins, and crystallized. Native and SeMet crystals were obtained by hanging drop vapor diffusion. The structure of PRD-C was determined at 1.8 Å resolution, using multi-wavelength anomalous dispersion phasing of SeMet-substituted protein. PRD-B was determined at 3.0 Å resolution using single-wavelength anomalous scattering data of the SeMet protein (Choi et al., 2002).

The DP PRD-B and PRD-C structures are very similar, as expected from their sequence homology. Each PRD has 4.5 copies of the PR (Fig. 3A). Unlike other repeat motif proteins that form extended structures, the PRD forms a globular structure with molecular dimensions of 30 × 34 × 48 Å. A full PR consists of β hairpin, followed by two antiparallel α helices, H1 and H2, of 8 and 14 amino acids, respectively. The helix H2 can be divided into two regions based on its intra- and interrepeat interactions: the first half forms part of the hydrophobic core within a repeat, whereas the second half participates in hydrophobic packing interactions with the next repeat. In the case of PR4, the second half of H2 is replaced with a loop containing hydrophilic residues, which does not contact the following half repeat PR5, which lacks H2. The conserved hydrophobic residues in PR5 interact with PR1 and the N-terminal and the C-terminal tails of the domain. The N-terminal and the C-terminal tails also pack against hydrophobic residues on H2 in PR1.

Superposition of the four PRs in each protein shows that PR1 and PR3 are more similar to each other than the others, with PR2 and PR4 having variations in the C-terminal half of H2: the C-terminal region of H2 in PR2 is bent, and H2 of PR4 is shorter. The relative orientation between PR1 and PR2 is different from that between PR2 and PR3. PR1 and PR2 are roughly parallel, whereas PR2 and PR3 are related by an approximate rotation of 120°. Moreover, there are interrepeat interactions between non-adjacent repeats in the PRD; for instance, PR1 makes contacts with PR3 and PR5 as well as with PR2. These specific, conserved variations in the PR sequences give rise to differences in the interrepeat packing to form an overall globular structure, and thereby constrain the domain to contain 4.5 repeats. The lack of regular interrepeat packing contrasts with other repeat motif structures, including those with HEAT, armadillo, leucine-rich, ankyrin, and WD40 repeats. In those structures, each repeat packs

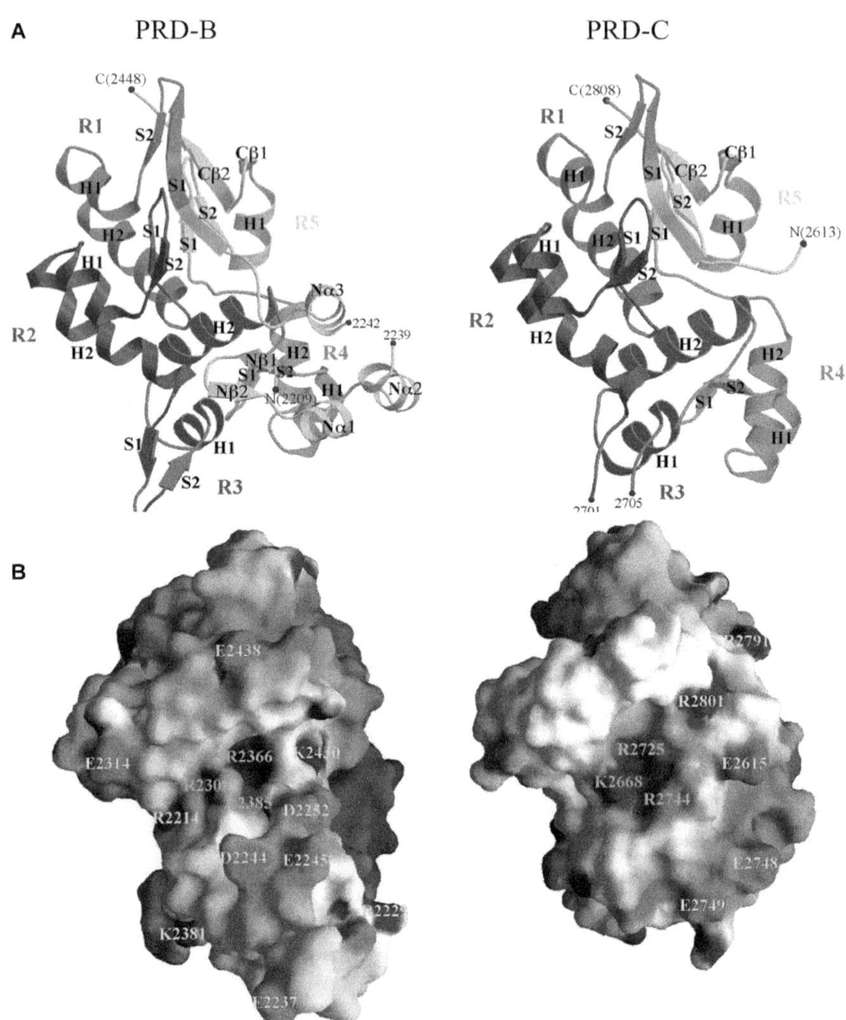

Figure 3 Crystal structures of PRD-B (left) and PRD-C (right). (A) Ribbon representation of the PRDs. Plakin repeats (PRs) 1–5 are colored in orange, magenta, blue, green, and yellow green. The N-terminal and C-terminal regions that flank the core PRD are shown in gray. The secondary structure elements of each PR are labeled as S1, S2, H1, and H2. Secondary structures in the N-terminal PR-like motif are labeled as Nβ1, Nβ2, Nα1, Nα2, and Nα3. The N- and C-termini, and the ends of breaks in the structure, are indicated with small black spheres. (B) Electrostatic surface representations of PRD-B and PRD-C. Negatively charged and positively charged regions are colored red and blue, respectively. Contoured at $\pm 10\ k_BT/e$. Conserved basic residues, which are labeled in yellow constitute positively charged groove in PRD. *Adapted from Choi et al. (2002).* (See the color plate.)

against the neighboring repeat in a regular way, to form either a circular (WD40 motif) or an elongated structure; thus, elongated structures can have variable numbers of repeats.

Although the overall structure of the PRD is known, it is not clear how multiple PRDs are arranged in DP. PRD-B contains an extra N-terminal PR-like motif consisting of a β hairpin followed by antiparallel α helices, but it has three α helices instead of the two found in the canonical PR. This region protrudes from the globular main body of PRD and could conceivably pack against PRD-A. Sequence analysis suggests this N-terminal PR-like motif is also present in PRD-A, but structural information is not yet available. The relative orientation between PRD-B and PRD-C is hard to predict given the long flexible linker between them. Other plakin family members also have multiple PRDs (Sonnenberg & Liem, 2007); for example, plectin contains five copies of PRD-B and one PRD-C. As an extreme case, epiplakin has 13 copies of PRD-B. The significance of the different numbers and combinations of PRD subtypes (A, B, and C) present in these proteins is not understood. The presence of multiple PRDs may alter IF affinity, as discussed in the following section.

3.3 Binding to Intermediate Filaments

The structures of the DP PRDs reveal a conserved, positively charged groove (Fig. 3B). Modeling showed that this groove would accommodate a helix, so given the helical coiled-coil structure of the IF building block (Koster et al., 2015; Parry & Steinert, 1999), it was proposed that the groove forms part of the IF-binding site (Choi et al., 2002). To date, no experimental verification of this model has been reported.

Assaying IF binding is challenging due to the tendency of IFs to aggregate and their generally poor solubility. In this regard, vimentin is a relatively well-behaved IF and allows binding to be assayed by cosedimentation. Commercially available monomeric vimentin (0.5 mg ml^{-1} in 5 mM PIPES, pH 7.0, and 1 mM EDTA) (Cytoskeleton, Inc.) is polymerized by adjusting the solution to a final concentration of 20 mM HEPES, pH 7.5, 150 mM NaCl, and incubated for 30 min at 37 °C. The protein of interest (10 mg ml^{-1}, 20 mM HEPES, pH 7.5, and 150 mM NaCl) is then mixed with 20 μl polymerized vimentin at molar ratios spanning 1:1 to 80:1 protein:vimentin; PRD-C saturated binding at a 40:1 ratio (Choi et al., 2002). The mixture is then centrifuged at 100,000 × g for 30 min. The supernatant is withdrawn and the pellet dissolved in SDS-PAGE sample buffer. Supernatants and pellets are run on SDS-PAGE to detect cosedimentation in the pellet. Several controls to assess background sedimentation are essential for these assays: sedimentation of the protein in the absence of vimentin, and inclusion of a non-IF-binding protein (typically BSA) to insure that any cosedimentation is not due to nonspecific trapping of the protein in the filament network.

Using this vimentin-binding assay, it was shown that the individual PRDs of DPCT bind weakly, but specifically, to vimentin (Choi et al., 2002). Sequences homologous to the first 114 residues of the 151 amino acid linker between PRD-B and PRD-C have been proposed to be important for IF binding (DiColandrea, Karashima, Määttä, & Watt, 2000; Nikolic, Nulty, Mir, & Wiche, 1996), although a construct comprising PRD-B and the following linker does not bind more strongly than PRD-B alone. A DP fragment comprising PRD-A and PRD-B also binds to vimentin weakly, whereas a fragment containing PRD-B, the following linker and PRD-C binds more strongly, as does a fragment spanning all three PRDs. These observations suggest that, at a minimum, the presence of multiple PRDs provides avidity enhancement, perhaps by interacting with more than one protofilament. The weak binding of the PRD-A–PRD-B construct, which has only a short sequence between PRDs, suggests that flexibility between domains is needed in order to allow simultaneous engagement of an IF. More structural and site-specific mutagenesis data will be needed to understand the specific interaction of DPCT and other plakins with IFs.

4. FUTURE DIRECTIONS

Despite the progress in understanding the structural motifs comprising DP, there are many important questions that remain unanswered. First, there has still not been a definitive analysis of the precise protein–protein interactions in the Dsg/Dsc/plakoglobin/PKP/DP complex that constitutes the membrane-proximal "outer dense plaque" of the desmosome (North et al., 1999). DPNT has been found to bind to each of the desmosomal cadherins, plakoglobin, and plakophilins. Many of these studies employed indirect methods to detect protein–protein interactions including co-immunoprecipitation, far western blotting, and yeast two hybrid analysis (Bonne et al., 2003; Chen, Bonné, Hatzfeld, van Roy, & Green, 2002; Kowalczyk et al., 1997, 1999; Smith & Fuchs, 1998). To date, only the interaction with the PKP1 head domain and a weaker interaction with the Dsg1 tail have been demonstrated with purified proteins (Al-Jassar et al., 2011; Kami et al., 2009), and no quantitative affinity data have been reported. Similarly, although binding of DPCT to IFs is established, there are neither quantitative binding data nor experimental data locating the IF-binding sites on the PRDs. A second, and potentially related, issue is the role of posttranslational modifications (PTMs) of DP and their effect on protein–protein interactions. The role of PTMs is only now beginning to be appreciated, as shown by the effect of arginine methylation and serine phosphorylation of the DPCT tail on IF association (Albrecht et al., 2015). In this regard, the methods described above for protein purification employ bacterial expression, and thus are unable to provide proteins with native PTMs. As these modifications are discovered, expression in mammalian systems and/or *in vitro* enzymatic modification will be needed to understand the roles of PTMs.

We anticipate that methods for purifying DP will enable a quantitative and structural description of the interactions of desmosomal components. Several missense mutations associated with ARVD/C appear to affect the interactions with other partners, and others may affect protein stability, so structural and affinity data will be needed to understand their effects at a molecular level (Al-Jassar, Bikker, et al., 2013; Yang et al., 2006). Finally, the molecular basis for the role of DP in transmitting mechanical force between desmosomal cadherins and IFs is still unclear. In particular, the presence of spectrin-like repeats of the plakin domain seem likely has fundamental roles in conferring mechanical elasticity to these junctions, but this hypothesis awaits experimental verification.

ACKNOWLEDGMENTS

This work was supported by the Interdisciplinary Research Program from the Research Institute for Basic Sciences, Seoul National University (H.-J.C.) and grant U01 GM094663 from the U.S. National Institutes of Health (W.I.W.).

REFERENCES

Albrecht, L. V., Zhang, L., Shabanowitz, J., Purevjav, E., Towbin, J. A., Hunt, D. F., et al. (2015). GSK3- and PRMT-1-dependent modifications of desmoplakin control desmoplakin-cytoskeleton dynamics. *Journal of Cell Biology*, *208*, 597–612.

Al-Jassar, C., Bernado, P., Chidgey, M., & Overduin, M. (2013). Hinged plakin domains provide specialized degrees of articulation in envoplakin, periplakin and desmoplakin. *PloS One*, *8*, e69767.

Al-Jassar, C., Bikker, H., Overduin, M., & Chidgey, M. (2013). Mechanistic basis of desmosome-targeted diseases. *Journal of Molecular Biology*, *425*(21), 4006–4022.

Al-Jassar, C., Knowles, T., Jeeves, M., Kami, K., Behr, E., Bikker, H., et al. (2011). The nonlinear structure of the desmoplakin plakin domain and the effects of cardiomyopathy-linked mutations. *Journal of Molecular Biology*, *411*, 1049–1061.

Bass-Zubek, A. E., Godsel, L. M., Delmar, M., & Green, K. J. (2009). Plakophilins: Multifunctional scaffolds for adhesion and signaling. *Current Opinion in Cell Biology*, *21*, 708–716.

Bernardo, P., Mylonas, E., Petoukhov, M. V., Blackledge, M., & Svergun, D. I. (2007). Structural characterization of flexible proteins using small-angle X-ray scattering. *Journal of the American Chemical Society*, *129*, 5656–5664.

Bonne, S., Gilbert, B., Hatzfeld, M., Chen, X., Green, K. J., & van Roy, F. (2003). Defining desmosomal plakophilin-3 interactions. *Journal of Cell Biology*, *161*, 403–416.

Bornslaeger, E. A., Corcoran, C. M., Stappenbeck, T. S., & Green, K. J. (1996). Breaking the connection: Displacement of the desmosomal plaque protein desmoplakin from cell-cell interfaces disrupts anchorage of intermediate filament bundles and alters intercellular junction assembly. *Journal of Cell Biology*, *134*, 985–1001.

Chen, X., Bonné, S., Hatzfeld, M., van Roy, F., & Green, K. J. (2002). Protein binding and functional characterization of plakophilin 2. *Journal of Biological Chemistry*, *277*, 10512–10522.

Choi, H.-J., Gross, J. C., Pokutta, S., & Weis, W. I. (2009). Interactions of plakoglobin and β-catenin with desmosomal cadherins: Basis of selective exclusion of α- and β-catenin from desmosomes. *Journal of Biological Chemistry*, *284*, 31776–31788.

Choi, H.-J., Park-Snyder, S., Pascoe, L. T., Green, K. J., & Weis, W. I. (2002). Structures of two intermediate filament-binding fragments of desmoplakin reveal a unique repeat motif structure. *Natural Structural Biology*, *9*, 612–620.

Choi, H.-J., & Weis, W. I. (2011). Crystal structure of a rigid four-spectrin-repeat fragment of the human desmoplakin plakin domain. *Journal of Molecular Biology*, *409*(5), 800–812.

DiColandrea, T., Karashima, T., Määttä, A., & Watt, F. M. (2000). Subcellular distribution of envoplakin and periplakin: Insights into their role as precursors of the epidermal cornified envelope. *Journal of Cell Biology*, *151*, 573–585.

Green, K. J., Parry, D. A. D., Stenert, P. M., Virata, M. L. A., Wagner, R. M., Angst, D. B., et al. (1990). Structure of the human desmoplakins. Implications for function in the desmosomal plaque. *The Journal of Biological Chemistry*, *265*, 2603–2612.

Huber, A. H., & Weis, W. I. (2001). The structure of the β-catenin/E-cadherin complex and the molecular basis of diverse ligand recognition by β-catenin. *Cell*, *105*, 391–402.

Jefferson, J. J., Ciatto, C., Shapiro, L., & Liem, R. K. (2007). Structural analysis of the plakin domain of bullous pemphigoid antigen1 (BPAG1) suggests that plakins are members of the spectrin superfamily. *Journal of Molecular Biology*, *366*(1), 244–257.

Kami, K., Chidgey, M., Dafforn, T., & Overduin, M. (2009). The desmoglein-specific cytoplasmic region is intrinsically disordered in solution and interacts with multiple desmosomal protein partners. *Journal of Molecular Biology*, *386*(2), 531–543.

Koster, S., Weitz, D. A., Goldman, R. D., Aebi, U., & Herrmann, H. (2015). Intermediate filament mechanics *in vitro* and in the cell: From coiled coils to filaments, fibers and networks. *Current Opinion in Cell Biology*, *32*, 82–91.

Kouklis, P. D., Hutton, E., & Fuchs, E. (1994). Making a connection: Direct binding between keratin intermediate filaments and desmosomal proteins. *Journal of Cell Biology*, *127*, 1049–1060.

Kowalczyk, A. P., Bornslaeger, E. A., Borgwardt, J. E., Palka, H. L., Bhaliwal, A. S., Corcoran, C. M., et al. (1997). The amino-terminal domain of desmoplakin binds to plakoglobin and clusters desmosomal cadherin-plakoglobin complexes. *Journal of Cell Biology*, *139*, 773–784.

Kowalczyk, A. P., & Green, K. J. (2013). Structure, function, and regulation of desmosomes. *Progress in Molecular Biology and Translational Science*, *116*, 95–118.

Kowalczyk, A. P., Hatzfeld, M., Bornslaeger, E. A., Kopp, D. S., Borgwardt, J. E., Corcoran, C. M., et al. (1999). The head domain of plakophilin-1 binds to desmoplakin and enhances its recruitment to desmosomes. *Journal of Biological Chemistry, 274*, 18145–18148.

Mathur, M., Goodwin, L., & Cowin, P. (1994). Interactions of the cytoplasmic domain of the desmosomal cadherin Dsg1 with plakoglobin. *Journal of Biological Chemistry, 269*, 14075–14080.

Meng, J.-J., Bornslaeger, E. A., Green, K. J., Steinert, P. M., & Ip, W. (1997). Two-hybrid analysis reveals fundamental differences in direct interactions between desmoplakin and cell type-specific intermediate filaments. *Journal of Biological Chemistry, 272*, 21495–21503.

Nikolic, B., Nulty, E. M., Mir, B., & Wiche, G. (1996). Basic amino acid residue cluster within nuclear targeting sequence motif is essential for cytoplasmic plectin-vimentin network junctions. *Journal of Cell Biology, 134*, 1455–1467.

North, A. J., Bardsley, W. G., Hyam, J., Bornslaeger, E. A., Cordingley, H. C., Trinnaman, B., et al. (1999). Molecular map of the desmosomal plaque. *Journal of Cell Science, 112*(Pt 23), 4325–4336.

Parry, D. A., & Steinert, P. M. (1999). Intermediate filaments: Molecular architecture, assembly, dynamics and polymorphism. *Quarterly Reviews of Biophysics, 32*, 99–187.

Smith, E. A., & Fuchs, E. (1998). Defining the interactions between intermediate filaments and desmosomes. *Journal of Cell Biology, 141*, 1229–1241.

Sonnenberg, A., & Liem, R. K. H. (2007). Plakins in development and disease. *Experimental Cell Research, 313*, 2189–2203.

Sonnenberg, A., Rojas, A. M., & de Pereda, J. M. (2007). The structure of a tandem pair of spectrin repeats of plectin reveals a modular organization of the plakin domain. *Journal of Molecular Biology, 368*, 1379–1391.

Stappenbeck, T. S., & Green, K. G. (1992). The desmoplakin carboxyl terminus coaligns with and specifically disrupts intermediate filament networks when expressed in cultured cells. *Journal of Cell Biology, 116*, 1197–1209.

Wahl, J. D., Sacco, P. A., McGranahan-Sadler, T. M., Sauppe, L. M., Wheelock, M. J., & Johnson, K. R. (1996). Plakoglobin domains that define its association with the desmosomal cadherins and the classical cadherins: Identification of unique and shared domains. *Journal of Cell Science, 109*, 1143–1154.

Witcher, L. L., Collins, R., Puttagunta, S., Mechanic, S. E., Munson, M., Gumbiner, B., et al. (1996). Desmosomal cadherin binding domains of plakoglobin. *Journal of Biological Chemistry, 271*, 10904–10909.

Yang, Z., Bowles, N. E., Scherer, S. E., Taylor, M. D., Kearney, D. L., Ge, S., et al. (2006). Desmosomal dysfunction due to mutations in desmoplakin causes arrhythmogenic right ventricular dysplasia/cardiomyopathy. *Circulation Research, 99*, 646–655.

Yu, R. C., Jahn, R., & Brünger, A. T. (1999). NSF N-terminal domain crystal structure: Models of NSF action. *Molecular Cell, 4*, 97–107.

CHAPTER TWELVE

Degradation of the Intermediate Filament Family by Gigaxonin

Pascale Bomont[*,†,1]
[*]Atip-Avenir Team, Montpellier, France
[†]INM Inserm U1051, Université de Montpellier, Montpellier, France
[1]Corresponding author: e-mail address: pascale.bomont@inserm.fr

Contents

Abstract

Intermediate filament turnover is a highly dynamic process required to maintain tissue integrity and is implicated in degenerative and regenerative processes. Despite these essential roles, little is known about the mechanisms that cause the degradation of intermediate filaments. Nevertheless, the last decade has seen the emergence of the ubiquitin proteasome system, in particular E3 ubiquitin ligases, as important regulators. Here, we will focus on the first identified factor controlling the degradation of the entire intermediate filament family, the gigaxonin-E3 ligase. We will present the scientific achievements and the methodologies to study gigaxonin and its crucial role in intermediate filament turnover.

1. INTRODUCTION

Analysis of the turnover of neurofilaments (NFs) in mice (using *in vivo* radioisotope labeling or an inducible system) (Millecamps, Gowing, Corti, Mallet, & Julien, 2007; Nixon & Logvinenko, 1986) revealed that NFs undergo degradation along the entire axons and that this is regulated by

Methods in Enzymology, Volume 569
ISSN 0076-6879
http://dx.doi.org/10.1016/bs.mie.2015.07.009

the NF network itself (Rao, Yuan, Campbell, Kumar, & Nixon, 2012). The importance of the local proteolysis of NFs has been revealed in both neurodegenerative and -regenerative processes (Cassereau et al., 2013; Wang, Medress, & Barres, 2012).

While it is established that a turnover of NFs exists *in vivo*, very little is known about the mechanisms that control their degradation. Mild proteolysis of NFs can be caused by various proteases (Perrot, Berges, Bocquet, & Eyer, 2008). To date, there is no direct evidence of a role of autophagy in the degradation of NFs. Supporting the initial finding that purified NFs from bovine spinal cord are ubiquitinated and can be deubiquitinated (Gou & Leterrier, 1995), three E3 ubiquitin ligases (Trim2, gigaxonin, and DCAF8) have been identified in neurodegenerative diseases presenting an abnormal aggregation of NFs (Bomont et al., 2000; Klein et al., 2014; Ylikallio et al., 2013). While these studies suggest that NF accumulation could result from impaired degradation induced by mutations in the E3 ligases, more direct evidences showed that Trim2 selectively binds and ubiquitinates the NF-L subunit (Balastik et al., 2008), but without affecting other intermediate filaments (IFs) (Ylikallio et al., 2013).

We present here gigaxonin, the substrate adaptor of an E3 ubiquitin ligase as the first regulator of the IF family, controlling the turnover of NFs but also of other IF types. Gigaxonin is mutated in giant axonal neuropathy (GAN) (Bomont et al., 2000), an extremely severe and rare neurodegenerative disorder characterized by a general deterioration of the nervous system and an extensive aggregation of IFs throughout the body (Kuhlenbaumer, Timmerman, & Bomont, 1993). The analysis of nerve biopsies from patients revealed that concomitantly to axonal degeneration and an abnormal accumulation of NFs (Asbury, Gale, Cox, Baringer, & Berg, 1972), many other IFs become similarly disorganized in other tissues. Thus, keratin, desmin, vimentin, GFAP, and NFs were shown to aggregate in GAN (Prineas, Ouvrier, Wright, Walsh, & McLeod, 1976), which therefore represents to our knowledge the only disease affecting all IFs, and hence supporting the key fundamental role(s) of gigaxonin in sustaining IF organization. In this review, we will present the studies which revealed that gigaxonin controls the degradation of the IF family, as well as the methodologies to study gigaxonin and its role in IF turnover.

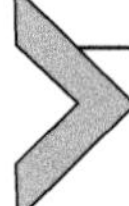

2. GIGAXONIN EXPRESSION AND LOCALIZATION

2.1 Broad and Low Expression

The *GAN* mRNA (4.7 kb) is transcribed ubiquitously at a low level, as shown by the results of Northern blot analysis on human tissues and

RT-PCR on multiple mouse tissues (Bomont et al., 2000). While the corresponding protein with an approximate molecular weight of 65 kDa can be easily detected in an overexpression system by Western blot and immunoprecipitation analysis (Bomont & Koenig, 2003), it is much more difficult to detect endogenous gigaxonin. We have made considerable efforts in the past years to detect endogenous gigaxonin, combining the generation of 32 mono/polyclonal antibodies (from multiple human antigens: the N-terminal BTB portion of gigaxonin, its full-length counterpart, or distinct peptides) with the creation of internal controls for their specificity, i.e., lymphoblast cell lines of GAN patients and a GAN knockout model. Thus, using our monoclonal antibody GigA on immunoblot (Cleveland, Yamanaka, & Bomont, 2009), we demonstrated that the molecular weight of gigaxonin is lower than expected (65 kDa) and is expressed ubiquitously but preferentially in the nervous system (Fig. 1A and B; Ganay et al., 2011). The very low amount of gigaxonin, estimated to be 1.25×10^{-3}% of the total amount of proteins in detergent-soluble lysates of mouse brain, and 7500 molecules in human lymphoblasts (Cleveland et al., 2009) accounts for the difficulty to identify the specific protein using (non)-commercially available antibodies with a high background reactivity (see notes in Section 2.1.2 and Fig. 1C).

2.1.1 Protocol

2.1.1.1 Tissues

1. The volume of lysis buffer is calculated as a *v*/*w* ratio of 5:1 (*v*: total volume in ml of the lysis buffer; *w*: weight in gram of the tissue).
2. Snap-frozen tissues are first homogenized on ice with a 1 × lysis buffer without detergent and with proteases inhibitors (150 m*M* NaCl; 50 m*M* Tris, pH 7.5; 0.1 m*M* DTT; 10 μg/ml aprotinin; 10 μg/ml leupeptin; 10 μg/ml pepstatin A; 10 μg/ml chymostatin; 1 m*M* PMSF).
3. Subsequently, 1 × detergent is added to the samples (from a 10 × stock solution of 10% Triton-X100; 10% sodium deoxycholate; 1% SDS), and samples are kept on ice for 30 min.
4. The samples are centrifuged at 12,000 × *g* for 10 min at 4 °C and the supernatants are recovered. Protein content is determined by Bradford prior to storage (preferably in small aliquot at −20 °C).

2.1.1.2 Cells

1. Cells are lysed in 50 m*M* Tris, pH 7.5; 150 m*M* NaCl; 1% Triton X-100; 5 m*M* EDTA; 10 μg/ml leupeptin; 10 μg/ml pepstatin A; 10 μg/ml chymostatin.

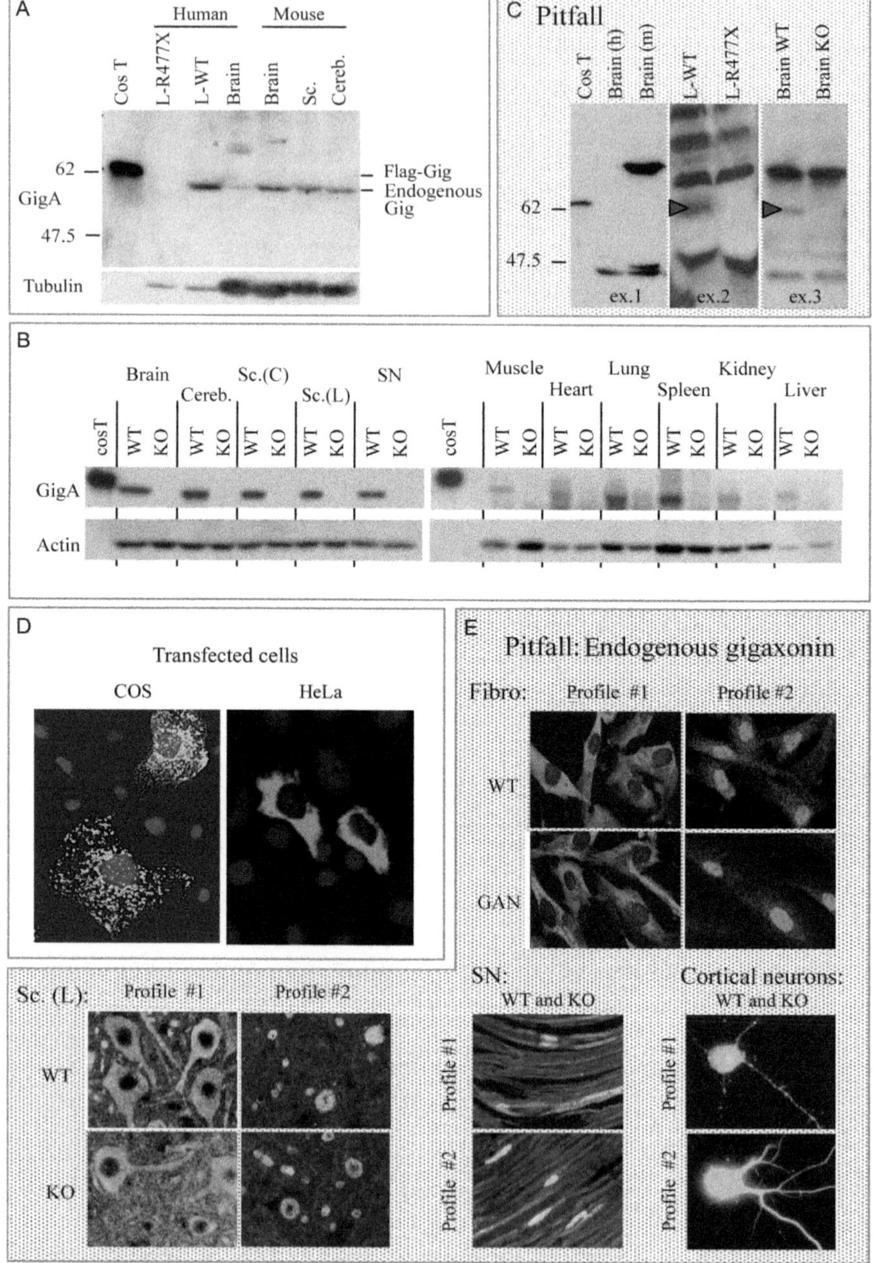

Figure 1 Broad and low abundance of gigaxonin. (A) The monoclonal GigA antibody detects endogenous gigaxonin at a molecular weight lower than expected (a 65 kDa band with a Flag-GAN construction transfected in Cos cells: Cos-T), in human samples ("L"-human lymphoblast cell line, "Brain") and mouse tissues ("Brain," spinal cord "Sc.,"

2. Proteins, their amounts being estimated by Bradford, are denatured for 5 min at 95 °C, in sample buffer (final concentration: 62.5 m*M* Tris, pH 6.8; 2% SDS; 10% glycerol; 2% β-mercaptoethanol; 0.01% bromophenol blue).
3. Samples are loaded on a 10% SDS-PAGE.
4. Wet transfer or semi-dry systems (like i-blot from Invitrogen) can be used to transfer proteins to membranes. The buffer for wet transfer is 25 m*M* Tris, pH 8.3; 192 m*M* glycine; 20% ethanol.

2.1.2 Notes and Potential Pitfalls

A. Endogenous gigaxonin is highly sensitive to the effect of freeze–thaw cycles of the samples. Accordingly, it is important to perform Western blot analysis directly after cell lysis and to aliquot the remaining proteins after determining their protein concentration.

B. 80 μg of protein is needed to perform a Western blot on endogenous protein using wet transfer, but 50 μg is sufficient for the semi-dry system. Transfer with ethanol is more efficient than with methanol in detecting endogenous gigaxonin.

C. So far, the quantification of gigaxonin in human and mouse samples by Western blot has not revealed variations in the level of the following standard normalizers: tubulin, GAPDH, or actin.

Figure 1—Cont'd and cerebellum "Cereb."). GigA antibody is described in Cleveland et al. (2009). Specificity of detection is verified in a lymphoblast cell line derived from a patient bearing a homozygous nonsense mutation at amino acid position 477 (L-R477X). (B) Ubiquitous but preferential expression of gigaxonin in mouse neuronal tissues. Control for specificity is demonstrated by the corresponding tissues from the GAN knockout model "KO.," "Sc. (C)," "Sc. (L)," and "SN" stand for cervical, lumbar sections of the spinal cord, and sciatic nerve, respectively. (C) Examples of aspecific detection using other polyclonal antibodies (ex. 1–3 correspond, respectively, to Ding et al. (2002), Cullen et al. (2004), and Sigma #SAB4200104). "Brain (h) and (m)" correspond to human and mouse brain samples, respectively. The arrowhead points to the endogenous gigaxonin. Cellular distribution of gigaxonin. (D) A fine or large granular distribution in the cytoplasm of transfected HeLa and Cos cells defines ectopic expression of gigaxonin. (E) Absence of specific detection of the endogenous gigaxonin with available gigaxonin antibodies. Examples presented include primary fibroblasts "Fibro," lumbar sections of the spinal cord "Sc. (L)" and longitudinal sections of sciatic nerves "SN" of mouse, as well as the derived E15.5 cortical neurons. As demonstrated by the internal controls (GAN fibroblasts and GAN KO tissues/cells), the two distinct profiles revealed by the screening of all 32 gigaxonin antibodies are not specific. *Panel (A) adapted from Cleveland et al. (2009) by permission of Oxford University Press; Panel (B) adapted from Ganay, Boizot, Burrer, Chauvin, and Bomont (2011). Copyright (2011) with permission from Biomed Central.* (See the color plate.)

D. It is crucial to add GAN mutant samples as internal controls (GAN patient cells, GAN KO mouse tissues, or RNAi samples), as many Gig-antibodies (from commercial and noncommercial providers) detect many aspecific bands (see three different examples in Fig. 1C).

2.2 Gigaxonin Localization

2.2.1 Transfected Cells

Ectopically overexpressed gigaxonin is easily detected. Its distribution is in a fine or large granules in the cytoplasm of several cell types, including HeLa and Cos cells (Bomont & Koenig, 2003; Fig. 1D). The aggregate pattern of overexpressed gigaxonin does not become colocalized with early endosome, endoplasmic reticulum, and Golgi, as assessed by coimmunostaining with the EEA1, PDI, and GM130 antigens, respectively (personal, unpublished results).

2.2.2 Endogenous Protein

Regarding the endogenous gigaxonin, rare studies reported its localization in rat/human tissues or mouse neurons (Cullen et al., 2004; Ding et al., 2002) but the aspecific band(s) revealed by the antibodies used in Western blotting (Fig. 1C), combined with the lack of internal controls for specificity seriously hampers a proper conclusion. To address this important issue, we systematically screened our 32 monoclonal/polyclonal gigaxonin antibodies as well as commercially available reagents in WT and gigaxonin-depleted samples. In spite of extensive efforts in adjusting the methodologies, we were unable to reveal any specific staining by immunofluorescence in skin-derived fibroblasts, primary motor, and cortical neurons, or spinal cord and sciatic nerves of mice (Fig. 1E).

While we could not exclude that a residual mutated gigaxonin persists in patient fibroblasts and that an isoform of gigaxonin could be upregulated in the KO mouse, our data suggests that gigaxonin is too weakly expressed to be detected with the methodologies applied.

2.2.2.1 Pitfall

It is essential to add internal controls for specificity, using either patient-derived fibroblasts, KO tissues or alternatively, RNAi-transfected cells.

2.3 Substrate Adaptor of E3 Ubiquitin Ligases

The *GAN* gene encodes for gigaxonin, a new BTB-Kelch protein (Bomont et al., 2000), which constitutes a key subunit of Cul3-E3 ubiquitin ligases. Indeed, as the substrate adaptor of this enzymatic complex (Furukawa, He,

Borchers, & Xiong, 2003), gigaxonin would promote the specificity of ubiquitination on its targets and degradation by the proteasome. Accordingly, gigaxonin has been shown, in an overexpression system, to promote the ubiquitination and degradation of three microtubule-associated/chaperone proteins (MAP1B, MAP1S, and TBCB), via interaction with the Kelch domain (Allen et al., 2005; Ding et al., 2006; Wang et al., 2005). These findings, pointing to a possible regulation of microtubule components by gigaxonin are now challenged by the modest or absent increase of these proteins in both GAN patients and mouse gigaxonin-depleted samples (Cleveland et al., 2009; Ganay et al., 2011; Mahammad et al., 2013). Moreover, the modest contribution of the known factors causing neuronal death (Allen et al., 2005; Ding et al., 2006) and IF aggregation in human fibroblasts (more details in Section 3.2) (Cleveland et al., 2009) strongly suggests the existence of other factors more relevant for the disease.

3. ANALYZING GIGAXONIN ROLE IN IF ORGANIZATION

The fundamental role of gigaxonin in regulating the IF family emerged in the 1970s, when multiple analyses of patient biopsies revealed that not only NFs but also other IF types aggregate in the disease (Asbury et al., 1972; Prineas et al., 1976). Forty years later, key steps have been achieved, i.e., the identification of the *GAN* gene, the generation of important tools to study gigaxonin, and the creation of several GAN mouse models, allowing to reveal the broad disorganization of type I/II (keratin), type III (vimentin, desmin, GFAP, peripherin), and type IV (NFs, α-internexin) filaments in GAN. In this review, we will focus on the most extensively studied IFs in GAN, namely the NFs and vimentin, which identified gigaxonin as the first factor able to control the degradation of the IF family.

3.1 Neuronal IFs: GAN Mouse

The NF network is abnormally compact and disorganized within giant axons throughout the nervous system of GAN patients. To study these alterations, we and others have independently generated a knockout model for GAN, deleting either the promotor—$GAN^{\Delta ex1/\Delta ex1}$—(Dequen, Bomont, Gowing, Cleveland, & Julien, 2008) or early exons—$GAN^{\Delta ex3-5/\Delta ex3-5}$—(Ding et al., 2006; Ganay et al., 2011) of the *GAN* murine gene. While in none of these models, the pathology is as severe as in human patients—late onset, mild sensory/motor deficits with no overt neurodegeneration—(Dequen et al., 2008; Ganay et al., 2011), they present an IF disorganization reminiscent of GAN patients. The ultrastructural examination of the axoplasms of GAN nerves revealed an alteration of the spatial distribution of NFs in sciatic nerves, ventral, and dorsal roots of 48-week-old animals (Ganay et al., 2011). As in patients, NFs are not regularly orientated along the axis of the axon and can exhibit multiple orientations, as measured by the decreased CIRC value of all individual NF within a cross-section (Fig. 2A and Section 3.1.1 for details on the CIRC measurement).

While in these models, NFs are not as dramatically compact as in patients, they exhibit another feature of the human pathology. Indeed, like in patients (Donaghy, King, Thomas, & Workman, 1988), the mean diameter of NFs is increased in GAN peripheral nerves (Ganay et al., 2011), which may reflect impaired phosphorylation of the NFs, as demonstrated by the hypophoshorylation of NF-H in the brain of the GAN mouse (Dequen et al., 2008).

Concomitant to the alteration of their spatial distribution, the levels of NFs are increased in the nerves of GAN mouse (Fig. 2B). We showed that the alteration in NF content is comparable in the three GAN models and is more pronounced in the brain for the three subunits NF-L, NF-M, and NF-H. Most spectacularly, NF-L abundance increases four- to sevenfold in the brain of 48-week-old GAN mice (Ganay et al., 2011) and, in the sciatic nerve, is in the twofold range that was revealed in a GAN patient (Ionasescu et al., 1983). Independently, another study revealed the increased levels of the NF triplets, α-internexin, and vimentin as early as 3 months of age, in the brain, cerebellum, and spinal cord of $GAN^{\Delta ex1/\Delta ex1}$ mice (Dequen et al., 2008). Interestingly, the histological analysis of different structures of the central nervous system (Fig. 2C) showed an abnormal accumulation of NF-H and α-internexin in the cortex and the thalamus, NF-L in spinal motor neurons, and peripherin in the DRG of $GAN^{\Delta ex1/\Delta ex1}$ mice.

3.1.1 Protocol

The quantification of the alteration of NFs (Fig. 2A) was performed on electron micrographs by Image J, using a plug-in we developed to measure the circularity—CIRC—of individual NF. $CIRC = 4\pi * area/perimeter^{\wedge}2$. A value of 1 indicated a perfect circle. A value approaching 0 indicates an increasing elongated shape, and therefore an orientation not along the axis of the axons but in a perpendicular axis. The decreased CIRC value in KO mouse cannot be attributed to an improper cross-sectioning, as the standard deviation of the average circularity—which is representative of the variations in the orientation of individual NF within each micrograph—is increased in KO tissues. The determination of gigaxonin abundance in GAN tissues (Fig. 2B) is performed as described in Section 2.1.1.

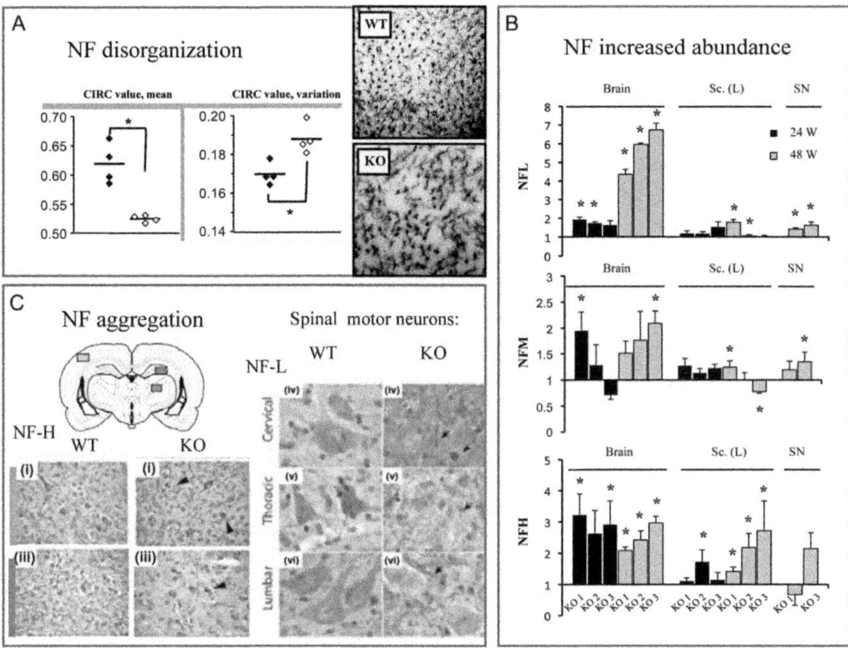

Figure 2 Severe alterations of NFs in GAN mice. (A) Disorganization of the spatial distribution of NFs in gigaxonin-depleted nerves. Electron micrographs of the axoplasms from transversal sections of GAN nerves show an abnormal orientation of the filaments concomitant to normally oriented NF along the axons. Illustrated here for the sciatic nerve, the quantification of NF-altered orientation is obtained by the measurement of the circularity of individual neurofilament (CIRC = 1 and CIRC < 1 representing a perfect circle and an elongated shape, respectively). The left panel displays the average circularity scores, measuring the general orientation of the neurofilaments in individual mice (the mean score per genotype is represented by a bar). The right panels show the standard deviations of the average circularity scores, a measure that is representative of the variations in the orientation of individual neurofilaments within each tissue section. (B) Increased abundance of NF subunits in the GAN models. The relative increase of protein content was established by comparing the mean abundance in each of the three GAN models (KO1 from Ganay et al., 2011; KO2 from Ding et al., 2006; KO3 from Dequen et al., 2008) with WT mice. Expression levels were quantified using anti-NF-L, NF-M, and NF-H antibodies and after normalization with GAPDH. (C) Accumulation of neuronal IFs in the central nervous system of GAN mice. Examples of an immunohistochemical analysis, showing the aggregation of NF-H in the cortex and thalamus and the accumulation of NF-L in the spinal motor neurons of the GAN$^{\Delta ex1/\Delta ex1}$ (KO3) mouse. *Panels (A) and (B) adapted from Ganay et al. (2011). Copyright (2011) with permission from Biomed Central; Panel (C) adapted from Dequen et al. (2008). Copyright (c) [John Wiley & Sons].*

3.2 Nonneuronal IFs: Patient's Skin Fibroblasts

Skin-derived primary fibroblasts from GAN patients were used as a model to study the aggregation of type III IF vimentin in disease (Pena, 1981). Since the initial description of the concentric bundle of vimentin in patients, few studies have revealed the high variability in the percentage of vimentin aggregation. The latter can differ between patients, but the results obtained with cells from the same patients may differ between different laboratories. This suggests a great sensibility to culture conditions and/or the age of the cells.

With the discovery of the *GAN* gene, we were able to push this analysis forward and study multiple fibroblasts of GAN patients with identified mutations (Bomont & Koenig, 2003). In normal conditions (10% serum), the vimentin network is disorganized and usually forms a single ovoid bundle located in the close proximity of the nucleus (Fig. 3A). The dense aggregates are highly resistant to detergent and locally distend the cells, as shown in Nomarski contrast. This phenotype is partial as the aggregates co-exist with a well-formed vimentin network within a cell, and only 3–15% of the fibroblasts exhibit these aggregates. Interestingly, the phenotype can be exacerbated and reversed by culture conditions such as a low serum content or at confluence (during 3 days), without being related to the stationary phase G0 of the cell cycle (Bomont & Koenig, 2003). In these conditions, not only vimentin aggregation can reach a 5–20-fold increase (with 48–88% of aggregation) but more pronounced compaction is also formed, ranging from two ovoid bundles to a perinuclear ring with all intermediate forms (Fig. 3A). Altogether, this study reveals that while gigaxonin depletion generates compact and insoluble bundling of vimentin, its aggregation process is highly dynamic.

The identification of microtubule-associated protein as substrate for ubiquitination by gigaxonin in an overexpressing system has misled the field, postulating that the excess of TBCB—due to impaired degradation by gigaxonin depletion—destroys the microtubule network and could therefore be responsible for the aggregation of IFs in GAN (Wang et al., 2005). Several studies now prove that, unlike the tubulin chaperone TBCE, which completely upsets the structure of the MT network, overexpression of TBCB is mildly affecting MTs. Indeed, the MT network is not perturbed in $64 \pm 4\%$ of the cells and is at best slightly reduced in intensity in remaining cells (Cleveland et al., 2009; Kortazar et al., 2007). Additionally, the TBCB level appeared not to be elevated in fibroblasts of several GAN patients (Cleveland et al., 2009) or to be reduced upon gigaxonin overexpression (Mahammad et al., 2013). Moreover, the MT density/distribution does not

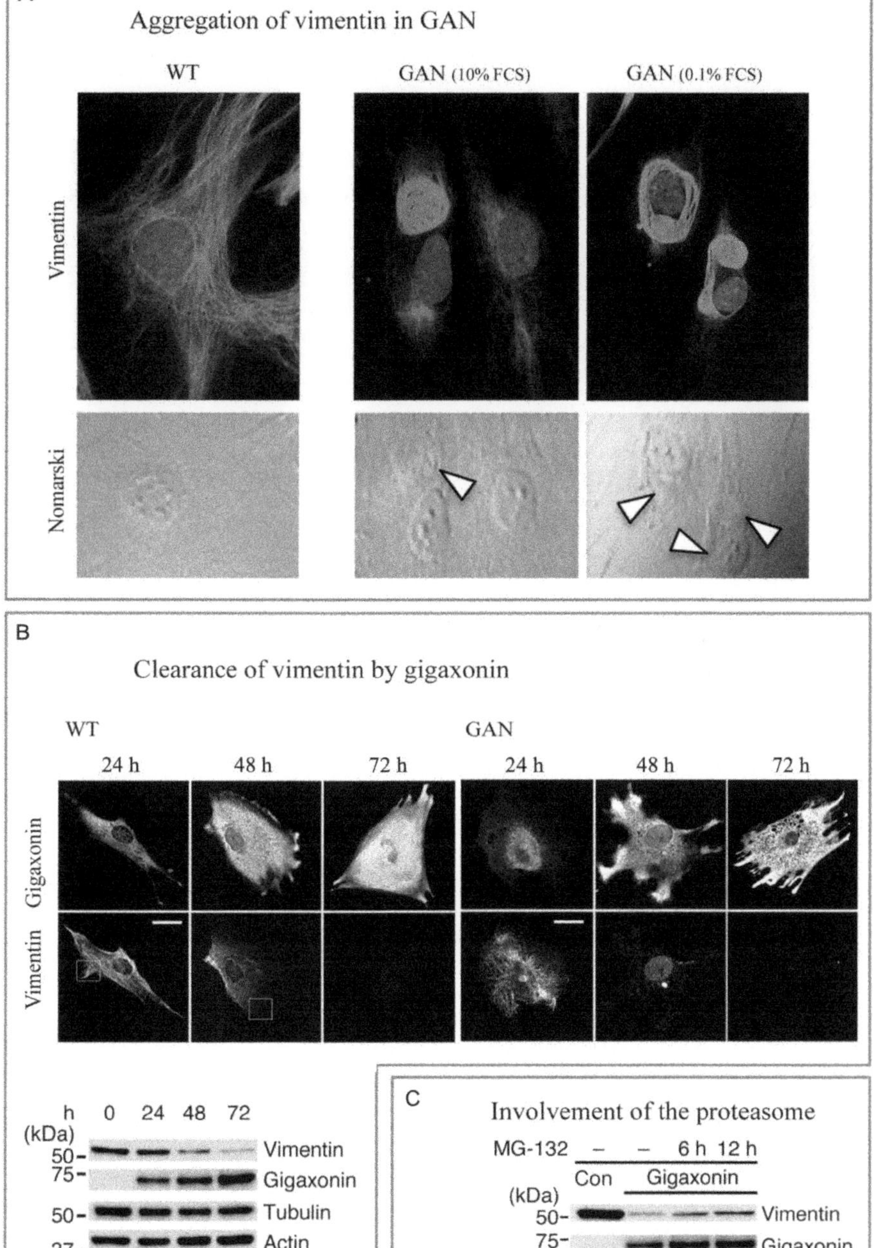

Figure 3 See legend on next page.

noticeably differ between control and patient fibroblasts, Finally, we also showed that the destruction of MTs does not generate the GAN-like aggregate in control cells but exacerbates vimentin aggregation in patients (Bomont & Koenig, 2003; Cleveland et al., 2009), therefore rather suggesting that the MT network is protective from the total collapse of vimentin in GAN. Thus, the evidence makes the proposal of microtubule instability from elevated TBCB in the absence of gigaxonin an unlikely explanation for the effect of gigaxonin on IF arrays, especially outside the nervous system, e.g., in skin-derived fibroblasts. In neuronal tissues, the reduced MT density observed by several groups in GAN KO mice (Ganay et al., 2011; Wang et al., 2005) may result from the aggregated NFs, which have been shown to bind soluble tubulin and therefore regulate locally the polymerization of microtubules (Bocquet et al., 2009).

Thus, GAN is a disease of IFs that has an impact on the microtubule network in neuronal systems through a yet unknown mechanism.

3.2.1 Protocol

Primary fibroblasts are maintained in DMEM with 10 or 0.1% fetal calf serum (FCS), supplemented with 40 μg/ml gentamycin and 1% penicillin/streptomycin.

Figure 3 Regulation of vimentin by gigaxonin in health and disease. (A) Aggregation of vimentin in GAN fibroblasts is partial, conditional, and highly dynamic. Vimentin forms dense perinuclear ovoid bundles in GAN fibroblasts, which are visible by Nomarski contrast (arrowheads) and co-exist with a well-organized vimentin network within the same cell. The phenotype is partial as only a subset of cells contains such aggregates but depends on the conditions, as it is exacerbated upon serum deprivation (0.1% FCS). Under this condition, not only the proportion of cells which aggregate increases drastically, but the aggregates are also more pronounced and form dense perinuclear rings. (B) Clearance of vimentin upon gigaxonin overexpression. The time course of vimentin degradation in human fibroblasts upon viral-mediated delivery of Flag-tagged gigaxonin. The progressive destruction of vimentin, assessed both by Western blot on total cell extracts and immunofluorescence, begins 48 h after gigaxonin surexpression and is effective on vimentin aggregates at 72 h posttransduction in GAN fibroblasts. (C) Involvement of the proteasome in the gigaxonin-mediated clearance of vimentin. Human fibroblasts deprived of vimentin by the expression of gigaxonin for 72 h are treated with the proteasome inhibitor MG-132 for 6–12 h. A slight increase in the amount of vimentin is obtained upon treatment, indicating a partial blocking of vimentin degradation by the proteasome in primary fibroblasts. *Panel (A) adapted from Bomont and Koenig (2003) by permission of Oxford University Press; Panels (B) and (C) reproduced from Mahammad et al. (2013) with permission of J Clin Invest.* (See the color plate.)

1. The cells are fixed for 10 min in 4% paraformaldehyde (PFA in PBS 1 ×).
2. Three washing of 5 min with PBS 1 × are performed.
3. The cells are incubated for 1 h in the blocking buffer (0.1% Triton-X100, 3% BSA in PBS 1 ×).
4. Primary antibodies prepared in the blocking buffer are incubated O/N at 4 °C.
5. Three washing with PBS 1 × are performed.
6. Secondary antibodies are added in the blocking buffer and incubated for 1 h.
7. Three washing with PBS 1 × are performed.
8. The cells are counterstained with Hoechst (Sigma, 5 mg/ml) for 10 s.
9. The cells are washed with PBS 1 ×.
10. And mounted with antifade reagent.

3.2.2 *Notes*

Primary fibroblasts usually exhibit a high phenotypic variability, but culture conditions have been shown to greatly influence the degree of vimentin disorganization in GAN. Trypsin, serum percentage, confluence, and number of passages are parameters that influence the aggregation. Thus, besides the variability that exists between patients, laboratories have found a great discrepancy in the percentage of cells presenting aggregates from the same patients. Accordingly, it is essential to simultaneously compare multiple controls and multiple patients.

3.3 IF Degradation Is Controlled by Gigaxonin

Different hypotheses have been proposed for the mechanisms responsible for IF aggregation in GAN, including impaired assembly, distribution, transport, or degradation. The identification of gigaxonin as the substrate adaptor of an E3 ligase complex provided the concept that gigaxonin controls the degradation either of a regulator of IF dynamics or of the IFs themselves. Supporting the latter proposal, increased levels of neuronal IFs have been revealed in GAN mouse tissues (Dequen et al., 2008; Ganay et al., 2011).

With the goal to reverse the vimentin aggregation in patient fibroblasts, a recent study demonstrated the formidable and unique control that gigaxonin exerts on IFs. Indeed, ectopic expression of gigaxonin using a lentivirus expression system induces a complete clearance of the vimentin network, not only in control healthy cells but also in GAN fibroblasts and MEFs derived from the GAN mouse model (Mahammad et al., 2013). The kinetics of vimentin destruction targets first the well-formed network at 48 h of gigaxonin expression to dissolve the aggregates after 72 h (Fig. 3B). Importantly, this total destruction, visible both by immunofluorescence and Western blot on total cell extracts also extends to other IFs, including peripherin and NF-L in differentiated neuronal cells. The study of several deletion mutants of vimentin shows that gigaxonin is able to degrade vimentin fragments containing the central ROD domain common to all IFs, and that their incorporation onto a filament is not required for this (Mahammad et al., 2013). This degradation is mediated by an interaction between gigaxonin and vimentin, that may be direct or indirect, and that involves the proteasome. Indeed, while the ubiquitination of vimentin has not been established yet, the screening of lysosome, autophagy, or proteasome inhibitors shows a partial restoration of vimentin upon MG132 treatment (Fig. 3C).

Altogether, this study establishes gigaxonin as the first factor known to date able to control the degradation of several IFs.

3.3.1 Protocol

1. Gigaxonin lentiviruses are produced by cotransfection of the plasmid pLex-MCS-Flag-gigaxonin with the helper plasmids pVSVG and pAX2 into 293FT cells, using the Xfect transfection reagent.
2. Two days after transfection, the culture supernatants are collected.
3. Control and GAN primary fibroblasts are incubated with viral supernatant with 8 μg/ml polybrene for 4–8 h, after which the virus-containing medium is replaced by fresh medium.
4. At different times after viral transduction (24, 48, 72 h), cells are processed for immunofluorescence and Western blot, as described in Sections 2.1.1 and 3.2.1.
5. For proteasome inhibition (Fig. 3C), control fibroblasts are incubated with viral supernatant as described above, cultured for 72 h to induce vimentin clearance, and subsequently treated with 12 μ*M* MG-132 for 6 and 12 h.

4. FUTURE PERSPECTIVES

In this review, we have described the function of gigaxonin as the substrate adaptor of an E3 ubiquitin ligase, and its role in regulating the IF network in neuronal and nonneuronal systems. While much has been achieved, several important questions remain to be answered. One major interest is to develop new tools for defining the subcellular localization of gigaxonin.

Confirming a direct interaction of gigaxonin with IF proteins at a physiological level would be very important to establish the mechanisms of IF degradation induced by gigaxonin. Although the proteasome has been shown to participate in the gigaxonin-mediated degradation of vimentin, the precise mechanisms responsible for it remain to be defined. In particular, demonstrating the ubiquitination of IFs by gigaxonin is essential, as well as developing more robust cellular systems to confirm their proteasome-dependent degradation. Additionally, it would be interesting to identify which kind of IFs is a target for degradation: soluble IFs or intermediate form of polymers (short IFs, unit length filaments, dimers, etc.). Beyond IFs, identifying the other targets of gigaxonin will be particularly essential to study the neurodegenerative pathways engaged in GAN. Indeed, the modest contribution of known partners of gigaxonin to neuronal death strongly suggests the existence of other target proteins more relevant for the disease. In the absence of a strong mouse model for GAN, this aspect will require the development of a new animal model, which mimics the severity of the human pathology. With regard to the ongoing gene therapy approach for GAN, determining the exact dosage of reintroduced gigaxonin that can reverse IF aggregation without completely destroying the network in the long term would be essential.

Forty years after the initial description of GAN and the observation of the extensive disorganization of IFs throughout the body, it is now established that gigaxonin is the first factor controlling the degradation of the IF family. Thus, studying a rare and neglected disorder allowed to shed light onto the degradation process of IFs, a poorly explored area of IF biology. Pursuing the study of gigaxonin in the future may be full of promises and will most certainly have impact on other IF-related diseases, in particular the common neurodegenerative diseases characterized by NF aggregation.

ACKNOWLEDGMENTS

This work is funded by l'Institut National de la Santé et de la Recherche Médicale (INSERM, P. B.) and grants from the Atip-Avenir program (CNRS-INSERM), the Région Languedoc-Roussillon, and the Association Française contre les Myopathies (AFM).

REFERENCES

Allen, E., Ding, J., Wang, W., Pramanik, S., Chou, J., Yau, V., et al. (2005). Gigaxonin-controlled degradation of MAP1B light chain is critical to neuronal survival. *Nature*, *438*(7065), 224–228.

Asbury, A. K., Gale, M. K., Cox, S. C., Baringer, J. R., & Berg, B. O. (1972). Giant axonal neuropathy—A unique case with segmental neurofilamentous masses. *Acta Neuropathologica*, *20*(3), 237–247.

Balastik, M., Ferraguti, F., Pires-da Silva, A., Lee, T. H., Alvarez-Bolado, G., Lu, K. P., et al. (2008). Deficiency in ubiquitin ligase TRIM2 causes accumulation of neurofilament light chain and neurodegeneration. *Proceedings of the National Academy of Sciences of the United States of America, 105*(33), 12016–12021.

Bocquet, A., Berges, R., Frank, R., Robert, P., Peterson, A. C., & Eyer, J. (2009). Neurofilaments bind tubulin and modulate its polymerization. *The Journal of Neuroscience, 29*(35), 11043–11054.

Bomont, P., Cavalier, L., Blondeau, F., Ben Hamida, C., Belal, S., Tazir, M., et al. (2000). The gene encoding gigaxonin, a new member of the cytoskeletal BTB/kelch repeat family, is mutated in giant axonal neuropathy. *Nature Genetics, 26*(3), 370–374.

Bomont, P., & Koenig, M. (2003). Intermediate filament aggregation in fibroblasts of giant axonal neuropathy patients is aggravated in non dividing cells and by microtubule destabilization. *Human Molecular Genetics, 12*(8), 813–822.

Cassereau, J., Nicolas, G., Lonchampt, P., Pinier, M., Barthelaix, A., Eyer, J., et al. (2013). Axonal regeneration is compromised in NFH-LacZ transgenic mice but not in NFH-GFP mice. *Neuroscience, 228*, 101–108.

Cleveland, D. W., Yamanaka, K., & Bomont, P. (2009). Gigaxonin controls vimentin organization through a tubulin chaperone-independent pathway. *Human Molecular Genetics, 18*(8), 1384–1394.

Cullen, V. C., Brownlees, J., Banner, S., Anderton, B. H., Leigh, P. N., Shaw, C. E., et al. (2004). Gigaxonin is associated with the Golgi and dimerises via its BTB domain. *Neuroreport, 15*(5), 873–876.

Dequen, F., Bomont, P., Gowing, G., Cleveland, D. W., & Julien, J. P. (2008). Modest loss of peripheral axons, muscle atrophy and formation of brain inclusions in mice with targeted deletion of gigaxonin exon 1. *Journal of Neurochemistry, 107*(1), 253–264.

Ding, J., Allen, E., Wang, W., Valle, A., Wu, C., Nardine, T., et al. (2006). Gene targeting of GAN in mouse causes a toxic accumulation of microtubule-associated protein 8 and impaired retrograde axonal transport. *Human Molecular Genetics, 15*(9), 1451–1463.

Ding, J., Liu, J. J., Kowal, A. S., Nardine, T., Bhattacharya, P., Lee, A., et al. (2002). Microtubule-associated protein 1B: A neuronal binding partner for gigaxonin. *The Journal of Cell Biology, 158*(3), 427–433.

Donaghy, M., King, R. H., Thomas, P. K., & Workman, J. M. (1988). Abnormalities of the axonal cytoskeleton in giant axonal neuropathy. *Journal of Neurocytology, 17*(2), 197–208.

Furukawa, M., He, Y. J., Borchers, C., & Xiong, Y. (2003). Targeting of protein ubiquitination by BTB-Cullin 3-Roc1 ubiquitin ligases. *Nature Cell Biology, 5*(11), 1001–1007.

Ganay, T., Boizot, A., Burrer, R., Chauvin, J. P., & Bomont, P. (2011). Sensory-motor deficits and neurofilament disorganization in gigaxonin-null mice. *Molecular Neurodegeneration, 12*(6), 25.

Gou, J. P., & Leterrier, J. F. (1995). Possible involvement of ubiquitination in neurofilament degradation. *Biochemical and Biophysical Research Communications, 217*(2), 529–538.

Ionasescu, V., Searby, C., Rubenstein, P., Sandra, A., Cancilla, P., & Robillard, J. (1983). Giant axonal neuropathy: Normal protein composition of neurofilaments. *Journal of Neurology, Neurosurgery and Psychiatry, 46*(6), 551–554.

Klein, C. J., Wu, Y., Vogel, P., Goebel, H. H., Bonnemann, C., Zukosky, K., et al. (2014). Ubiquitin ligase defect by DCAF8 mutation causes HMSN2 with giant axons. *Neurology, 82*, 873–878.

Kortazar, D., Fanarraga, M. L., Carranza, G., Bellido, J., Villegas, J. C., Avila, J., et al. (2007). Role of cofactors B (TBCB) and E (TBCE) in tubulin heterodimer dissociation. *Experimental Cell Research, 313*, 425–436.

Kuhlenbaumer, G., Timmerman, V., & Bomont, P. (1993). Giant axonal neuropathy. In R. A. Pagon, M. P. Adam, H. H. Ardinger, S. E. Wallace, A. Amemiya, L. J. H. Bean,

T. D. Bird, C. R. Dolan, C. T. Fong, R. J. H. Smith, K. Stephens, (Eds.). GeneReviews® [Internet]. Seattle (WA): University of Washington, Seattle; 1993–2015.

Mahammad, S., Murthy, S. N., Didonna, A., Grin, B., Israeli, E., Perrot, R., et al. (2013). Giant axonal neuropathy-associated gigaxonin mutations impair intermediate filament protein degradation. *The Journal of Clinical Investigation*, *123*(5), 1964–1975.

Millecamps, S., Gowing, G., Corti, O., Mallet, J., & Julien, J. P. (2007). Conditional NF-L transgene expression in mice for *in vivo* analysis of turnover and transport rate of neurofilaments. *The Journal of Neuroscience*, *27*(18), 4947–4956.

Nixon, R. A., & Logvinenko, K. B. (1986). Multiple fates of newly synthesized neurofilament proteins: Evidence for a stationary neurofilament network distributed nonuniformly along axons of retinal ganglion cell neurons. *The Journal of Cell Biology*, *102*(2), 647–659.

Pena, S. D. (1981). Giant axonal neuropathy: Intermediate filament aggregates in cultured skin fibroblasts. *Neurology*, *31*(11), 1470–1473.

Perrot, R., Berges, R., Bocquet, A., & Eyer, J. (2008). Review of the multiple aspects of neurofilament functions, and their possible contribution to neurodegeneration. *Molecular Neurobiology*, *38*(1), 27–65.

Prineas, J. W., Ouvrier, R. A., Wright, R. G., Walsh, J. C., & McLeod, J. G. (1976). Gian axonal neuropathy—A generalized disorder of cytoplasmic microfilament formation. *Journal of Neuropathology and Experimental Neurology*, *35*(4), 458–470.

Rao, M. V., Yuan, A., Campbell, J., Kumar, A., & Nixon, R. A. (2012). The C-terminal domains of NF-H and NF-M subunits maintain axonal neurofilament content by blocking turnover of the stationary neurofilament network. *PLoS One*, 7(9), e44320.

Wang, W., Ding, J., Allen, E., Zhu, P., Zhang, L., Vogel, H., et al. (2005). Gigaxonin interacts with tubulin folding cofactor B and controls its degradation through the ubiquitin-proteasome pathway. *Current Biology*, *15*(22), 2050–2055.

Wang, J. T., Medress, Z. A., & Barres, B. A. (2012). Axon degeneration: Molecular mechanisms of a self-destruction pathway. *The Journal of Cell Biology*, *196*(1), 7–18.

Ylikallio, E., Poyhonen, R., Zimon, M., De Vriendt, E., Hilander, T., Paetau, A., et al. (2013). Deficiency of the E3 ubiquitin ligase TRIM2 in early-onset axonal neuropathy. *Human Molecular Genetics*, *22*(15), 2975–2983.

PART III

Functional and Genetic Analysis of the Plakin and Other Cytoskeletal Cross-Linkers

CHAPTER THIRTEEN

Functional and Genetic Analysis of Plectin in Skin and Muscle

Günther A. Rezniczek*, Lilli Winter†, Gernot Walko‡, Gerhard Wiche§,1

*Department of Obstetrics & Gynecology, Marien Hospital Herne, Ruhr-Universität Bochum, Herne, Germany

†Institute of Neuropathology, University Hospital Erlangen, Erlangen, Germany

‡Centre for Stem Cells & Regenerative Medicine, Faculty of Life Sciences and Medicine, King's College London, London, United Kingdom

§Department of Biochemistry & Cell Biology, Max F. Perutz Laboratories, University of Vienna, Vienna, Austria

[1]Corresponding author: e-mail address: gerhard.wiche@univie.ac.at

Contents

Methods in Enzymology, Volume 569
ISSN 0076-6879
http://dx.doi.org/10.1016/bs.mie.2015.05.003

Abstract

Plectin is a large cytoskeletal linker protein with a multitude of functions affecting various cellular processes. It is expressed as several different isoforms from a highly complex gene. Both, this transcript diversity (mainly caused by short 5′-sequences contained in alternative first exons) and the size (>500 kDa) of the resulting proteins, present considerable challenges to plectin researchers. In this chapter, we will consider these problems and offer advice on how to tackle them best. As plectin has been studied most extensively in skin and muscle, we will focus on these types of tissues and describe some selected methods in detail. Foremost, however, we aim to give the readers some good pointers to available tools and into the existing literature.

1. INTRODUCTION

Plectin is a large multimodular and multifunctional cytolinker of the plakin protein family (Bouameur, Favre, & Borradori, 2014). Having diverse cellular functions, it can sustain multiple interactions with other proteins, form compact oligomeric structures by self-association, serve as scaffold for signaling proteins, and interact with and act as organizer of all three cytoskeletal filament systems: actin, intermediate filaments (IFs), and microtubules (MTs; Castañón, Walko, Winter, & Wiche, 2013). Arguably, its most important functions are within the context of the various IF systems where it interacts with IFs by directly binding to their subunit proteins and anchoring them at sites of strategic importance for the organization and performance of a particular cell type (Wiche, Osmanagic-Myers, & Castañón, 2015). These include peripheral cell junctions such as hemidesmosomes in basal keratinocytes of skin, focal adhesion in fibroblasts, abaxonal membrane complexes in Schwann cells, and costameres and neuromuscular junctions in muscle cells. Furthermore, plectin links IFs to intracellular structures and organelles, such as the nucleus and mitochondria, and the contractile apparatus (sarcomeric Z-disks) in skeletal and heart muscle (Walko, Castañón, & Wiche, 2014; Wiche et al., 2015).

1.1 Gene and Protein Structure

Plectin is expressed as several protein isoforms from a single gene (*PLEC*) that is located on chromosome 8q24 in humans (mouse chromosome 15, rat chromosome 7). Detailed analysis of the mouse gene (Fuchs et al., 1999) revealed well over 40 exons (spanning over 62 kb; Fig. 1A) and an unusual 5′ transcript complexity (Fig. 1B) resulting in the tissue- and cell type-specific expression of several plectin isoforms (Fig. 1C) with a deduced

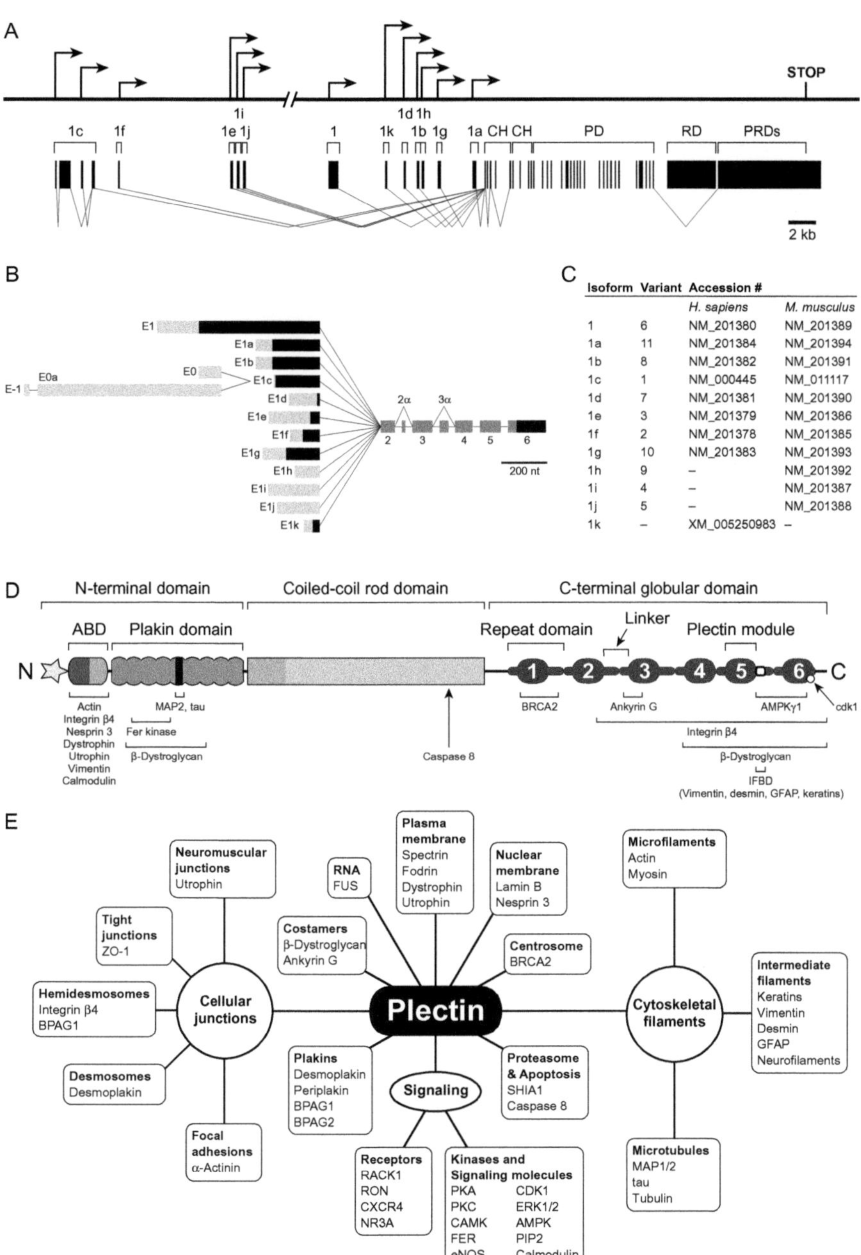

Isoform	Variant	Accession # *H. sapiens*	*M. musculus*
1	6	NM_201380	NM_201389
1a	11	NM_201384	NM_201394
1b	8	NM_201382	NM_201391
1c	1	NM_000445	NM_011117
1d	7	NM_201381	NM_201390
1e	3	NM_201379	NM_201386
1f	2	NM_201378	NM_201385
1g	10	NM_201383	NM_201393
1h	9	–	NM_201392
1i	4	–	NM_201387
1j	5	–	NM_201388
1k	–	XM_005250983	–

Figure 1 (A) Genomic organization of the plectin gene. Alternative transcription start sites are indicated by arrows. Exons are represented by boxes. Splice events in the 5′ region before exon 2 and alternative splice events following exon 2 are indicated by lines connecting the boxes. Scale is approximate. (B) Variable 5′ ends of plectin

(Continued)

molecular mass for full-length plectin ranging from 499 to 533 kDa, depending on the particular isoform (Rezniczek, Abrahamsberg, Fuchs, Spazierer, & Wiche, 2003). Rotary shadowing electron microscopy of purified plectin molecules revealed a dumbbell-like structure comprising a central 200-nm-long rod domain flanked by large globular domains (Fig. 1D). The N-terminal domain harbors a conventional actin-binding domain (ABD) preceding the plakin domain (PD). The ABD is composed of two calponin homology (CH) domains closely resembling the ABD of fimbrin (Sevcík, Urbániková, Kost'an, Janda, & Wiche, 2004). Of note, two short sequences encoded by the optionally spliced exons 2α and 3α are inserted in the first CH domain in certain plectin isoforms (Fuchs et al., 1999). The PD, common to all isoforms, comprises nine spectrin repeats with one Src-homology 3 domain inserted in repeat 5 (Sonnenberg, Rojas, & de Pereda, 2007). The 130-kDa-rod domain is encoded by exon 31 and is missing in the so-called rodless isoforms (Elliott et al., 1997; Fuchs, Spazierer, & Wiche, 2005). It contains an almost continuous 1127-residue-long, mainly α-helical coiled coil showing long stretches of heptad repeats with a staggered charge periodicity of 10.4 for both acidic and basic residues. The C-terminal domain contains six plectin repeat domains (PRDs; each made of a conserved core region/module and connected by short linker sequences) and a 70-amino acid-long serine-rich terminal tail (Janda, Damborsky, Rezniczek, & Wiche, 2001). The PRDs are most likely tightly packed into a compact globular domain through interlinkage by hydrogen bonds and disulfide bridges (Janda et al., 2001; Spurny et al., 2007).

1.2 Expression, Subcellular Localization, and Molecular Interactions

Plectin is particularly abundant in tissues subjected to great mechanical stress, such as stratified and simple epithelia, skeletal and heart muscle, and blood vessels (Castañón et al., 2013). The extensive use of alternative first exons

Figure 1—Cont'd transcripts shown in detail. Noncoding (parts of) exons are light gray; coding (parts of) exons are black. Dark gray (parts of) exons are coding if preceded by one of the coding first exons (1, 1a–g, and 1k) and noncoding otherwise (1h–j). (C) Accession numbers of plectin transcript variants and isoforms. (D) Schematic representation of the protein. See Section 1.1 for details. The different N-termini of the isoforms are indicated by a star. Positions of the mapped binding regions to various proteins are indicated below the scheme (IFBD, IF-binding domain). (E) Schematic representation of proteins known to interact directly with plectin. Plectin is shown in the center with the interacting proteins grouped by their cellular location or functional categories.

with their separate promotors allows for cell type-specific expression of plectin isoforms differing only in their short N-terminal sequences. These short sequences, in turn, determine the cellular location and function of plectin. Plectin isoform 1 (P1), for example, is targeted to the nucleus/ER membrane, P1a to hemidesmosomes, P1b to mitochondria, P1c to MTs, P1d to Z-disks, and P1f to focal adhesions and costameres (Wiche et al., 2015; see Fig. 2B). Tissue-specific expression was characteristic also of exons 2α and 3α, with isoforms containing exon 2α being expressed in brain, heart, and skeletal muscle and exon 3α being brain specific (Fuchs et al., 1999). Furthermore, the alternative first exons, and specifically their 5′-untranslated regions, were found to determine the stability of gene products by controlling initiation of translation (Rezniczek et al., 2003).

The multiple functional domains of plectin provide for its ability to interact with a vast array of different proteins and thereby fulfill its major roles as mechanical reinforcer and/or scaffold for signaling molecules. Figure 1E gives an overview of proteins known to interact directly with plectin (reviewed in Castañón et al., 2013). Plectin's roles as major component of hemidesmosomes (Walko et al., 2014) and networking and anchoring element of IFs (Wiche et al., 2015) have been the highlight of recent reviews. Some specific binding domains that have been mapped on plectin are indicated in Fig. 1D.

1.3 Extracellular Plectin

Normally, plectin is exclusively cytoplasmic. However, it has recently been discovered that the aberrant expression and mislocalization in pancreatic ductal adenocarcinoma (PDAC) led to the inclusion of plectin in exosomes released into the extracellular space (Shin et al., 2013).

1.4 Plectin and Human Diseases

Plectin gene mutations result in multiple diseases manifesting with muscular dystrophy (MD), skin blistering, and signs of neuropathy (Winter & Wiche, 2013). Their specific symptoms and severity depend on the mutation(s) present in the patient.

The most common manifestation is epidermolysis bullosa simplex (EBS) with MD, a rare autosomal-recessive disorder characterized by neonatal skin blistering and delayed progressive MD (OMIM #226670). Others include EBS-MD combined with myasthenic syndrome, EBS with pyloric atresia (OMIM #612138), and EBS-Ogna (OMIM #131950). Distinct and specific disease phenotypes have been expected for mutations in the alternative first exons, and recently such mutations have been identified in patients: a homozygous 9 bp deletion in exon 1f (encoding an important muscle isoform) leading to limb-girdle MD type 2Q (LGMD2Q, OMIM #613723) without skin involvement (Gundesli et al., 2010) and a homozygous nonsense mutation in exon 1a (predominantly expressed in keratinocytes) causing EBS without extracutaneous involvement (Gostyńska et al., 2015).

In recent years, plectin has been found to be deregulated in carcinoma cells and tumor tissues, potentially serving as a biomarker. Furthermore, some specific functions of plectin in the cancer context have emerged, such as modulating the invasiveness of bladder carcinoma cells via invadopodia, affecting migration and invasion of head and neck squamous cell carcinoma cells, and playing a role in PDAC tumor pathogenesis (reviewed in Wiche et al., 2015).

2. MAIN CHALLENGES

When working with plectin, researchers face a number of challenges. One is plectin's size. While some of the size-imposed experimental limitations can be overcome by carefully selected conditions and protocols, plectin fragments of smaller size will usually have to be employed. Most others are related to the splicing complexity of plectin transcripts. As discussed above, different plectin isoforms fulfill distinct and specific roles within a cell. Thus, major questions arising when investigating the function(s) of plectin are: Which isoform am I dealing with, and does it make a difference?

2.1 How to Distinguish Isoforms

The most desirable tools to identify isoforms *in situ* are isoform-specific antibodies. While such antibodies have been successfully raised for a number but not all of the plectin isoforms (see Supplementary Table 1 on http://dx.doi.org/10.1016/bs.mie.2015.05.003), none are available commercially. For some sequences, like those encoded by exons 1d and 2α, both only five amino acids long, so far it has proven impossible to obtain specific antibodies. In mouse and rat (not yet confirmed in humans), plectin transcript variants 4, 5, and 9 (see Fig. 1C) start with noncoding first exons. The putative single protein species (P1h, P1i, P1j) that results from translation of these different transcripts starting at the first ATG within exon 6 (Rezniczek et al., 2003) has no known features that would enable their distinction from other isoforms (except size). If sufficient and pure enough protein can be obtained, protein sequencing and/or mass spectroscopy will give the answer. On the transcript level, reverse transcription (RT)-PCR can be used to obtain (limited) information about plectin isoform expression (see Section 4.1.1).

2.2 How to Study Isoform Functions

One should always bear in mind that the "ubiquitous plectin" is the combination of a multitude of distinctly behaving isoforms. In the following sections, we will discuss available tools and present a selection of *in vitro* and *in vivo* approaches that have been used to study plectin. Supplementary Table 2 on http://dx.doi.org/10.1016/bs.mie.2015.05.003 provides an extensive methodological survey of the plectin literature.

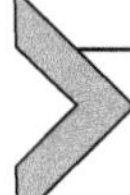

3. TOOLS

3.1 Antibodies

A list of antibodies to plectin is shown in Supplementary Table 1 on http://dx.doi.org/10.1016/bs.mie.2015.05.003. Note that when the whole molecule is used as an immunogen, most resulting antibodies will be reactive with the highly antigenic rod domain (due to its α-helical nature and surface charge). Thus, in order to raise antibodies against specific plectin domains, immunize animals with recombinant proteins or peptides. To generate antibodies recognizing short sequences (like those encoded by the first exons), using short glutathione-*S*-transferase (GST)-tagged target sequences as immunogen and maltose-binding protein (MBP) fusions of the same sequences immobilized on CNBr-Sepharose for subsequent affinity purification has proved useful (Andrä et al., 2003).

3.2 PCR Primers

Supplementary Table 3 on http://dx.doi.org/10.1016/bs.mie.2015.05.003 lists selected primer pairs (mouse and human) that have been used to identify plectin isoform transcripts.

3.3 Mouse Models and Cell Lines

Several genetically altered mouse lines have been generated to study the function of plectin and its isoforms. Conditional (tissue-restricted) and isoform-specific knockout and knock-in mice, mimicking the clinical features of plectinopathies, are outstanding models to study functional and molecular consequences of plectin gene defects and to develop treatment and therapy strategies. Table 1 gives an overview of the available mouse

Table 1 Mouse Models

Type/Designation	Affected Tissues	Phenotype	Ref.[a]
Full knockout			
Plectin-null	All	Skin blistering, skeletal and heart muscle abnormalities, death at age 1–2 days	3
Conditional knockout			
K5-Cre (inducible/ noninducible)	Epidermis	Skin blistering, death at age 1–2 days	2
MCK-Cre	Skeletal muscle and heart	Progressive striated muscle degeneration, intermediate filament (desmin) aggregation	5
Pax7-Cre	Satellite cells	Muscle weakness, neuromuscular junction defects, reduced life span, desmin aggregation	6
P0-Cre	Schwann cells	Myelin sheath deformation	10
nes-Cre (nestin)	Neuronal precursor cells	Decreased motor nerve conduction velocity (MNCV)	4
Isoform-specific knockout			
P1	Fibroblasts, leukocytes, muscle	Impaired migration potential of fibroblasts/T cells, reduced leukocyte infiltration in wound healing	1
P1b	Mitochondria	Mitochondrial shape changes	11
P1c	Neurons, basal keratinocytes	Reduced MNCV and microtubule dynamics; cell migration, glucose uptake, and mitotic spindle aberrations	4, 8
P1d	Skeletal muscle and heart	Striated muscle degeneration, localized desmin aggregation	5
Double knockout			
Dystrophin (*mdx*)/ plectin (MCK-Cre)	Skeletal muscle and heart	Restoration of sarcolemmal integrity of *mdx* skeletal muscle	7

Table 1 Mouse Models—cont'd

Type/Designation	Affected Tissues	Phenotype	Ref.
Desmin/P1d	Skeletal muscle and heart	Massive skeletal muscle degeneration	5
Knock-in			
Dominant EBS-Ogna mutation	Epidermis	Skin blistering, no muscular manifestation	9

[a]References: [1]Abrahamsberg et al. (2005); [2]Ackerl et al. (2007); [3]Andrä et al. (1997); [4]Fuchs et al. (2009); [5]Konieczny et al. (2008); [6]Mihailovska et al. (2014); [7]Raith et al. (2013); [8]Valencia et al. (2013); [9]Walko et al. (2011); [10]Walko et al. (2013); and [11]Winter, Abrahamsberg, and Wiche (2008).

models (for detailed descriptions, see Castañón et al., 2013; Winter & Wiche, 2013).

Another valuable tool are plectin-deficient cell lines lacking the expression of total plectin or individual isoforms. Such cell lines have successfully been derived after crossing the corresponding knockout mouse line into a $p53^{-/-}$ background. Absence of p53 facilitates immortalization of cells that otherwise are difficult to immortalize (Metz, Harris, & Adams, 1995). So far, the following plectin-deficient (total, i.e., all isoforms) cell lines have been established: basal keratinocytes (Andrä et al., 2003), dermal fibroblasts (Osmanagic-Myers & Wiche, 2004), and skeletal muscle myoblasts (Winter et al., 2014). Alternatively, immortalization of plectin-deficient endothelial cells has been achieved by polyoma middle T-expressing retrovirus transduction (Spurny et al., 2007).

4. METHODS, PEARLS, AND PITFALLS

4.1 Determining Plectin Isoform Expression Profiles

On the transcript level, the determination of isoform expression profile and relative abundance of mRNAs is straightforward. The easiest method is to perform RT-PCR analysis (end-point or real-time). For accurate transcript quantitation, RNase protection assays may be more suitable (a detailed protocol can be found in Rezniczek, Janda, & Wiche, 2004). On the protein level, mass spectrometry has been used successfully to detect differential plectin expression in samples; however, isoform profiling has yet to be established. Sodium dodecyl sulfate (SDS)-polyacrylamide gel electrophoresis (PAGE) coupled with immunoblot analysis (and possibly isoform-specific antibodies) is an easy-to-perform method.

4.1.1 Quantitative Real-Time PCR

Extract total RNA using suitable reagents/kits. Measure RNA quantity and optionally check RNA integrity on a 1% agarose/2.2 *M* formaldehyde denaturing gel. Synthesize cDNA from 1 μg of total RNA using any cDNA-synthesis kit with random hexamer primers. Do not use oligo(dT) primers, as plectin transcripts are >15 kb long and, therefore, potentially fragmented. Prepare and run PCRs using suitable plectin primers (Supplementary Table 3 on http://dx.doi.org/10.1016/bs.mie.2015.05.003). For absolute quantitation, generate standard curves from serial dilutions of linearized plasmids carrying the target sequences. Supplementary Fig. 1 on http://dx.doi.org/10.1016/bs.mie.2015.05.003 shows expression levels of different plectin transcripts in tissue samples from epidermis and muscle. When performing relative quantification (ΔΔCt method), note that specific plectin isoforms can be quantitatively compared between different samples but not multiple isoforms within the same sample (at least not without additional considerations and work).

4.1.2 SDS-PAGE and Western Blotting

SDS-5% polyacrylamide gels with 4% stacking gels offer a good compromise between good resolution of plectin isoforms and ease of handling. Recommended sample buffer: 50 m*M* Tris/HCl, pH 6.8, 100 m*M* dithiothreitol (DTT), 2% SDS, 10% glycerol, 0.1% bromophenol blue (prepare as 2–5 × stock). To transfer plectin onto nitrocellulose membranes, we recommend using a tank blot system. Equilibrate the gel as well as filter paper, pads, and membrane in transfer solution (3.05 g/l Tris, 14.4 g/l glycine; without methanol or SDS) and assemble the gel/membrane sandwich. To transfer low percentage acrylamide gels, instead of lifting them with your fingers risking tears, "pour" them from plastic trays. Blot overnight at 25 V at 4 °C. Carry out standard immunodetection using suitable antibodies.

4.2 Preparation of Cell and Tissue Lysates and Purification of Plectin

4.2.1 Direct Cell Lysis, Triton X-100-Based Extraction, and Digitonin-Based Fractionation

The following basic protocols are suitable for most adherent cultures. Unless otherwise noted, amounts apply to a standard 10-cm (58 cm^2) confluent culture plate.

4.2.1.1 Direct Lysis

To prepare total cell lysates suitable for SDS-PAGE and immunoblotting, wash cells twice with phosphate buffered saline (PBS) and overlay them with 10 μl/cm^2 2 × sample buffer (see Section 4.1.2). Use a cell scraper to facilitate lysis. Pass the lysate through a blunt 27-G needle several times to shear DNA and to reduce viscosity.

4.2.1.2 Detergent Extraction

Scrape cells into PBS, collect (5 min/200 × *g*/4 °C) and gently resuspend them in 700 μl of ice-cold 0.5% (v/v) Triton X-100, 100 m*M* NaCl, 50 m*M* HEPES/HCl, pH 7.0, 5 m*M* $MgCl_2$, 1 m*M* EGTA, and 0.1 m*M* DTT containing a protease/phosphatase inhibitor mix. Incubate with end-over-end rotation for 45 min/4 °C. After centrifugation (30 min/15,800 × *g*/4 °C), collect the supernatant (detergent-soluble extract) and dissolve the pellet (detergent-insoluble extract) in 125 μl 8 *M* urea with shaking for 30 min/4 °C. To prepare the extracts for SDS-PAGE/immunoblotting, add the appropriate volume of sample buffer (see Section 4.1.2). Heat the detergent-soluble fraction for 5 min/95 °C. The detergent-insoluble fraction should be passed through a 27-G needle and incubated for 5 min/60 °C.

4.2.1.3 Cellular Fractionation

Collect cells (5 min/200 × *g*/4 °C) and gently resuspend them in 150 μl of ice-cold 0.01% digitonin, 10 m*M* piperazine-N,N′-bis(2-ethanesulfonic acid) (PIPES), pH 6.8, 300 m*M* sucrose, 100 m*M* NaCl, 3 m*M* $MgCl_2$, 5 m*M* EDTA, and 0.2 m*M* DTT containing a protease/phosphatase inhibitor mix. Incubate with end-over-end rotation for 10 min/4 °C. Centrifuge (1 min/15,800 × *g*) and collect the supernatant (cytosolic fraction). Resuspend the pellet in 150 μl 0.5% Triton X-100, 10 m*M* PIPES, pH 7.4, 300 m*M* sucrose, 100 m*M* NaCl, 3 m*M* $MgCl_2$, 3 m*M* EDTA, and 0.2 m*M* DTT containing a protease inhibitor mixture. After 20 min of end-over-end rotation, centrifuge (1 min/15,800 × *g*) to obtain the membrane fraction (supernatant) and cytoskeleton fraction (pellet). Resuspend the cytoskeleton fraction in 40 μl 8 *M* urea and incubate on ice for 30 min (with frequent vortexing). Store aliquots at −20 or −80 °C and/or process for immunoblotting.

4.2.2 Preparation of Protein Extracts from Adult Mouse Tail Skin Epidermis

Solubilizing epidermal proteins can be quite challenging. The following protocol should provide good results for mouse tissues.

4.2.2.1 Isolation of Epidermal Tissue

Sacrifice a mouse and rinse once with 70% ethanol. Optionally, cut the tail tip to be used for genotyping. Using sterile scissors remove the complete tail and pull away the skin, starting at the base and continuing toward the tip. Cut the resulting "skin-tube" into ~1-cm-broad pieces, open each tube with a longitudinal cut, and flatten the skin pieces dermis-side down on the culture surface of a cell culture dish. Cut the skin pieces into ~5 × 5 mm pieces. Next, prepare a 6-cm culture dish containing 5 ml Dispase II (Roche) solution (5 U/ml) in keratinocyte growth medium (KGM; Lonza) without additives or serum, supplemented with gentamycin and amphotericin B. Grasp opposite ends of one side of a skin piece with forceps, lift it up, and transfer it to the Dispase-dish so that the dermis side of the skin piece touches the Dispase II solution. Then, lower the rest of the skin piece and release the forceps. Small skin pieces will generally float perfectly flat on the Dispase II solution. If necessary, flatten curled under edges using the rough side of one tip of forceps placed under the skin piece and gliding away from it while steadying the skin piece with the other forceps placed gently on the center of the piece. Incubate the dish at 37 °C for 1 h to digest the basement membrane. After digestion is complete, grasp each floating skin piece as before and transfer it, epidermis-side down, onto the dry inner surface of a culture dish lid. Spread the skin pieces gently to allow the epidermis to make full contact with the plastic. Next, lift the dermis up straight above the epidermis, holding the epidermis down with the other forceps, if necessary.

4.2.2.2 Tissue Extraction

Immediately shock-freeze epidermal tissue pieces in liquid nitrogen. Grind the frozen tissue under permanent cover of liquid nitrogen to a fine powder, using pestle and mortar. Transfer the pulverized tissue to a preweighed polypropylene round bottom tube, weigh, and add 400 μl per 100 mg of tissue of 20 m*M* Tris/HCl, pH 7.5, 150 m*M* NaCl, 1% (v/v) Triton X-100, 0.5% (v/v) NP-40, 1 m*M* EDTA, 1 m*M* phenylmethylsulfonyl fluoride (PMSF) (extraction buffer), supplemented with protease/phosphatase inhibitor mix.

Homogenize the pulverized tissue at 4 °C using an Ultra-Turrax equipped with a small rotor (to fit into the tube) at 12,000 rpm (move up/down 20 ×, cool on ice for 1 min after five up/down movements). Transfer the tissue slurry to a small glass-in-glass Dounce homogenizer and homogenize further by moving the pestle up/down 10 ×. Transfer the final homogenate to a micro test tube and incubate for 30 min/4 °C on a shaking platform at full speed. Mix the homogenate with an equal volume of 5 × sample buffer (see Section 4.1.2) and incubate for 5 min/95 °C. Centrifuge (15 min/ >10,000 × *g*/room temperature) and carefully transfer the supernatant to a fresh tube using a 27-G blunt needle attached to a 1-ml syringe. Aliquot and store at −80 °C.

4.2.3 Purification from Cultured Cells

Triton X-100/high salt-resistant cytoskeletal preparations typically contain around 35% of the total cellular protein and 75% of cellular plectin. Gel permeation chromatography of urea- or sodium *N*-laurosylsarcosine-solubilized IF preparations yields purified plectin fractions that elute with the void volume of the column or shortly afterwards. Rat glioma C6 cells are an excellent source for plectin (for a detailed protocol see Rezniczek et al., 2004), but other cells containing plectin and vimentin can be used as well.

4.3 Isolation and Culture of Plectin-Deficient Cells from Skin and Muscle

To perform detailed analyses on the cellular level, isolated cells from plectin-deficient mice cultured *ex vivo* are essential. Usually, primary cell lines are preferred, but not suitable for all cell types and analyses. Immortalized cell lines were established from plectin$^{-/-}$/p53$^{-/-}$ double knockout mice, as described above (Section 3.3).

4.3.1 Keratinocytes

4.3.1.1 Isolation

Sacrifice neonatal mice (1–2 days old) by decapitation and rinse twice with 70% ethanol. Amputate tail and limbs using sharp(!) sterile scissors and forceps. Immediately process limbs for the preparation of tissue sections. Insert small sharp scissors through the hole at the tail and cut the skin along the dorsal midline to the neck; be careful not to cut into the body cavity. Gently pull the skin away from the body, and layer it onto Dispase II solution in a 6-cm culture dish as described in Section 4.2.2. Incubate for 60 min/ 37 °C. Next, flatten skin epidermis-side down onto the dry, clean lid of a

10-cm culture dish and gently peel back the dermis using forceps. Place the epidermal sheet onto a sterile glass slide and use sterile razor blades to mince it into pieces small enough to enter the tip of a plastic 5-ml pipette. Add 3 ml T/E (0.25% trypsin/0.02% EDTA in Ca^{2+}/Mg^{2+}-free PBS) and use the pipette to gently resuspend the epidermal pieces. Transfer the suspension to a 15-ml tube and incubate in a water bath for 10 min/37 °C; occasionally shake the tube. Then, pipette up/down 50 times with a 1-ml pipette to release epidermal cells from the tissue pieces. Filter the resulting cell suspension through a 40-μm cell strainer into a 50-ml tube filled with 40 ml of cold adhesion medium. Centrifuge (5 min/150 × *g*), resuspend the cell pellet in warm KGM w/o Ca^{2+} (Lonza) supplemented with 1% (v/v) insulin/transferrin/selenium (ITS; Life Technologies) and 8% Chelex-100-treated fetal calf serum (FCS) and Ca^{2+} adjusted to 0.1 m*M* (adhesion medium; for guidelines on how to remove calcium from serum see Lichti, Anders, & Yuspa, 2008) and plate onto 3.5-cm cell culture dishes (four dishes per neonatal mouse skin, two dishes in case of plectin-null skin) coated with collagen IV (Corning) or 804G cell-derived extracellular matrix (ECM; Langhofer, Hopkinson, & Jones, 1993). Maintain the dishes in a cell culture incubator (37 °C/5% CO_2) until keratinocytes have attached (usually after 2 h on collagen IV-coated dishes, shorter when cells are plated on 804G cell-derived ECM). Once cells have attached, wash the dishes twice with KGM w/o Ca^{2+} supplemented with 1% ITS, 2% Chelex-100-treated FCS and Ca^{2+} adjusted to 0.05 m*M* (growth medium) to remove differentiated and other unattached cells. Note, cell yields will be smaller for plectin-null keratinocytes due to mechanical damage and cell death occurring during the isolation procedure. Growing keratinocytes in adhesion medium on 804G cell-derived ECM facilitates formation of hemidesmosome-like protein complexes.

4.3.1.2 Maintenance/Differentiation

Replace the growth medium every second day. Confluence will normally be reached 4–5 days after plating. Note that primary mouse keratinocytes cannot be passaged, whereas immortalized keratinocyte cell lines ($p53^{-/-}$) have been used up to passage number 20 without displaying signs of transformation (Osmanagic-Myers et al., 2006). Similar to primary cells, the immortalized cell line faithfully recapitulates key normal keratinocyte functions, including formation of hemidesmosome-like protein complexes, and cells can be induced to undergo stratification and terminal differentiation (Kostan, Gregor, Walko, & Wiche, 2009; Osmanagic-Myers et al., 2006). To initiate stratification, grow keratinocytes to confluence in growth medium and raise the Ca^{2+} level from 0.05 to 1.8 m*M*; culture for up to 96 h.

4.3.2 Myoblasts

4.3.2.1 Isolation

Sacrifice neonatal (2–3 days old) mice, rinse twice with 70% ethanol, and amputate front and hind limbs. For enzymatic dissociation, put limbs into 3 ml 0.2% collagenase I in DMEM containing 100 n*M* nonessential amino acids (NEAA), 2 m*M* L-glutamine, 50 U/ml penicillin, and 50 μg/ml streptomycin (P/S) (all reagents from Gibco/Life Technologies) for 1.5–2 h/37 °C with gentle agitation. Then, pour the digested tissue into a 6-cm cell culture dish containing 5 ml prewarmed medium (serum-free DMEM supplemented with NEAA, P/S, and L-glutamine). Gently triturate with a 10-ml pipette to release single muscle fibers from the bones. Transfer the muscle slurry into a 15-ml tube and centrifuge (3 min/15 × *g*). Wash the pellet twice with PBS (each time centrifuge as before). Resuspend the fibers in 10 ml DMEM and allow large fiber bundles to settle by gravity for 1 min. Transfer the supernatant (containing single fibers) into a fresh tube and centrifuge (5 min/50 × *g*). Finally, resuspend the pellet in DMEM containing 20% FCS, 10% horse serum, 1% chicken embryo extract, and P/S and plate on Matrigel™ (BD Biosciences; diluted 1:10 in DMEM)-coated 10-cm dishes for 4–24 h at 37 °C/5% CO_2. Primary myoblasts can be maintained for three to four passages, whereas immortalized myoblast cell lines ($p53^{-/-}$) have been used with passage number 40 and beyond.

4.3.2.2 Maintenance of Myoblasts and Differentiation into Multinucleated Myotubes

Myoblasts are routinely maintained in Ham's F10 (Gibco) supplemented with 20% FCS, 2.5 ng/ml basic fibroblast growth factor (Promega), and P/S (proliferation medium) on collagen-coated (0.01% collagen in PBS) culture dishes at 37 °C/5% CO_2. When reaching 80% confluence, cells have to be passaged. To split cultures, wash cells with PBS and incubate with 0.25% trypsin/0.02% EDTA in Ca^{2+}/Mg^{2+}-free PBS. After a few minutes, cells detach from the culture dish. Add 7 ml proliferation medium to inhibit trypsin. Transfer the cell suspension into a 15-ml tube and centrifuge (3 min/200 × *g*). Resuspend cells in proliferation medium, preplate on uncoated cell culture dishes for 1–2 h (to remove contaminating fibroblasts; perform this preplating procedure routinely at every splitting), and finally transfer to collagen-coated dishes. To induce differentiation, switch

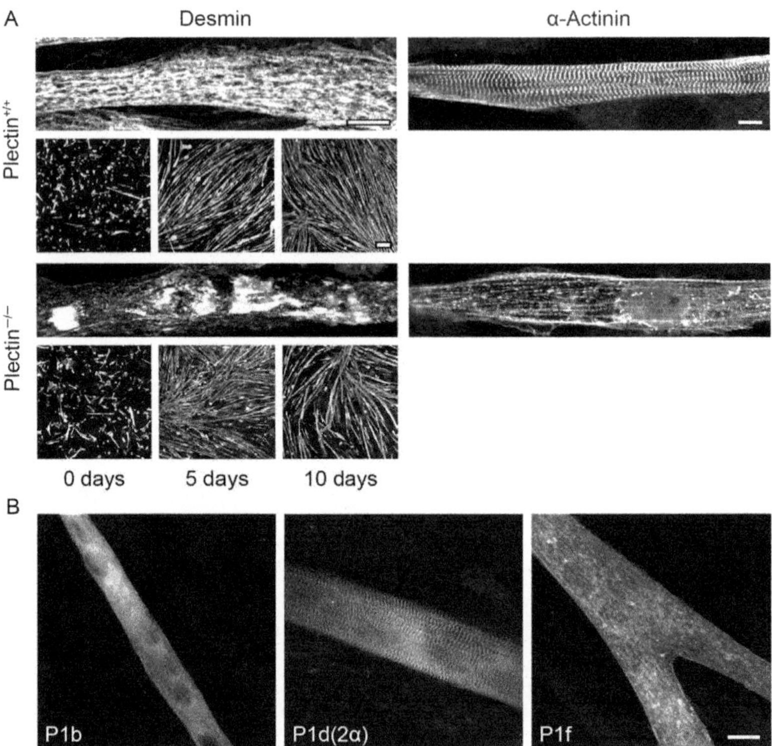

Figure 2 (A) Plectin$^{+/+}$ and plectin$^{-/-}$ myoblasts differentiate into multinucleated myotubes in culture. Myotubes differentiated for 10 days were subjected to immunostaining using antibodies to desmin and α-actinin. Note the collapse of the desmin intermediate filament network and the generation of desmin-positive protein aggregates in plectin$^{-/-}$ myotubes. Moreover, plectin$^{-/-}$ myotubes displayed dissolution of sarcomeric structures, while plectin$^{+/+}$ myotubes developed intact Z-striations, as visualized by staining for α-actinin. Bars, 10 μm. (B) Expression of EGFP-tagged full-length plectin isoforms (P1b, P1d with 2α, P1f) in mouse myotubes differentiated *in vitro*. Bar, 20 μm.

confluent (90–100%) myoblast cultures to DMEM containing 5% horse serum and P/S. Immortalized myoblasts with or without plectin differentiate into multinucleated myotubes over the course of 10 days (Fig. 2A).

4.4 Expression and Purification of Recombinant Plectin

4.4.1 Expression in Escherichia coli

While expression of full-length plectin in bacteria is not possible, smaller fragments representing functional and structural subdomains have been expressed on a wide scale. However, these often accumulate in inclusion bodies and lack biological activity. In such cases, fine-tuning the expression conditions (concentration of inducer, stringent reducing conditions, or temperature of induction) can increase both the solubility of these otherwise insoluble plectin fragments and the chances of them retaining their activity. Fusion to different tags can facilitate both purification and solubility/activity. Purification tags that have been used for the expression of plectin fragments are His-tag, GST, and MBP. The His-tag is usually the favored tag, since it is small, and purification under both denaturing and physiological conditions is rapid and robust. The larger GST and MBP tags are usually only used when either very short plectin sequences are expressed, or the fragments exhibit very low solubility. For instance, plectin's ABD (especially versions containing the exons 2α- and 3α-encoded sequences) had low solubility under conditions suitable for interaction with actin or integrin β4 in overlay and cosedimentation assays. This problem could be solved by using MBP-fusions of the ABD with comparatively much higher solubility. Basic protocols for the purification of His-tagged proteins have been published previously (Rezniczek et al., 2004).

4.4.2 Expression in Baculovirus

The advantage of this method is that larger proteins can be expressed than in the *E. coli* system. For example, the full-length rod domain of plectin (amino acid residues 1375–2501 in NP_035247) and even some larger versions (e.g., 898–2501 and 1245–2668) have been expressed in soluble form. Recombinant baculovirus were generated using the Bac-to-Bac expression system (Invitrogen). To facilitate purification, recombinant proteins carried either a His- or a GST-tag.

4.4.2.1 Procedure

For protein expression, grow Sf9 cells at 27 °C in Insect Express medium (Lonza) supplemented with 10% FCS and 2 m*M* L-glutamine. Infect cells when they have reached 90% confluence ($4–5 \times 10^7$ cells in 175 cm^2 flasks) at a multiplicity of infection of four to five and harvest 72 h postinfection. Wash cells twice with PBS and resuspend the cell pellet in 1 ml lysis buffer (20 m*M* HEPES, pH 7.4, 1% Triton X-100, 1% glycerol, 1 m*M* PMSF). Combine the cell pellets from at least four infected flasks and lyse cells on ice for 15 min with occasional vortexing. Centrifuge (15 min/13,000 × *g*/4 °C), collect the supernatant, and bring up to a final volume of 20 ml with binding buffer (50 m*M* Na-phosphate, pH 7.4, 300 m*M* NaCl, 1 m*M* PMSF). Protein purification is done by affinity chromatography using 1 ml HiTrap Chelating HP or GSTrap HP columns with the Äkta FPLC system (GE Healthcare). His-tagged proteins are bound to Co^{2+}-charged columns and eluted with a linear gradient of 0–500 m*M* imidazole in binding buffer. GST-tagged proteins are bound to glutathione sepharose and eluted with binding buffer containing 10 m*M* reduced glutathione. Typical yield from four 175 cm^2 flasks expressing the His-tagged full-length rod is 2–3 mg of protein; in general yields for the GST-tagged versions are slightly lower. The plectin rod and the variants described here are very stable and can be stored at 4 °C for several months (add 10% glycerol for freezing). Plectin rod variants form oligomers and polymorphic structures through lateral association of dimeric coiled coils (Walko et al., 2011; Fig. 3).

4.4.3 Ectopic Expression of Plectin in Mammalian Cells

Ectopic (over)expression of proteins or protein domains (often fused to fluorescent tags) is a commonly used method to obtain information about their

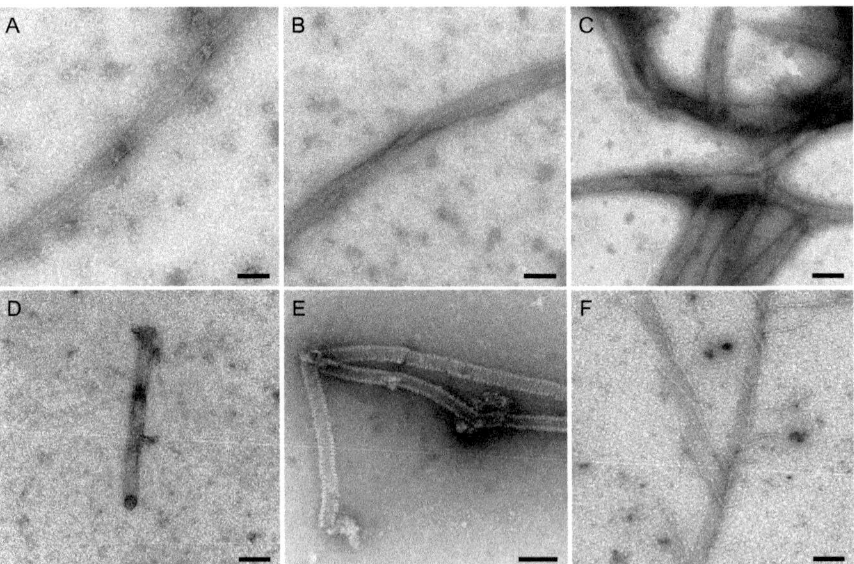

Figure 3 Gallery of electron micrographs depicting polymeric plectin rod structures assembled by lateral self-association. Conditions used for self-assembly were identical or similar to those described in Walko et al. (2011). The structures shown resemble (A) filamentous flat sheets with periodic restrictions, (B) twisted sheets/ribbons, (C) networks of seemingly interconnected filaments, (D) tube-like filaments, (E) paracrystalline tubes, and (F) uncoiling sheets/tubes. Bars, 100 nm.

subcellular localization and function in mammalian cells. This has frequently been used to study plectin, often in combination with plectin-deficient cells, either from patients with gene defects in plectin leading to the loss of plectin expression (e.g., EBS-MD keratinocytes) or from plectin-knockout mice. However, potential pitfalls of ectopic overexpression, such as changes in the cell's biochemistry in response to the transfection agents, nonphysiological (too high) concentrations of the expressed protein, and aggregation, have to be considered. Furthermore, when expressing N-terminal fragments of plectin (with C-terminal tags), it was found that in some cases the fidelity of protein translation, i.e., the recognition of the correct start codon, was suboptimal. This was not the case when the respective 5′-UTR sequences were included in the cDNA constructs (Rezniczek et al., 2003).

For cloning and transfection of full-length plectin cDNAs (insert sizes of up to 15.2 kb), standard techniques can be used. Transfection of full-length plectin expression plasmids into a number of different cell types was successfully achieved using common liposomal and nonliposomal transfection agents such as Lipofectamine (Life Technologies) or FuGENE 6 (Roche), albeit with varying transfection efficiencies. When using keratinocytes and myocytes, transfect before initiating differentiation. Especially in the case of myocytes, expect the number of myotubes expressing recombinant full-length plectin to be low (Fig. 2B).

4.5 Elimination/Suppression of Plectin (Isoforms)

4.5.1 Gene Targeting in Mice

To establish plectin's function in an animal model, plectin-null mice were generated initially. Due to their early death (2–3 days after birth), the usefulness of these mice was limited. To avoid these limitations and multisystemic effects caused by total loss of plectin, two strategies were employed. First, mice were generated bearing "floxed" plectin alleles (i.e., the to-be-removed sequences were flanked by loxP elements). This allowed tissue-restricted deletion (by crossing mice to so-called Cre-deleter mice expressing the Cre recombinase in a cell type- or tissue-specific manner) or inducible deletion (e.g., after activation of Cre-ER^{T} with tamoxifen) of plectin. Second, isoform-specific plectin-knockout mice were generated by targeting specific first exons. This has been done for P1, P1b, P1c, and P1d. For an overview of available plectin-knockout mouse lines, see Table 1.

4.5.2 Knockdown Strategies

RNAi has been used to successfully knockdown plectin in cell lines. Lentiviral transduction of shRNA ensures high efficiency. Lentivirus vectors encoding shRNA against human plectin (as well as other species) are available from a number of commercial sources with target sequences common to all plectin variants. Plectin isoform-specific target sequences have also been used to knockdown human P1a (5′-AGGGCCTCTGAGGGCAAGAAA-3′), P1c (5′-TCTCTGAAGATGTCTCCAATG-3′), and P1f (5′-AGGTGCGC-GAGAAGTACAAAG-3′; Shin et al., 2013).

4.6 Visualization of Plectin in Cells

4.6.1 Immunofluorescence Microscopy

To study the subcellular localization of plectin, the use of confocal laser scanning microscopy is highly recommended. Adherent cells grown in culture can be fixed with either ice-cold methanol for 90 s or 4% paraformaldehyde in PBS for 15 min with subsequent permeabilization (5 min, 0.05% Triton X-100 in PBS). After fixation, apply blocking solution (optional) and antibodies. For tissue sections of frozen or paraffin-embedded material, monoclonal antibodies usually perform better than sera.

4.6.2 Life Cell Imaging

The possibility to express or inject proteins with fluorescent tags into (plectin-deficient) cells makes time-lapse video microscopy an excellent tool to study plectin in dynamic cellular processes (Mihailovska et al., 2014; Valencia et al., 2013; Walko et al., 2011; Winter et al., 2014). When using laser scanning microscopes, additional possibilities become available, such as fluorescence recovery after photobleaching (Winter et al., 2014).

4.6.3 Plectin-Targeting Peptide

In a screen for peptides specifically binding to cell surface antigens of PDAC versus normal pancreatic duct cells, the sequence KTLLPTP was found and shown to bind to plectin (Kelly et al., 2008). Such plectin-targeted peptides, conjugated to fluorescent tags or magnetofluorescent particles together with, e.g., intravital confocal microscopy or magnetic resonance imaging, can be used to detect extracellular plectin. These peptides could also be useful for *in vitro* assays.

4.7 Binding Assays

Various binding assays have been used to study plectin (Rezniczek et al., 2004). For overlay assays, the main concerns are solubility of the proteins (discussed above, Section 4.4) and detection of binding. Choices for detection include radioactive labeling, fluorescent labeling (including Eu^{3+}-chelates), and antibodies. Yeast two-hybrid assays have been used to screen for new interaction partners of plectin as well as to map and characterize known interactions.

4.7.1 Cosedimentation

Cosedimentation assays are particularly useful and important when studying cytoskeletal filament-associated proteins. We will not provide protocols here, as the methods are not specific for plectin. What should be considered, however, are the domains of plectin and the sequences specific for certain plectin isoforms, as these can affect the results. For example, in some studies (Fontao et al., 2001; Geerts et al., 1999; Litjens et al., 2003), versions of the ABD fused to MBP (N-terminal) were used, while others worked with His-tagged proteins (García-Alvarez, Bobkov, Sonnenberg, & de Pereda, 2003). Fontao et al. reported an apparent K_d of approx. 0.3 μM for plectin–actin binding, which was in excellent agreement with the K_d reported previously for a plectin fragment encoded by exons 1–24 (Andrä, Nikolic, Stöcher, Drenckhahn, & Wiche, 1998). The different K_d of 22 μM observed by Garcia-Alvarez et al. was very likely due to the absence of additional sequences N-terminal of the ABD. This is also in line with the finding that stress fiber association of plectin 1d (exon 1d encodes only five amino acids) was less pronounced in comparison to other isoforms in transfected cells (Rezniczek et al., 2003).

4.7.2 Coimmunoprecipitation

Coimmunoprecipitation and pull-down assays are another important class of assays applicable to plectin. The main difficulty lies, again, in the size of the protein and its relative insolubility due to its numerous associations with cytoskeletal and membrane structures. The right balance between detergent solubilization and resulting stringency of the conditions and conservation of labile complexes must be found. Below, we provide two basic protocols that have been used successfully to precipitate signaling complexes associated with plectin from cultured cells (Osmanagic-Myers & Wiche, 2004). When working with mouse cells or tissues, ideally a negative control from a plectin-deficient animal should be included in the analysis.

4.7.2.1 Coimmunoprecipitation from Total Cell Lysates

Grow cells to the desired density, wash twice with ice-cold PBS, and collect the cells as described in Section 4.2.1 (detergent extraction). To ensure optimal activity of the nucleases, incubate the tubes at room temperature for 10 min. Then, add Triton X-100 to a final concentration of 1%. Optionally, SDS can be added to 0.1% at this point. Leave at room temperature for another 5 min before removing insoluble matter by centrifugation (1 min/>10,000 × *g*). Transfer the supernatant to a fresh tube. A 30-min preclearing step using protein A (or G) sepharose (35 μl of a 10% solution per 500 μl supernatant) equilibrated in lysis buffer is essential to prevent high background. After removal of the protein A (or G) beads (2 min/ >10,000 × *g*), incubate the lysates with antibodies by rotation either for 3 h/room temperature or for overnight/4 °C. To recover antibody complexes, add 50 μl of a 50% protein A (or G) sepharose slurry (per 500 μl lysate) and incubate by rotation for 3–6 h/4 °C. Recover the beads (centrifuge for 1 min with a slow increase from 0 to 3000 rpm over 30 s) and wash 3× with 1 ml of 50 m*M* HEPES, pH 7.0, 5 m*M* $MgCl_2$, 1 m*M* EGTA, 100 m*M* NaCl, 1% Triton X-100, at 4 °C.

4.7.2.2 Coimmunoprecipitation from Cell Fractions

Perform cell fractionation as described in Section 4.2.1. Use the cytosolic and membrane fractions for coimmunoprecipitation. Before adding primary antibodies, an optional preclearing step can be performed. Add primary antibodies and incubate 3 h to overnight with end-over-end rotation. Add protein A (or G) sepharose and incubate for 3 h to overnight (incubation times need to be individually balanced between specificity and efficiency of precipitation). Remove the beads by careful centrifugation (see above) and wash 3–4 × with the respective lysis solution (inhibitors are not necessary).

5. CONCLUSIONS

Plectin was first described 35 years ago (Pytela & Wiche, 1980). Since then, plectin has fulfilled and exceeded its initially thought role as an organizer of the cytoskeleton. Beyond mere structural responsibilities, plectin has been established as a scaffolding platform, bringing together the right molecules at the right time and location to perform their tasks, and to be itself involved in many and diverse regulatory processes. This diversity is owed to a large part of an elaborate fine-tuning mechanism of plectin provided by the alternative splicing of its gene, leading to subtle, yet functionally significant variations in the expressed isoforms. While our understanding of plectin isoforms has greatly increased in recent years, future plectin research will have to remain focused on the specific roles of these isoforms. It was the aim of this chapter to spur new interest for plectin and provide a methodical starting point for researchers new to this protein.

ACKNOWLEDGMENTS

We thank Maria-Josefa Castañón, Irmgard Fischer, and Selma Osmanagic-Myers for providing protocols and data.

REFERENCES

Abrahamsberg, C., Fuchs, P., Osmanagic-Myers, S., Fischer, I., Propst, F., Elbe-Bürger, A., et al. (2005). Targeted ablation of plectin isoform 1 uncovers role of cytolinker proteins in leukocyte recruitment. *Proceedings of the National Academy of Sciences of the United States of America, 102*(51), 18449–18454.

Ackerl, R., Walko, G., Fuchs, P., Fischer, I., Schmuth, M., & Wiche, G. (2007). Conditional targeting of plectin in prenatal and adult mouse stratified epithelia causes keratinocyte fragility and lesional epidermal barrier defects. *Journal of Cell Science, 120*(14), 2435–2443.

Andrä, K., Kornacker, I., Jörgl, A., Zörer, M., Spazierer, D., Fuchs, P., et al. (2003). Plectin-isoform-specific rescue of hemidesmosomal defects in plectin (−/−) keratinocytes. *The Journal of Investigative Dermatology, 120*(2), 189–197.

Andrä, K., Lassmann, H., Bittner, R., Shorny, S., Fässler, R., Propst, F., et al. (1997). Targeted inactivation of plectin reveals essential function in maintaining the integrity of skin, muscle, and heart cytoarchitecture. *Genes & Development, 11*(23), 3143–3156.

Andrä, K., Nikolic, B., Stöcher, M., Drenckhahn, D., & Wiche, G. (1998). Not just scaffolding: Plectin regulates actin dynamics in cultured cells. *Genes & Development, 12*(21), 3442–3451.

Bouameur, J.-E., Favre, B., & Borradori, L. (2014). Plakins, a versatile family of cytolinkers: Roles in skin integrity and in human diseases. *The Journal of Investigative Dermatology, 134*(4), 885–894.

Castañón, M. J., Walko, G., Winter, L., & Wiche, G. (2013). Plectin-intermediate filament partnership in skin, skeletal muscle, and peripheral nerve. *Histochemistry and Cell Biology, 140*(1), 33–53.

Elliott, C. E., Becker, B., Oehler, S., Castañón, M. J., Hauptmann, R., & Wiche, G. (1997). Plectin transcript diversity: Identification and tissue distribution of variants with distinct first coding exons and rodless isoforms. *Genomics, 42*(1), 115–125.

Fontao, L., Geerts, D., Kuikman, I., Koster, J., Kramer, D., & Sonnenberg, A. (2001). The interaction of plectin with actin: Evidence for cross-linking of actin filaments by dimerization of the actin-binding domain of plectin. *Journal of Cell Science, 114*(Pt. 11), 2065–2076.

Fuchs, P., Spazierer, D., & Wiche, G. (2005). Plectin rodless isoform expression and its detection in mouse brain. *Cellular and Molecular Neurobiology, 25*(7), 1141–1150.

Fuchs, P., Zörer, M., Rezniczek, G. A., Spazierer, D., Oehler, S., Castañón, M. J., et al. (1999). Unusual 5′ transcript complexity of plectin isoforms: Novel tissue-specific exons modulate actin binding activity. *Human Molecular Genetics, 8*(13), 2461–2472.

Fuchs, P., Zörer, M., Reipert, S., Rezniczek, G. A., Propst, F., Walko, G., et al. (2009). Targeted inactivation of a developmentally regulated neural plectin isoform (plectin 1c) in mice leads to reduced motor nerve conduction velocity. *The Journal of Biological Chemistry, 284*(39), 26502–26509.

García-Alvarez, B., Bobkov, A., Sonnenberg, A., & de Pereda, J. M. (2003). Structural and functional analysis of the actin binding domain of plectin suggests alternative mechanisms for binding to F-actin and integrin beta4. *Structure, 11*(6), 615–625.

Geerts, D., Fontao, L., Nievers, M. G., Schaapveld, R. Q., Purkis, P. E., Wheeler, G. N., et al. (1999). Binding of integrin alpha6beta4 to plectin prevents plectin association with F-actin but does not interfere with intermediate filament binding. *The Journal of Cell Biology*, *147*(2), 417–434.

Gostyńska, K. B., Nijenhuis, M., Lemmink, H., Pas, H. H., Pasmooij, A. M., Lang, K. K., et al. (2015). Mutation in exon 1a of PLEC, leading to disruption of plectin isoform 1a, causes autosomal-recessive skin-only epidermolysis bullosa simplex. *Human Molecular Genetics*, *24*(11), 3155–3162.

Gundesli, H., Talim, B., Korkusuz, P., Balci-Hayta, B., Cirak, S., Akarsu, N. A., et al. (2010). Mutation in exon 1f of PLEC, leading to disruption of plectin isoform 1f, causes autosomal-recessive limb-girdle muscular dystrophy. *American Journal of Human Genetics*, *87*(6), 834–841.

Janda, L., Damborsky, J., Rezniczek, G. A., & Wiche, G. (2001). Plectin repeats and modules: Strategic cysteines and their presumed impact on cytolinker functions. *Bioessays*, *23*(11), 1064–1069.

Kelly, K. A., Bardeesy, N., Anbazhagan, R., Gurumurthy, S., Berger, J., Alencar, H., et al. (2008). Targeted nanoparticles for imaging incipient pancreatic ductal adenocarcinoma. *PLoS Medicine*, *5*(4), e85.

Konieczny, P., Fuchs, P., Reipert, S., Kunz, W. S., Zeöld, A., Fischer, I., et al. (2008). Myofiber integrity depends on desmin network targeting to Z-disks and costameres via distinct plectin isoforms. *The Journal of Cell Biology*, *181*(4), 667–681.

Kostan, J., Gregor, M., Walko, G., & Wiche, G. (2009). Plectin isoform-dependent regulation of keratin-integrin alpha6beta4 anchorage via Ca^{2+}/calmodulin. *The Journal of Biological Chemistry*, *284*(27), 18525–18536.

Langhofer, M., Hopkinson, S. B., & Jones, J. C. (1993). The matrix secreted by 804G cells contains laminin-related components that participate in hemidesmosome assembly in vitro. *Journal of Cell Science*, *105*(Pt. 3), 753–764.

Lichti, U., Anders, J., & Yuspa, S. H. (2008). Isolation and short-term culture of primary keratinocytes, hair follicle populations and dermal cells from newborn mice and keratinocytes from adult mice for in vitro analysis and for grafting to immunodeficient mice. *Nature Protocols*, *3*(5), 799–810.

Litjens, S. H., Koster, J., Kuikman, I., van Wilpe, S., de Pereda, J. M., & Sonnenberg, A. (2003). Specificity of binding of the plectin actin-binding domain to beta4 integrin. *Molecular Biology of the Cell*, *14*(10), 4039–4050.

Metz, T., Harris, A. W., & Adams, J. M. (1995). Absence of p53 allows direct immortalization of hematopoietic cells by the myc and raf oncogenes. *Cell*, *82*(1), 29–36.

Mihailovska, E., Raith, M., Valencia, R. G., Fischer, I., Al Banchaabouchi, M., Herbst, R., et al. (2014). Neuromuscular synapse integrity requires linkage of acetylcholine receptors to postsynaptic intermediate filament networks via rapsyn–plectin 1f complexes. *Molecular Biology of the Cell*, *25*(25), 4130–4149.

Osmanagic-Myers, S., Gregor, M., Walko, G., Burgstaller, G., Reipert, S., & Wiche, G. (2006). Plectin-controlled keratin cytoarchitecture affects MAP kinases involved in cellular stress response and migration. *The Journal of Cell Biology*, *174*(4), 557–568.

Osmanagic-Myers, S., & Wiche, G. (2004). Plectin-RACK1 (receptor for activated C kinase 1) scaffolding: A novel mechanism to regulate protein kinase C activity. *The Journal of Biological Chemistry*, *279*(18), 18701–18710.

Pytela, R., & Wiche, G. (1980). High molecular weight polypeptides (270,000–340,000) from cultured cells are related to hog brain microtubule-associated proteins but copurify with intermediate filaments. *Proceedings of the National Academy of Sciences of the United States of America*, 77(8), 4808–4812.

Raith, M., Valencia, R. G., Fischer, I., Orthofer, M., Penninger, J. M., Spuler, S., et al. (2013). Linking cytoarchitecture to metabolism: sarcolemma-associated plectin affects glucose uptake by destabilizing microtubule networks in mdx myofibers. *Skeletal Muscle, 3*, 14.

Rezniczek, G. A., Abrahamsberg, C., Fuchs, P., Spazierer, D., & Wiche, G. (2003). Plectin 5′-transcript diversity: Short alternative sequences determine stability of gene products, initiation of translation and subcellular localization of isoforms. *Human Molecular Genetics, 12*(23), 3181–3194.

Rezniczek, G. A., Janda, L., & Wiche, G. (2004). Plectin. *Methods in Cell Biology, 78*, 721–755.

Sevcík, J., Urbániková, L., Kost'an, J., Janda, L., & Wiche, G. (2004). Actin-binding domain of mouse plectin. Crystal structure and binding to vimentin. *European Journal of Biochemistry, 271*(10), 1873–1884.

Shin, S. J., Smith, J. A., Rezniczek, G. A., Pan, S., Chen, R., Brentnall, T. A., et al. (2013). Unexpected gain of function for the scaffolding protein plectin due to mislocalization in pancreatic cancer. *Proceedings of the National Academy of Sciences of the United States of America, 110*(48), 19414–19419.

Sonnenberg, A., Rojas, A. M., & de Pereda, J. M. (2007). The structure of a tandem pair of spectrin repeats of plectin reveals a modular organization of the plakin domain. *Journal of Molecular Biology, 368*(5), 1379–1391.

Spurny, R., Abdoulrahman, K., Janda, L., Rünzler, D., Köhler, G., Castañón, M. J., et al. (2007). Oxidation and nitrosylation of cysteines proximal to the intermediate filament (IF)-binding site of plectin: Effects on structure and vimentin binding and involvement in IF collapse. *The Journal of Biological Chemistry, 282*(11), 8175–8187.

Valencia, R. G., Walko, G., Janda, L., Novacek, J., Mihailovska, E., Reipert, S., et al. (2013). Intermediate filament-associated cytolinker plectin 1c destabilizes microtubules in keratinocytes. *Molecular Biology of the Cell, 24*(6), 768–784.

Walko, G., Castañón, M. J., & Wiche, G. (2014). Molecular architecture and function of the hemidesmosome. *Cell and Tissue Research, 360*(2), 363–378.

Walko, G., Vukasinovic, N., Gross, K., Fischer, I., Sibitz, S., Fuchs, P., et al. (2011). Targeted proteolysis of plectin isoform 1a accounts for hemidesmosome dysfunction in mice mimicking the dominant skin blistering disease EBS-Ogna. *PLoS Genetics*, 7(12), e1002396.

Walko, G., Wögenstein, K. L., Winter, L., Fischer, I., Feltri, M. L., & Wiche, G. (2013). Stabilization of the dystroglycan complex in Cajal bands of myelinating Schwann cells through plectin-mediated anchorage to vimentin filaments. *Glia, 61*(8), 1274–1287.

Wiche, G., Osmanagic-Myers, S., & Castañón, M. J. (2015). Networking and anchoring through plectin: A key to IF functionality and mechanotransduction. *Current Opinion in Cell Biology, 32C*(32), 21–29.

Winter, L., Abrahamsberg, C., & Wiche, G. (2008). Plectin isoform 1b mediates mitochondrion-intermediate filament network linkage and controls organelle shape. *The Journal of Cell Biology, 181*(6), 903–911.

Winter, L., Staszewska, I., Mihailovska, E., Fischer, I., Goldmann, W. H., Schröder, R., et al. (2014). Chemical chaperone ameliorates pathological protein aggregation in plectin-deficient muscle. *The Journal of Clinical Investigation, 124*(3), 1144–1157.

Winter, L., & Wiche, G. (2013). The many faces of plectin and plectinopathies: Pathology and mechanisms. *Acta Neuropathologica, 125*(1), 77–93.

CHAPTER FOURTEEN

Functional and Genetic Analysis of Epiplakin in Epithelial Cells

Sandra Szabo, Karl L. Wögenstein, Peter Fuchs[1]
Department of Biochemistry and Cell Biology, Max F. Perutz Laboratories, University of Vienna, Vienna Biocenter (VBC), Vienna, Austria
[1]Corresponding author: e-mail address: peter.fuchs@univie.ac.at

Contents

Abstract

Epiplakin is a large member (>700 kDa) of the plakin protein family and exclusively expressed in epithelial cell types. Compared to other plakin proteins epiplakin exhibits an unusual structure as it consists entirely of a variable number of consecutive plakin repeat domains (13 in humans, 16 in mice). The only binding partners of epiplakin identified so far are keratins of simple as well as of stratified epithelia. Epiplakin-deficient mice show no obvious spontaneous phenotype. However, *ex vivo* studies using epiplakin-deficient primary cells indicated protective functions of epiplakin in response to stress. Recent studies using stress models for organs of the gastrointestinal tract revealed that epiplakin-deficient mice develop more pronounced pancreas and liver injuries than their wild-type littermates. In addition, impaired stress-induced keratin network reorganization was observed in the affected organs, and primary epiplakin-deficient

Methods in Enzymology, Volume 569
ISSN 0076-6879
http://dx.doi.org/10.1016/bs.mie.2015.06.018

hepatocytes showed reduced tolerance for forced keratin overexpression which could be rescued by a chemical chaperone. These findings indicate protective functions of epiplakin in chaperoning disease-induced keratin reorganization. In this review, we describe some of the methods we used to analyze epiplakin's function with the focus on biochemical and *ex vivo* techniques.

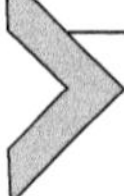

1. INTRODUCTION

1.1 Historical Perspective

The cytoskeleton consists of a network of actin filaments, intermediate filaments (IFs), and microtubules. Each of these components has unique functions while for certain cellular tasks or biological processes a coordinated action of at least two filament types is required. The molecular bases of such cytoskeletal cross-talks are proteins that mediate numerous interactions between different cytoskeletal networks and specific cellular compartments. Some of these so called cytoskeletal linker proteins are members of the plakin protein family (for review see Bouameur, Favre, & Borradori, 2014). This protein family is defined by the presence of several domains such as the plakin domain and/or the plakin repeat domain (PRD), the actin-binding domain, a rod domain, a spectrin repeat-containing rod, and a microtubule-binding domain. Many plakins are expressed in tissues that experience mechanical stress, such as epithelia and muscle, where they play a vital role in maintaining tissue integrity by cross-linking cytoskeletal filaments and anchoring them to membrane complexes. Hence, both inherited and autoimmune diseases that affect plakins can lead to disorders involving tissue fragility and skin blistering. Seven plakin family members have been identified: desmoplakin, plectin, bullous pemphigoid antigen 1 (BPAG1), microtubule actin cross-linking factor 1 (MACF1), envoplakin, periplakin, and epiplakin.

Epiplakin was originally identified as an autoantigen that reacted with serum from a patient who suffered from a subepidermal blistering disease (Fujiwara, Kohno, Iwamatsu, Naito, & Shinkai, 1996; Fujiwara, Kohno, Iwamatsu, & Shinkai, 1994). The protein is expressed in epithelial tissues, like those of skin, intestine, liver, and pancreas (Fujiwara et al., 2001; Spazierer et al., 2003). Compared to other members of the plakin protein family epiplakin has an unusual structure as it consists entirely of PRDs (Fujiwara et al., 2001; Spazierer et al., 2003). IF proteins, mainly keratins, are the only known cytoskeletal binding partners of epiplakin (Jang, Kalinin, Takahashi, Marekov, & Steinert, 2005; Spazierer, Raberger, Gross, Fuchs, & Wiche, 2008; Szabo et al., 2015; Wang, Sumiyoshi,

Yoshioka, & Fujiwara, 2006), which is an untypical characteristic for a plakin protein as many other protein family members are cytoskeletal linker proteins connecting different filament systems. Epiplakin-deficient mice did not show any discernible phenotype (Spazierer et al., 2006), except for a slight acceleration of keratinocyte migration (Goto et al., 2006). Surprisingly, even when the skin of knock-out mice was challenged by tape stripping or wounding, no differences to wild-type mice were observed (Spazierer et al., 2006). These findings are in clear contrast to other proteins belonging to the plakin protein family such as plectin, desmoplakin, and BPAG1, which play an important role in mechanically strengthening the skin as shown by phenotypes of knock-out mice (Andrä et al., 1997; Gallicano et al., 1998; Guo et al., 1995).

1.2 Gene Organization, Structural Properties, Expression Pattern, and Molecular Interactions

The murine epiplakin gene encodes a large protein with a molecular mass of 725 kDa consisting entirely of 16 PRDs (Fig. 1A) (Spazierer et al., 2003) whereas the human protein was reported to be smaller (~550 kDa) and to comprise 13 PRDs (Fujiwara et al., 2001). Murine and human epiplakin are both encoded by a single remarkably large exon consisting of 0.8–1.5 kb long DNA repeats (Fujiwara et al., 2001; Spazierer et al., 2003; Takeo et al., 2003). Eight of 16 of these repeats present in murine epiplakin are virtually identical both in regard to their encoding DNA and their amino-acid sequence (Fig. 1A). Epiplakin's expression is restricted to epithelial tissues and was found to be diverse in their various cell types (Fujiwara et al., 2001; Kokado et al., 2013; Matsuo et al., 2011; Spazierer et al., 2003; Szabo et al., 2015; Wögenstein et al., 2014; Yoshida et al., 2008, 2010). Its subcellular localization is dependent on keratin filaments (Szabo et al., 2015; Wögenstein et al., 2014) and it either perfectly colocalizes with filaments or decorates them in a punctuated pattern (Jang et al., 2005; Spazierer et al., 2008). All but 1 of the 16 murine epiplakin PRDs were shown to bind to keratins of basal keratinocytes (Spazierer et al., 2008). Keratin 8 and 18, the main keratins of most simple epithelia, were found to bind to 12 of the 16 PRDs of mouse epiplakin (Szabo et al., 2015).

1.3 Recent Findings from Analyses of Epiplakin-Deficient Mice and Cells

After the initial generation and phenotyping of two independent epiplakin-deficient mouse lines (Goto et al., 2006; Spazierer et al., 2006), these mice

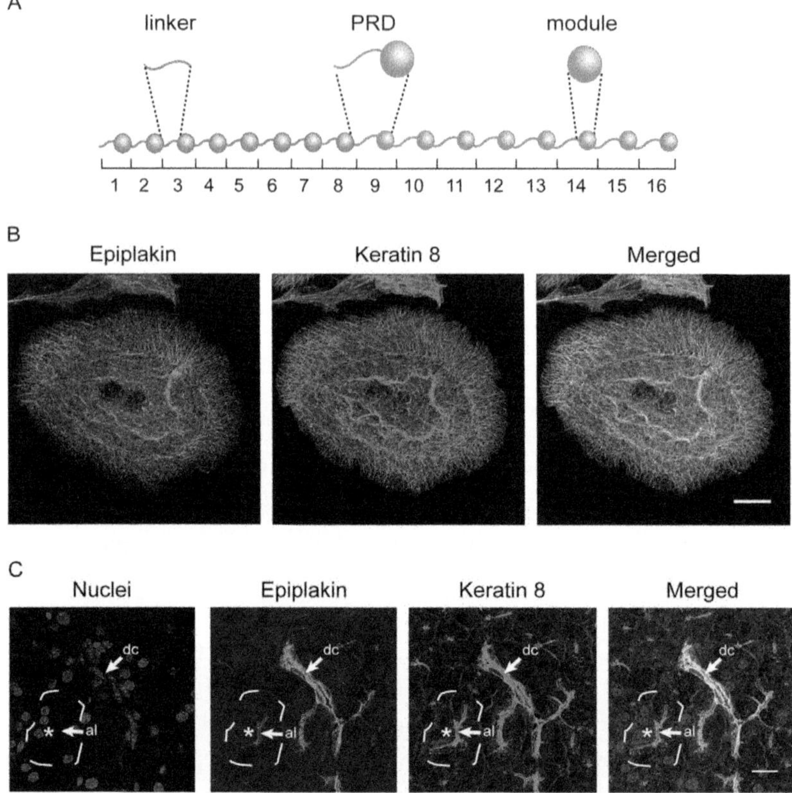

Figure 1 *Schematic illustration of murine epiplakin and immunofluorescence microscopy of epiplakin and keratin 8 in murine cells and tissue.* (A) Domain structure of mouse epiplakin showing 16 PRDs consisting of linkers (curved lines) and modules (circles). Modules that are virtually identical in amino acid sequence are shown in red (modules 9 to 16), those with lower homology in blue (modules 1 to 8). (B) Immunofluorescence microscopy of epiplakin and keratin 8 in primary hepatocytes. Cells were fixed with methanol and subjected to immunofluorescence microscopy using antibodies to epiplakin and K8. Epiplakin exhibits a dotted staining pattern lining keratin filaments. Scale bars, 20 μm. (C) Paraffin sections, prepared from pancreata of adult wild-type mice, were subjected to immunofluorescence microscopy using antibodies to epiplakin and K8. Epiplakin is found primarily in ductal cells (dc) and the apicolateral compartment (al) of acinar cells, where it colocalizes with K8-positive filaments. K8-positive filaments are also seen in perinuclear areas of acinar cells. Nuclear staining is shown for easier distinction of ductal and acinar cells. Whole acini are outlined by dashed lines for orientation. Asterisks indicate cytosolic/perinuclear areas. Scale bar, 10 μm. *Panel (A) illustration was originally published in* Journal of Hepatology; *Szabo et al. (2015). Panel (C) image was originally published in* PLoS One; *Wögenstein et al. (2014).* (See the color plate.)

were investigated in more detail, and studies using epiplakin-deficient primary cells as well as knock-down experiments were performed. In differentiated primary keratinocytes phosphatase, inhibitor-induced filament disruption was shown to proceed faster in epiplakin-deficient compared to wild-type cells (Spazierer et al., 2008). This observation suggested that epiplakin plays a role in keratin filament reorganization in response to stress. Another finding in epiplakin-deficient keratinocytes was that they exhibited reduced diameters of keratin filaments indicating that epiplakin accelerates keratin bundling in proliferating keratinocytes during wound healing (Ishikawa et al., 2010). Kokado and colleagues reported that loss of epiplakin affects cell differentiation, proliferation activity, and structural integrity of the corneal epithelium (Kokado et al., 2013). Similar to epiplakin-deficient skin, these authors also detected promoted migration-dependent wound healing in epiplakin-deficient corneal epithelium (Goto et al., 2006). These findings were supported by epiplakin knock-down experiments in HeLa cells which showed significantly elevated cellular motility suggesting that epiplakin affects rearrangement of IFs (Shimada et al., 2013). Recent studies using stress models for organs of the gastrointestinal tract showed that lack of epiplakin leads to more pronounced pancreas and liver injuries and to impaired stress-induced keratin network reorganization manifested by increased numbers of cells comprising keratin aggregates (Szabo et al., 2015; Wögenstein et al., 2014). These granules were also found in stressed epiplakin-deficient primary hepatocytes and the formation of these aggregates could be prevented by a chemical chaperone. These findings indicate that epiplakin has protective functions during certain injuries of the gastrointestinal tract by chaperoning disease-induced keratin reorganization.

2. MAIN CONSIDERATIONS

Studying epiplakin's functions issues challenges to scientists. First, epiplakin represents one of the proteins with the largest molecular sizes found in vertebrates. This fact considerably complicates biochemical experiments with full-length epiplakin proteins, such as immunoblotting, although these difficulties can mostly be overcome by using customized protocols. Second, murine epiplakin consists of eight PRDs (PRDs 9–16) almost identical in regard to nucleic acid and amino acid sequence. This particular feature exacerbates all kinds of cloning work with DNA encoding these repeats due to recombination events occurring between these almost identical DNA sequences. Thus, the coding cDNA for more than one of these repeats and for full-length epiplakin cDNA can only be cloned and propagated in *E. coli* with the aid of bacterial artificial chromosomes (BACs) as vector systems (Spazierer et al., 2003).

3. METHODS: PEARLS AND PITFALLS

3.1 Visualization of Epiplakin in Cells and Tissue

For studying the localization of epiplakin in cells or tissue, standard immunofluorescence techniques can be performed. Additionally, the expression pattern of epiplakin can be analyzed using standard immunohistochemistry techniques.

3.1.1 Immunofluorescence Staining in Primary Cells (*Fig. 1B*)

1. For fixation of cells, culture medium is aspirated and cells are washed twice with PBS.
2. Cells are incubated for 90 s with –20 °C methanol on a cooling block followed by two washing steps with PBS. Upon fixation, cells can be stored in PBS at 4 °C or directly subjected to immunofluorescence staining for epiplakin.
3. Selected areas on the culture dishes are encircled with a hydrophobic pen and endogenous unspecific antigen interactions are blocked with 2% BSA in PBS for 30 min at room temperature (RT).
4. An epiplakin antibody is diluted in blocking solution as recommended and applied to the culture dishes for 60 min at RT, followed by washing with PBS.
5. Sections are incubated with adequate fluorophore-conjugated IgGs diluted in PBS for 60 min at RT, followed by washing with PBS.
6. Nuclei are stained using Hoechst #33258 (Sigma–Aldrich, cat. no. 861405) diluted in PBS (3.3 μg/ml) for 5 min at RT, followed by washing with PBS.
7. In order to avoid formation of PBS-derived precipitates, cells are additionally washed with ddH_2O.
8. Slides are mounted with cover slips using Mowiol (preparation described below).
9. Stainings are analyzed using confocal laser scanning microscopy.

Pearls and pitfalls

- Optimal staining results are obtained when performing steps 3–6 in a humid chamber.
- In case of unspecific binding of antibodies, higher antibody dilutions should be tested.

Preparation of Mowiol

1. 6 g of glycerol and 2.4 g Mowiol 4–88 (Calbiochem, cat. no. 475904) are mixed with 6 ml of ddH_2O and stirred gently for 2 h at RT.
2. 12 ml of 0.2 *M* Tris–HCl, pH 8.5 are added and the mixture is stirred gently for 10 min, followed by heating to 50 °C.
3. After a centrifugation step at 3000 × *g*, the supernatant is filtered (pore size 40 μm) and used for mounting or stored at −20 °C.

3.1.2 *Immunofluorescence Staining on Paraffin Sections* (*Fig. 1C*)

1. Organs are isolated and fixed in 4% paraformaldehyde in PBS at 4 °C for 24 h, followed by dehydration, paraffin embedding, and sectioning of tissues according to standard protocols.
2. Paraffin sections are deparaffinized in xylene for 2 × 20 min and rehydrated by incubation of slides in 96% ethanol (twice), 70% ethanol, and 50% ethanol for 2 min each.
3. Sections are washed with ddH_2O before being boiled in antigen retrieval buffer (10 m*M* Tris–HCl, pH 9.0; 1 m*M* EDTA, pH 8.0) using a steam cooker.
4. Slides are allowed to cool for 30 min and are subsequently washed with ddH_2O.
5. Sections are encircled with a hydrophobic pen and endogenous unspecific antigen interactions are blocked by 2% BSA in PBS for 30 min at RT.
6. An epiplakin antibody is diluted in blocking solution as recommended and applied to the sections for 60 min at RT, followed by washing with PBS.
7. Sections are incubated with adequate fluorophore-conjugated anti-rabbit IgGs diluted in PBS for 60 min at RT, followed by washing with PBS.
8. Nuclei are stained using Hoechst #33258 (Sigma–Aldrich, cat. no. 861405) diluted in PBS (3.3 μg/ml) for 5 min at RT and slides are subsequently washed with PBS.

9. In order to avoid formation of PBS-derived precipitates, cells are additionally washed with ddH_2O.
10. Slides are mounted with cover slips using Mowiol (see Section 3.1.1).
11. Stained sections are analyzed using confocal laser scanning microscopy.

Pearls and pitfalls

- Optimal staining results are obtained when performing steps 5–8 in a humid chamber.
- In case of unspecific binding of antibodies, higher antibody dilutions should be tested. Alternatively, antigen retrieval using citrate buffer (1.8 m*M* citric acid; 8.2 m*M* sodium citrate, pH 6.0) might prove useful.

3.1.3 Immunohistochemistry on Paraffin Sections *(Fig. 2A)*

1. Paraffin sections are deparaffinized in xylene for 2 × 20 min and rehydrated by incubation of slides in 96% ethanol (twice), 70% ethanol, and 50% ethanol for 2 min each.
2. Sections are washed with ddH_2O before being boiled in antigen retrieval buffer (10 m*M* Tris–HCl, pH 9.0; 1 m*M* EDTA, pH 8.0) for 60 min using a steam cooker.
3. Slides are allowed to cool for 30 min at RT and are subsequently washed with ddH_2O.
4. Sections are encircled with a hydrophobic pen and endogenous peroxidases are blocked using Peroxidase Blocking Reagent (DAKO, cat. no. S2001) for 15 min at RT.
5. Endogenous unspecific antigen interactions are blocked with 2% BSA in PBS for 30 min at RT.
6. An epiplakin antibody is diluted in blocking solution as recommended and applied to the sections for 60 min at RT, followed by washing with PBS.
7. Sections are incubated with adequate biotinylated anti-rabbit IgGs (e.g., DAKO, cat. no. E0432) diluted in PBS for 30 min at RT, followed by washing with PBS.
8. Sections are incubated with horseradish peroxidase (HRPO)-coupled streptavidin (DAKO, cat. no. P0397) for 30 min at RT, followed by washing with PBS.
9. Bound antibodies are visualized using the Liquid DAB+ Substrate Chromogen System (DAKO, cat. no. K3468) according to the manufacturer's guidelines.
10. Sections are washed with PBS and counterstained with hematoxylin (Merck, cat. no. 109253) for 20–60 s.

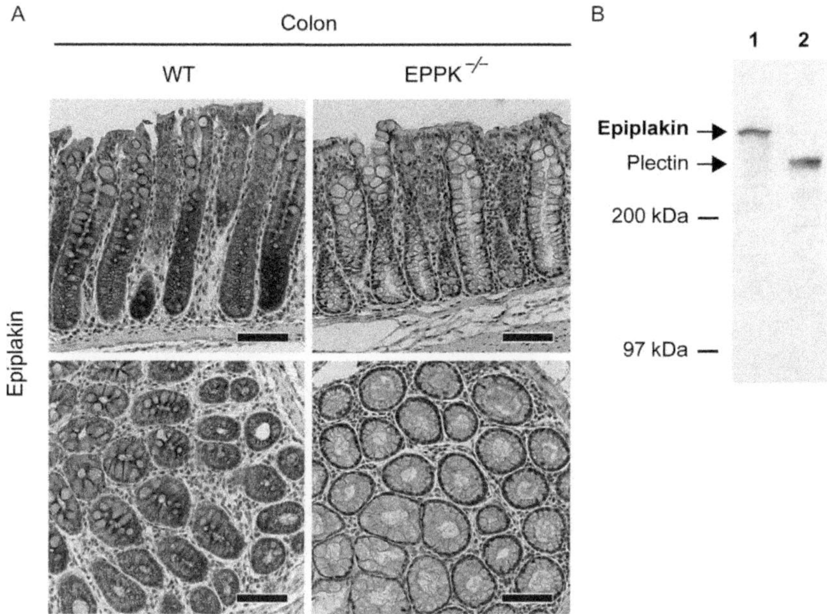

Figure 2 *Immunohistochemistry staining and immunoblots of epiplakin.* (A) Representative immunohistochemistry staining depicting the expression of epiplakin in murine colons from WT and epiplakin-deficient mice. Longitudinal (upper panels) and cross sections (lower panels) of colon crypts are shown. Epiplakin is expressed in epithelial cells of WT colon whereas the protein is completely absent in epiplakin-deficient colon, demonstrating the specificity of the antibody used. Scale bars, 50 μm. (B) Immunoblotting of epiplakin and plectin. A protein extract from newborn mouse was separated via SDS-PAGE, transferred to a nitrocellulose membrane and analyzed using antibodies to epiplakin (*lane 1*) and to plectin (*lane 2*). The positions of additional size markers are indicated. A 3.75% polyacrylamide gel was used. *Panel (B) image was originally published in* The Journal of Biological Chemistry; *Spazierer et al. (2003). © the American Society for Biochemistry and Molecular Biology.* (See the color plate.)

11. Sections are washed with tap water and briefly immersed in 0.5% HCl in 70% EtOH.
12. After purging of slides with warm tap water for 15 min, sections are dehydrated by incubation with 50% ethanol, 70% ethanol, and 96% ethanol for 2 min each.
13. Slides are incubated with *n*-Butyl acetate (Merck, cat. no. 109652) for 5 min before mounting with cover slips using Entellan mounting medium (Merck, cat. no. 107961).
14. Stained sections are analyzed using brightfield light microscopy.

Pearls and pitfalls

- Optimal staining results are obtained when performing steps 4–8 in a humid chamber.
- In case of unspecific binding of antibodies, higher antibody dilutions should be tested. Alternatively, antigen retrieval using citrate buffer (1.8 m*M* citric acid; 8.2 m*M* sodium citrate, pH 6.0) might prove useful.

3.2 Preparation of Protein Lysates

3.2.1 Preparation of Total Protein Lysates from Tissue

1. Organs are dissected and briefly washed with ice-cold PBS and snap frozen in liquid nitrogen.
2. Organs are pulverized under permanent liquid N_2-cover using a mortar and pistil and further homogenized in lysis buffer [10 m*M* Tris pH 7.5; 150 m*M* NaCl; 5 m*M* EDTA; 1% Triton X-100; 0.1% SDS; 1% deoxycholate; freshly supplemented with cOmplete Mini Protease Inhibitors (Roche Applied Science, cat no. 05892791001); Phosphatase Inhibitor Cocktails 2 and 3 (Sigma–Aldrich, cat no. P5726 and P0044); 100 µg/ml DNAse I; 100 µg/ml RNAse A; 1 m*M* PMSF] using a glass dounce homogenizer. Per 10 mg of organ, 100 µl of lysis buffer are used.
3. The resulting suspension is incubated for 10 min at RT with periodic mixing to guarantee efficient cell lysis and ribonuclease digestion.
4. Two parts of the lysate are mixed with one part sample buffer (400 m*M* Tris, pH 6.8; 10% SDS; 50% glycerol; 500 m*M* DTT; 0.1% Bromophenol blue) by brief vortexing, subsequently boiled for 5 min at 95 °C and vortexed again for 30 s.
5. The lysate is cleared by centrifugation at 13,800 × *g* and the supernatant is transferred to a fresh tube and either directly used for downstream analyses or stored at −80 °C.

3.2.2 Preparation of Total Protein Lysates from Cultured Cells

1. Dishes or flasks with cells are placed on ice and the cells are washed twice with ice-cold PBS.
2. The cells are overlaid with boiling sample buffer (400 m*M* Tris pH 6.8; 10% SDS; 50% glycerol; 500 m*M* DTT; 0.1% Bromophenol blue) and scraped into a microcentrifuge tube. Per 1×10^6 cells, 100 µl sample buffer are used.
3. The lysates are incubated at 95 °C for 5 min and subsequently repeatedly passed through a 27 gauge needle to tear genomic DNA.

4. The lysate is cleared by centrifugation at 13,800 × g and the supernatant is transferred to a fresh tube and either directly used for downstream analyses or stored at −80 °C.

Pearls and pitfalls

- To prepare protein lysates for immunoblot analysis of epiplakin, all experimental steps need to be performed fast and efficiently. The lysates need to be prepared freshly and immediately subjected to electrophoretic separation since upon storage (even at −80 °C), epiplakin frequently becomes degraded, resulting in a ladder-like signal pattern upon immunoblotting.
- In organs exhibiting low expression levels of epiplakin, tissue/buffer ratios have to be modified in order to obtain epiplakin concentrations high enough for detection via immunoblotting. High amounts of genomic DNA in concentrated lysates can lead to high viscosity, which can be reduced by repeatedly passing the lysate through a 27 gauge needle thereby tearing DNA or sonication. Alternatively, larger volumes of sample buffer can be added.

3.2.3 High Salt Extraction Method: Analysis of Epiplakin Solubility

In untreated keratinocytes, epiplakin can be detected in the cytosolic (soluble) as well as in the cytoskeletal (insoluble) cellular subfractions. In order to investigate potential alterations in epiplakin's solubility (e.g., during stress conditions), soluble and insoluble fractions are prepared from cultured primary cells or cell lines.

1. Cells are washed twice with PBS and pelleted at 300 × g for 2 min.
2. The cell pellet is lysed with (1 ml per ~0.5 cm^3 cells) 150 m*M* NaCl, 1.5 *M* KCl, 1 m*M* EDTA, 0.5% Triton X-100, 10 m*M* Tris–HCl, pH 7.5 (high-salt buffer), freshly supplemented with cOmplete Mini protease inhibitors (Roche Applied Science, cat. no. 05892791001) for 30 min at 4 °C.
3. After centrifugation at 16,000 × g for 10 min, the supernatant representing the cytosolic (soluble) fraction is mixed with SDS-containing sample buffer, incubated at 95 °C for 5 min and subjected to immunoblot analysis.
4. The pellet, representing the cytoskeletal (insoluble) fraction is solubilized with 8 *M* urea (~200 μl per ~0.5 cm^3 originally used cells) by vortexing or shaking on RT for 20 min.
5. The sample is mixed with 5 × SDS sample buffer in a 1:2 (sample buffer/urea) ratio, incubated at 65 °C for 10 min and centrifuged for 2 min at 16000 × g and RT.
6. The supernatant is transferred into a fresh tube and subjected to immunoblot analysis.

3.3 SDS-PAGE and Immunoblotting *(Fig. 2B)*

For SDS-polyacrylamide gel electrophoresis (PAGE) and subsequent blotting of epiplakin, we used standard protocols. However, due to epiplakin's susceptibility to degradation, all experimental steps need to be carried out consecutively without interruption. Either 4% or 6% SDS polyacrylamide gels with 4% stacking gels were used to separate epiplakin-containing lysates. While the lower percentage gels provide better resolution, they are also more challenging to handle. To achieve better resolution when using 6% gels, electrophoresis can be continued for up to 1 h after the dye front has migrated out of the gel. Notably, most commercially available length markers mixes lack proteins at the size of epiplakin, thus one might alternatively use large purified proteins (e.g., plectin) as size marker.

1. Samples are incubated at 95 °C for 5 min, subsequently placed on ice and loaded on the acrylamide gel. Separating gel: 4–6% acrylamide mix (Gerbu, cat no. 1108); 375 m*M* Tris–HCl pH 8.8; 0.1% SDS; 0.1% ammonium persulfate; 0.08% TEMED; stacking gel: 4% acrylamide mix; 125 m*M* Tris–HCl; pH 6.8; 0.1% SDS; 0.1% ammonium persulfate; 0.1% TEMED; running buffer: 25 m*M* Tris; 250 m*M* glycine; 0.1% SDS.
2. Electrophoresis is performed at 20 mA/gel until the bromophenol blue front migrates out of the gel and continued for additional 30 min (4% gels) or 60 min (6% gels).
3. For blotting of epiplakin, wet blotting onto a nicrocellulose membrane is recommended. To this cause, the gel and all components of the blotting sandwich, such as nitrocellulose membrane, filters pads, and sponges are equilibrated in transfer buffer (25 m*M* Tris; 250 m*M* glycine) and subsequently assembled.
4. Blotting is performed overnight at 25 V and 4 °C in a wet chamber. Next day, transferred proteins are reversibly stained with Ponceau-S solution (0.2% Ponceau S; 30% trichloracetic acid; 30% sulfonsalicylic acid; 40% ddH_2O).
5. To block unspecific binding sites, the nitrocellulose membrane is incubated in 5% milk powder in PBS for 1 h at RT. Incubation with the primary epiplakin antibody, diluted as recommended in PBS supplemented with 0.1% Tween-20 (PBS-T), is carried out overnight at 4 °C.
6. Immunodetection is performed using suitable secondary antibodies.

3.4 Protein–Protein Interaction Studies

The availability of purified proteins is a prerequisite for in vitro protein–protein interaction assays. Unfortunately, as the size of epiplakin exceeds the limitations of bacterial protein expression, expression of the full-length protein using standard bacterial expression systems is not recommended. However, as certain subdomains of epiplakin (i.e., PRDs) are almost identical in amino acid sequence (Spazierer et al., 2003), analysis of the function of individual PRDs might provide valuable insight into the function of full-length epiplakin. For the protein–protein interaction studies described, purified glutathione *S*-transferase (GST)-tagged PRD fusion proteins (Spazierer et al., 2008; Szabo et al., 2015) were used.

3.4.1 Blot Overlay with Individual Epiplakin PRDs (Fig. 3A)

In order to identify the major binding partners of epiplakin, i.e., epidermal as well as simple epithelial keratins, blot overlays have been predominantly used (Spazierer et al., 2008; Szabo et al., 2015). This interaction assay can be employed for analysis of direct binding of epiplakin to any protein of interest.

1. After gel electrophoresis, the proteins of interest (e.g., liver high-salt keratin extracts) are immobilized to a nitrocellulose membrane which is subsequently stained with Ponceau Red and cut between each protein lane using a scalpel.
2. Membrane stripes are blocked in 5% milk powder in 0.1% PBS-T for 3 h at RT.
3. Each membrane stripe is separately incubated with 0.1% PBS-T and 5 μg/ml of individual recombinant GST-tagged epiplakin PRDs for 1 h at 4 °C.
4. Membrane stripes are washed with 0.1% PBS-T for 3×10 min, followed by incubation with an anti-GST antibody for 1 h at RT or overnight at 4 °C.
5. Membrane stripes are washed with 0.1% PBS-T for 3×10 min.
6. Bound PRD–GST fusion proteins are visualized using adequate secondary antibodies (conjugated to either alkaline phosphatase or horseradish peroxidase) and the respective detection system.

Pearls and pitfalls

- In case of low affinity of the PRD–GST fusion protein to binding partners, longer incubation times should be tested. Alternatively, the amount of PRD–GST fusion protein employed may be increased.

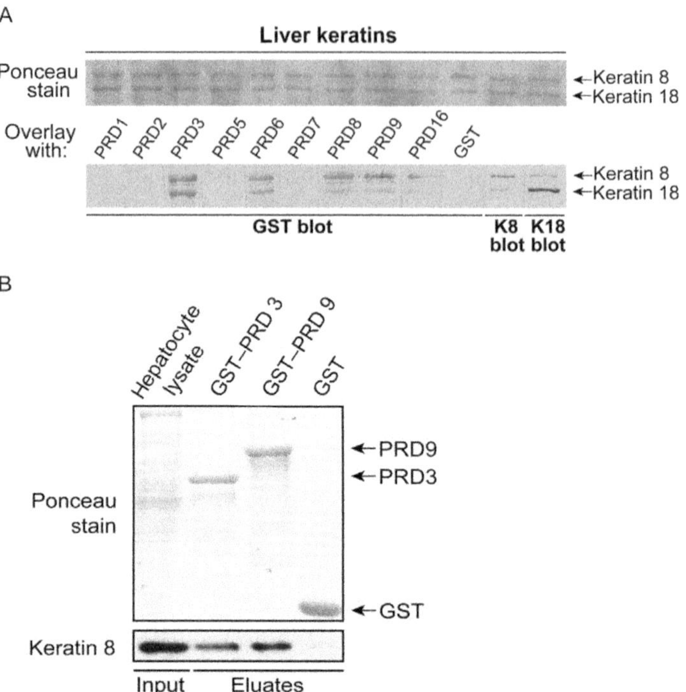

Figure 3 *Epiplakin directly binds to simple epithelial keratins via multiple plakin repeat domains.* (A) Blot overlay of keratins isolated from mouse livers with 5 μg/ml of the PRD–GST fusion protein indicated. Overlay of keratins with GST was used as negative control; the Ponceau stain demonstrates equal loading of hepatic keratins. Proteins bound to keratins were detected using a GST antibody whereas the position of keratins was visualized using antibodies recognizing K8 or K18, respectively. (B) Co-sedimentation of keratin 8 with two epiplakin PRD–GST fusion proteins. Proteins from a primary hepatocyte lysate (input) and proteins binding to PRD–GST–sepharose beads (eluates) were analyzed by immunoblotting using antibodies to K8; GST coupled to sepharose beads was used as negative control. The Ponceau stain demonstrates similar concentrations of the GST–PRD3, GST–PRD9, and GST proteins employed. Immunoblot analysis revealed specific co-sedimentation of K8 with PRD3 and PRD9, whereas no interaction was found with GST alone. *Both images were originally published in* Journal of Hepatology; *Szabo et al. (2015).*

- In case of unspecific protein interactions, the stringency of the binding buffer can be increased. Alternatively, the amount of PRD–GST fusion protein employed may be reduced.
- In order to exclude false-positive results as a consequence of unspecific binding of the GST-tag to the potential binding partners, adequate negative controls should be included in the experimental setup (e.g., recombinant GST).

3.4.2 Co-Sedimentation Assay: PRD–GST Pulldown *(Fig. 3B)*

As an alternative technique for studying interaction of epiplakin PRDs with proteins, cosedimentation assays have been used to confirm the interaction of certain PRDs with epidermal or simple epithelial keratins, respectively (Spazierer et al., 2008; Szabo et al., 2015).

1. Cells are scraped into 1 ml of PBS supplemented with 5 m*M* EDTA, followed by pelleting of the cells at 335 × *g* for 2 min.
2. The pellet is lysed with pre-chilled 2% Empigen BB Detergent (Calbiochem, cat. no. 324690) in PBS, freshly supplemented with cOmplete Mini protease inhibitor tablets (Roche Applied Science, cat. no. 05892791001) and Phosphatase Inhibitor Cocktails 2 and 3 (Sigma–Aldrich, cat. no. P5726 and P0044), 5 m*M* EDTA and 1 m*M* PMSF for 1 h at 4 °C.
3. The cell lysate is cleared by centrifugation at 16,000 × *g* for 20 min and the supernatant is transferred into a fresh tube and stored on ice until incubation with recombinant GST-tagged epiplakin PRDs.
4. 50% slurry of glutathione-Sepharose 4B beads (GE Healthcare, cat. no. 17-0756-01) in PBS is prepared according to the manufacturer's instructions.
5. 20 μg of GST–PRD fusion proteins are mixed with 50 μl of slurry and PBS to a total volume of 600 μl, followed by incubation at 4 °C for 1 h.
6. After centrifugation for 3 min at 1000 × *g* and 4 °C, the supernatant is removed and unspecific binding sites on the beads are blocked by incubation with 1 ml of 4% BSA in PBS for 1 h at 4 °C.
7. After centrifugation for 3 min at 1000 × *g* and 4 °C, the supernatant is discarded.
8. 600 μl of the freshly prepared cell lysate are added to the beads, followed by overnight incubation at 4 °C.
9. After centrifugation for 3 min at 1000 × *g* and 4 °C, the supernatant is discarded.
10. The beads are incubated with 1 ml of RIPA buffer (50 m*M* Tris–HCl, pH 7.5; 1% Triton X 100; 0.5% deoxycholate; 0.1% SDS; 150 m*M* NaCl and 1 m*M* EGTA) for 15 min at 4 °C, followed by repelleting of the beads for 3 min at 1000 × *g* and 4 °C. This washing step is repeated once.
11. After an additional washing step with PBS, 50–100 μl of 5 × SDS sample buffer are added to the beads and incubated at 95 °C for 5 min.
12. For elution of the bound proteins, the beads are centrifuged for 2 min at 16,000 × *g* and the supernatant is transferred into a fresh tube and subjected to immunoblot analysis using antibodies recognizing the protein of interest.

Pears and Pitfalls

- To avoid unspecific protein interactions, the stringency of the washing buffer may be increased. Alternatively, the amount of PRD–GST fusion protein employed may be reduced.
- In case of low affinity of the PRD–GST fusion protein to the protein of interest, the amount of PRD–GST fusion protein employed may be increased.
- In order to exclude false-positive results as a consequence of unspecific binding of the GST-tag to the protein of interest, adequate negative controls should be included in the experimental setup (e.g., recombinant GST).

3.5 Investigation of Epiplakin's Function Using Primary Hepatocytes

Not all of epiplakin's functions can be sufficiently investigated *in vivo*. For a more detailed investigation of the consequences of epiplakin deficiency, cell culture models proved to be highly useful (Spazierer et al., 2008; Szabo et al., 2015). Hence, we routinely isolated primary hepatocytes using a two-step collagenase perfusion protocol (adapted from Weerasinghe, Ku, Altshuler, Kwan, & Omary, 2014).

3.5.1 Overexpression of Fluorescently Tagged Epiplakin PRD-Fusion Proteins in Primary Hepatocytes

To study the cellular localization of single epiplakin PRDs, individual cDNAs coding for PRDs (Spazierer et al., 2008) are cloned into the mammalian expression vectors pEGFP-C2 or pEGFP-N2 (Clontech Laboratories, Inc.) employing standard cloning methods and are then transfected into cells.

1. $\sim 8 \times 10^4$ primary WT hepatocytes are seeded per 35 mm dish and instantly transfected with the individual DNA constructs using the transfection reagent Nanofectin (PAA, cat. no. Q051-005).
2. For transfection of one 35 mm dish, 1.6 μg plasmid DNA is mixed with 83 μl diluent and 5.5 μl of Nanofectin are mixed with 83 μl diluent. The mixtures containing the plasmid or Nanofectin are combined, immediately vortexed, and incubated at RT for 20 min.
3. The transfection mixture is pipetted onto the cells and evenly distributed by gentle shaking.

4. To analyze the intracellular distribution of individual PRDs as well as their potential colocalization with keratin filaments, cells are fixed 24 h after transfection with methanol as described in Section 3.1.1.
5. Cells are stained with an antibody recognizing K8 and an adequate fluorophore-coupled secondary antibody, followed by mounting with Mowiol as described in the Section 3.1.1.
6. Stained cells are subjected to confocal fluorescence microscopy and examined carefully regarding the localization pattern of individual EGFP-tagged PRDs.

3.5.2 Okadaic Acid-Treatment of Primary Hepatocytes

Phosphorylation of keratins is dramatically increased in response to stress, as shown in cultured cells and mouse models (Fausther, Villeneuve, & Cadrin, 2004; Ridge et al., 2005; Toivola, Zhou, English, & Omary, 2002). Consequently, phosphatase inhibitors, such as okadaic acid (OA), which cause hyperphosphorylation of keratins (Blankson, Holen, & Seglen, 1995; Chou & Omary, 1994; Kasahara et al., 1993; Yatsunami et al., 1993) represent a useful tool to mimic cellular stress accompanied by keratin network reorganization (Strnad, Windoffer, & Leube, 2001, 2002).

1. 4×10^5 wild-type and epiplakin-deficient hepatocytes each are seeded as a mixed culture on collagen-coated 35 mm dishes and cultivated for 18 h.
2. Cells are incubated with medium (Williams E Medium; 50 U/ml Penicillin; 50 µg/ml Streptomycin) containing 30 n*M* OA (Merck Millipore, cat no. 495609) for 20, 40, 60, and 90 min. Cells are immediately fixed with −20 °C cold methanol (see Section 3.1.1).
3. Immunofluorescence staining is performed using antibodies to epiplakin and keratin (see Section 3.1.1) and the cells are evaluated regarding their keratin distribution pattern.

Pearls and pitfalls

- In our hands treatment of primary hepatocytes with 30 n*M* OA induced efficient phosphorylation of keratins already after 20 min of incubation. Disassembly of keratin filaments as well as formation of keratin granules started approximately after 50 min of incubation.

3.5.3 Forced Overexpression of Keratin 8 in Primary Hepatocytes

To investigate the tolerance of epiplakin-deficient primary hepatocytes for high keratin levels, these cells can be challenged by forced overexpression of fluorescently tagged keratin 8 (Szabo et al., 2015). Transfected hepatocytes are analyzed regarding their keratin distribution pattern by immunofluorescence microscopy.

1. 4×10^5 wild-type and epiplakin-deficient primary hepatocytes in 1.5 ml medium are seeded on collagen-coated 35 mm dishes as a mixed culture and instantly transfected with a plasmid coding for full-length K8 tagged with a fluorophore using the Nanofectin transfection kit (PAA, cat no. Q051-005).
2. For transfection of one 35 mm dish, 1.6 μg plasmid DNA is mixed with 83 μl diluent. 5.5 μl of Nanofectin are mixed with 83 μl diluent. The mixtures containing the plasmid or Nanofectin respectively are combined, immediately vortexed, and incubated at RT for 20 min.
3. The transfection mixture is pipetted onto the cells and evenly distributed by gentle shaking.
4. After 24 h, the cells are fixed with 4% formaldehyde diluted in PBS and immunofluorescently labeled for epiplakin to distinguish between WT and epiplakin-deficient cells (see Section 3.1.1). Fluorescence from the K8-fusion protein is used to assess the pattern of exogenous keratins and the number of ectopically K8-expressing hepatocytes from each genotype is counted.

Pearls and pitfalls

- WT and epiplakin-deficient cells should be cultivated in the same dishes as a mixed culture to minimize variations in the treatment of the two genotypes, which can be distinguished easily by the use of anti-epiplakin antibodies.

3.5.4 Rescue Experiments Using the Chemical Chaperone TMAO

The low molecular weight chemical chaperone trimethylamine N-oxide (TMAO) was shown to be able to prevent the formation of keratin aggregates in cultured cells (Chamcheu et al., 2009). TMAO can be used to rescue the impaired keratin reorganization in primary epiplakin-deficient hepatocytes upon OA-treatment and forced keratin 8 overexpression (Szabo et al., 2015).

3.5.4.1 Rescue of OA-Induced Keratin Aggregate Formation

Primary hepatocytes isolated from WT and epiplakin-deficient livers are cultivated in plating medium containing 100 μ*M* TMAO (Sigma–Aldrich, cat no. 317594) for 18 h before being incubated with serum-free medium containing 30 μ*M* OA in combination with TMAO for 60 min. The protocol is continued according to Section 3.5.2.

3.5.4.2 Rescue of Aggregate Formation Induced by Keratin Overexpression

Primary hepatocytes isolated from WT and epiplakin-deficient livers are seeded on collagen-coated dishes as a mixed culture of epiplakin-deficient and WT cells as described in Section 3.5.3. in medium containing 100 μ*M* TMAO and instantly transfected with a plasmid coding for full-length K8 tagged with a fluorophore. The protocol is continued according to Section 3.5.3.

3.6 Antibodies

Both antibodies produced in our lab were generated by using short glutathion-*S*-transferase (GST)-tagged target sequences as immunogens which were injected into rabbits (Spazierer et al., 2003). Resulting sera were subsequently affinity purified by the use of maltose-binding protein fusions of the same sequences immobilized on CNBr-Sepharose. We have compiled a table of epiplakin-specific antibodies and some of their features and applications as described in literature (Table 1).

3.7 Generation of Epiplakin-Deficient Mice and their Phenotyping Using Stress Models

In our lab, a mouse line deficient for epiplakin has been created through targeted inactivation of its gene by replacing the first ~2.5 kb of epiplakin's single coding exon (Spazierer et al., 2006). These mice did not show any discernible phenotype unless certain stress models affecting organs comprising epithelia were applied which are listed in Table 2. Protocols for disease models resulting in phenotypes in this epiplakin-deficient mouse line are listed below.

3.7.1 Induction of Pancreatitis by Injection of Caerulein

Induction of experimental pancreatitis in mice and rat using caerulein represents a well established and acknowledged method (Adler, Hahn, Kern, & Rao, 1985).

1. Mice are subjected to 7 hourly intraperitoneal injections of 50 μg/kg caerulein-diethylamine salt: pGlu-QD[Y(SO_3H)]TGWMDF-NH_2 (Caslo, custom-ordered) diluted in PBS or of just PBS as control.
2. Mice are sacrificed at predefined time points by decapitation and their blood is collected to obtain serum for measuring disease-associated enzyme levels. To this cause, the blood is allowed to

Table 1 Epiplakin Antibodies
Commercially Available

Designation	Species	Type	Epitope	Applications
Santa Cruz Biotechnology				
Epiplakin 1 (C-13)	Rabbit	Polyclonal	Human C-terminal	IB, ImFl, ELISA
Epiplakin 1 (T-16)	Rabbit	Polyclonal	Human N-terminal	IB, ImFl, ELISA
MyBioSource.com				
Rabbit anti-human epiplakin polyclonal antibody	Rabbit	Polyclonal	Human	IB, ELISA

Raised by Scientific Working Groups

Citation	Species	Type	Epitope	Applications
Fujiwara et al. (2001)	Rabbit	Peptide	Human PRD 9	IB, ImFl
Fujiwara et al. (2001)	Rat	Polyclonal	Human linker PRD 9	IB, ImFl
Spazierer et al. (2003)	Rabbit	Polyclonal	Murine linker PRD 4	IB
Spazierer et al. (2003)	Rabbit	Polyclonal	Murine linker PRD 16	IB, ImFl, IHC
Jang et al. (2005)	Rabbit	Peptide	Several human linkers	IB, ImFl
Goto et al. (2006)	Rabbit	Polyclonal	Murine linker PRD 16	ImFl, IHC
Yoshida et al. (2010)	Rabbit	Polyclonal	Murine linker PRD 16	ImFl
Shimada et al. (2013)	Rat	Polyclonal	Human linker PRD 9	IB, ImFl, IHC

IB, immunoblotting; IHC, immunohistochemistry; ImFl, immunofluorescence; PRD, plakin repeat domain.

coagulate for 30 min and then centrifuged at $3000 \times g$ for 10 min. The supernatant is transferred into a fresh tube and subjected to further analyses.

3. The pancreas is isolated and one third is fixed in 4% buffered formaldehyde overnight and subsequently embedded in paraffin. The remaining organ parts are used to generate protein lysates and cDNA.

Pearls and Pitfalls

– It is essential to use sex matched littermate mice as non-related individuals frequently show great differences regarding their susceptibility to caerulein-induced pancreatitis.

Table 2 Stress Models Resulting in Phenotypes of the Epiplakin-Deficient Mouse Line First Described in Spazierer et al. (2006)

Stress Model	Resulting Disease/ Impairments	Phenotypes Observed	Citation
Pancreas			
Caerulein injection	Acute pancreatitis Edema formation, cell death, inflammatory infiltration	More severe course of disease Increased number of acinar cells displaying keratin aggregates	Wögenstein et al. (2014)
Liver			
CBDL	Increased biliary pressure Ductular rupture Necrosis of hepatocytes	More severe course of disease Increased number of hepatocytes displaying keratin aggregates	Szabo et al. (2015)
DDC intoxication	Steatohepatitis-like liver damage Formation of MDBs	More severe course of disease Increased number of hepatocytes displaying keratin aggregates	Szabo et al. (2015)

CBDL, common bile duct ligation; DDC, 3,5-diethoxy-carbonyl-1,4-dihydrocollidine; MDBs, Mallory–Denk bodies.

3.7.2 Common Bile Duct Ligation

Common bile duct ligation is a well-established liver stress model resulting in increased biliary pressure, followed by ductular rupture subsequently leading to necrosis of the surrounding hepatocytes (Georgiev et al., 2008).

1. Animals are anesthetized with ketamine and xylazine.
2. After midline laparotomy, the common bile duct is exposed and double ligated with 6-0 silk sutures.
3. The abdomen is closed in layers, and the animals are allowed to recover on a heat pad.
4. After 5 days, mice are anesthetized with 150 μl isoflurane and sacrificed for collection of blood and liver samples.
5. Successful bile duct ligation is verified by macroscopic evaluation and an increased, dilated gallbladder.

Pearls and Pitfalls

– It is essential to use sex matched littermate mice as non-related individuals frequently show great differences regarding their susceptibility to common bile duct ligation-induced liver injury.

3.7.3 DDC Treatment

Administration of the porphyrinogenic liver toxin 3,5-diethoxy-carbonyl-1,4-dihydrocollidine (DDC) leads to steatohepatitis-like liver damage combined with the formation of cytoplasmic keratin-containing aggregates in hepatocytes referred to as Mallory–Denk bodies (MDBs) (Fickert et al., 2007).

1. Mice are fed with a diet supplemented with 0.1% DDC (#137030, Sigma–Aldrich) for 4–12 weeks.
2. Control animals are exposed to the corresponding DDC-free diet for the same period.
3. After 4–12 weeks, animals are sacrificed for collection of blood and liver samples.

Pearls and Pitfalls

– It is essential to use sex matched littermate mice as non-related individuals frequently show great differences regarding their susceptibility to DDC-induced liver disease.

4. CONCLUSIONS

After epiplakin was first identified 21 years and described 14 years ago early studies focused on revealing its unusual structure, its specific expression pattern, and its interaction partners. After the generation of epiplakin-deficient mouse lines and initial phenotyping, it soon became clear that even though this protein is closely related to other plakins (especially plectin), epiplakin's functions are distinctly different from those of most other protein family members. Several more recent findings obtained from studies using different epiplakin-deficient epithelial cell types and organs are now clearly pointing towards functions of epiplakin in keratin reorganization and remodeling after disease-induced stress. Moreover, experiments using chemical chaperones indicate a chaperone-like role of epiplakin tailored for keratins.

Future studies will investigate epiplakin's close interactions with keratins in more detail and it will be interesting to see whether epiplakin fulfills different functions in specific epithelial tissues and cell types considering the diversity of keratin variants and their multifactorial and tissue-specific roles.

Future investigations will additionally have to shed light on the molecular details of how epiplakin interacts with keratins and how it fulfills its keratin chaperone-like functions. However, the findings obtained until now already revealed originally unexpected functions for this unusual plakin protein family member.

ACKNOWLEDGMENTS

Studies carried out in the authors' laboratory were supported by grants of the Austrian Science Research Fund.

REFERENCES

Adler, G., Hahn, C., Kern, H. F., & Rao, K. N. (1985). Cerulein-induced pancreatitis in rats: Increased lysosomal enzyme activity and autophagocytosis. *Digestion*, *32*(1), 10–18.

Andrä, K., Lassmann, H., Bittner, R., Shorny, S., Fässler, R., Propst, F., et al. (1997). Targeted inactivation of plectin reveals essential function in maintaining the integrity of skin, muscle, and heart cytoarchitecture. *Genes & Development*, *11*(23), 3143–3156.

Blankson, H., Holen, I., & Seglen, P. O. (1995). Disruption of the cytokeratin cytoskeleton and inhibition of hepatocytic autophagy by okadaic acid. *Experimental Cell Research*, *218*(2), 522–530 (Research support, Non-U.S. Gov't).

Bouameur, J. E., Favre, B., & Borradori, L. (2014). Plakins, a versatile family of cytolinkers: Roles in skin integrity and in human diseases. *The Journal of Investigative Dermatology*, *134*(4), 885–894.

Chamcheu, J. C., Lorie, E. P., Akgul, B., Bannbers, E., Virtanen, M., Gammon, L., et al. (2009). Characterization of immortalized human epidermolysis bullosa simplex (KRT5) cell lines: Trimethylamine N-oxide protects the keratin cytoskeleton against disruptive stress condition. *Journal of Dermatological Science*, *53*(3), 198–206.

Chou, C. F., & Omary, M. B. (1994). Mitotic arrest with anti-microtubule agents or okadaic acid is associated with increased glycoprotein terminal GlcNAc's. *Journal of Cell Science*, *107*(Pt 7), 1833–1843.

Fausther, M., Villeneuve, L., & Cadrin, M. (2004). Heat shock protein 70 expression, keratin phosphorylation and Mallory body formation in hepatocytes from griseofulvin-intoxicated mice. *Comparative Hepatology*, *3*(1), 5.

Fickert, P., Stoger, U., Fuchsbichler, A., Moustafa, T., Marschall, H. U., Weiglein, A. H., et al. (2007). A new xenobiotic-induced mouse model of sclerosing cholangitis and biliary fibrosis. *The American Journal of Pathology*, *171*(2), 525–536.

Fujiwara, S., Kohno, K., Iwamatsu, A., Naito, I., & Shinkai, H. (1996). Identification of a 450-kDa human epidermal autoantigen as a new member of the plectin family. *The Journal of Investigative Dermatology*, *106*(5), 1125–1130.

Fujiwara, S., Kohno, K., Iwamatsu, A., & Shinkai, H. (1994). A new bullous pemphigoid antigen. *Dermatology*, *189*(Suppl. 1), 120–122.

Fujiwara, S., Takeo, N., Otani, Y., Parry, D. A., Kunimatsu, M., Lu, R., et al. (2001). Epiplakin, a novel member of the plakin family originally identified as a 450-kDa human epidermal autoantigen. Structure and tissue localization. *The Journal of Biological Chemistry*, *276*(16), 13340–13347.

Gallicano, G. I., Kouklis, P., Bauer, C., Yin, M., Vasioukhin, V., Degenstein, L., et al. (1998). Desmoplakin is required early in development for assembly of desmosomes and cytoskeletal linkage. *The Journal of Cell Biology*, *143*(7), 2009–2022.

Georgiev, P., Jochum, W., Heinrich, S., Jang, J. H., Nocito, A., Dahm, F., et al. (2008). Characterization of time-related changes after experimental bile duct ligation. *The British Journal of Surgery*, *95*(5), 646–656.

Goto, M., Sumiyoshi, H., Sakai, T., Fassler, R., Ohashi, S., Adachi, E., et al. (2006). Elimination of epiplakin by gene targeting results in acceleration of keratinocyte migration in mice. *Molecular and Cellular Biology*, *26*(2), 548–558.

Guo, L., Degenstein, L., Dowling, J., Yu, Q. C., Wollmann, R., Perman, B., et al. (1995). Gene targeting of BPAG1: Abnormalities in mechanical strength and cell migration in stratified epithelia and neurologic degeneration. *Cell*, *81*(2), 233–243.

Ishikawa, K., Sumiyoshi, H., Matsuo, N., Takeo, N., Goto, M., Okamoto, O., et al. (2010). Epiplakin accelerates the lateral organization of keratin filaments during wound healing. *Journal of Dermatological Science*, *60*(2), 95–104.

Jang, S. I., Kalinin, A., Takahashi, K., Marekov, L. N., & Steinert, P. M. (2005). Characterization of human epiplakin: RNAi-mediated epiplakin depletion leads to the disruption of keratin and vimentin IF networks. *Journal of Cell Science*, *118*(Pt 4), 781–793.

Kasahara, K., Kartasova, T., Ren, X. Q., Ikuta, T., Chida, K., & Kuroki, T. (1993). Hyperphosphorylation of keratins by treatment with okadaic acid of BALB/MK-2 mouse keratinocytes. *The Journal of Biological Chemistry*, *268*(31), 23531–23537.

Kokado, M., Okada, Y., Goto, M., Ishikawa, K., Miyamoto, T., Yamanaka, O., et al. (2013). Increased fragility, impaired differentiation, and acceleration of migration of corneal epithelium of epiplakin-null mice. *Investigative Ophthalmology & Visual Science*, *54*(5), 3780–3789.

Matsuo, A., Yoshida, T., Yasukawa, T., Miki, R., Kume, K., & Kume, S. (2011). Epiplakin1 is expressed in the cholangiocyte lineage cells in normal liver and adult progenitor cells in injured liver. *Gene Expression Patterns*, *11*(3-4), 255–262.

Ridge, K. M., Linz, L., Flitney, F. W., Kuczmarski, E. R., Chou, Y. H., Omary, M. B., et al. (2005). Keratin 8 phosphorylation by protein kinase C delta regulates shear stress-mediated disassembly of keratin intermediate filaments in alveolar epithelial cells. *The Journal of Biological Chemistry*, *280*(34), 30400–30405.

Shimada, H., Nambu-Niibori, A., Wilson-Morifuji, M., Mizuguchi, S., Araki, N., Sumiyoshi, H., et al. (2013). Epiplakin modifies the motility of the HeLa cells and accumulates at the outer surfaces of 3-D cell clusters. *The Journal of Dermatology*, *40*(4), 249–258.

Spazierer, D., Fuchs, P., Proll, V., Janda, L., Oehler, S., Fischer, I., et al. (2003). Epiplakin gene analysis in mouse reveals a single exon encoding a 725-kDa protein with expression restricted to epithelial tissues. *The Journal of Biological Chemistry*, *278*(34), 31657–31666.

Spazierer, D., Fuchs, P., Reipert, S., Fischer, I., Schmuth, M., Lassmann, H., et al. (2006). Epiplakin is dispensable for skin barrier function and for integrity of keratin network cytoarchitecture in simple and stratified epithelia. *Molecular and Cellular Biology*, *26*(2), 559–568 (Research support, Non-U.S. Gov't).

Spazierer, D., Raberger, J., Gross, K., Fuchs, P., & Wiche, G. (2008). Stress-induced recruitment of epiplakin to keratin networks increases their resistance to hyperphosphorylation-induced disruption. *Journal of Cell Science*, *121*(Pt 6), 825–833 (Research support, Non-U.S. Gov't).

Strnad, P., Windoffer, R., & Leube, R. E. (2001). In vivo detection of cytokeratin filament network breakdown in cells treated with the phosphatase inhibitor okadaic acid. *Cell and Tissue Research*, *306*(2), 277–293.

Strnad, P., Windoffer, R., & Leube, R. E. (2002). Induction of rapid and reversible cytokeratin filament network remodeling by inhibition of tyrosine phosphatases. *Journal of Cell Science*, *115*(Pt 21), 4133–4148.

Szabo, S., Wögenstein, K. L., Österreicher, C. H., Guldiken, N., Chen, Y., Doler, C., et al. (2015). Epiplakin attenuates experimental mouse liver injury by chaperoning keratin reorganization. *Journal of Hepatology*, *62*(6), 1357–1366.

Takeo, N., Wang, W., Matsuo, N., Sumiyoshi, H., Yoshioka, H., & Fujiwara, S. (2003). Structure and heterogeneity of the human gene for epiplakin (EPPK1). *The Journal of Investigative Dermatology*, *121*(5), 1224–1226.

Toivola, D. M., Zhou, Q., English, L. S., & Omary, M. B. (2002). Type II keratins are phosphorylated on a unique motif during stress and mitosis in tissues and cultured cells. *Molecular Biology of the Cell*, *13*(6), 1857–1870.

Wang, W., Sumiyoshi, H., Yoshioka, H., & Fujiwara, S. (2006). Interactions between epiplakin and intermediate filaments. *The Journal of Dermatology*, *33*(8), 518–527.

Weerasinghe, S. V., Ku, N. O., Altshuler, P. J., Kwan, R., & Omary, M. B. (2014). Mutation of caspase-digestion sites in keratin 18 interferes with filament reorganization, and predisposes to hepatocyte necrosis and loss of membrane integrity. *Journal of Cell Science*, *127*(Pt 7), 1464–1475.

Wögenstein, K. L., Szabo, S., Lunova, M., Wiche, G., Haybaeck, J., Strnad, P., et al. (2014). Epiplakin deficiency aggravates murine caerulein-induced acute pancreatitis and favors the formation of acinar keratin granules. *PLoS One*, *9*(9), e108323.

Yatsunami, J., Komori, A., Ohta, T., Suganuma, M., Yuspa, S. H., & Fujiki, H. (1993). Hyperphosphorylation of cytokeratins by okadaic acid class tumor promoters in primary human keratinocytes. *Cancer Research*, *53*(5), 992–996 (Research support, Non-U.S. Gov't).

Yoshida, T., Guo, X., Namekata, K., Mitamura, Y., Kume, S., & Harada, T. (2010). Expression of Epiplakin1 in the developing and adult mouse retina. *Japanese Journal of Ophthalmology*, *54*(1), 85–88.

Yoshida, T., Shiraki, N., Baba, H., Goto, M., Fujiwara, S., Kume, K., et al. (2008). Expression patterns of epiplakin1 in pancreas, pancreatic cancer and regenerating pancreas. *Genes to Cells*, *13*(7), 667–678.

CHAPTER FIFTEEN

In Vitro Model of the Epidermis: Connecting Protein Function to 3D Structure

Christopher Arnette*,[1], Jennifer L. Koetsier*,[1], Paul Hoover†, Spiro Getsios†,‡,§, Kathleen J. Green*,†,§,[2]

*Department of Pathology, Northwestern University Feinberg School of Medicine, Chicago, Illinois, USA
†Department of Dermatology, Northwestern University Feinberg School of Medicine, Chicago, Illinois, USA
‡Department of Cell and Molecular Biology, Northwestern University Feinberg School of Medicine, Chicago, Illinois, USA
§Robert H. Lurie Comprehensive Cancer Center, Northwestern University, Chicago, Illinois, USA
[2]Corresponding author: e-mail address: kgreen@northwestern.edu

Contents

Abstract

Much of our understanding of the biological processes that underlie cellular functions in humans, such as cell–cell communication, intracellular signaling, and transcriptional and posttranscriptional control of gene expression, has been acquired from studying

[1] These authors contributed equally to this work.

Methods in Enzymology, Volume 569
ISSN 0076-6879
http://dx.doi.org/10.1016/bs.mie.2015.07.015

cells in a two-dimensional (2D) tissue culture environment. However, it has become increasingly evident that the 2D environment does not support certain cell functions. The need for more physiologically relevant models prompted the development of three-dimensional (3D) cultures of epithelial, endothelial, and neuronal tissues (Shamir & Ewald, 2014). These models afford investigators with powerful tools to study the contribution of spatial organization, often in the context of relevant extracellular matrix and stromal components, to cellular and tissue homeostasis in normal and disease states.

1. INTRODUCTION

In this chapter, we will review methodology associated with one of the longest standing and best-characterized examples of three-dimensional (3D) culture models—that of the living skin equivalent. Epidermal organotypic "raft" cultures, grown on an air–liquid interface, were originally developed in the 1980s as an *ex vivo* approach for understanding human skin function (Asselineau & Prunieras, 1984; Prunieras, Regnier, & Woodley, 1983) and continue to provide an important alternative to animal models (Getsios et al., 2009). The cultures are typically generated using primary keratinocytes isolated from readily available human foreskins; however, adult keratinocytes and keratinocytes derived from iPS cells (Bilousova & Roop, 2013; Kogut, Roop, & Bilousova, 2014) are emerging as additional cell sources. These isolated keratinocytes can be passaged multiple times in culture and genetically manipulated using viral vectors or RNAi. Outcomes of genetic or pharmacologic manipulation can be assessed in weeks rather than months, and cultures are easily harvested and analyzed. Here, we discuss how this 3D tissue model can be used to interrogate the role of the intermediate filament (IF)–desmosome scaffold in cytoarchitectural and signaling pathways driving epidermal differentiation.

1.1 The Skin

3D organotypic cultures of human epidermis mimic the outermost layer of the skin. The epidermis is a constantly renewing, multilayered epithelium composed primarily of keratinocytes, which derive their name from the word keratin, the major cellular component that provides building blocks for IFs (Coulombe & Fuchs, 1990). Keratinocytes begin their life in the basal proliferative layer of the epidermis. Upon committing to undergo differentiation, they stop dividing and leave the basal layer, transiting through the

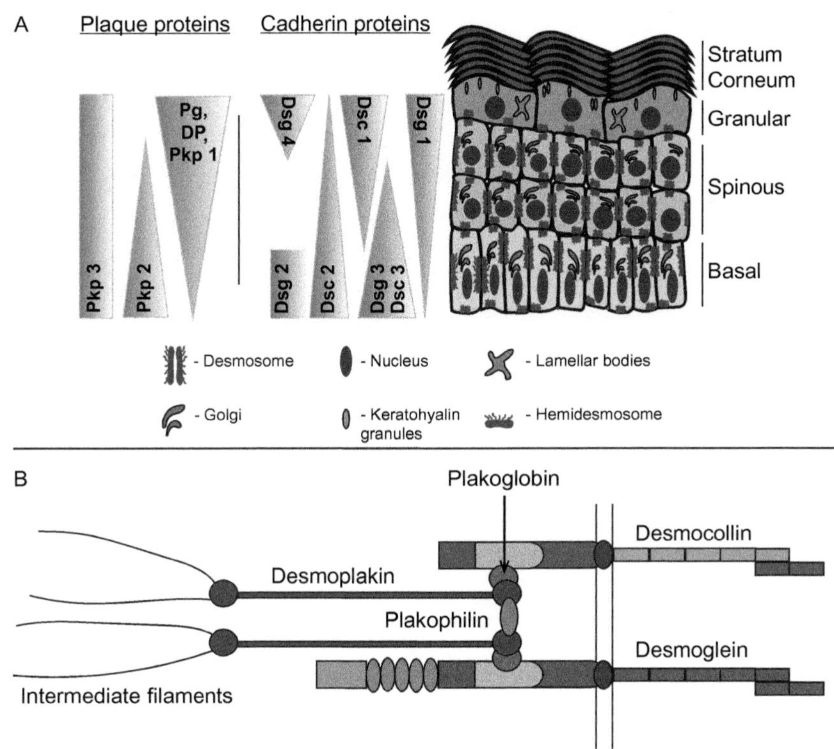

Figure 1 Organization and expression of desmosomal components within the skin. (A) There are four layers within the stratified epidermis: basal, spinous, granular, and cornified. Each layer contains a complement of differentiation-specific junctional proteins critical to support epidermal cytoarchitecture and drive tissue morphogenesis. (B) Desmosomes are specialized, calcium-dependent adhesive structures that are composed of three families of proteins: desmosomal cadherins, armadillo proteins, and plakin proteins. These structures link to the intermediate filament cytoskeleton to maintain tissue integrity. (See the color plate.)

spinous, granular, and, ultimately, cornified layers (Fig. 1). This process results in the creation of a barrier that protects against insults from the environment and prevents water loss. Throughout the differentiating tissue, the keratinocyte–IF scaffold associates with calcium-dependent, cell–cell adhesive structures known as desmosomes (see Fig. 1B; Harmon & Green, 2013; Osmani & Labouesse, 2015). Interaction of desmosomal components and IFs is dictated by a tightly regulated transcriptional program that results in differentiation-dependent patterning of both desmosomal components and IFs within the epidermis (Desai, Harmon, & Green, 2009; Kowalczyk & Green, 2013). As cells commit to terminal differentiation, they switch from expression of basal keratins, keratin 5/keratin 14 (Fuchs & Green, 1980; Nelson & Sun, 1983), to more suprabasal keratins, keratin 1/keratin 10 (Fuchs & Green, 1980; Kim, Marchuk, & Fuchs, 1984). Likewise, while desmogleins 2 and 3 concentrate within the proliferating basal layer, desmoglein 1 becomes concentrated in the suprabasal layers of the epidermis (see Fig. 1A) (Green & Simpson, 2007).

Tissue level integrity provided by IFs relies on connections to intercellular desmosomes (Saito, Tucker, Kohlhorst, Niessen, & Kowalczyk, 2012). This supracellular IF–desmosome scaffold allows the epidermis to maintain its structural stability in the face of constant remodeling, resulting in a highly dynamic but organized tissue that balances epidermal cell renewal with cell removal. Here, we discuss how 3D organotypic cultures provide an important tool for investigating the functions of desmosomes not only as mediators of mechanical tissue integrity but also as spatially distinct regulators of keratinocyte differentiation.

2. HISTORICAL PERSPECTIVE

2.1 Technical History of Model Development

Up until 1975, attempts to serial culture keratinocytes were hindered by a lack of understanding of the necessary components to prolong cellular lifetimes in culture. Rheinwald and Green overcame these conditions, establishing a method by which isolated keratinocytes could be grown as sheets on a feeder layer of lethally irradiated 3T3 fibroblasts to support and maintain keratinocyte colony growth and stratification (Rheinwald & Green, 1975). Both the ability to isolate and stratify keratinocytes in a 3D culture provided a new avenue for probing questions related to skin biogenesis.

In 1976, Freeman and colleagues successfully grew 3D organotypic cultures of keratinocytes on raised metal grids to support maintenance of keratinocytes at the air–liquid interface to induce differentiation and stratification. Unlike Rheinwald and Green, Freeman et al. used porcine skin dermis as the extracellular matrix (ECM) component; however, an artifact of this *in vitro* method was the presence of membrane granules, which are not found in native tissue. Nonetheless, the composition of the cultures closely resembled normal skin, allowing for successful grafting back into human patients (Freeman, Igel, Herrman, & Kleinfeld, 1976).

Building upon the use of alternative ECM substrates, Prunieras, Regnier, and Woodley turned to a more physiological substrate by growing keratinocytes on de-epidermized dermis derived from human skin (Prunieras et al., 1983). While this method allowed for differentiation, it lacked proper keratin patterning. Then, in 1984, Asselineau and Prunieras developed a system using a collagen–fibroblast plug as the substrate to support keratinocyte growth (Asselineau & Prunieras, 1984). While the cells on this collagen–fibroblast stratified, keratinocyte-specific differentiation markers were misexpressed, Kopan and colleagues found that they could induce proper expression of differentiation-specific keratins by removing vitamin A from the medium of submerged cultures or growing 3D organotypic cultures in the absence of retinoic acid (Kopan, Traska, & Fuchs, 1987) permitting proper stratification and differentiation marker expression.

Continuous improvements on the original method have made this a common technique and also led to the development of alternative systems, such as the transwell system described below. While both of these methods (grid and transwell) are viable options for creating 3D raft cultures, the transwell system has the advantage of being contained and not requiring a physical lifting of the collagen plug. However both have improved our understanding of the organization of the skin, as well as the function of the cells within it.

3. CELLULAR CONSTITUENTS OF 3D RAFT CULTURES

3.1 Cell Lines Used

The protocol described below utilizes keratinocytes and fibroblasts to make 3D raft cultures. Normal Human Epidermal Keratinocytes (NHEKs) are typically isolated from human neonatal foreskin but can also be isolated from biopsies taken from patient skin of nonpathological or pathological origin. The lethally irradiated J2 clone of 3T3 mouse fibroblasts, which serve as a feeder layer and secrete paracrine factors to support keratinocyte growth (Rheinwald & Green, 1975), will be used. The methods to isolate and culture these two cell types are summarized below Table 1.

3.2 Isolation of Keratinocytes from Neonatal Foreskin

Note: Under Institutional Review Board (IRB) approval, neonatal foreskins are collected in transport media (Hanks Balanced Salt Solution containing no calcium or magnesium). The foreskin can be stored at 4 °C and processed within 48 h after receipt.

Table 1 Cell Lines Used

Cell Line	Media	Media Supplements	Source
Normal human keratinocytes	Medium 154 (M154)	Human keratinocyte growth supplement (HKGS), 0.07 m*M* Ca^{2+}, and antibiotics (1 × Gentamicin/ Amphotericin B)	Human neonatal foreskin
3T3 J2 fibroblasts	DMEM	10% Fetal bovine serum and 1% Penicillin/Streptomycin	Originally derived from Howard Green (Rheinwald & Green, 1975)

Materials Needed

- Hanks' Balanced Salt Solution (HBSS)
- Dispase solution (2.4 U/mL) (Roche, Catalog #04942078001)
- 0.05% Trypsin/1 m*M* EDTA (Corning, Catalog #30-002-CI)
- Fetal bovine serum (FBS)
- Medium 154 (M154; Life Technologies, Catalog #M-154CF-500)
- Sterile forceps and scissors
- 40 μm nylon cell strainers (Corning, Catalog #352340)

3.2.1 Isolation Protocol

Day 1

1. Carefully remove HBSS and replace it with phosphate-buffered saline (PBS) containing no Ca^{2+} or Mg^{2+}.
2. Rinse foreskin with PBS by pipetting up and down to remove any excess blood (2 ×).
3. Using sterile forceps, transfer foreskin to a 10-cm dish containing 3 mL of PBS.
4. Cut foreskin on one side so that it lays flat with the epidermis side down.
5. Remove excess fat and vessels using sterile forceps and scissors.
6. Cut remaining tissue into small pieces (0.8 cm × 0.8 cm).
7. Place tissue epidermal side up in 4 mL of dispase solution (2.4 U/mL; prepared according to manufacturer's instructions) in a 60 mm dish.
8. Incubate overnight at 4 °C.

Day 2

1. Using sterile forceps, separate the epidermis from the dermis from each piece of foreskin.
2. Transfer epidermal pieces to a 10-cm dish containing 3–4 mL of 0.25% Trypsin/1 m*M* EDTA.

3. Incubate 10–15 min at 37 °C to dissociate the cells.
4. Inactivate trypsin by adding 0.5 mL FBS to the dish.
5. With sterile forceps, scrape any remaining large epidermal pieces vigorously against the bottom of the dish for 10–30 s/piece. This further helps create a single cell suspension by breaking up any remaining larger sheets of cells as leaving the cells in trypsin too long can be detrimental to the cells causing cell lysis and a lower cell yield.
6. Using a pipet, transfer the cell suspension and debris through a 40-μm nylon cell strainer placed into a 50-mL conical tube. Rinse the scraped dish once with 5 mL of PBS and also filter through the same 40-μm cell strainer used for filtering the cell suspension. By running the cells and the rinse through a cell strainer, the larger debris are removed.
7. Transfer the cell suspension and the rinse (collected in same tube after straining) to a 15-mL conical tube.
8. Centrifuge at 200–250 × *g* for 5 min at room temperature to pellet cells.
9. Discard supernatant. Resuspend the cells in 7–8 mL of M154 and plate in a 10-cm dish. Cell yield will vary based on the size of the tissue.
10. Change the media 24 h later. Media should be changed every other day.

3.3 Culturing NHEKs

NHEKs are cultured in M154 (Life Technologies, Catalog #M-154CF-500) supplemented with Human Keratinocyte Growth Supplement (HKGS; Life Technologies, Catalog# S-001-5), Gentamicin/Amphotericin solution (Life Technologies, Catalog# R-015-10), and a final Ca^{2+} concentration of 0.07 m*M*. Maintaining the culture in 0.07 m*M* Ca^{2+} will keep the cells in a relatively "undifferentiated" state with limited expression of differentiation markers such as desmoglein 1, keratin 10, or loricrin (Denning et al., 1998; Hohl, Lichti, Breitkreutz, Steinert, & Roop, 1991). It must be noted that primary keratinocytes have a finite lifespan in culture that limits their proliferative capacity; however, methods to extend their lifetime in culture, such as the presence of a fibroblast feeder layer, have been explored (Green, Rheinwald, & Sun, 1977; Ramirez et al., 2001). Recently, it was reported that addition of the Rho kinase inhibitor, Y-27632, in combination with a fibroblast feeder layer, increases NHEK proliferation and results in efficient immortalization of NHEKs genetically indistinguishable from their freshly isolated counterparts (Chapman, Liu, Meyers, Schlegel, & McBride, 2010).

Note: After isolating NHEKs from foreskin, cultures should be checked daily for contamination. Like any cultured cells, they too are susceptible to bacterial and fungal contamination.

Comment: NHEK cultures will contain a population of dermal fibroblasts. To remove fibroblasts from the culture following attachment to the culture dish, add 0.02% EDTA, incubate at room temperature for 1–5 min, and then vigorously pipet against the dish surface to dislodge fibroblasts. Remove EDTA and add fresh medium to keratinocytes. Alternatively, removal of dermal fibroblasts from the culture typically occurs after two to three passages of NHEKs, as keratinocyte growth will outpace fibroblast growth in M154.

3.3.1 NHEK Passage Protocol

1. Wash cells 1 × with PBS.
2. Add 1–2 mL of 0.05% Trypsin/1 m*M* EDTA and incubate at 37 °C for 5–10 min.

   Note: Different cell lines require different trypsinization times. To avoid over-trypsinization, which can reduce cell viability, it is essential to monitor the dish frequently (every few minutes) and collect the cells as quickly as possible following their release from the substrate.
3. To inactivate trypsin, add 2–3 × the volume of DMEM supplemented with 10% FBS (the FBS will neutralize the Trypsin).
4. Centrifuge at 200–250 × *g* for 5 min to pellet the cells.
5. Discard the supernatant.
6. Resuspend the cells in M154 and replate $0.5–1.0 \times 10^6$ cells per 10 cm. Cell number will depend on experimental conditions.

3.4 3T3 Fibroblasts

The J2 clone of 3T3 mouse fibroblasts can be used as a feeder layer when culturing NHEKs and as a component of the collagen plug scaffold for organotypic raft cultures (Rheinwald & Green, 1975). The use of a feeder layer provides a longer keratinocyte lifespan by secretion of paracrine factors necessary to support growth (Barreca et al., 1992) and through reorganization of ECM components within the plug.

3.4.1 Generation of 3T3 Feeder Layers for Growth of Human Keratinocytes

Materials Needed

- Dulbecco's Modification of Eagle's Medium (DMEM, 1 × with 4.5 g/L glucose, L-glutamine, and sodium pyruvate; Corning, Reference #10-013-CV)

- PBS (Corning Reference #21-040-CV)
- Trypsin (Corning, Catalog #30-002-C1)
- Mitomycin C (VWR, Catalog#80055-338)
- FBS
- Penicillin/Streptomycin (Corning, Reference #30-002-CI)

1. Grow 3T3 cells to 75–80% confluence using DMEM supplemented with 10% FBS and 1% Penicillin/Streptomycin. Never allow 3T3 cells to become confluent.
2. Treat with 4 μg/mL of mitomycin C for 2–4h at 37 °C.
   Note: mitomycin C is a cytotoxic agent and appropriate personal protective equipment (PPE) should be worn when handling it.
3. Aspirate off mitomycin C-containing medium and wash cells 3 × with PBS to remove any trace of this cytotoxic agent.
4. Trypsinize the 3T3 cells for ~2 min and neutralize with DMEM containing 10% FBS. Centrifuge cells at 960 × *g* for 5 min to pellet and resuspend in FAD+ keratinocyte growth media (see Table 2).
5. Plate 5.0×10^4 cells/10-cm dish with resuspended keratinocytes (see Section 3.3.1).
   Note: NHEKs can also be cultured using a feeder layer of 3T3 fibroblasts and maintained in FAD+ media (see Table 2) or alone in the presence of M154. See Section 3.3 for NHEK culture information using M154.

4. 3D RAFT CULTURES

4.1 Utilization of 3D Raft Cultures

3D *in vitro* models offer an important complement to two-dimensional (2D) models, providing a more physiologically relevant context, while lending itself to numerous applications for studying biological processes (Elliott & Yuan, 2011).

The epidermal organotypic cultures discussed in this chapter bear a striking resemblance to normal human epidermis (Fig. 2), replicating the polarized protein distribution and architectural features characteristic of each cell layer seen *in vivo* (see Fig. 2B). While cellular architecture is readily reproduced in the 3D cultures, a potential limitation is that a well-defined basement membrane is not always fully formed (S. Getsios, personal communication, June 26, 2015). Nevertheless, 3D organotypic raft cultures are a powerful tool to study the epidermis and are amenable to genetic manipulation by CRISPR/Cas9-mediated gene ablation (Lopez-Pajares

Table 2 Media Recipes

E-Media Recipe

Reagent	Vendor	Catalog	Storage	Volume
DMEM/F12	Corning	10-090-CV	4 °C	500 mL
DMEM	Corning	10-013-CV	4 °C	500 mL
Note: This give 3:1 DMEM:F12 Combine and remove 83 mL				
Gentamycin/ Amphotericin B	Life Technologies	R01510	−20 °C	0.500 mL
E-cocktail (see below)			−20 °C	10 mL
0.4 μg/mL Hydrocortisone	Sigma	H0888	Room temperature	
10 ng/mL Cholera toxin	Sigma	C8052	4 °C	

E-Cocktail Recipe

Reagent	Vendor	Catalog #	Storage Temperature	How to Prepare
180 μm Adenine	Sigma	A2786	Room temperature	486 mg in 20 mL of deionized water + 500 μL 6 *N* HCl
5 μg/mL human recombinant insulin	Sigma	I2643 or I6634 (bovine insulin)	−20 °C	100 mg in 20 mL 0.1 *N* HCl
5 μg/mL human apo-transferrin	Sigma	T1147	4 °C	100 mg in 20 mL PBS
5 μg/mL triiodothyronine (T3)	Sigma	T6397	−20 °C	**1.** Dissolve 6.8 mg of T3 in 50 mL 0.02 *N* NaOH. This equals a concentration of 2×10^{-4} *M*. Store at −20 °C. **2.** Dilute 0.1 mL of the 2×10^{-4} *M* stock in 9.9 mL of PBS. The concentration is now 2×10^{-6} *M*. **3.** Dilute 0.2 mL of 2×10^{-6} *M* in 19.8 mL of

Table 2 Media Recipes—cont'd
E-Cocktail Recipe

Reagent	Vendor	Catalog #	Storage Temperature	How to Prepare
				PBS. This gives a final concentration of 2×10^{-8} *M* and will be used in the E-cocktail.

1. Filter-sterilize each component individually.
2. Combine all reagents and bring to a final volume of 200 mL in PBS. Sterile filter.
3. Make 10 mL aliquots and store at −20 °C.
4. One 10 mL aliquot is added when 1 L of E-Media is made.

FAD+ Media (250 mL)

Reagent	Amount
E-media	237.5 mL
Supplement with extra 5% FBS	12.5 mL
10 ng/mL EGF	2.5 μL of 1 mg/mL stock

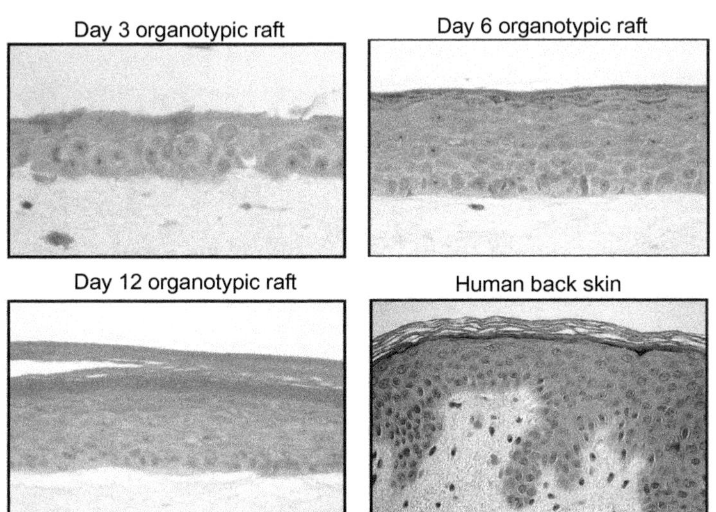

Figure 2 Comparison to human epidermis and time course of 3D raft development. Keratinocytes were seeded on collagen plugs and were grown submerged conditions for 2 days. The cultures were then lifted to the air–liquid interface. Organotypic rafts were harvested 3, 6, or 12 days after being lifted to air. *Representative images of keratinocyte stratification in 3D raft cultures (provided by Paul Hoover of the Northwestern University Skin Disease Research Core).* (See the color plate.)

et al., 2015), RNA interference (Simpson, Kojima, & Getsios, 2010), or viral delivery of silencing constructs or expression vectors, which allows one to analyze how loss of protein function contributes to a process or to mimic disease conditions (Brooks et al., 2014). Furthermore, keratinocytes derived from patient biopsies can be expanded in culture and utilized in 3D models, affording the opportunity to ask directly how genetic defects impact skin structure and function (Brooke et al., 2014).

Here, we describe the protocol for a transwell system for generating 3D organotypic cultures (Fig. 3). We outline genetic manipulation and analysis tools that allow for an understanding of the biology of the 3D culture. As an illustration of the utility of these methods, we present data showing how desmoglein 1 knockdown via retroviral transduction impacts differentiation of 3D organotypic cultures.

Materials Needed

- Transwell plates (Corning, Catalog #355467)
- Transwell inserts (3 μm pore size) (Corning, Catalog #353091). One insert is placed into one well of the transwell plate.
- Media: E-Media: see Table 2 for recipe.

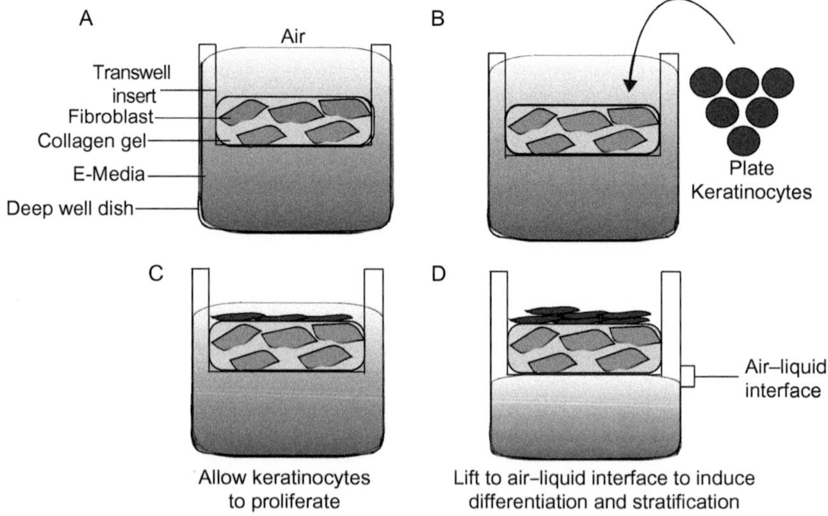

Figure 3 Generation of 3D organotypic cultures using the transwell method. (A) A collagen–fibroblast plug is made in the upper chamber (transwell). Media are supplied from the top and bottom. (B and C) Keratinocytes are seeded onto the collagen–fibroblast plug and kept under submerged conditions: media are supplied from the top and bottom chambers. Keratinocytes are allowed to proliferate. (D) Organotypic rafts are raised to air–liquid interface. Media are only supplied from the bottom chamber.

- Collagen plug: Rat Tail Collagen Type I, high concentration (100 mg/bottle; Corning, Catalog#354249)
- Reconstitution buffer: (1.1 g $NaHCO_3$, 2.3 g HEPES resuspended in 50 mL of 0.05 *N* NaOH, and sterile filtered)
- 10 × DMEM (Sigma, Catalog #D2429)
- 0.5 *N* Sodium hydroxide (NaOH)
- Trypsin (Corning, Catalog #30-002-CI)
- DMEM (Corning, Catalog #10-013-CV)

4.2 3D Organotypic Raft Culture Protocol

4.2.1 Making the Collagen–Fibroblast Plugs

The transwell system requires 2.0 mL collagen mix/raft and 7.5×10^5 J2 3T3 fibroblasts per six well plug. Each collagen plug contains 1×10^6 NHEKs (see Section 4.2.3 for plating NHEKs onto the plug).

4.2.2 Generation of Collagen Plugs Protocol (for 10 Plugs)

1. Trypsinize 3T3 cells in 2 mL of 0.05% Trypsin/1 m*M* EDTA to dissociate the cells.
2. Inactivate trypsin by adding serum-containing media (e.g. DMEM).
3. Count the number of cells using a hemocytometer.
4. Plate the exact number of cells required for rafts (e.g., for 10 rafts, one would need 7.5×10^6 cells) by centrifugation at 960 × *g* for 5 min.
5. Discard supernatant.
6. Resuspend the 3T3 cells in 2 mL (1/10 of the final volume) of 10 × reconstitution buffer.
7. Add 2 mL (1/10 of final volume) of 10 × DMEM.
8. Add collagen to a final concentration of 4 mg/mL.
9. Add sterile water to bring the final volume to 20 mL for 10 plugs.
10. Neutralize pH with 0.5 *N* NaOH and mix thoroughly without generating air bubbles.

    Note: The ideal color of the collagen plug is faint red/light pink. If orange, it is too acidic. If bright pink, it is too basic.
11. Pipet 2.0 mL of the collagen–fibroblast mixture into the top chamber of the transwell.
12. Incubate dishes in a 37 °C, humidified CO_2 incubator for 30 minutes to allow the collagen plug to polymerize.
13. Add 2 mL of DMEM supplemented with 10% FBS and antibiotics to top chamber.

14. Add 13 mL of DMEM supplemented with 10% FBS and antibiotics to the bottom chamber. Incubate for 24–48 h at 37 °C prior to plating NHEKs on top. This process will allow the formation of a smooth, flat surface for when the keratinocytes are seeded.

4.2.3 Seeding NHEKs onto Collagen–Fibroblast Plugs and Creating an "Air–Liquid" Interface

Note: NHEKs are cultured in a keratinocyte growth media (e.g., M154) prior to seeding onto the collagen plugs. Once on the collagen plugs, the cells are cultured in E-Media supplemented with EGF (5 ng/mL) (Keratinocyte Growth Media; Wu et al., 1982). Once introduced to the air–liquid interface, the raft cultures are fed with only E-Media that has not been supplemented with EGF.

1. Trypsinize NHEKs as described in Section 3.3.1.
2. Count the cells using a hemocytometer. For an experiment with 10 plugs, 10×10^6 cells would be required for the experiment.
3. Remove the media from both the top and bottom chamber of the transwell plate.
4. Plate 1.0×10^6 NHEKS per plug in 2 mL of E-Media supplemented with EGF (5 ng/mL) in the top chamber of the transwell. The addition of EGF to the media supports cell proliferation to better ensure a confluent keratinocyte sheet when cells are introduced to the air–liquid interface.
5. Fill the bottom chamber with 13 mL of E-Media supplemented with EGF (5 ng/mL).
6. Allow cells to grow to confluency as described above.
7. To create an air–liquid interface, remove the media from the top chamber of the transwell without disturbing the collagen plug or the NHEK sheet.
8. Remove the media from the bottom chamber and replace with E-Media (not supplemented with EGF). Approximately 9–10 mL are required to fill the lower chamber up to the bottom of the insert. The media level should not exceed past the insert.
9. Remove the media from the bottom chamber every 2 days and replace with fresh E-Media containing no EGF. Take care not to disturb the top chamber of the transwell when the media in the bottom chamber is replaced

Comment: Culture methods typically allow for maintenance of NHEKs at the air–liquid interface for up to 3 weeks. Stratification can be observed within 1–2 days of introduction to an air–liquid interface (Sugihara, Toda, Yonemitsu, & Watanabe, 2001) and morphological differentiation can be seen as early as 1 week (Popov, Kovalski, Grandi, Bagnoli, & Amieva, 2014). Maintenance of 3D raft cultures at the air–liquid interface for 2 weeks or longer is typically necessary for development of a more mature raft with well-defined layers (Hildebrand, Hakkinen, Wiebe, & Larjava, 2002).

5. EXPERIMENTAL MANIPULATION OF THE ORGANOTYPIC RAFT CULTURES

5.1 Genetic Manipulation of Raft Cultures

3D organotypic cultures provide researchers with a more complex model of the skin that is amenable to genetic manipulation. For example, to identify the function of Dsg1 in normal skin processes and how those functions may be altered in disease states, such as pemphigus foliaceus, Dsg1 or its associated binding partner, Erbin, was genetically silenced in keratinocytes. 3D organotypic rafts made from these infected cells allowed Getsios et al. and Harmon et al. to understand the normal functions of Dsg1 within the epidermis, as silencing resulted in impaired morphogenesis and differentiation of the 3D cultures as compared to control (Getsios et al., 2009; Harmon et al., 2013). Alternatively, the use CRISPR/Cas9-mediated gene ablation technique in primary keratinocytes is also an option. Using this gene-targeting approach, the Khavari group showed that loss of transcription factors, MAF and MAFB, severely impaired expression of differentiation markers and differentiation genes in 3D organotypic raft cultures (Lopez-Pajares et al., 2015). While this gene-editing technique represents a powerful tool, editing in primary cells is generally more challenging, likely due to factors such as reduced transfection efficiency, differences in promoter activity, or DNA repair mechanisms (Hendel et al., 2015). Ultimately, the ability to readily silence proteins within this system provides a practical tool to understand how the 3D structure is connected to protein function.

5.1.1 Knockdown Using RNA Interference via Electroporation

While several methods exist to silence gene function in 3D cultures, the use of electroporation has become a common method for this purpose.

Electroporation can increase transfection efficiency for difficult cell types (Escobar-Chavez, Bonilla-Martinez, Villegas-Gonzalez, & Revilla-Vazquez, 2009). However, lack of optimization for specific cell types can result in the formation of pores that may become too large or fail to close after membrane discharge, resulting in cellular damage or rupture (Weaver, 1995), thus, care must be taken to find the proper conditions. Once optimized, electroporation represents a quick and efficient option for gene silencing.

Materials Needed

- Lonza Amaxa Nucleofector™ 2b Device (Lonza, Catalog #AAB-1001)
- Cell line Nucleofector™ Kit V (Lonza, Catalog #VCA-1003; Kit includes: Buffer V, electroporation cuvettes, and cell droppers)
- Medium 154 (M154; Life Technologies, Catalog #M-154CF-500)
- 0.05% Trypsin (Corning, Catalog #30-002-CI)/1 m*M* EDTA
- Dulbecco's Modification of Eagle's Medium (DMEM, 1× with 4.5 g/L glucose, L-glutamine, and sodium pyruvate; Corning, Catalog #10-013-CV)
- siRNA (commonly used vendors: Dharmacon, Sigma, IDT)

  Note: Prior to beginning, allow Buffer V and M154 to warm to room temperature.

1. Grow NHEKS to ~90% confluence in a 10 cm culture dish.
2. Aspirate media and wash cells 3× in PBS.
3. Add 1–2 mL of 0.05% Trypsin/1 m*M* EDTA and incubate cells at 37 °C for 5–10 min to dissociate cells.
4. To inactivate Trypsin/EDTA, add 2–3× the volume of DMEM supplemented with 10% FBS (the FBS will neutralize the Trypsin).
5. Pipette dissociated cell solution up and down to break apart any cellular clumps that may have formed during trypsinization.
6. Count the number of cells in solution using a hemocytometer. Typically, a 10-cm dish will yield ~3 million cells at confluence; however, this can vary between 2 and 4 million depending on the clone.
7. Centrifuge the correct number of cells needed for the experiment at 200–250 × *g* for 5 min to pellet the cells.

   Note: To seed cells at confluency, plate 1.0×10^6 cells per collagen plug. Scale up or down as determined by total number of cells required for experiment.
8. Aspirate off supernatant and resuspend cells in Buffer V solution by adding 100 μL per 1 million cells
9. Add appropriate amount of siRNA (concentration range varies between 10 and 50 n*M*, but concentration for each siRNA should

be empirically determined) per 1 million cells to the cell/Buffer V solution and mix well.

10. Add 100 μL of the cell/siRNA/Buffer V solution to each electroporation cuvette.
11. Select the correct electroporation program on the Lonza Amaxa nucleoporator and place the cuvette into the machine.

    Note: The correct program for specific cell types will need to be determined for each application; however, recommendations for programs can be found on the Lonza website: http://bio.lonza.com/resources/product-instructions/protocols/#ops_k. For Buffer V, the program is X-001.
12. Run electroporation program.
13. Remove electroporated cells from the cuvette using the cell dropper and add to fresh media.

    Note: Cuvettes can be pooled if necessary. If 4 million cells are needed, use two cuvettes with 2 million cells/cuvette and plate in the same dish.
14. Allow cells to attach for 24 h, and then replace media with fresh M154 the following day.

5.1.2 Knockdown and Overexpression Using Retroviral Systems

Utilization of viral systems to silence or overexpress a particular gene represents a feasible alternative to traditional transfection or siRNA methods in 3D organotypic rafts (Getsios et al., 2009) (Fig. 4), including electroporation. Concentrated virus can be added directly to NHEK media to infect them prior to 3D organotypic raft formation. For a more detailed description of designing plasmids and viral production, see Simpson et al. (2010).

Materials Needed

- Medium 154 (Life Technologies, Catalog# M-154CF-500)
- Viral supernatant (Simpson et al., 2010)
- PBS
- 10% bleach solution

1. Grow NHEKs to ~20% to 30% confluence in a 10-cm dish prior to infection.

   Note: This is a BSL-2 level procedure. Proper handling of cultures must be followed according to institutional rDNA protocols. PPE should be worn at all times.
2. Aspirate off M154 media and wash cells 2 × with PBS.
3. Add 4 mL of fresh M154 to cells.

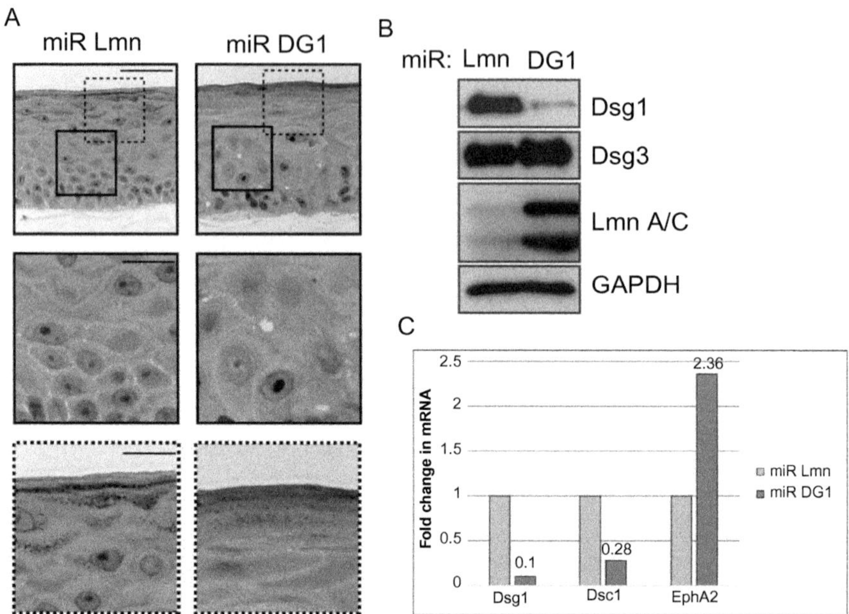

Figure 4 Manipulation and analysis of Dsg1 in 3D organotypic raft cultures. (A) H&E-stained sections of 6-day-old rafts expressing either miR Lmn or miR DG1. Silencing of DG1 results in altered suprabasal morphology (insets with continuous lines) and impaired differentiation (insets with dashed lines) compared to control. (B) Western blot analysis of desmosomal proteins, desmoglein 1 and 3, lamin A/C (Lmn A/C), and GAPDH (glyceraldehyde-3-phosphate dehydrogenase) in 6-day-old 3D raft cultures. Either Dsg1 or Lamin expression was silenced in raft cultures using retroviruses engineered to express microRNA (miRNA)-like sequences specific for each protein. KD reveals ~90% reduction of protein levels under each condition. (C) Real-time PCR analysis show decreased levels of desmocollin 1 mRNA levels following Dsg1 knockdown in 3D raft cultures, while a gene target of EGFR, EphA2, mRNA levels are increased. Note: *Figure originally published in Getsios et al. (2009) (© 2009* The Journal of Cell Biology. *doi: 10.1083/jcb.200809044); used with permission.* (See the color plate.)

4. Carefully add viral supernatant to the dish.
5. Incubate cells at 32 °C, which is a permissible temperature for retroviral infection, for 90 min.
6. Pipette off viral supernatant-containing media into a 10% bleach solution.
7. Wash cells 2 × with PBS, discarding each wash in a 10% bleach solution.
8. Add 7–8 mL of M154 and incubate at 37 °C.
9. When cells have reached desired confluency, prepare cells for 3D organotypic raft formation as described in Section 4.2.3.

Comment: Use of replication-deficient viruses for infection creates a more stable, long-term knockdown or overexpression in NHEKs.

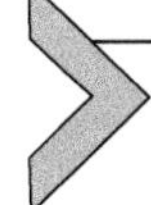

6. ANALYSIS OF ORGANOTYPIC RAFT CULTURES AND OTHER METHODS

Standard analysis techniques commonly used for 2D cultures can also be applied to 3D raft cultures. 3D raft cultures can be harvested from the collagen plug wholly or in part for protein and/or RNA analysis (See Fig. 4B and C). Protein–protein interactions *in situ* can also be assessed by Bio-ID screens and proximity ligation assays or via immunoprecipitation. Additionally, 3D organotypic cultures also allow for the addition of pharmacological agents directly to the media to interfere with protein function or to modulate signaling pathways. The effect of modulating these signaling cascades, such as alterations in proliferation, can be measured through 5-bromo-2′-deoxyuridine incorporation (Getsios et al., 2009). Standard histological methods and analysis can also be utilized to stain proteins of interest. The flexibility and ability to investigate several questions at once utilizing the 3D organotypic culture system make them an attractive tool for investigating protein function.

7. CONCLUSIONS

The evolution of 3D organotypic raft culture technology has impacted a range of research areas from cancer biology to pharmacology to basic cell biology. One potential reason for this is that such a model lends itself to a number of applications. While our focus has been on the use of 3D models to study structural components of the skin, these 3D cultures are being used to model disease (Shamir & Ewald, 2014), study drug delivery (Nam, Smith, Lone, Kwon, & Kim, 2015), and investigate mechanisms of human development (Ader & Tanaka, 2014). Continued improvements of the system are likely to emerge in the future with the engineering of novel biocompatible scaffolding materials, as well as incorporation of increasingly complex cellular components to more accurately mimic the skin.

ACKNOWLEDGMENTS

This work was supported in part by grants to K.J.G. from the National Institute of Arthritis and Musculoskeletal and Skin Diseases (R37 AR043380 and RO1 AR041836), the National Cancer Institute (R01 CA122151), the Joseph L. Mayberry Endowment, and in part by a grant to S.G. from the National Institute of Arthritis and Musculoskeletal and Skin Diseases (RO1 AR062110) and the Northwestern University Skin Disease Research Core (P30 AR057216).

REFERENCES

Ader, M., & Tanaka, E. M. (2014). Modeling human development in 3D culture. *Current Opinion in Cell Biology*, *31*, 23–28.

Asselineau, D., & Prunieras, M. (1984). Reconstruction of 'simplified' skin: Control of fabrication. *The British Journal of Dermatology*, *111*(Suppl. 27), 219–222.

Barreca, A., De Luca, M., Del Monte, P., Bondanza, S., Damonte, G., Cariola, G., et al. (1992). In vitro paracrine regulation of human keratinocyte growth by fibroblast-derived insulin-like growth factors. *Journal of Cellular Physiology*, *151*(2), 262–268.

Bilousova, G., & Roop, D. R. (2013). Generation of functional multipotent keratinocytes from mouse induced pluripotent stem cells. *Methods in Molecular Biology*, *961*, 337–350.

Brooke, M. A., Etheridge, S. L., Kaplan, N., Simpson, C., O'Toole, E. A., Ishida-Yamamoto, A., et al. (2014). iRHOM2-dependent regulation of ADAM17 in cutaneous disease and epidermal barrier function. *Human Molecular Genetics*, *23*(15), 4064–4076.

Brooks, Y. S., Ostano, P., Jo, S. H., Dai, J., Getsios, S., Dziunycz, P., et al. (2014). Multifactorial ERbeta and NOTCH1 control of squamous differentiation and cancer. *The Journal of Clinical Investigation*, *124*(5), 2260–2276.

Chapman, S., Liu, X., Meyers, C., Schlegel, R., & McBride, A. A. (2010). Human keratinocytes are efficiently immortalized by a Rho kinase inhibitor. *The Journal of Clinical Investigation*, *120*(7), 2619–2626.

Coulombe, P. A., & Fuchs, E. (1990). Elucidating the early stages of keratin filament assembly. *The Journal of Cell Biology*, *111*(1), 153–169.

Denning, M. F., Guy, S. G., Ellerbroek, S. M., Norvell, S. M., Kowalczyk, A. P., & Green, K. J. (1998). The expression of desmoglein isoforms in cultured human keratinocytes is regulated by calcium, serum, and protein kinase C. *Experimental Cell Research*, *239*(1), 50–59.

Desai, B. V., Harmon, R. M., & Green, K. J. (2009). Desmosomes at a glance. *Journal of Cell Science*, *122*(Pt. 24), 4401–4407.

Elliott, N. T., & Yuan, F. (2011). A review of three-dimensional in vitro tissue models for drug discovery and transport studies. *Journal of Pharmaceutical Sciences*, *100*(1), 59–74.

Escobar-Chavez, J. J., Bonilla-Martinez, D., Villegas-Gonzalez, M. A., & Revilla-Vazquez, A. L. (2009). Electroporation as an efficient physical enhancer for skin drug delivery. *Journal of Clinical Pharmacology*, *49*(11), 1262–1283.

Freeman, A. E., Igel, H. J., Herrman, B. J., & Kleinfeld, K. L. (1976). Growth and characterization of human skin epithelial cell cultures. *In Vitro*, *12*(5), 352–362.

Fuchs, E., & Green, H. (1980). Changes in keratin gene expression during terminal differentiation of the keratinocyte. *Cell*, *19*(4), 1033–1042.

Getsios, S., Simpson, C. L., Kojima, S., Harmon, R., Sheu, L. J., Dusek, R. L., et al. (2009). Desmoglein 1-dependent suppression of EGFR signaling promotes epidermal differentiation and morphogenesis. *The Journal of Cell Biology*, *185*(7), 1243–1258.

Green, H., Rheinwald, J. G., & Sun, T. T. (1977). Properties of an epithelial cell type in culture: The epidermal keratinocyte and its dependence on products of the fibroblast. *Progress in Clinical and Biological Research*, *17*, 493–500.

Green, K. J., & Simpson, C. L. (2007). Desmosomes: New perspectives on a classic. *The Journal of Investigative Dermatology*, *127*(11), 2499–2515.

Harmon, R. M., & Green, K. J. (2013). Structural and functional diversity of desmosomes. *Cell Communication & Adhesion*, *20*(6), 171–187.

Harmon, R. M., Simpson, C. L., Johnson, J. L., Koetsier, J. L., Dubash, A. D., Najor, N. A., et al. (2013). Desmoglein-1/Erbin interaction suppresses ERK activation to support epidermal differentiation. *The Journal of Clinical Investigation*, *123*(4), 1556–1570.

Hendel, A., Bak, R. O., Clark, J. T., Kennedy, A. B., Ryan, D. E., Roy, S., et al. (2015). Chemically modified guide RNAs enhance CRISPR-Cas genome editing in human primary cells. *Nature Biotechnology*.

Hildebrand, H. C., Hakkinen, L., Wiebe, C. B., & Larjava, H. S. (2002). Characterization of organotypic keratinocyte cultures on de-epithelialized bovine tongue mucosa. *Histology and Histopathology*, *17*(1), 151–163.

Hohl, D., Lichti, U., Breitkreutz, D., Steinert, P. M., & Roop, D. R. (1991). Transcription of the human loricrin gene in vitro is induced by calcium and cell density and suppressed by retinoic acid. *The Journal of Investigative Dermatology*, *96*(4), 414–418.

Kim, K. H., Marchuk, D., & Fuchs, E. (1984). Expression of unusually large keratins during terminal differentiation: Balance of type I and type II keratins is not disrupted. *The Journal of Cell Biology*, *99*(5), 1872–1877.

Kogut, I., Roop, D. R., & Bilousova, G. (2014). Differentiation of human induced pluripotent stem cells into a keratinocyte lineage. *Methods in Molecular Biology*, *1195*, 1–12.

Kopan, R., Traska, G., & Fuchs, E. (1987). Retinoids as important regulators of terminal differentiation: Examining keratin expression in individual epidermal cells at various stages of keratinization. *The Journal of Cell Biology*, *105*(1), 427–440.

Kowalczyk, A. P., & Green, K. J. (2013). Structure, function, and regulation of desmosomes. *Progress in Molecular Biology and Translational Science*, *116*, 95–118.

Lopez-Pajares, V., Qu, K., Zhang, J., Webster, D. E., Barajas, B. C., Siprashvili, Z., et al. (2015). A LncRNA-MAF:MAFB transcription factor network regulates epidermal differentiation. *Developmental Cell*, *32*(6), 693–706.

Nam, K. H., Smith, A. S., Lone, S., Kwon, S., & Kim, D. H. (2015). Biomimetic 3D tissue models for advanced high-throughput drug screening. *Journal of Laboratory Automation*, *20*(3), 201–215.

Nelson, W. G., & Sun, T. T. (1983). The 50- and 58-kdalton keratin classes as molecular markers for stratified squamous epithelia: Cell culture studies. *The Journal of Cell Biology*, *97*(1), 244–251.

Osmani, N., & Labouesse, M. (2015). Remodeling of keratin-coupled cell adhesion complexes. *Current Opinion in Cell Biology*, *32*, 30–38.

Popov, L., Kovalski, J., Grandi, G., Bagnoli, F., & Amieva, M. R. (2014). Three-dimensional human skin models to understand *Staphylococcus aureus* skin colonization and infection. *Frontiers in Immunology*, *5*, 41.

Prunieras, M., Regnier, M., & Woodley, D. (1983). Methods for cultivation of keratinocytes with an air-liquid interface. *The Journal of Investigative Dermatology*, *81*(Suppl. 1), 28s–33s.

Ramirez, R. D., Morales, C. P., Herbert, B. S., Rohde, J. M., Passons, C., Shay, J. W., et al. (2001). Putative telomere-independent mechanisms of replicative aging reflect inadequate growth conditions. *Genes & Development*, *15*(4), 398–403.

Rheinwald, J. G., & Green, H. (1975). Serial cultivation of strains of human epidermal keratinocytes: The formation of keratinizing colonies from single cells. *Cell*, *6*(3), 331–343.

Saito, M., Tucker, D. K., Kohlhorst, D., Niessen, C. M., & Kowalczyk, A. P. (2012). Classical and desmosomal cadherins at a glance. *Journal of Cell Science*, *125*(Pt. 11), 2547–2552.

Shamir, E. R., & Ewald, A. J. (2014). Three-dimensional organotypic culture: Experimental models of mammalian biology and disease. *Nature Reviews. Molecular Cell Biology*, *15*(10), 647–664.

Simpson, C. L., Kojima, S., & Getsios, S. (2010). RNA interference in keratinocytes and an organotypic model of human epidermis. *Methods in Molecular Biology*, *585*, 127–146.

Sugihara, H., Toda, S., Yonemitsu, N., & Watanabe, K. (2001). Effects of fat cells on keratinocytes and fibroblasts in a reconstructed rat skin model using collagen gel matrix culture. *The British Journal of Dermatology*, *144*(2), 244–253.

Weaver, J. C. (1995). Electroporation theory. Concepts and mechanisms. *Methods in Molecular Biology*, *55*, 3–28.

Wu, Y. J., Parker, L. M., Binder, N. E., Beckett, M. A., Sinard, J. H., Griffiths, C. T., et al. (1982). The mesothelial keratins: A new family of cytoskeletal proteins identified in cultured mesothelial cells and nonkeratinizing epithelia. *Cell*, *31*(3 Pt. 2), 693–703.

CHAPTER SIXTEEN

Functional Analysis of Periplakin and Envoplakin, Cytoskeletal Linkers, and Cornified Envelope Precursor Proteins

Veronika Boczonadi*, Arto Määttä[†,1]
*Institute of Genetic Medicine, Newcastle University, Newcastle-upon-Tyne, United Kingdom
[†]School of Biological and Biomedical Sciences, Durham University, Durham, United Kingdom
[1]Corresponding author: e-mail address: arto.maatta@durham.ac.uk

Contents

Abstract

Envoplakin and periplakin are the two smallest plakin family cytoskeletal linker proteins that connect intermediate filaments to cellular junctions and other membrane locations. These two plakins have a structural role in the assembly of the cornified envelope (CE), the terminal stage of epidermal differentiation. Analysis of gene-targeted mice lacking both these plakins and the third initial CE scaffold protein, involucrin, demonstrate the importance of the structural integrity of CE for a proper epidermal barrier function. It has emerged that periplakin, which also has a wider tissue distribution than envoplakin,

Methods in Enzymology, Volume 569
ISSN 0076-6879
http://dx.doi.org/10.1016/bs.mie.2015.06.019

has additional, independent roles. Periplakin participates in the cytoskeletal organization also in other tissues and interacts with a wide range of membrane-associated proteins such as kazrin and butyrophilin BTN3A1. This review covers methods used to understand periplakin and envoplakin functions in cell culture models, including siRNA ablation of periplakin expression and the use of tagged protein domain constructs to study localization and interactions. In addition, assays that can be used to analyze CEs and epidermal barrier function in gene-targeted mice are described and discussed.

1. INTRODUCTION

Envoplakin and periplakin are the two smallest plakin cytoskeletal linker proteins related to desmoplakin, plectin, and bullous pemphigoid antigen-1 (Bouameur, Favre, & Borradori, 2014). However, they are nonetheless relatively large polypeptides, 210 and 195 kDa, respectively (Ruhrberg, Hajibagheri, Simon, Dooley, & Watt, 1996; Ruhrberg, Hajibagheri, Parry, & Watt, 1997). The two plakins were originally found in a proteomic search for constituents of cornified envelopes (CEs) (Simon & Green, 1984), the covalently cross-linked protein compartment of the epidermal barrier and the final endpoint of keratinocyte differentiation (Kalinin, Kajava, & Steinert, 2002). Biochemical evidence based on analysis of cross-linked peptides isolated from mature and developing envelopes (Steinert & Marekov, 1999) and subcellular targeting of plakin protein domains (DiColandrea, Karashima, Määttä, & Watt, 2000) indicated that these two plakins together with a third CE precursor protein, involucrin, constitute a scaffold beneath the plasma membrane upon which the late differentiation proteins, such as small proline-rich proteins and late envelope proteins, are subsequently assembled (Kalinin et al., 2002).

The role of the periplakin and envoplakin in the acquisition of the epidermal barrier was confirmed by analysis of gene-targeted mice. The methods to analyze the mouse models are covered in Section 6. Intriguingly, single knockout mice of either envoplakin (Määttä, DiColandrea, Groot, & Watt, 2001), periplakin (Aho et al., 2004), or involucrin (Djian, Easley, & Green, 2000) demonstrated only minor changes, such as a developmental delay in the barrier formation in the case of the mice lacking envoplakin (Määttä et al., 2001). Breeding the individual knockout strains to generate a triple-KO strain lacking all three CE scaffold proteins (Sevilla et al., 2007), for brevity referred to as EPI −/− mice (Cipolat, Hoste, Natsuga, Quist, & Watt, 2014), showed abnormalities in several aspects of the barrier homeostasis, namely, abnormal stratum corneum lipid distribution, imbalance in the epidermal proteolytic balance, and dermal infiltration of T-lymphocytes and decreased number of dendritic epidermal T cells (DETC) consistent with an immune response to a compromised barrier (Sevilla et al., 2007). Recently the EPI −/−mice were analyzed further to reveal a connection of the epidermal barrier function and skin carcinogenesis (Cipolat et al., 2014). The barrier defect was associated with resistance to papilloma formation via a mechanism involving an interplay between the epidermis and the immune system. EPI −/− mice have an exaggerated response to the tumor promoter TPA resulting in immune infiltration and elevated serum levels of thymic stromal lymphopoietin (Cipolat et al., 2014). This exciting result further emphasizes the need to understand the functions and interactions of periplakin and envoplakin.

In the skin, both envoplakin and periplakin are also implicated in an autoimmune blistering condition paraneoplastic pemphigus (Borradori et al., 1998; Kiyokawa et al., 1998; Mahoney, Aho, Uitto, & Stanley, 1998). In this condition, a subset of patients with an underlying neoplasm, such as non-Hodgkin lymphoma or leukemia, develop circulating antibodies against plakin family members resulting in mucosal and epidermal blistering with only 49% 1-year survival rate (Leger et al., 2012). The diagnostic differentiation from other blistering diseases is difficult and the assays used are more relevant for clinical laboratories rather than basic research (Kelly et al., 2015).

Of the two proteins, periplakin shows a wider tissue distribution and participates in cytoskeletal organization and interactions with membrane proteins in several other cell types in addition to the epidermis. For example, periplakin function has been investigated in simple epithelial cells (Boczonadi, McInroy, & Määttä, 2007; Long, Boczonadi, McInroy, Goldberg, & Määttä, 2006), eye lens (Yoon & FitzGerald, 2009), neuronal tissue (Feng et al., 2003; Francke et al., 2006; Murdoch et al., 2005; Ward, Jenkins, & Milligan, 2009), the liver (Ito et al., 2013), activation of PBMC-derived human γδT cells in co-culture experiments (Rhodes et al., 2015), and carcinomas (Matsumoto et al., 2014; Nishimori et al., 2006; Otsubo et al., 2015). Owing, perhaps, to the restricted tissue distribution of envoplakin and its dependence on periplakin for correct subcellular targeting (DiColandrea et al., 2000), much less studies have been carried out on envoplakin than periplakin. This said, the approaches and methods discussed in this review can equally well be adapted to study either of the proteins. In addition to the gene-targeted animal models, a major input to our knowledge about functions of especially periplakin has been achieved by elucidating interaction partners and by ablating periplakin expression in cultured cells by RNA interference. The studies of protein-binding partners and role in subcellular localization have been most often carried by separately investigating the easily defined protein domains of periplakin and envoplakin rather than the full-length proteins.

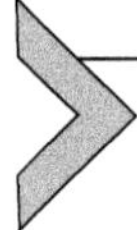

2. IMPLICATIONS OF PLAKIN DOMAIN STRUCTURE TO FUNCTIONAL ANALYSIS

Envoplakin and periplakin share the domain structure of the major desmosomal protein desmoplakin with an N-terminal plakin domain that is targeted to cell junctions and other membrane locations and a C-terminus that interacts with intermediate filament proteins. The central rod domains of envoplakin and periplakin are unique among the plakin family members by being able to mediate heterodimer formation (Kalinin et al., 2004; Ruhrberg et al., 1997). The domain structure is illustrated in the Fig. 1. The key development in understanding of the N-terminal plakin domain came after the crystal structure of plectin N-terminus was shown to comprise nine spectrin repeats with a Src homology three domain inserted in the central repeat (Sonnenberg, Rojas, & de Pereda, 2007), with the SH3 domain interacting with the fourth repeat and not likely to recognize canonical proline-rich motifs (Ortega, Buey, Sonnenberg, & de Pereda, 2011).

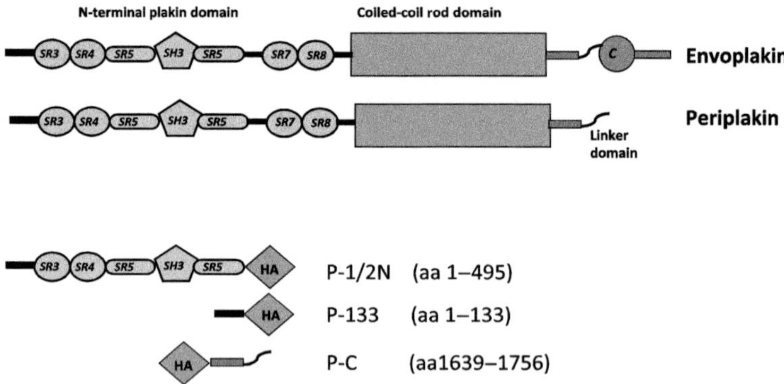

Figure 1 Domain structure of periplakin and envoplakin. The N-terminal plakin domain comprises spectrin repeats, a Src homology three domain, and an N-terminal extension (131 amino acids in periplakin and 144 in envoplakin). The C-terminal linker domain of periplakin is only 117 amino acids but sufficient for binding to intermediate filaments. Below, examples of widely used periplakin domain constructs. N-terminal constructs on pCI-neo backbone use a HA-tag in their C-terminus (FLAG tag for envoplakin), whereas the linker domain constructs have HA or eGFP tag at the N-terminus.

Periplakin and envoplakin N-terminal domains contain repeats SR3–SR8 with the SH3 domain again embedded in the SR5 (Al-Jassar, Bernadó, Chidgey, & Overduin, 2013; Sonnenberg et al., 2007). In addition, the SR6 repeat is missing in envoplakin and periplakin and replaced by a shorter helix (Al-Jassar et al., 2013). The extreme N-termini of the proteins (132 first aa for periplakin and 143 for envoplakin) are predicted to be unstructured (Al-Jassar et al., 2013) but contribute to the interactions with binding partners such as kazrin (Groot, Sevilla, Nishi, DiColandrea, & Watt, 2004). The C-terminal domains of envoplakin and periplakin are shorter than those of the other plakin family members. However, even the periplakin C-terminus (so-called linker domain) is sufficient for binding to intermediate filament proteins (Karashima & Watt, 2002; Kazerounian, Uitto, & Aho, 2002).

Examples of expression constructs used to study localization and interactions of periplakin and envoplakin domains are shown in Fig. 1. Many N-terminal constructs such as those described in DiColandrea et al. (2000) were originally designed before the spectrin-repeat structure was elucidated. However, the most widely used constructs such as the P-1/2-N (DiColandrea et al., 2000) map very well to the actual organization of the N-terminus (Fig. 1) and these constructs, or further deletion variants of them, have been successfully used for immunofluorescence (Fig. 2A.) and co-immunoprecipitation studies (Boczonadi & Määttä, 2012; Boczonadi et al., 2007; Groot et al., 2004; Rhodes et al., 2015). Tagged periplakin constructs used in these studies were generated on a pCI-neo (Promega, Cat. no. E1841) backbone (DiColandrea et al., 2000).

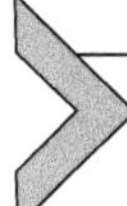

3. ANALYSIS OF THE PLAKIN EXPRESSION AND LOCALIZATION IN CELL CULTURE MODELS

3.1 Antibody Resources for Periplakin and Envoplakin

Several proprietary antibodies are available for both periplakin and envoplakin. These are listed in Table 1. All these antibodies are suitable for immunoblotting and immunofluorescence applications. An example of TD2 antibody staining result can be seen in Fig. 2A demonstrating cell border staining in MCF-7 cells. In addition to antibodies that are listed in the Table 1, several commercial vendors provide periplakin antibodies. These are mostly rabbit polyclonal antibodies but some of them have been only tested for either immunoblotting of immunohistochemistry. The original AE11 monoclonal antibody (Ma & Sun, 1986) is available from Santa

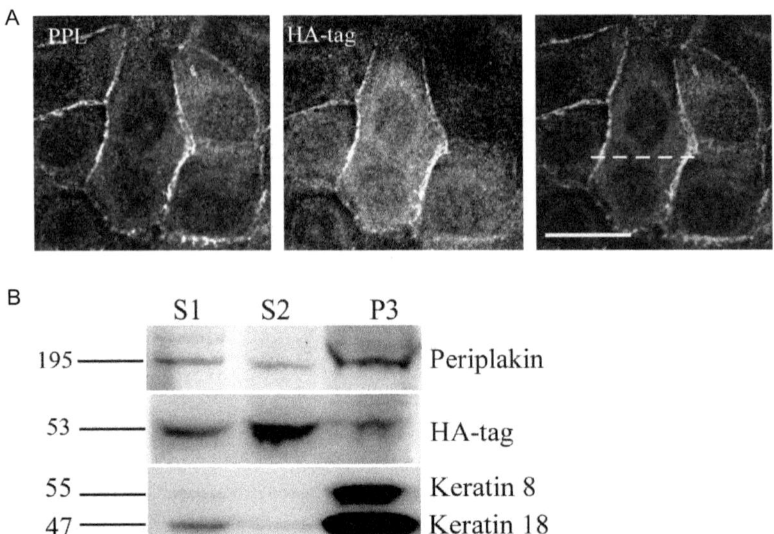

Figure 2 (A) Stably transfected MCF-7 cells expressing p1/2-HA periplakin construct. The N-terminal construct shows targeting to cell membrane (anti-HA tag immunofluorescence mAb Y-11, 1:200 dilution) and colocalization with endogenous periplakin (PPL panel, immunofluorescence with TD2, 1:100 dilution). Merged panels at right (TD2 green (labeled PPL), HA-tag red (note localization at cell borders)). Scale bar = 20 μm. (B) Subcellular fractionation of stably transfected MCF-7 cells expressing periplakin N-terminus. S1 is the saponin-soluble fraction and S2 denotes the Triton X-100-soluble fraction. Immunoblots of both endogenous periplakin (using TD2 antibody) and transfected 1/2 N-terminus (HA-tag) are shown. Keratin 8/18 immunoblot is a marker for the Triton-insoluble (P3) pellet fraction. *Reproduced with permission from Boczonadi et al. (2007).* (See the color plate.)

Cruz Biotechnology (Cat. no. sc-80605). Currently, four companies provide antibodies against envoplakin. A monoclonal antibody, 5G96, is available from United Sates Biological, Cat. no. E3317, and Santa Cruz Biotechnology, sc-71051.

To carry out immunofluorescence of either endogenous plakin proteins or transfected, epitope-tagged protein domains, cultured cells can be fixed with ice-cold 1:1 methanol–acetone (or ice-cold pure methanol) on ice for 5 min or with 4% paraformaldehyde for 10 min at room temperature followed by permeabilization with 0.1% Triton X-100 (Sigma-Aldrich Cat. no. T8787) for 5 min at room temperature. DiColandrea et al. (2000) found that the latter protocol is best suited for visualization of membrane-associated periplakin and envoplakin, whereas the methanol–acetone fixation is better for visualizing intermediate filament cytoskeleton and associated plakin proteins.

Table 1 Antibodies Against Periplakin and Envoplakin

Ab	Mono/Poly	Epitope	Reactivity	References
Periplakin antibodies				
AE11	mAb	Not mapped	Human	Ma and Sun (1986)
PPL448	mAb	aa 1548–1583	Human	Aho (2004)
CR3	pAb	aa 1639–1664	Human, mouse	Ruhrberg et al. (1997)
TD2	pAb	aa 1646–1694	Human, mouse	Määttä et al. (2001)
r-pAb	pAb	aa 817–838	Human	Aho (2004)
BOCZ1	pAb	aa 1663–1677	Human, mouse	Boczonadi and Määttä (2012)
Envoplakin antibodies				
CR1	pAb	aa 2019–2033	Human	Ruhrberg et al. (1996)
CR5	pAb	aa 222–793	Human	Ruhrberg et al. (1997)
aEnvopl	pAb	aa 2019–2033	Human	Steinert and Marekov (1999)

Note that reactivity that is based on published data is likely to be wider.

For immunofluorescence of frozen tissue sections, it is recommended that the sections are blocked with 0.02% fish skin gelatin (Sigma-Aldrich Cat. no. G7765), 2% bovine serum albumin, and 10% goat serum followed by incubation with primary antibody (30–60 min room temperature or overnight at +4 °C). It should be noted that in some situations, there is evidence of potential epitope masking. It was noticed (Aho, 2004) that the monoclonal ab PPL448 preferentially stains basal aspect of basal keratinocytes in human frozen skin sections if a routine 5 min permeabilization step was used but had a similar suprabasal cell border staining as the other periplakin antibodies after a prolonged, 20 min incubation with 0.1% Triton X-100.

3.2 Transient and Stable Transfections of Plakin Protein Domain Constructs

(1) Cells are plated on sterile coverslips in 24-well plates so that on the day of the transfection they have reached about 50% confluency. For MCF-7 (ATCC number HTB-22), HeLa (ATCC number CCL-2) cells, or HaCaT (a keratinocyte cell line, can be obtained from CLS Cell

Lines Service using a Materials Transfer Agreement), 30,000 cells per well are seeded the day before transfection. Primary human keratinocytes (e.g., from Life Technologies, C-005-5C; Lonza, 192906; or ATCC, PCS-200-011) are maintained in keratinocyte serum-free media (KSFM) with supplements (Life Technologies Cat. no. 17005-075) and plated at a density of 50,000 cells per well the day before transfection.

(2) For established epithelial cell lines, virtually all commercial transfection regents result in good expression of plakin protein domains. Primary keratinocytes can be transfected using SuperFect reagent (Qiagen, Cat. no. 301305) as follows.

(3) For each well, a mixture of 0.5 μg of expression plasmid diluted in 60 μL of KSFM without growth supplements and antibiotics and 2.5 μL SuperFect reagent is prepared. The mixture is vortexed for 10 s and left to stand in room temperature for 5 min. Meanwhile, old media is aspirated from the cells and the wells are washed once with PBS.

(4) A further 350 μL of KSFM with growth supplements and antibiotics is added to the transfection mixture that is immediately transferred on the cells by spreading the mixture drop wise over the cells. Cells are incubated for 2 h with the transfection mixture, washed with PBS, and grown in KSFM with growth supplements and antibiotics for 24 or 48 h before fixing the cells on coverslips for immunofluorescence analysis.

To generate stably transfected MCF-7 cell lines expressing periplakin protein domains (Boczonadi et al., 2007; Long et al., 2006), the same expression vectors as for the transient transfections were utilized. The protocol resulted in stable expression of the N-terminal periplakin constructs that was retained for several passages and enabled cells to be expanded also for proteomic experiments (Boczonadi & Määttä, 2012; Boczonadi et al., 2007). An example of membrane-targeted expression of HA-tagged periplakin N-terminus is shown in Fig. 2A. Co-localization with the endogenous protein demonstrates the importance of the N-terminal domain in targeting of the protein.

(1) Cells are grown to 50–60% confluency on 6-well plates and transfected with 1 μg of the pCI-neo-based expression vectors using, e.g., GeneJuice (Merck Millipore Cat. no. 70967) in a ratio of 3 μL reagent for 1 μg of plasmid.

(2) After 48 h, neomycin-resistant cells are selected with 500 μg/mL neomycin (G418, LifeTechnologies, Cat. no. 10130135) changing the media every second day for 3–4 weeks until only resistant individual colonies remain.

(3) Colonies are isolated by trypsinizing them individually inside cloning rings and expanded first on 96-well plates and then on 24-well plates. Cells on replicate wells are grown on coverslips to check the expression level by indirect immunofluorescence.

(4) The selected and expanded clones can be maintained in 250 μg/mL neomycin.

3.3 Detergent Subcellular Fractionation

Periplakin can be targeted to several different subcellular localizations depending on the interactions of its N-terminal and C-terminal domains (Aho, 2004; Boczonadi & Määttä, 2012; Boczonadi et al., 2007; DiColandrea et al., 2000; Long et al., 2006; Rhodes et al., 2015). A quantitative way to assess the distribution of periplakin is to carry out subcellular fractionation. The following protocol was originally used to investigate plakoglobin (Palka & Green, 1997) and subsequently adopted for periplakin (Boczonadi & Määttä, 2012; Long et al., 2006). Usually, the majority of periplakin is found in the detergent-insoluble fractions but, for example, N-terminal domains that lack the C-terminal linker domain are found in the cytoskeletal pellet to a lesser extent. An example of fractionation experiment is shown in Fig. 2B. The endogenous periplakin is mostly present in the Triton X-100 insoluble pellet fraction (P3) with less protein found in the saponin-soluble cytosolic and Triton-soluble fractions, whereas the transfected N-terminus that lacks the intermediate filament-binding domain is preferentially found in the Triton-soluble membrane fraction. As a control that changes the solubility of intermediate filaments and associated proteins, cells can be treated with 500 n*M* protein phosphatase-2 inhibitor okadaic acid (OA) (Merck Millipore Cat. no. 495604) for 2 h before extraction. OA application results in hyperphosphorylation and subsequent collapse of keratin cytoskeleton (Long et al., 2006).

(1) Confluent epithelial or epidermal cells on 100 mm cell culture dishes (e.g., 2–3 $\times 10^6$ HaCaT or MCF-7 cells) are extracted on plates with 450 μL of saponin buffer (0.01% w/v saponin, 10 m*M* Tris, pH 7.5, 140 m*M* NaCl, 5 m*M* EDTA, 2 m*M* EGTA, and protease inhibitors) on ice for 10 min, scraped, and centrifuged at 14,000 $\times g$ for 30 min at 4 °C.

(2) The supernatant fraction is saved and the pellet is extracted by vortexing 1 min using 450 μL of ice-cold Triton buffer (using 1% v/v Triton

X-100 instead of saponin in the extraction buffer), centrifuged as above leaving behind the cytoskeletal pellet fraction.

(3) The Triton X-100 insoluble proteins are solubilized in 600 μL of SDS/urea buffer [1% SDS, 8 *M* urea, 10 m*M* Tris–HCl, pH 7.5, 5 m*M* EDTA, 2 m*M* EGTA, and protease inhibitors (LifeTechnologies Cat. no. 78443)] while the volume of saponin and Triton-extracted fractions is equalized to 600 μL with 4 × Laemmli sample buffer.

Saponin extraction can be also used in conjunction with immunofluorescence to evaluate the association of the full-length endogenous protein or transfected C-terminal domains with cytoskeleton (Karashima & Watt, 2002). Detergent-labile components of the cytoskeleton can be removed by extracting the cells with 0.6% saponin in 50 m*M* imidazole, pH 6.8, 50 m*M* KCl, 0.5 m*M* $MgCl_2$, 0.1 m*M* EDTA, 1 m*M* EGTA, and 4% polyethylene glycol PEG8000 for 30 min prior to fixation with 4% paraformaldehyde for 30 min at +4 °C (Karashima & Watt, 2002).

4. ABLATION OF PERIPLAKIN EXPRESSION BY RNA INTERFERENCE

The functional analysis of periplakin in cultured cell lines has benefited from the fact that the protein is a relatively straightforward target for knock down of expression by RNA interference (Long et al., 2006; Rhodes et al., 2015). Both transiently transfected siRNAs and stably integrated viral constructs expressing shRNA have been utilized. Table 2 lists sequence targets in the periplakin mRNA that have been successfully used for knockdown experiments. Up to 95% depletion of periplakin protein levels can be achieved 48 h posttransfection using the following approach (Long et al., 2006).

Table 2 siRNA and shRNA Target Sequences in Periplakin mRNA

Sequence	siRNA/shRNA	References
5′-ATGTATAAAATGCTTGGCCTG	siRNA	Long et al. (2006)
5′-TGCTCGTATTTCCGGTTGGTG	siRNA	Long et al. (2006)
5′-GAGGGTATGTATAAAATGCTT	siRNA	Long et al. (2006)
5′-ACCAGAAGAACCTGCTAGA	shRNA	Rhodes et al. (2015)
5′-GTGGAAGTCAAAGAGGTGA	shRNA	Rhodes et al. (2015)
5′GACTGATCGAAAGGTGTGA	shRNA	Rhodes et al. (2015)

(1) Cells are plated as described above. The day before transfections, the culture medium is changed to 500 μL/well of serum-free and antibiotics-free medium.

(2) For each well of a 24-well plate to be transfected, 60 pmol of siRNA is diluted in 50 μL of serum-free DMEM medium. Separately, 3 μL of Oligofectamine reagent (Life Technologies Cat. no. 12252-011) is mixed with 12 μL of serum-free DMEM and incubated for 5 min at ambient room temperature. Next, the diluted siRNA and Oligofectamine/DMEM are mixed gently and incubated for a further 20 min at room temperature to allow formation of complexes between Oligofectamine and siRNA before applying the mixture on the cells.

(3) 6–24 h later, the transfection mixture is removed from the cells and replaced with DMEM medium (with 10% FCS and antibiotics). For experiments that require confluent cultures, the cells can be left for several days with the downregulation lasting to at least until 96 h time point. For example, scratch wounding fro migration assays by a 200 μL pipette tip was usually carried out when cells had formed a confluent monolayer 50–72 h after the transfection.

In a recent study where stable short-hairpin RNA ablation of periplakin was achieved (Rhodes et al., 2015), the DNA encoding shRNA (Table 2) was subcloned into *Bam*H1/*Eco*R1 site pHR-SIREN/puro lentiviral vector backbone (Rhodes et al., 2015) and viral particles were produced in the HEK 293T packaging cell line (ATCC number CRL-11268) by co-transfection with packaging, pCMV8.91 gag pol, and viral envelope, pMDG VSV-G env, plasmids. Viral particles were then used to transduce HeLa cells followed by selection with 1 μg/mL puromycin.

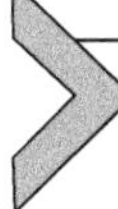

5. IDENTIFICATION OF PERIPLAKIN INTERACTION PARTNERS

5.1 Co-immunoprecipitation Using HA-Tagged Periplakin N-Terminus

One approach to identify interacting partners for the periplakin N-terminus has been co-immunoprecipitation of proteins binding to HA-tagged periplakin constructs in stably transfected MCF-7 cell lines followed by 2D-protein gel electrophoresis (Boczonadi & Määttä, 2012; Boczonadi et al., 2007). For most consistent results, commercial kit with immobilized anti-HA tag antibody (Pierce HA-tag IP/Co-IP, Life Technologies Cat. no.

26180) was used. It is advisable to include both parental cell and empty vector transfected cells as negative controls for stably transfected lines.

(1) Cells are washed three times in ice-cold PBS and extracted with M-PER lysis reagent (Life Technologies Cat. no. 78501). For routine experiments, an IP-lysis buffer (50 m*M* Tris–HCl, pH 7.5, 150 m*M* NaCl, 1% Nonident P-40 (Sigma-Aldrich 21-3277), 0.5% Na-deoxycholate (Sigma-Aldrich D6750), and protease inhibitor cocktail) can equally well be used. The samples are cleared from cell debris by pelleting in a microcentrifuge at +4 °C for 20 min.

(2) For the immunoprecipitation, 200 μL of cell lysate is mixed with 20 μL of immobilized anti-HA agarose (Pierce HA-tag IP/Co-IP kit, Life Technologies Cat. no. 26180) and incubated overnight at +4 °C with end-over-end rotation.

(3) Samples are washed three times with TTBS (0.05% Tween 20 in 25 m*M* Tris–HCl, pH 7.2, 150 m*M* NaCl) using spin columns provided with the kit and eluted in 2 × nonreducing sample buffer.

(4) For 1D-gel electrophoresis and subsequent immunoblotting the samples are reduced with boiling the sample in 3 μL of 1 *M* dithiothreitol for 25 μL of sample. The samples for 2D electrophoresis are diluted with 100 μL of distilled water and precipitated with 400 μL of acetone for 30 min at room temperature to remove charged detergents.

(5) For isoelectric focusing, the acetone-precipitated pellets are suspended in 120 μL of 9 *M* urea, 2 *M* thiourea, 4% CHAPS, 1% DTT, and 2% ampholytes with a trace of bromophenol blue and a further 2.5 μL of 50% DTT is added. For most of the experiments, a wide (pH 3–10) range can be used. Immobilized pH gradient strips (70 mm; ReadyStrip, BioRad Cat. no. 163-2000) are rehydrated with the sample overnight in wells covered with paraffin oil before loading of the samples. The run conditions for IEF were 2 mA current and 5 W power.

(6) After the first dimension run, IPG strips are rinsed in distilled water and equilibrated in 50 m*M* Tris–HCl, 10 mg/mL DTT, 10% SDS, 6 *M* urea, and 30% glycerol for 15 min, rinsed in distilled water, and placed in equilibration buffer containing 48 mg/mL iodoacetamide for a further 15 min. The second dimension run is carried out on 70 × 80 mm-sized 10% SDS-PAGE gels (without a stacking gel, but the IPG strips immobilized on the top of the separating gel with the help of 1% agarose). The second dimension is run at 200 V for 2 h.

For MALDI-TOF MS identification of the protein spots, Boczonadi and coworkers (Boczonadi & Määttä, 2012; Boczonadi et al., 2007) used a ProGest Workstation (Genomic Solutions) and ProGest Trypsin kit for in-gel digestion (Genomic Solutions Cat. no. 0080-0210) to generate tryptic peptides that were lyophilized, resuspended in 10 mL of 0.1% formic acid, and analyzed by a Voyager DE-STR mass spectrometer. The peptide fingerprints were matched to theoretical digests from a nonredundant NCBL database using MASCOT search algorithm (matrixscience.com).

5.2 Identification of Periplakin Interactions by Yeast Two-Hybrid Analysis

Another approach to find out interaction partners for periplakin has been yeast two-hybrid assay. For example, Groot et al. (2004) employed MATCHMAKER protocol (Clontech Cat. No. 630489) and used the 133 first amino acids of the periplakin N-terminus (i.e., the region that is not part of the spectrin repeats) as a bait cloned into pGBKT7 plasmid to screen MATCHMAKER keratinocyte cDNA library in pGAD10 vector (note that currently a Universal Human Library, Clontech Cat. no. 630481, is recommended). This screen identified Kazrin, a conserved desmosomal protein that regulates junctional assembly and can promote keratinocyte differentiation (Groot et al., 2004; Nachat et al., 2009; Sevilla, Nachat, Groot, & Watt, 2008). The interaction was confirmed by immunoprecipitation of FLAG-tagged Kazrin and HA-tagged periplakin N-terminus using transiently transfected COS-7 (ATCC number CRL-1651) cells (Groot et al., 2004).

Likewise, the C-terminal linker domain has been used as a bait. Aho's research group (Kazerounian & Aho, 2003; Kazerounian et al., 2002) used amino acids 1584–1756 to probe Clontech human keratinocyte library to confirm binding of the C-terminus to vimentin and keratin 8 and to identify periphilin, a nuclear protein in undifferentiated keratinocytes that co-localizes with periplakin at cell borders upon differentiation (Kazerounian & Aho, 2003).

Notably, periplakin has been identified in several other Y2H screens that have studied interactions of various membrane-associated proteins (see, e.g., Beekman, Bakema, van de Winkel, & Leusen, 2004; Murdoch et al., 2005; Rhodes et al., 2015). These interactions underscore the versatility of plakin family proteins as organizers of cytoskeletal connections to a number of unrelated proteins.

6. ENVOPLAKIN AND PERIPLAKIN GENE-TARGETED MICE

6.1 Envoplakin Gene Targeting and Visualization of the Knockin LacZ Activity

To generate envoplakin gene knockout mice, the targeting vector deleted coding regions of the eight first exons of the mouse *Evpl* gene from a *Sty*I restriction site in the 5′ noncoding region to a *Avr*II site in the eighth intron replacing the coding exons with an internal ribosome entry-site driven LacZ and a neomycin resistance gene. (Määttä et al., 2001) Absence of envoplakin mRNA expression in the homozygous mice was confirmed by RT-PCR using a forward primer 5′-ACCATGTTCAAGGGGCTGAGCAA from the first exon and a reverse primer 5′-TCAGGGTCTGGTGCTGGGAT from the 12th exon of the gene. In addition, the following primers from the last exon of the gene were used for RT-PCR to confirm absence of any truncated or alternative transcripts 5′-GCTGGTACAGCACCTGACT GGG (forward) and 5′-GTGGGCCATCTCCTGCTCGAAGTC (reverse).

The IRS-LacZ cassette driven by the endogenous Evpl promoter allows detection of the promoter activity to identify the tissues envoplakin would have been expressed (Määttä et al., 2001). Since prolonged formaldehyde fixation will result in reduced LacZ enzyme activity (Ma, Rogers, Zbar, & Schmidt, 2002), the following protocol is recommended.

(1) Tissues are fixed for 2 h in paraformaldehyde. Longer fixation times will reduce the staining intensity. Optionally, the tissues can be impregnated with sucrose by incubating them in 15% sucrose in PBS for 4 h at +4 °C and then in 30% sucrose in PBS for overnight at +4 °C.

(2) The lightly fixed tissue pieces are embedded in OCT compound and frozen on dry ice.

(3) 10 μm cryostat sections are brought to room temperature and washed three times for 5 min with PBS.

(4) Staining with X-Gal solution [1 mg/mL of 5-bromo-4-chloro-3-indolyl β-D-galactoside (X-Gal), 2 m*M* $MgCl_2$, 5 m*M* potassium ferrocyanide, 5 m*M* potassium ferricyanide, 0.01% sodium deoxycholate, and 0.02% Nonident P-40] is carried out at +37 °C overnight in humidified conditions.

(5) The sections are washed three times with distilled water, counterstained with Hematoxylin, dehydrated in an ascending series of ethanol (3 min each in 50%, 70%, 95%, and 100% ethanol) followed by 3 min in 1:1 xylene: ethanol and 2 × 3 min in xylene followed by mounting with a coverslip.

6.2 Periplakin Gene Targeting

The gene targeting of periplakin involved generating a deletion from a *Xho*I site in the 16th intron into a *Xba*I site in the 22nd exon replacing exons 17–21 and the 5′-end of the exon 22 with a neomycin selection cassette. The targeted allele can be detected by PCR using exon 16-specific forward primer 5′-CTC ATA CGA GAA CAG GCT G-3′ and a reverse primer 5′-CTG CTT GGC CAC CTG TAG-3′ from the 17th exon to produce a 982-bp PCR product from the wild-type allele or a PGK promoter-specific reverse primer, 5′-CCA GAG GCC ACT TGT GTA G-3′ to generate a 602-bp band from the targeted allele.

The experimental protocols to analyze the stratum corneum and barrier function of either the single knockout mouse strains or the triple-KO mice (EPI −/−) have included isolation of detergent-resistant corneocytes from the epidermis and analysis of the barrier functionality in embryos by a dye-exclusion assay.

6.3 Cornified Envelope Isolation and Analysis

Sufficient number of terminally differentiated keratinocytes, also called corneocytes or CEs, can be isolated from small (5 × 5 mm) pieces of skin. Shaved back skin or an ear tip is a preferred source compared to chemically depilated skin due to high concentration of the reducing agent thioglycolate in the commercial hair removal products.

(1) Skin samples are boiled for 20 min in 20 m*M* Tris–HCl, pH 7.5, 5 m*M* EDTA, 10 m*M* DTT, and 2% SDS. Remnants of intact skin are removed by forceps.

(2) Corneocytes are pelleted by a centrifugation at 5000 × *g* for 5 min, washed twice with a buffer containing 0.2% SDS (otherwise the same as the isolation buffer), and pelleted again.

(3) Washed corneocytes can be stored in 500 μL of the washing buffer at +4 °C. For imaging, aliquots of isolated corneocytes were pelleted and resuspended in 50 μL of PBS.

(4) Immature corneocytes that look thin and fragile can be distinguished from mature, hexagonal, and thicker ones by observing under a phase-contrast microscope. Alternatively, corneocytes can be stained with 0.5 mg/mL nile red (Life Technologies Cat. no. N-1142) for 10 min, pelleted and spotted in a minimal volume of PBS on glass slides, mounted and observed with a fluorescence microscope. Figure 3A shows how mature and immature corneocytes can be identified after nile red staining or in phase-contrast images.

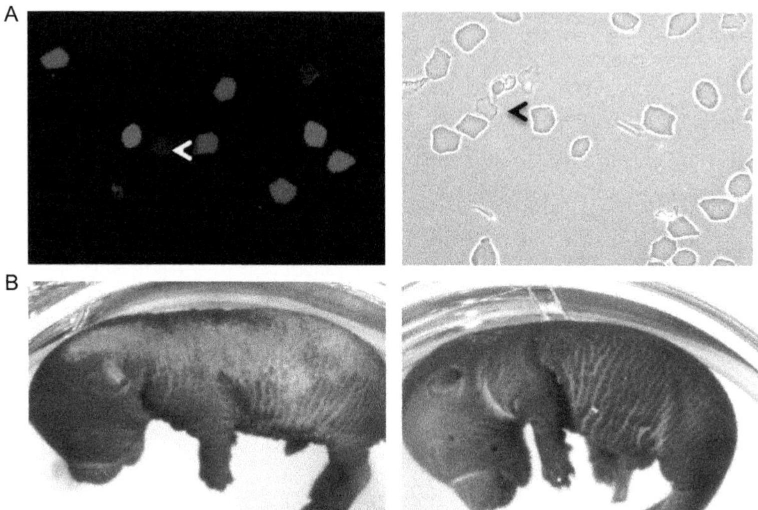

Figure 3 (A) Corneocytes isolated from mouse skin and stained with nile red (left-hand panel) or unstained and visualized by phase-contrast light microscopy (right-hand panel). Intensely stained hexagonal envelopes are mature whereas faintly stained, irregularly shaped envelopes (arrowheads) represent earlier stages of the terminal differentiation. (B) Toluidine blue assay for barrier development. E16.5 mouse embryos of heterozygous (left) or envoplakin −/− homozygous (right). (See the color plate.)

(5) The mechanical resilience of the isolated corneocytes can be evaluated by sonication. Wild-type, mature corneocytes remain largely intact during a short burst of sonication (5–10 s using a probe sonicator or 20–40 s on bath sonicator), whereas almost all the EPI −/− corneocytes are broken (Sevilla et al., 2007).

6.4 Dye-Exclusion Assay for Epidermal Barrier

Dye-exclusion assay can be used to evaluate developmental status of the barrier in mouse embryos (Hardman, Sisi, Banbury, & Byrne, 1998) or to localize regions of barrier disruption newborn or adult epidermis. The epidermal barrier develops between days E15.5 (no barrier, dye impregnates the skin) and E17.5 (skin impermeable to the dye) as a moving front starting from the head (Hardman et al., 1998). Both envoplakin single knockout mice and EPI −/− mice demonstrated a delay in their barrier acquisition (Määttä et al., 2001; Sevilla et al., 2007). Figure 3B shows Evpl −/− and +/− embryos stained with toluidine blue to demonstrate the developmental delay showing how the barrier has started to develop in the head and back skin of the heterozygous littermate, whereas the skin of the knockout mice is still fully permeable.

(1) Unfixed mouse embryos of known gestational age (generated by timed matings with a 5-h mating window) are dehydrated with ascending series of methanol (1 min in each of 25%, 50%, 75%, and 100% methanol) and then rehydrated using a descending series of the same concentrations.

(2) The samples are washed with PBS and stained for 1 min with 0.0125% toluidine blue (Sigma-Aldrich T3260).

(3) Excess dye is washed with PBS before the samples are photographed using a camera attached to a stereomicroscope.

(4) For adult mouse skin, the following modification of the protocol can be used (Ambler & Watt, 2010). Small (2 cm^2) pieces of shaved back skin are fixed overnight in 4% paraformaldehyde in PBS, dehydrated and rehydrated in methanol as described in step 1 above, and stained in 1% toluidine blue before destaining for 5–10 min in PBS. The stained skin can then be processed to thick (25 μm) frozen sections that are analyzed and photographed using a dissection microscope.

7. CONCLUSIONS AND OPEN QUESTIONS

Since their original identification as CE precursor proteins, envoplakin, and periplakin (Ruhrberg et al., 1996, 1997) have been studied in both animal and cell culture models. Investigation of the epidermis of the gene-targeted mice, especially of EPI −/− animals that lack both plakins and involucrin, has confirmed that the scaffolding proteins that constitute the initial protein assembly underneath the plasma membrane are required for the integrity of the epidermal barrier (Cipolat et al., 2014; Sevilla et al., 2007). However, the wide-ranging phenotype seen in the affected epidermis, ranging from processing defect of profilaggrin to disorganization of stratum corneum lipids (Sevilla et al., 2007), means that molecular mechanism leading to the development of the barrier defect has not yet been fully explained. For example, it is not yet clear which aspect(s) of the stratum corneum phenotype is behind the barrier "leakiness" that recruits the immune cell infiltrate. Given how the immune surveillance inhibits carcinogenesis in these mice (Cipolat et al., 2014) and the recently found role for periplakin in the BTN3A1-mediated activation of γδT cells by phosphoantigens (Rhodes et al., 2015), this area merits further research by utilizing additional mouse models and co-culture systems with 3D epidermal and full-thickness skin equivalents.

Several periplakin-interacting proteins have been identified by yeast two-hybrid assays and co-immunoprecipitation (Boczonadi et al., 2007;

Boczonadi & Määttä, 2012; Groot et al., 2004). Conversely, periplakin N-terminus has been identified by Y2H as a partner for several other membrane-associated proteins (Rhodes et al., 2015). This said, compared to other plakin family members, little is known how interactions of periplakin envoplakin with intermediate filaments and other interaction partners are regulated. The growing number of proteins interacting with periplakin promotes questions on whether some interactions are preferential over the others. Likewise, the internal hinge in the plakin head domains located before the SR7 repeat allows different flexibility for envoplakin and periplakin heads compared to desmoplakin but the functional significance of this observation is not yet understood (Al-Jassar et al., 2013). Many of the molecular interactions and localization of periplakin domains have so far been assayed only at static time points and it would be interesting to see dynamics of periplakin and envoplakin domains tagged with fluorescent proteins in live cell experiments. So far, only the C-terminal domain of periplakin has been investigated with a GFP tag and only in conjunction with OA-mediated collapse of keratin network (Long et al., 2006). Likewise, it is already detailed knowledge on how posttranslational modifications such as phosphorylation regulate interactions of other closely related plakins, e.g., desmoplakin (Albrecht et al., 2015; Fontao et al., 2003; Kröger et al., 2013). Therefore, mapping the phosphorylation of periplakin and envoplakin and consequences of these modifications should be of high priority for future research.

ACKNOWLEDGMENTS

Our research on plakin proteins has been supported by BBSRC and British Skin Foundation.

REFERENCES

Aho, S. (2004). Many faces of periplakin: Domain-specific antibodies detect the protein throughout the epidermis, explaining the multiple protein-protein interactions. *Cell and Tissue Research, 316*, 87–97.

Aho, S., Li, K., Ryoo, Y., McGee, C., Ishida-Yamamoto, A., Uitto, J., et al. (2004). Periplakin gene targeting reveals a constituent of the cornified cell envelope dispensable for normal mouse development. *Molecular and Cellular Biology, 24*, 6410–6418.

Albrecht, L. V., Zhang, L., Shabanowitz, J., Purevjav, E., Towbin, J. A., Hunt, D. F., et al. (2015). GSK3- and PRMT-1-dependent modifications of desmoplakin control desmoplakin-cytoskeleton dynamics. *Journal of Cell Biology, 208*, 597–612.

Al-Jassar, C., Bernadó, P., Chidgey, M., & Overduin, M. (2013). Hinged plakin domains provide specialized degrees of articulation in envoplakin, periplakin and desmoplakin. *PLoS One, 29*, e69767.

Ambler, C. A., & Watt, F. M. (2010). Adult epidermal Notch activity indces dermal accumulation of T cells and neural crest derivatives through upregulation of jagged 1. *Development*, *137*, 3569–3579.

Beekman, J. M., Bakema, J. E., van de Winkel, J. G., & Leusen, J. H. (2004). Direct interaction between FcgammaRI (CD64) and periplakin controls receptor endocytosis and ligand binding capacity. *Proceedings of the National Academy of Sciences of the United States of America*, *101*, 10392–10397.

Boczonadi, V., & Määttä, A. (2012). Annexin A9 is a periplakin interacting partner in membrane-targeted cytoskeletal linker protein complexes. *FEBS Letters*, *586*, 3090–3096.

Boczonadi, V., McInroy, L., & Määttä, A. (2007). Cytolinker cross-talk: Periplakin N-terminus interacts with plectin to regulate keratin organisation and epithelial migration. *Experimental Cell Research*, *313*, 3579–3591.

Borradori, L., Trüeb, R. M., Jaunin, F., Limat, A., Favre, B., & Saurat, J. H. (1998). Autoantibodies from a patient with paraneoplastic pemphigus bind periplakin, a novel member of the plakin family. *Journal of Investigative Dermatology*, *111*, 338–340.

Bouameur, J. E., Favre, B., & Borradori, L. (2014). Plakins, a versatile family of cytolinkers: Roles in skin integrity and in human diseases. *Journal of Investigative Dermatology*, *134*, 885–894.

Cipolat, S., Hoste, E., Natsuga, K., Quist, S. R., & Watt, F. M. (2014). Epidermal barrier defects link atopic dermatitis with altered skin cancer susceptibility. *eLife*, *3*, e01888.

DiColandrea, T., Karashima, T., Määttä, A., & Watt, F. M. (2000). Subcellular distribution of envoplakin and periplakin: Insights into their role as precursors of the epidermal cornified envelope. *Journal of Cell Biology*, *151*, 573–586.

Djian, P., Easley, K., & Green, H. (2000). Targeted ablation of the murine involucrin gene. *Journal of Cell Biology*, *151*, 381–388.

Feng, G. J., Kellett, E., Scorer, C. A., Wilde, J., White, J. H., & Milligan, G. (2003). Selective interactions between helix VIII of the human mu-opioid receptors and the C terminus of periplakin disrupt G protein activation. *Journal of Biological Chemistry*, *278*, 33400–33407.

Fontao, L., Favre, B., Riou, S., Geerts, D., Jaunin, F., Saurat, J. H., et al. (2003). Interaction of the bullous pemphigoid antigen 1 (BP230) and desmoplakin with intermediate filaments is mediated by distinct sequences within their COOH terminus. *Molecular Biology of the Cell*, *14*, 1978–1992.

Francke, F., Ward, R. J., Jenkins, L., Kellett, E., Richter, D., Milligan, G., et al. (2006). Interaction of neurochondrin with the melanin-concentrating hormone receptor 1 interferes with G protein-coupled signal transduction but not agonist-mediated internalization. *Journal of Biological Chemistry*, *281*, 32496–32507.

Groot, K. R., Sevilla, L. M., Nishi, K., DiColandrea, T., & Watt, F. M. (2004). Kazrin, a novel periplakin-interacting protein associated with desmosomes and the keratinocyte plasma membrane. *Journal of Cell Biology*, *166*, 653–659.

Hardman, M. J., Sisi, P., Banbury, D. N., & Byrne, C. (1998). Patterned acquisition of skin barrier function during development. *Development*, *125*, 1541–1552.

Ito, S., Satoh, J., Matsubara, T., Shah, Y. M., Ahn, S. H., Anderson, C. R., et al. (2013). Cholestasis induces reversible accumulation of periplakin in mouse liver. *BMC Gastroenterology*, *13*, 116.

Kalinin, A. E., Idler, W. W., Marekov, L. N., McPhie, P., Bowers, B., Steinert, P. M., et al. (2004). Co-assembly of envoplakin and periplakin into oligomers and $Ca(^{2+})$-dependent vesicle binding: Implications for cornified cell envelope formation in stratified squamous epithelia. *Journal of Biological Chemistry*, *279*, 22773–22780.

Kalinin, A. E., Kajava, A. V., & Steinert, P. M. (2002). Epithelial barrier function: Assembly and structural features of the cornified cell envelope. *Bioessays*, *24*, 789–800.

Karashima, T., & Watt, F. M. (2002). Interaction of periplakin and envoplakin with intermediate filaments. *Journal of Cell Science, 115*, 5027–5037.

Kazerounian, S., & Aho, S. (2003). Characterization of periphilin, a widespread, highly insoluble nuclear protein and potential constituent of the keratinocyte cornified envelope. *Journal of Biological Chemistry, 278*, 36707–36717.

Kazerounian, S., Uitto, J., & Aho, S. (2002). Unique role for the periplakin tail in intermediate filament association: Specific binding to keratin 8 and vimentin. *Experimental Dermatology, 11*, 428–438.

Kelly, S., Culican, S., Silvestrini, R. A., Vu, J., Schifter, M., Fulcher, D. A., et al. (2015). Comparative study of five serological assays for the diagnosis of paraneoplastic pemphigus. *Pathology, 47*, 58–61.

Kiyokawa, C., Ruhrberg, C., Nie, Z., Karashima, T., Mori, O., Nishikawa, T., et al. (1998). Envoplakin and periplakin are components of the paraneoplastic pemphigus antigen complex. *Journal of Investigative Dermatology, 111*, 1236–1238.

Kröger, C., Loschke, F., Schwarz, N., Windoffer, R., Leube, R. E., & Magin, T. M. (2013). Keratins control intercellular adhesion involving PKC-α-mediated desmoplakin phosphorylation. *Journal of Cell Biology, 201*, 681–692.

Leger, S., Picard, D., Ingen-Housz-Oro, S., Arnault, J. P., Aubin, F., Carsuzaa, F., et al. (2012). Prognostic factors of paraneoplastic pemphigus. *Archives of Dermatology, 148*, 1165–1172.

Long, H. A., Boczonadi, V., McInroy, L., Goldberg, M., & Määttä, A. (2006). Periplakin-dependent re-organisation of keratin cytoskeleton and loss of collective migration in keratin-8-downregulated epithelial sheets. *Journal of Cell Science, 119*, 5147–5159.

Ma, W., Rogers, K., Zbar, B., & Schmidt, L. (2002). Effect of different fixatives on beta-galactosidase activity. *Journal of Histochemistry and Cytochemistry, 50*, 1421–1424.

Ma, A. S., & Sun, T. T. (1986). Differentiation-dependent chnages in the solubility of a 195-kD protein in human epidermal keratinocytes. *Journal of Cell Biology, 103*, 41–48.

Määttä, A., DiColandrea, T., Groot, K., & Watt, F. M. (2001). Gene targeting of envoplakin, a cytoskeletal linker protein and precursor of the epidermal cornified envelope. *Molecular and Cellular Biology, 21*, 7047–7053.

Mahoney, M. G., Aho, S., Uitto, J., & Stanley, J. R. (1998). The members of the plakin family of proteins recognized by paraneoplastic pemphigus antibodies include periplakin. *Journal of Investigative Dermatology, 111*, 308–313.

Matsumoto, K., Ikeda, M., Sato, Y., Kuruma, H., Kamata, Y., Nishimori, T., et al. (2014). Loss of periplakin expression is associated with pathological stage and cancer-specific survival in patients with urothelial carcinoma of the urinary bladder. *Biomedical Research, 35*, 201–206.

Murdoch, H., Feng, G. J., Bächner, D., Ormiston, L., White, J. H., Richter, D., et al. (2005). Periplakin interferes with G protein activation by the melanin-concentrating hormone receptor-1 by binding to the proximal segment of the receptor C-terminal tail. *Journal of Biological Chemistry, 280*, 8208–8220.

Nachat, R., Cipolat, S., Sevilla, L. M., Chhatriwala, M., Groot, K. R., & Watt, F. M. (2009). KazrinE is a desmosome-associated liprin that colocalises with acetylated microtubules. *Journal of Cell Science, 122*, 4035–4041.

Nishimori, T., Tomonaga, T., Matsushita, K., Oh-Ishi, M., Kodera, Y., Maeda, T., et al. (2006). Proteomic analysis of primary esophageal squamous cell carcinoma reveals downregulation of a cell adhesion protein, periplakin. *Proteomics, 6*, 1011–1018.

Ortega, E., Buey, R. M., Sonnenberg, A., & de Pereda, J. M. (2011). The structure of the plakin domain of plectin reveals a non-canonical SH3 domain interacting with its fourth spectrin repeat. *Journal of Biological Chemistry, 286*, 12429–12438.

Otsubo, T., Hagiwara, T., Tamura-Nakano, M., Sezaki, T., Miyake, O., Hinohara, C., et al. (2015). Aberrant DNA hypermethylation reduces the expression of the desmosome-related molecule periplakin in esophageal squamous cell carcinoma. *Cancer Medicine*, *4*, 415–425. http://dx.doi.org/10.1002/cam4.369.

Palka, H. L., & Green, K. J. (1997). Roles of plakoglobin end domains in desmosome assembly. *Journal of Cell Science*, *110*, 2359–2371.

Rhodes, D. A., Chen, H. C., Price, A. J., Keeble, A. H., Davey, M. S., James, L. C., et al. (2015). Activation of human γδ T cells by cytosolic interactions of BTN3A1 with soluble phosphoantigens and the cytoskeletal adaptor periplakin. *The Journal of Immunology*, *194*, 2390–2398.

Ruhrberg, C., Hajibagheri, M. A., Parry, D. A., & Watt, F. M. (1997). Periplakin, a novel component of cornified envelopes and desmosomes that belongs to the plakin family and forms complexes with envoplakin. *Journal of Cell Biology*, *139*, 1835–1849.

Ruhrberg, C., Hajibagheri, M. A., Simon, M., Dooley, T. P., & Watt, F. M. (1996). Envoplakin, a novel precursor of the cornified envelope that has homology to desmoplakin. *Journal of Cell Biology*, *134*, 715–729.

Sevilla, L. M., Nachat, R., Groot, K. R., Klement, J. F., Uitto, J., Djian, P., et al. (2007). Mice deficient in involucrin, envoplakin, and periplakin have a defective epidermal barrier. *Journal of Cell Biology*, *179*, 1599–1612.

Sevilla, L. M., Nachat, R., Groot, K. R., & Watt, F. M. (2008). Kazrin regulates keratinocyte cytoskeletal networks, intercellular junctions and differentiation. *Journal of Cell Science*, *121*, 3561–3569.

Simon, M., & Green, H. (1984). Participation of membrane-associated proteins in the formation of the cross-linked envelope of the keratinocyte. *Cell*, *36*, 827–834.

Sonnenberg, A., Rojas, A. M., & de Pereda, J. M. (2007). The structure of a tandem pair of spectrin repeats of plectin reveals a modular organization of the plakin domain. *Journal of Molecular Biology*, *368*, 1379–1391.

Steinert, P. M., & Marekov, L. N. (1999). Initiation of assembly of the cell envelope barrier structure of stratified squamous epithelia. *Molecular Biology of the Cell*, *10*, 4247–4261.

Ward, R. J., Jenkins, L., & Milligan, G. (2009). Selectivity and functional consequences of interactions of family A G protein-coupled receptors with neurochondrin and periplakin. *Journal of Neurochemistry*, *109*, 182–192.

Yoon, K. H., & FitzGerald, P. G. (2009). Periplakin interactions with lens intermediate and beaded filaments. *Investigative Ophthalmology & Visual Science*, *50*, 1283–1289.

CHAPTER SEVENTEEN

Microtubule-Actin Cross-Linking Factor 1: Domains, Interaction Partners, and Tissue-Specific Functions

Dmitry Goryunov, Ronald K.H. Liem[1]
Department of Pathology and Cell Biology, Taub Institute for Research on Alzheimer's Disease and the Aging Brain, Columbia University College of Physicians and Surgeons, New York, USA
[1]Corresponding author: e-mail address: rkl2@cumc.columbia.edu

Contents

Abstract

The cytoskeleton of most eukaryotic cells is composed of three principal filamentous components: actin filaments, microtubules (MTs), and intermediate filaments. It is a highly dynamic system that plays crucial roles in a wide range of cellular processes, including migration, adhesion, cytokinesis, morphogenesis, intracellular traffic and signaling, and structural flexibility. Among the large number of cytoskeleton-associated

Methods in Enzymology, Volume 569
ISSN 0076-6879
http://dx.doi.org/10.1016/bs.mie.2015.05.022

proteins characterized to date, microtubule-actin cross-linking factor 1 (MACF1) is arguably the most versatile integrator and modulator of cytoskeleton-related processes. MACF1 belongs to the plakin family of proteins, and within it, to the spectraplakin subfamily. These proteins are characterized by the ability to bridge MT and actin cytoskeletal networks in a dynamic fashion, which underlies their involvement in the regulation of cell migration, axonal extension, and vesicular traffic. Studying MACF1 functions has provided insights not only into the regulation of the cytoskeleton but also into molecular mechanisms of both normal cellular physiology and cellular pathology. Multiple MACF1 isoforms exist, composed of a large variety of alternatively spliced domains. Each of these domains mediates a specific set of interactions and functions. These functions are manifested in tissue and cell-specific phenotypes observed in conditional *MACF1* knockout mice. The conditional models described to date reveal critical roles of MACF1 in mammalian skin, nervous system, heart muscle, and intestinal epithelia. Complete elimination of MACF1 is early embryonic lethal, indicating an essential role for MACF1 in early development. Further studies of MACF1 domains and their interactions will likely reveal multiple new roles of this protein in various tissues.

1. INTRODUCTION

1.1 The Plakin Family

The microtubule-actin cross-linking factor 1 (MACF1) is a member of the plakin family of proteins (Jefferson, Leung, & Liem, 2004; Ruhrberg & Watt, 1997; Sonnenberg & Liem, 2007; Suozzi, Wu, & Fuchs, 2012). The plakins are a group of large cytoskeletal linkers with multidomain structures that mediate a wide range of diverse cellular functions. The mammalian plakins include MACF1, bullous pemphigoid antigen 1 (BPAG1), desmoplakin, plectin, periplakin, envoplakin, and epiplakin (Jefferson et al., 2004; Leung, Green, & Liem, 2002; Ruhrberg & Watt, 1997; Suozzi et al., 2012). All the plakins have a conserved plakin domain (described below). In addition, MACF1 and BPAG1 have a series of rod domain consisting of a series of spectrin repeats. These two proteins are therefore also called spectraplakins (Bosher et al., 2003; Roper & Brown, 2003; Suozzi et al., 2012). Plakin family members have the ability to cross-link actin filaments, microtubules (MTs), and intermediate filaments (IFs) to one another and to various cell junction complexes. In addition, various plakins are involved in processes regulating cell migration, differentiation, and vesicular traffic (Boyer, Bernstein, & Boudreau-Lariviere, 2010; Jefferson et al., 2004; Ruhrberg & Watt, 1997; Sonnenberg & Liem, 2007; Suozzi et al., 2012).

Spectraplakins are phylogenetically conserved. *Kakapo/shortstop/shot* is a *Drosophila* gene closely related to both *MACF1* and *BPAG1* and described in more detail in Chapter 18 (Spectraplakins in *Drosophila*, by A. Prokop). Mutations in *shot* result in multiple defects, including axonal extension, dendrite morphology, epidermal muscle attachment, and tendon cell differentiation (Roper & Brown, 2003; Subramanian et al., 2003). Shot recruits EB1/APC to promote MT assembly at the muscle–tendon junction. The mutant phenotype is characterized by a failure to extend both motor and sensory axons to reach their targets (Subramanian et al., 2003). In *C. elegans*, the *vab-10* gene encodes a protein that contains many of the same domains as the two mammalian spectraplakins. *Vab-10* mutants exhibit abnormal body morphology and show motility defects (Bosher et al., 2003; Jorgensen et al., 2014). The spectraplakin subfamily also includes a single gene in zebrafish, *magellan*. Mutations in *magellan* result in defects in animal–vegetal polarity of the oocyte and egg (Dosch et al., 2004; Gupta et al., 2010).

1.2 MACF1

A partial *MACF1* cDNA was initially isolated as a clone encoding actin cross-linking protein 7 (ACF7), based on its homology to previously characterized actin-binding domains (ABDs) (Byers, Beggs, McNally, & Kunkel, 1995). Analysis of a longer, though still incomplete, *ACF7* cDNA revealed that it is related most closely to *BPAG1*, which was also only partially cloned and characterized at the time (Bernier, Mathieu, De Repentigny, Vidal, & Kothary, 1996). Three isoforms of ACF7 were partially characterized in mice, and two of them contained putative ABDs (Bernier et al., 1996). Two groups later reported the cloning of a high-molecular-weight protein in humans with close similarity to plectin and BPAG1 (Okuda et al., 1999; Sun et al., 1999). The two groups called the protein trabeculin-α (Sun et al., 1999) and macrophin (Okuda et al., 1999). It was eventually demonstrated that both these proteins were encoded by the *MACF1* gene (Leung, Sun, Zheng, Knowles, & Liem, 1999).

Two major MACF1 isoforms, MACF1a (~608 kDa) and MACF1b (~809 kDa), result from alternative splicing (Leung, Sun, Zheng, et al., 1999; Lin, Chen, Leung, Parry, & Liem, 2005). The two isoforms consist of a variety of domains, described in more detail below and shown in Fig. 1. In addition to the N-terminal ABD, MACF1 has a novel microtubule-binding domain (MTBD) at its C-terminus, thereby making it microtubule-actin cross-linking factor. Alternative initiation of both

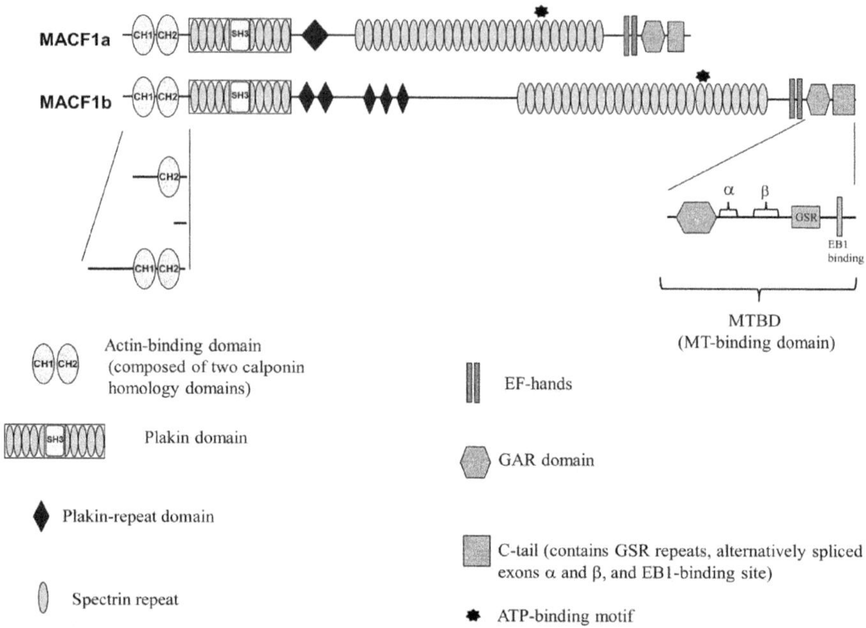

Figure 1 The *MACF1* gene is alternatively spliced, resulting in two major isoforms, MACF1a and MACF1b. Both isoforms contain an actin-binding domain (ABD) composed of two calponin-homology domains (CH1 and CH2), a plakin domain, three (MACF1a) or five (MACF1b) plakin-repeat domains (PRDs), a series of 28 spectrin repeats (spectrin rod), a pair of EF-hand motifs, and a C-terminal MT-binding domain (MTBD). The plakin domain comprises nine spectrin repeats, with an SH3 domain located in the middle of the fifth repeat. Embedded in the spectrin rod is an ATP-binding motif required for intrinsic ATPase activity, which is stimulated by F-actin. The MTBD is composed of a Gas2/GAR22-related (GAR) subdomain and a region containing several GSR (Gly-Ser-Arg) motifs. The MTBD contains two alternatively spliced exons and a single SxIP motif that functions as a binding site for the EB1 family of MT tip-interacting proteins (TIPs). Alternative initiation of both *MACF1a* and *MACF1b* can result in additional isoforms, some lacking one or both CH parts of the ABD (Bernier et al., 1996; Gong, Besirli, & Lomax, 2001; Jefferson, Leung, & Liem, 2006; Kumar & Wittmann, 2012; Leung, Sun, Zheng, et al., 1999; Lin et al., 2005; Ortega, Buey, Sonnenberg, & de Pereda, 2011; Poliakova et al., 2014; Slep et al., 2005; Sonnenberg & Liem, 2007; Sonnenberg, Rojas, & de Pereda, 2007; Wu, Kodama, & Fuchs, 2008).

MACF1a and *MACF1b* can result in additional isoforms (Bernier et al., 1996; Gong et al., 2001; Jefferson et al., 2006). Splice variants of the MACF1 C-tail have also been described (Poliakova et al., 2014).

MACF1 is central to the regulation of cell migration. In keratinocytes, MACF1 associated preferentially with MTs that aligned with actin filaments near the cell periphery as well as with focal adhesions (FAs) (Karakesisoglou, Yang, & Fuchs, 2000). It functions as an organizer of polarity factors at the leading edge of migrating cells and acts as a MT-actin cross-linker to coordinate MT dynamics in cell migration (Kodama, Karakesisoglou, Wong, Vaezi, & Fuchs, 2003). MACF1 also interacts with ELMO (engulfment and motility) proteins, inducing the formation of membrane protrusions during integrin-mediated cell spreading. ELMO1 recruits MACF1 to the plasma membrane, where MACF1 promotes MT capture and stability (Margaron, Fradet, & Cote, 2013). The role of MACF1 in cell migration impinges on its ability to function as a MT plus-end-tracking protein (+TIP), enabling it to coordinate the interactions of growing MT ends with the actin cytoskeleton at the plasma membrane (Gupta et al., 2010). MACF1 also binds directly to the +TIP protein EB1 and mediates the cortical localization of another MT +TIP, CLASP2 (Drabek et al., 2006; Slep et al., 2005). In migrating epithelial cells, GSK3β phosphorylates the MACF1 MTBD, resulting in the dissociation of MACF1 from the MTs. Phosphorylation of MACF1 is necessary for MT growth and for skin stem cell migration (Wu et al., 2011).

The ability of MACF1 to orchestrate interactions of different cytoskeletal elements and membrane-associated complexes is central to its role in neuronal extension (Goryunov, He, Lin, Leung, & Liem, 2010; Ka, Jung, Mueller, & Kim, 2014; Sanchez-Soriano et al., 2009). In cultured mouse neurons, MACF1 regulates the organization of neuronal MTs via a mechanism that is dependent on both its ABD and MTBD. It can also control the formation of filopodia through regulation of F-actin. This latter function is dependent on the presence of the C-terminal EF-hand motifs and an interaction with the translational regulator Krasavietz/eIF5C. Together, these functions enable MACF1 to coordinate the organization of both actin and MT networks during axonal growth (Sanchez-Soriano et al., 2009). Consistent with these findings, elimination of MACF1a in the mouse nervous system causes multiple brain developmental abnormalities indicative of defects in axonal growth (Goryunov et al., 2010). The involvement of MACF1 in neuronal migration is cell-autonomous rather than mediated by radial glia or neuronal progenitor cells (Ka et al., 2014).

Increasing evidence also points to MACF1 involvement in various aspects of intracellular vesicular traffic (Burgo et al., 2012; Kakinuma, Ichikawa, Tsukada, Nakamura, & Toh, 2004; Lin et al., 2005). MACF1 appears to play a key role in the molecular mechanisms that allow secretory vesicles to load onto MTs and reach the cell surface during exocytosis (Burgo et al., 2012; Kakinuma et al., 2004). Through a sequence in its C-terminus, MACF1 interacts with the *trans*-Golgi network (TGN) protein p230/Golgin245. This interaction is critical for the transport of GPI-anchored proteins along MTs from the TGN to the cell periphery (Kakinuma et al., 2004). During this process, MACF1 and GolginA4 bind activated Rab21-GTP, an event required for directed movement of TI-VAMP/VAMP7 vesicles from the cell center to the cell periphery along the MTs (Burgo et al., 2012). MACF1b has also been shown to interact with the Golgi apparatus via its most N-terminal plakin repeats. This association appears to be crucial for the integrity of the Golgi complex as RNAi knockdown of MACF1 results in its dispersal (Lin et al., 2005).

2. MACF1 DOMAINS

The MACF1 proteins are composed of an unusually large variety of structural domains and sequence motifs, which might explain the extraordinary multi-functionality of these spectraplakins. Our laboratory was the first to demonstrate that, in addition to the ABD and plakin domain, MACF1 contains multiple plakin-repeat domains (PRDs), a series of spectrin repeats, two putative EF-hand calcium-binding motifs, and a C-terminal MTBD (Leung, Sun, & Liem, 1999; Leung, Sun, Zheng, et al., 1999). Each of these domains imparts a new function on the protein, and together they provide for a unique combination of structural, signaling, and mechanical properties that enable MACF1 to serve as an integrator of numerous cellular processes. The organization of the domains, as well as the various isoforms of MACF1a/b is shown in Fig. 1.

2.1 ABD

Both MACF1a and MACF1b contain a well-characterized ABD at their N-termini (Fig. 1). The MACF1 ABD consists of two calponin-homology domains (CH1 and CH2), the tandem sequence motifs that are preserved throughout the spectrin superfamily of proteins and confer on them the ability to bind actin filaments (Stradal, Kranewitter, Winder, & Gimona, 1998). Although the CH1 domain of BPAG1 can bind F-actin on its own, this binding is significantly augmented by the CH2 domain (Jefferson et al., 2004). The ABD of MACF1 has been shown to associate with F-actin both *in vitro* and *in vivo* (Karakesisoglou et al., 2000; Leung, Sun, Zheng, et al., 1999; Yang et al., 1996). Isoforms of MACF1 have been described that only contain the CH2 domain or lack the ABD altogether (Goryunov et al., 2010; Karakesisoglou et al., 2000; see Fig. 1). The ABD of MACF1 also interacts with rapsyn, suggesting that acetylcholine receptors in mammalian muscle are immobilized in the sarcolemmal membrane through MACF1-mediated anchoring to the actin cytoskeleton (Antolik et al., 2007).

2.2 Plakin Domain

The plakin domain is the defining feature of the plakin family, including MACF1. Detailed structural studies of the plakin domain have shown that it contains a putative Src-homology 3 (SH3) domain positioned in the middle of a spectrin repeat that is itself flanked on each side by four complete spectrin repeats (Jefferson, Ciatto, Shapiro, & Liem, 2007; Ortega et al., 2011; Sonnenberg et al., 2007). Most plakin domains studied interact with components of cellular junctions, although so far this has not been demonstrated for the MACF1 plakin domain. The plakin domains of plectin and BPAG1 interact with β4-integrin and the cytoplasmic domain of BPAG2 in hemidesmosomes (Rezniczek, de Pereda, Reipert, & Wiche, 1998). The plakin domain of desmoplakin interacts with the armadillo repeat family members plakoglobin (γ-catenin), plakophilin-1, and plakophilin-2, as well as with the desmosomal cadherin desmocollin (Green & Gaudry, 2000). The plakin domain of MACF1 has been reported to interact weakly with MTs (Karakesisoglou et al., 2000).

2.3 Plakin-Repeat Domains

All MACF1 isoforms known to date contain several PRDs located between the plakin domain and the spectrin-repeat rod (Fig. 1). MACF1a contains three PRDs; MACF1b contains an additional two that mediate association with the Golgi apparatus (Jefferson et al., 2004; Lin et al., 2005). Each PRD is composed of a conserved central plectin module and flanking linker sequences (Janda, Damborsky, Rezniczek, & Wiche, 2001). The plectin module is a globular domain composed of 4.5 tandem repeats of 38 amino acids, which are called plectin or plakin repeats (Jefferson et al., 2004). The function of the MACF1 PRD has not yet been determined, although the PRD containing regions of other plakins have been found to bind to IFs (Lapouge et al., 2006; Nikolic, Mac Nulty, Mir, & Wiche, 1996; Wiche, Gromov, Donovan, Castanon, & Fuchs, 1993).

2.4 Spectrin Repeats

Both major MACF1 isoforms contain 28 spectrin repeats in the middle of the molecule, between the PRDs and the C-terminal MTBD. Each spectrin repeat consists of 106 amino acids, together comprising three α-helices

(Speicher & Marchesi, 1984). Together, the spectrin repeats serve as a physical spacer as well as a structural element that renders MACF1 flexible, a property critical for its functions as a mechanical connector of membrane-bound and cytoskeletal structures. A putative SH3 domain has been identified in the ninth spectrin repeat of α-spectrin, suggesting that the spectrin-repeat rods of plakins may harbor functions other than purely structural or mechanical (Sonnenberg et al., 2007). The spectrin rod contains an ATP-binding domain that is required for intrinsic ATPase activity of MACF1, which is stimulated by F-actin *in vitro* (Wu et al., 2008).

2.5 MTBD

The structure and functions of the MACF1 MTBD are considerably more complicated than the name suggests. Originally, it was shown to be composed of two regions, each of which was capable of binding or bundling MTs on its own (Leung, Sun, Zheng, et al., 1999; Sun, Leung, & Liem, 2001). The more N-terminal region is called the Gas2/GAR22-related (GAR) domain. The more C-terminal region, or C-tail, contains several GSR (Gly-Ser-Arg) motifs and carries a positive overall charge. When combined, the two regions bind MTs more strongly than either one does in isolation (Sun et al., 2001; Suozzi et al., 2012). The MACF1 MTBD contains at least two alternatively spliced exons, which may play a role in modulation of the MT-binding capacity of MACF1 isoforms in different tissues (Poliakova et al., 2014). In addition, the C-tail of MACF1 contains a single SxIP motif. These motifs function as binding sites for the EB1 family of MT tip-interacting proteins (Kumar & Wittmann, 2012; Slep et al., 2005). Consistent with this, the C-tail of MACF1 has been shown to bind EB1 and EB3 (Poliakova et al., 2014).

3. METHODS TO STUDY PLAKIN DOMAIN INTERACTIONS

3.1 *In Vitro* Interactions

3.1.1 Spin-Down Assay

In vitro binding assays are often used to show that MACF1 domains can interact with MTs, IFs, or actin filaments. One commonly employed assay involves incubation of preassembled filaments with a protein containing a potential binding domain, followed by high-speed centrifugation to pellet the filaments (Goriounov, Leung, & Liem, 2003; Sun et al., 2001). The presence of the studied protein in the pellet indicates filament binding, whereas its presence in the supernate shows lack of binding. The highest sensitivity can be achieved if the protein being studied is labeled radioactively, normally through *in vitro* transcription/translation. (The procedures described in this section are intended to demonstrate the overall logic of each assay rather than to be used as detailed protocols. The specific technical details will vary depending on the reagents used as well as the investigator's purpose in performing the experiment.)

1. Radioactively labeled proteins can be prepared using commercial kits for coupled *in vitro* transcription/translation. Such kits are available from multiple manufacturers, including Promega, Life Technologies, and EMD-Millipore (formerly Novagen). The single-tube protein system 3 is designed for *in vitro* synthesis of proteins directly from supercoiled or linear DNA templates containing a T7 or Sp6 promoter. The system is based on a two-step protocol in which transcription with T7 or Sp6 polymerase is directly followed by translation in an optimized rabbit reticulocyte lysate. Linearization is essential for efficient transcription in this system.
2. To test labeled proteins for the ability to bind actin filaments, we used the Actin-Binding Protein Spin-Down Biochem Kit from Cytoskeleton, Inc. This kit provides an easy and convenient way to prepare F-actin, which is then used to cosediment any binding proteins. A 250-μg aliquot of monomeric actin (kit component AKL99) is thawed on ice, 250 μl of general actin buffer (BSA01; 5 m*M* Tris–HCl, pH 8.0, and 0.2 m*M* $CaCl_2$) is added, and the mixture is pipetted up and down several times. The resulting actin concentration is 1 mg/ml. The actin preparation is incubated on ice for 30 min, followed by addition of 25 μl of actin polymerization buffer (BSA02; 500 m*M* KCl, 20 m*M* $MgCl_2$, and 10 m*M* ATP in 100 m*M* Tris, pH 7.5) and incubation at room temperature for 1 h. This resulting F-actin stock (23 μ*M*) can be stored at 4 °C for up to a week.
3. To test for protein binding to actin filaments, 40 μl of F-actin stock is combined with 10 μl of labeled protein (prespun at 150,000 × *g* for 20 min at 4 °C) and mixed by stirring gently with the pipette tip. α-actinin or BSA can be used as positive and negative binding controls, respectively. The protein mixtures are incubated at room temperature for 30 min, layered on top of 100 μl of cushion buffer (BSA03; 5 m*M* Tris, pH 8.0, 50 m*M* KCl, 2 m*M* $MgCl_2$, and 10% glycerol) and spun at 150,000 × *g* for 1.5 h at room temperature. An aliquot of the supernate and the actin pellet are boiled in SDS denaturing buffer and resolved on an SDS-PAGE gel. The gel is stained, dried in a vacuum dryer on a sheet of filter paper, and exposed to X-ray film for 24–72 h at −80 °C.

A similar assay can be used to test whether a protein is capable of binding MTs. The Microtubule Associated Protein Spin-Down Kit from Cytoskeleton, Inc., can be used with ^{35}S-labeled proteins prepared as described above. The basic assay includes incubation of the protein of interest with preassembled MTs, high-speed centrifugation to spin down the MTs and associated proteins through a buffered cushion, and analysis of the pellet and supernate fractions by SDS-PAGE followed by autoradiography.

1. To prepare MTs, tubulin monomer is incubated in MT cushion buffer at 35 °C for 20 min. In the meantime, taxol is added to tubulin dilution buffer (TDB) to a final concentration of 40 μ*M*, and the resulting solution is added to the cushion buffer with polymerized MTs for stabilization. The procedure produces a population of MTs that are 5–10 μm in length. The MTs will remain stable for up to a week when stored at 4 °C.
2. To assay for MT binding, 10 μl of radioactively labeled protein is mixed with 20 μl of MTs and 20 μl of TDB/taxol solution in a microfuge tube, and the mixture is incubated at room temperature for 20 min. The reaction is then loaded on top of 100 μl of cushion buffer and centrifuged at 100,000 × *g* for 40 min at room temperature. The supernate and pellet fractions are resolved on an SDS-PAGE gel as described above for the actin-binding assay, and the labeled protein is detected by autoradiography. MAP2 and BSA can be used as positive and negative binding controls, respectively.

It should be noted that the spin-down assay described here leaves open the possibility that an observed interaction is not direct but instead mediated by a component of the transcription/translation mix. To demonstrate unequivocally that an interaction is direct, purified recombinant proteins can be used in conjunction with immunochemical analysis of the soluble and insoluble fractions by Western blotting. It is also the preferred method for determination of the binding constant of a putative filament-binding domain (Sun et al., 2001). Similar caveats apply to co-immunoprecipitation, another method often used to obtain evidence of protein–protein interactions. Unless purified or recombinant proteins are used, it is always possible that another protein present in the cell lysate functions as a mediator of the interaction.

3.1.2 Far-Western Blotting

The Far-Western blotting assay, also called the overlay binding assay, is another *in vitro* approach to determine if a MACF1-like domain interacts with another protein, including cytoskeletal proteins. The assay is essentially a modification of the Western blot. The protein under study is resolved on an SDS-PAGE gel and transferred onto a membrane, and the other potential interacting protein is used to probe the membrane, much like the primary antibody in Western blotting. Far-Western assays have been used to study the interactions of plectin and desmoplakin with IF proteins (Meng, Bornslaeger, Green, Steinert, & Ip, 1997; Nikolic et al., 1996). Our laboratory used Far-Western blotting to demonstrate interactions between the IF protein peripherin and a BPAG1 domain (Leung, Sun, & Liem, 1999).

1. Histidine-tagged peripherin and BPAG1 protein containing a putative IF-binding domain were expressed in bacteria and purified by column chromatography. The BPAG1 construct also included an in-frame FLAG tag.
2. The peripherin protein was resolved by SDS-PAGE and transferred onto a nylon membrane, which was then washed to remove the SDS and allow for at least partial renaturation of the transferred peripherin.
3. The membrane was blocked with 5% BSA, incubated with 5 μg/ml purified BPAG1 protein, and washed. The BPAG1 protein that remained bound to the peripherin on the membrane was then detected with an anti-FLAG monoclonal antibody, which was in turn visualized with a secondary HRP-antibody conjugate.

Unlike peripherin, none of the neurofilament triplet proteins (NFL, NFM, and NFH) interacted with BPAG1 in this assay. One should keep in mind that the Far-Western assay detects interactions between monomeric proteins. To confirm that a putative interaction identified with this assay applies to filaments as well, a slot-blot assay can be used. In this assay, *in vitro* produced filaments are immobilized on a membrane and probed with the protein under study, similar to standard Far-Western blotting (Leung, Sun, & Liem, 1999).

3.2 *In Vivo* Interactions

3.2.1 Two-Hybrid Assay

The yeast two-hybrid assay is a classical method for identifying and characterizing protein–protein interactions *in vivo* (Fields & Song, 1989). The assay is based on reconstitution of a functional GAL4 transcriptional activator and consequent activation of a reporter gene under the control of a GAL4-responsive promoter. The target is cloned into a "bait" vector in frame with the GAL4 DNA-binding domain. The library to be screened (or the second of the two putative interacting proteins) is cloned in another vector in frame with the GAL4 activation domain. The two types of hybrid plasmids are cotransformed into yeast cells. If the two proteins interact, a functional GAL4 activator is reconstituted, and the expression of a reporter gene (lacZ in the case of SFY526 strain) is induced. Reporter activation can be assessed quantitatively by a β-galactosidase assay. Our laboratory has used the two-hybrid system to characterize interactions between an IF-binding domain of BPAG1 and neuronal IF proteins (Leung, Sun, & Liem, 1999). This approach was also used to identify MACF1 as a binding partner of p230/ Golgin-245, a TGN protein involved in transporting GPI-anchored membrane proteins to the cell periphery (Kakinuma et al., 2004).

As with Far-Western blotting, it is important to remember that the two-hybrid assay identifies interactions of monomeric proteins rather than filaments. The interaction of a protein domain with a cytoskeletal filament protein in the two-hybrid assay may be different from its interaction with the corresponding filaments in cells or may depend on a specific modification of the filaments.

3.2.2 Immunofluorescent Staining of Cultured Cells or Tissues

Co-localization of two proteins in cultured cells or tissues, typically revealed with immunostaining and fluorescent microscopy, can serve as indirect evidence of the proteins interacting *in vivo*. Again, the apparent interaction could be enabled by a third, often unknown protein that is not being visualized in the experiment. In the case of transiently transfected cells, overexpression of a protein under study can result in increased co-localization of two other components, each of which can be a monomeric protein, a filamentous structure, or a membrane-associated complex. Such enhanced co-localization provides strong, if indirect, evidence that the protein is an actual cytolinker or regulator (Goriounov et al., 2003; Karakesisoglou et al., 2000; Leung, Sun, Zheng, et al., 1999; Leung, Zheng, Prater, & Liem, 2001; Wu et al., 2008).

A similar approach using electron microscopy can yield much more direct evidence of interaction. In an elegant study utilizing electron microscopy with immunogold labeling, Svitkina and coworkers demonstrated that plectin side-arms directly connect IFs to MTs as well as actin filaments (Svitkina, Verkhovsky, & Borisy, 1996). Recently, 3D electron tomography was used to show that a protein that most likely represents MACF1 forms physical linkers between MTs and the F-actin network of the cuticular plate, near the apical surface of the mechanosensory hair cell, in zebrafish (Antonellis et al., 2014).

3.2.3 Interactions with Membranous Organelles

A density gradient fractionation assay that largely preserves *in vivo* interactions can be used to demonstrate the association of a protein or one of its domains with a membranous organelle (Gao, Alvarez, Nelson, & Sztul, 1998). Our laboratory employed this approach to show that MACF1b is associated with the Golgi apparatus (Lin et al., 2005).

1. Enriched Golgi fractions were isolated by sucrose step gradient centrifugation from H460 cells. Cells were suspended in 0.5 *M* sucrose in PMG buffer (0.1 *M* potassium phosphate, 5 m*M* $MgCl_2$, 1 m*M* dithiothreitol, and protease inhibitors (Roche), pH 7.0), homogenized with a Dounce homogenizer on ice, and centrifuged at 1000 × *g* for 10 min at 4 °C to remove the nuclei and cell debris.
2. The supernate was diluted to 0.25 *M* sucrose in PMG buffer, underlayered with 0.5 *M*, 0.86 *M*, and 1.3 *M* sucrose in PMG, and centrifuged at 25,000 rpm in a SW28 rotor for 2 h at 4 °C. The interfaces between 0.5 *M*/0.86 *M* sucrose-PMG, enriched for Golgi proteins, and the interface between 0.86 *M*/1.3 *M* sucrose-PMG, enriched for ER proteins, were isolated.
3. The two collected fractions were resolved on SDS-PAGE gels, transferred onto PVDF membranes, blocked with 5% milk in PBS-Tween (0.1%), and probed with antibodies against MACF1b and a secondary antibody conjugate. The supernates were similarly processed as controls. The MACF1b protein was visualized using enhanced chemiluminescence. MACF1b was detected in the Golgi-enriched fraction but not in the ER fraction. It was also shown by immunofluorescent staining of transfected cells that the MACF1b association with Golgi membranes was mediated by the two most N-terminal PRDs (Lin et al., 2005).

4. TISSUE-SPECIFIC FUNCTIONS OF MACF1 IN MAMMALS

In mammals, MACF1 is expressed ubiquitously both in the embryo and in the adult (Gong et al., 2001; Leung, Sun, Zheng, et al., 1999). Given its many functions and wide-spread expression pattern, it was no surprise that inactivation of the *Macf1* gene in mice by standard gene-targeting techniques resulted in embryonic lethality at the gastrulation stage (Chen et al., 2006; Kodama et al., 2003). The early developmental defect was shown to be related to the Wnt/β-catenin signaling pathway. In cultured wild-type cells, MACF1 associated with a complex containing several classic Wnt signaling molecules, including axin, APC, β-catenin, and GSK3β (Chen et al., 2006). MACF1 may be necessary for the translocation of axin, β-catenin, and GSK3β to the plasma membrane in response to Wnt activation of frizzled/LRP5/6 and for the subsequent degradation of axin (Chen et al., 2006). Another mechanism that may cause the embryonic lethal phenotype is likely related to the MACF1 role in the regulation of MT growth and stabilization at the plasma membrane. Primary cells derived from *Macf1* knockout embryos contained relatively unstable and excessively long MTs with skewed cytoplasmic trajectories (Kodama et al., 2003).

4.1 Skin

To investigate tissue-specific functions of MACF1, several conditional Cre-loxP knockout mice have been generated (Fassett et al., 2013; Goryunov et al., 2010; Ka et al., 2014; Liang et al., 2013; Wu et al., 2008). Targeting the *Macf1* gene in skin epidermis with a *Cre* transgene controlled by a keratin-14 promoter resulted in overtly normal animals. Their skin, however, was more susceptible to mechanical injury compared to that of wild-type mice and showed slowed wound healing (Wu et al., 2008). Keratinocytes derived from the mutant mice showed deficiency in MT tracking along F-actin to FAs, decreased FA-actin dynamics, and impaired epidermal migration. Protein constructs that contained only the ABD and the MTBD (including the EB1-binding site) were not sufficient to rescue the defects in FA-cytoskeletal dynamics and migration in these cells (Wu et al., 2008). A cryptic ATPase motif in MACF1 may be required for complete rescue. The intrinsic ATPase activity of MACF1 was stimulated by F-actin, suggesting that MACF1 may not only target MTs to FAs but could also mediate delivery of factors required for FA turnover, such as dynamin, along the MTs aligned with F-actin (Ezratty, Partridge, & Gundersen, 2005; Kaverina, Rottner, & Small, 1998). In keratinocytes and perhaps other migrating cells, MACF1 may therefore function as the functional equivalent of the yeast EB1–Kar9–Myo2 complex (Wu et al., 2008).

4.2 Nervous System

MACF1 is expressed at particularly high levels throughout the nervous system (Bernier et al., 1996; Gong et al., 2001; Leung, Sun, Zheng, et al., 1999). We targeted the *Macf1* gene in the entire developing nervous system with a *Cre* transgene under the control of the nestin promoter (Goryunov et al., 2010). The mutant mice died within 24–36 h after birth and showed a plethora of brain developmental abnormalities indicative of defects in neuronal migration and axonal extension. We observed a disorganized cerebral cortex with a mixed layer structure and pyramidal neuron heterotopia (misplacement) in the hippocampus. We also found thalamocortical and corticofugal fibers to be disorganized and the anterior and hippocampal commissures to be aplastic or missing entirely. Although the mutant brains contained no MACF1a and MACF1b proteins, they did express an apparent novel isoform identical to MACF1a but missing the ABD. Fortuitously, this finding provided direct evidence that the ability to cross-link MTs and actin filaments is critical for MACF1 role in neuronal development, likely reflecting its involvement in both neuronal cell migration and process outgrowth (Goryunov et al., 2010; Sanchez-Soriano et al., 2009; Wu et al., 2008). Although the nestin promoter drives Cre expression in the peripheral as well as central nervous systems (PNS and CNS), we did not detect any apparent peripheral abnormalities in the mutant mice. It is possible that MACF1 involvement in the function of the PNS is less critical in the early development or else the neuronal isoform of BPAG1 may play a bigger role in compensating for MACF1 in the PNS.

A recent study based on conditional *Macf1* inactivation specifically in neuronal cells further showed that the role of MACF1 in neuronal migration during brain development was cell-autonomous to neurons rather than radial glia or neuronal progenitor cells (Ka et al., 2014). This was achieved using two approaches. First, *in utero* electroporation was performed on *Macf1*$^{loxP/loxP}$ embryos with a construct that expressed Cre recombinase in neuronal populations but not in neural progenitors. The migration of MACF1-deficient pyramidal neurons both in the cortex and in the hippocampus was affected. Second, by crossing the *Macf1*$^{loxP/loxP}$ mouse with a mouse expressing Cre under the control of a nexin promoter, embryos were produced in which *Macf1* was inactivated exclusively in neurons of the dorsal telencephalon but not in dividing neural progenitors in the developing cerebral cortex. Similar to our findings, neuronal pyramidal layers were abnormally spaced in the *Macf1* nexin-Cre knockout cortex, suggesting that MACF1 plays a role in neuronal layer organization and migration (Goryunov et al., 2010; Ka et al., 2014). The defective migration of MACF1-deficient neurons was shown to be caused by MT destabilization and static centrosomes. Consistent with previous *in vitro* data, the role of MACF1 in correct positioning of migrating neurons was mediated by GSK-3 signaling (Chen et al., 2006; Ka et al., 2014; Wu et al., 2011).

Using a *Cre* transgene driven by a calmodulin kinase 2 promoter we also generated mice-lacking MACF1a/b in the pyramidal cell layer of the hippocampus, but in the adult rather than the embryo. These mice developed normally and were morphologically indistinguishable from their control or wild-type littermates. However, they displayed a pronounced deficiency in a behavioral fear conditioning test that requires hippocampus-mediated learning (contextual conditioning) but not in a similar test that is known to be hippocampus independent (cued conditioning) (Fig. 2; Fanselow & Poulos, 2005). In the contextual conditioning setup, each mouse was placed in a cage with a conductive metal-grid floor for 1 min, after which a mild yet painful electric shock was administered through the grid. The conditioning procedure was repeated five times. After 1 h, the mouse was returned to the cage but no shock was given. Instead, the fraction of time the mouse spent

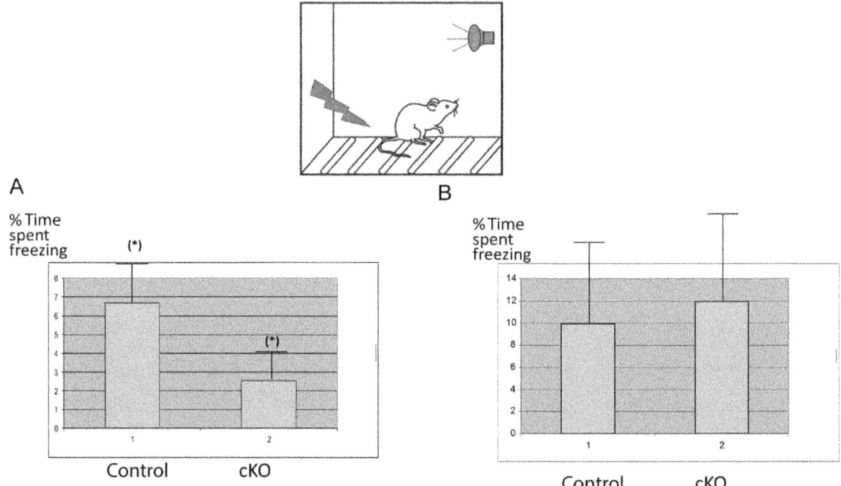

Figure 2 Elimination of MACF1a/b in the hippocampus causes deficits in contextual fear conditioning. In this experiment, a mouse is placed in a cage with a metal-grid floor. An aversive stimulus (mild electric shock) is administered with or without a preceding auditory signal (cue). In the first case, the environment of the cage itself is associated with the aversive stimulus (contextual conditioning). In the second case, it is the auditory stimulus that is associated with the shock (cued conditioning). The percentage of time a conditioned mouse spends "freezing" (staying motionless) after being exposed to the associative stimulus (placement back in the cage or the cue sound) is used as a measure of learning. *Macf1* CAMK-Cre knockout mice showed a significantly ($p=0.05$) decreased freezing time percentage in the hippocampus dependent, contextual test (A) but not in the hippocampus independent, cued conditioning test (B). At least five mice were tested in each condition. Significance was determined using the one-tailed *t*-test.

motionless (freezing), as if in anticipation of being shocked, was measured. Compared to control mice, *Macf1* mutant mice showed shorter "freezing" times, indicative of a learning deficit where the mouse fails to associate the metal-grid cage environment with the aversive stimulus (Fig. 2A). In the cued conditioning setup, each mouse was first pre-exposed to the metal-grid cage environment without electric shocking. It was then conditioned in a similar training procedure as the one used for contextual learning, except the shock was administered shortly after a loud tone sounded in the cage. After 1 h, the mouse was placed back in the cage, the tone sounded, and the "freezing" time fraction was measured. The mouse would spend more time motionless if it had learned to associate the sound with the aversive stimulus. No significant differences between control and mutant mice in the cued conditioning setup (Fig. 2B). These findings reveal a possible function of MACF1 in the processes of memory formation and consolidation in the hippocampus (Sanders, Wiltgen, & Fanselow, 2003).

4.3 Heart Muscle

The role of MACF1 in heart muscle has been studied in a conditional knockout mouse where Cre expression was controlled by a tamoxifen-inducible heart-specific myosin promoter (αMHC-MerCreMer) (Fassett et al., 2013). Under normal conditions, MACF1 expression in cardiomyocytes was dispensable for normal heart structure and function. However, under the stress conditions of aortic pressure overload, MACF1 expression in heart was increased, and tamoxifen-induced elimination of MACF1 significantly exacerbated the overload-induced left ventricle (LV) hypertrophy, dilation, and contractile dysfunction. Tubulin isolated from heart muscle of overload-stressed knockout animals redistributed into a Triton-soluble membrane fraction to a much larger extent compared to wild-type controls. Combined with a strong correlation with LV dysfunction, this finding suggests that MACF1 regulation of MT organization is essential for ventricular adaptation to hemodynamic overload (Fassett et al., 2013). The authors speculate that MACF1 expression in the heart is important for limiting aberrant MT-membrane interactions, which in turn can impair contractile function. *Macf1* disruption also resulted in reduction of membrane-associated caveolin 3 levels and increased the levels of membrane-associate PKCα and β1 integrin after pressure overload, suggesting MACF1 function is important for spatial regulation of signaling proteins in LV hypertrophy (Fassett et al., 2013).

4.4 Intestinal Epithelia

Macf1 has been inactivated in the intestinal mucosal epithelia of mice using a *Cre* transgene controlled by a villin promoter (Liang et al., 2013). MACF1 deficiency resulted in significant interstitial proliferation and columnar epithelial cell rearrangement. In addition, colonic paracellular permeability was found to be reduced compared to control intestines. Since the paracellular barrier in intestinal epithelia largely consists of tight junctions (TJ) between epithelial cells, the authors examined the expression levels of three major TJ components: ZO-1, occludin, and claudin-1. All three proteins were significantly downregulated in the mutant epithelia. It is therefore possible that MACF1 regulates cytoskeleton dynamics to alter mucosal epithelial arrangement and colonic paracellular permeability in part through disrupting TJ formation (Liang et al., 2013).

4.5 Human Disorders

Recently, the first instance of a *MACF1*-linked human disorder has been reported (Jorgensen et al., 2014). The patients, all members of a single family, were diagnosed with a novel type of myopathy characterized by periodic hypotonia, facial weakness, lax muscles, contractures, and muscle pains. The severity of the symptoms varied significantly among the affected individuals. No heart symptoms were evident, possibly because of the patients' inability to engage in strenuous physical activities. Routine laboratory tests and metabolic analyses revealed no abnormalities in any of the affected relatives. The cause of the disorder was traced to a duplication on chromosome 1p34.4, resulting in overall reduction of MACF1 protein levels. The patients showed no skeletal muscle abnormalities at the histological or immunohistochemical levels. However, membrane foldings were detected in the TJ membranes between endothelial cells as well as in membranes facing the muscle fibers. In the satellite cells, membrane foldings were mostly present in the cell periphery and appeared as collapsed or folded structures and membranous whorls (Jorgensen et al., 2014).

MACF1 mutations are likely to be involved in the pathogenesis of many other human diseases, as recent correlational studies have indicated. A multilayer-omics approach combining whole exome, methylome, and transcriptome analyses showed that genetic alterations of *MACF1* are associated with renal cell carcinomas (Arai et al., 2014). Genetic changes resulting in alternative splicing of *MACF1* exon 8 have been linked to non-small cell lung cancer as well as breast cancer (Misquitta-Ali et al.,

2011). *MACF1* has also been indirectly implicated in brain developmental disorders leading to autism and schizophrenia: MACF1 is part of the interactome of DISC1 (Disrupted in Schizophrenia 1), a protein mutations in which appear to be associated with susceptibility to these disorders (Camargo et al., 2007; Kamiya et al., 2005; Tomita, Kubo, Ishii, & Nakajima, 2011).

5. CONCLUSIONS

As an integrator of the cytoskeleton and multiple other cellular systems and processes, MACF1 is involved in a large variety of interactions. We have described some of the methods that can be used to study these interactions. It is important to remember that most of the methods are indirect or involve monomeric proteins as potential binding partners. Ultimately, functional *in vivo* studies should be performed to test the significance of any putative interactions identified in an *in vitro* system. Generation of conditional and inducible-conditional knockout mice is the definitive approach to verify the function of MACF1 in a particular mammalian cell type or tissue. As the number of such studies grows, the demonstrated repertoires of both physiological and pathological roles of MACF1 and other spectraplakins are bound to expand significantly.

REFERENCES

Antolik, C., Catino, D. H., O'Neill, A. M., Resneck, W. G., Ursitti, J. A., & Bloch, R. J. (2007). The actin binding domain of ACF7 binds directly to the tetratricopeptide repeat domains of rapsyn. *Neuroscience*, *145*(1), 56–65.

Antonellis, P. J., Pollock, L. M., Chou, S. W., Hassan, A., Geng, R., Chen, X., et al. (2014). ACF7 is a hair-bundle antecedent, positioned to integrate cuticular plate actin and somatic tubulin. *Journal of Neuroscience*, *34*(1), 305–312. http://dx.doi.org/10.1523/JNEUROSCI.1880-13.2014.

Arai, E., Sakamoto, H., Ichikawa, H., Totsuka, H., Chiku, S., Gotoh, M., et al. (2014). Multilayer-omics analysis of renal cell carcinoma, including the whole exome, methylome and transcriptome. *International Journal of Cancer*, *135*(6), 1330–1342. http://dx.doi.org/10.1002/ijc.28768.

Bernier, G., Mathieu, M., De Repentigny, Y., Vidal, S. M., & Kothary, R. (1996). Cloning and characterization of mouse ACF7, a novel member of the dystonin subfamily of actin binding proteins. *Genomics*, *38*(1), 19–29.

Bosher, J. M., Hahn, B. S., Legouis, R., Sookhareea, S., Weimer, R. M., Gansmuller, A., et al. (2003). The Caenorhabditis elegans vab-10 spectraplakin isoforms protect the epidermis against internal and external forces. *Journal of Cell Biology*, *161*(4), 757–768.

Boyer, J. G., Bernstein, M. A., & Boudreau-Lariviere, C. (2010). Plakins in striated muscle. *Muscle and Nerve*, *41*(3), 299–308. http://dx.doi.org/10.1002/mus.21472.

Burgo, A., Proux-Gillardeaux, V., Sotirakis, E., Bun, P., Casano, A., Verraes, A., et al. (2012). A molecular network for the transport of the TI-VAMP/VAMP7 vesicles from cell center to periphery. *Developmental Cell, 23*(1), 166–180. http://dx.doi.org/10.1016/j.devcel.2012.04.019.

Byers, T. J., Beggs, A. H., McNally, E. M., & Kunkel, L. M. (1995). Novel actin crosslinker superfamily member identified by a two step degenerate PCR procedure. *FEBS Letters, 368*(3), 500–504.

Camargo, L. M., Collura, V., Rain, J. C., Mizuguchi, K., Hermjakob, H., Kerrien, S., et al. (2007). Disrupted in Schizophrenia 1 Interactome: Evidence for the close connectivity of risk genes and a potential synaptic basis for schizophrenia. *Molecular Psychiatry, 12*(1), 74–86. http://dx.doi.org/10.1038/sj.mp.4001880.

Chen, H. J., Lin, C. M., Lin, C. S., Perez-Olle, R., Leung, C. L., & Liem, R. K. (2006). The role of microtubule actin cross-linking factor 1 (MACF1) in the Wnt signaling pathway. *Genes and Development, 20*(14), 1933–1945.

Dosch, R., Wagner, D. S., Mintzer, K. A., Runke, G., Wiemelt, A. P., & Mullins, M. C. (2004). Maternal control of vertebrate development before the midblastula transition: Mutants from the zebrafish I. *Developmental Cell, 6*(6), 771–780. http://dx.doi.org/10.1016/j.devcel.2004.05.002.

Drabek, K., van Ham, M., Stepanova, T., Draegestein, K., van Horssen, R., Sayas, C. L., et al. (2006). Role of CLASP2 in microtubule stabilization and the regulation of persistent motility. *Current Biology, 16*(22), 2259–2264.

Ezratty, E. J., Partridge, M. A., & Gundersen, G. G. (2005). Microtubule-induced focal adhesion disassembly is mediated by dynamin and focal adhesion kinase. *Nature Cell Biology*, 7(6), 581–590. http://dx.doi.org/10.1038/ncb1262.

Fanselow, M. S., & Poulos, A. M. (2005). The neuroscience of mammalian associative learning. *Annual Review of Psychology, 56*, 207–234. http://dx.doi.org/10.1146/annurev.psych.56.091103.070213.

Fassett, J. T., Xu, X., Kwak, D., Wang, H., Liu, X., Hu, X., et al. (2013). Microtubule actin cross-linking factor 1 regulates cardiomyocyte microtubule distribution and adaptation to hemodynamic overload. *PLoS One, 8*(9), e73887. http://dx.doi.org/10.1371/journal.pone.0073887.

Fields, S., & Song, O. (1989). A novel genetic system to detect protein–protein interactions. *Nature, 340*(6230), 245–246. http://dx.doi.org/10.1038/340245a0.

Gao, Y. S., Alvarez, C., Nelson, D. S., & Sztul, E. (1998). Molecular cloning, characterization, and dynamics of rat formiminotransferase cyclodeaminase, a Golgi-associated 58-kDa protein. *Journal of Biological Chemistry, 273*(50), 33825–33834.

Gong, T. W., Besirli, C. G., & Lomax, M. I. (2001). MACF1 gene structure: A hybrid of plectin and dystrophin. *Mammalian Genome, 12*(11), 852–861.

Goriounov, D., Leung, C. L., & Liem, R. K. (2003). Protein products of human Gas2-related genes on chromosomes 17 and 22 (hGAR17 and hGAR22) associate with both microfilaments and microtubules. *Journal of Cell Science, 116*(Pt. 6), 1045–1058.

Goryunov, D., He, C. Z., Lin, C. S., Leung, C. L., & Liem, R. K. (2010). Nervous-tissue-specific elimination of microtubule-actin crosslinking factor 1a results in multiple developmental defects in the mouse brain. *Molecular and Cellular Neuroscience, 44*(1), 1–14. http://dx.doi.org/10.1016/j.mcn.2010.01.010.

Green, K. J., & Gaudry, C. A. (2000). Are desmosomes more than tethers for intermediate filaments? *Nature Reviews Molecular Cell Biology, 1*(3), 208–216. http://dx.doi.org/10.1038/35043032.

Gupta, T., Marlow, F. L., Ferriola, D., Mackiewicz, K., Dapprich, J., Monos, D., et al. (2010). Microtubule actin crosslinking factor 1 regulates the Balbiani body and animal-vegetal polarity of the zebrafish oocyte. *PLoS Genetics, 6*(8), e1001073. http://dx.doi.org/10.1371/journal.pgen.1001073.

Janda, L., Damborsky, J., Rezniczek, G. A., & Wiche, G. (2001). Plectin repeats and modules: Strategic cysteines and their presumed impact on cytolinker functions. *BioEssays, 23*(11), 1064–1069. http://dx.doi.org/10.1002/bies.1151.

Jefferson, J. J., Ciatto, C., Shapiro, L., & Liem, R. K. (2007). Structural analysis of the plakin domain of bullous pemphigoid antigen1 (BPAG1) suggests that plakins are members of the spectrin superfamily. *Journal of Molecular Biology*, *366*(1), 244–257. http://dx.doi.org/10.1016/j.jmb.2006.11.036.

Jefferson, J. J., Leung, C. L., & Liem, R. K. (2004). Plakins: Goliaths that link cell junctions and the cytoskeleton. *Nature Reviews Molecular Cell Biology*, *5*(7), 542–553. http://dx.doi.org/10.1038/nrm1425.

Jefferson, J. J., Leung, C. L., & Liem, R. K. (2006). Dissecting the sequence specific functions of alternative N-terminal isoforms of mouse bullous pemphigoid antigen 1. *Experimental Cell Research*, *312*(15), 2712–2725. http://dx.doi.org/10.1016/j.yexcr.2006.04.025, S0014-4827(06)00169-8 [pii].

Jorgensen, L. H., Mosbech, M. B., Faergeman, N. J., Graakjaer, J., Jacobsen, S. V., & Schroder, H. D. (2014). Duplication in the microtubule-actin cross-linking factor 1 gene causes a novel neuromuscular condition. *Scientific Reports*, *4*, 5180. http://dx.doi.org/10.1038/srep05180.

Ka, M., Jung, E. M., Mueller, U., & Kim, W. Y. (2014). MACF1 regulates the migration of pyramidal neurons via microtubule dynamics and GSK-3 signaling. *Developmental Biology*, *395*(1), 4–18. http://dx.doi.org/10.1016/j.ydbio.2014.09.009.

Kakinuma, T., Ichikawa, H., Tsukada, Y., Nakamura, T., & Toh, B. H. (2004). Interaction between p230 and MACF1 is associated with transport of a glycosyl phosphatidyl inositol-anchored protein from the Golgi to the cell periphery. *Experimental Cell Research*, *298*(2), 388–398. http://dx.doi.org/10.1016/j.yexcr.2004.04.047.

Kamiya, A., Kubo, K., Tomoda, T., Takaki, M., Youn, R., Ozeki, Y., et al. (2005). A schizophrenia-associated mutation of DISC1 perturbs cerebral cortex development. *Nature Cell Biology*, 7(12), 1167–1178. http://dx.doi.org/10.1038/ncb1328.

Karakesisoglou, I., Yang, Y., & Fuchs, E. (2000). An epidermal plakin that integrates actin and microtubule networks at cellular junctions. *Journal of Cell Biology*, *149*(1), 195–208.

Kaverina, I., Rottner, K., & Small, J. V. (1998). Targeting, capture, and stabilization of microtubules at early focal adhesions. *Journal of Cell Biology*, *142*(1), 181–190.

Kodama, A., Karakesisoglou, I., Wong, E., Vaezi, A., & Fuchs, E. (2003). ACF7: An essential integrator of microtubule dynamics. *Cell*, *115*(3), 343–354.

Kumar, P., & Wittmann, T. (2012). +TIPs: SxIPping along microtubule ends. *Trends in Cell Biology*, *22*(8), 418–428. http://dx.doi.org/10.1016/j.tcb.2012.05.005.

Lapouge, K., Fontao, L., Champliaud, M. F., Jaunin, F., Frias, M. A., Favre, B., et al. (2006). New insights into the molecular basis of desmoplakin- and desmin-related cardiomyopathies. *Journal of Cell Science*, *119*(Pt. 23), 4974–4985. http://dx.doi.org/10.1242/jcs.03255.

Leung, C. L., Green, K. J., & Liem, R. K. (2002). Plakins: A family of versatile cytolinker proteins. *Trends in Cell Biology*, *12*(1), 37–45.

Leung, C. L., Sun, D., & Liem, R. K. (1999). The intermediate filament protein peripherin is the specific interaction partner of mouse BPAG1-n (dystonin) in neurons. *Journal of Cell Biology*, *144*(3), 435–446.

Leung, C. L., Sun, D., Zheng, M., Knowles, D. R., & Liem, R. K. (1999). Microtubule actin cross-linking factor (MACF): A hybrid of dystonin and dystrophin that can interact with the actin and microtubule cytoskeletons. *Journal of Cell Biology*, *147*(6), 1275–1286.

Leung, C. L., Zheng, M., Prater, S. M., & Liem, R. K. (2001). The BPAG1 locus: Alternative splicing produces multiple isoforms with distinct cytoskeletal linker domains, including predominant isoforms in neurons and muscles. *Journal of Cell Biology*, *154*(4), 691–697.

Liang, Y., Shi, C., Yang, J., Chen, H., Xia, Y., Zhang, P., et al. (2013). ACF7 regulates colonic permeability. *International Journal of Molecular Medicine*, *31*(4), 861–866. http://dx.doi.org/10.3892/ijmm.2013.1284.

Lin, C. M., Chen, H. J., Leung, C. L., Parry, D. A., & Liem, R. K. (2005). Microtubule actin crosslinking factor 1b: A novel plakin that localizes to the Golgi complex. *Journal of Cell Science*, *118*(Pt. 16), 3727–3738.

Margaron, Y., Fradet, N., & Cote, J. F. (2013). ELMO recruits actin cross-linking family 7 (ACF7) at the cell membrane for microtubule capture and stabilization of cellular protrusions. *Journal of Biological Chemistry*, *288*(2), 1184–1199. http://dx.doi.org/10.1074/jbc.M112.431825.

Meng, J. J., Bornslaeger, E. A., Green, K. J., Steinert, P. M., & Ip, W. (1997). Two-hybrid analysis reveals fundamental differences in direct interactions between desmoplakin and cell type-specific intermediate filaments. *Journal of Biological Chemistry*, *272*(34), 21495–21503.

Misquitta-Ali, C. M., Cheng, E., O'Hanlon, D., Liu, N., McGlade, C. J., Tsao, M. S., et al. (2011). Global profiling and molecular characterization of alternative splicing events misregulated in lung cancer. *Molecular and Cellular Biology*, *31*(1), 138–150. http://dx.doi.org/10.1128/MCB.00709-10.

Nikolic, B., Mac Nulty, E., Mir, B., & Wiche, G. (1996). Basic amino acid residue cluster within nuclear targeting sequence motif is essential for cytoplasmic plectin-vimentin network junctions. *Journal of Cell Biology*, *134*(6), 1455–1467.

Okuda, T., Matsuda, S., Nakatsugawa, S., Ichigotani, Y., Iwahashi, N., Takahashi, M., et al. (1999). Molecular cloning of macrophin, a human homologue of Drosophila kakapo with a close structural similarity to plectin and dystrophin. *Biochemical and Biophysical Research Communications*, *264*(2), 568–574. http://dx.doi.org/10.1006/bbrc.1999.1538.

Ortega, E., Buey, R. M., Sonnenberg, A., & de Pereda, J. M. (2011). The structure of the plakin domain of plectin reveals a non-canonical SH3 domain interacting with its fourth spectrin repeat. *Journal of Biological Chemistry*, *286*(14), 12429–12438. http://dx.doi.org/10.1074/jbc.M110.197467.

Poliakova, K., Adebola, A., Leung, C. L., Favre, B., Liem, R. K., Schepens, I., et al. (2014). BPAG1a and b associate with EB1 and EB3 and modulate vesicular transport, Golgi apparatus structure, and cell migration in C2.7 myoblasts. *PLoS One*, *9*(9), e107535. http://dx.doi.org/10.1371/journal.pone.0107535.

Rezniczek, G. A., de Pereda, J. M., Reipert, S., & Wiche, G. (1998). Linking integrin alpha6beta4-based cell adhesion to the intermediate filament cytoskeleton: Direct interaction between the beta4 subunit and plectin at multiple molecular sites. *Journal of Cell Biology*, *141*(1), 209–225.

Roper, K., & Brown, N. H. (2003). Maintaining epithelial integrity: A function for gigantic spectraplakin isoforms in adherens junctions. *Journal of Cell Biology*, *162*(7), 1305–1315. http://dx.doi.org/10.1083/jcb.200307089.

Ruhrberg, C., & Watt, F. M. (1997). The plakin family: Versatile organizers of cytoskeletal architecture. *Current Opinion in Genetics and Development*, 7(3), 392–397.

Sanchez-Soriano, N., Travis, M., Dajas-Bailador, F., Goncalves-Pimentel, C., Whitmarsh, A. J., & Prokop, A. (2009). Mouse ACF7 and Drosophila short stop modulate filopodia formation and microtubule organisation during neuronal growth. *Journal of Cell Science*, *122*(Pt. 14), 2534–2542. http://dx.doi.org/10.1242/jcs.046268.

Sanders, M. J., Wiltgen, B. J., & Fanselow, M. S. (2003). The place of the hippocampus in fear conditioning. *European Journal of Pharmacology*, *463*(1–3), 217–223.

Slep, K. C., Rogers, S. L., Elliott, S. L., Ohkura, H., Kolodziej, P. A., & Vale, R. D. (2005). Structural determinants for EB1-mediated recruitment of APC and spectraplakins to the microtubule plus end. *Journal of Cell Biology*, *168*(4), 587–598. http://dx.doi.org/10.1083/jcb.200410114.

Sonnenberg, A., & Liem, R. K. (2007). Plakins in development and disease. *Experimental Cell Research*, *313*(10), 2189–2203.

Sonnenberg, A., Rojas, A. M., & de Pereda, J. M. (2007). The structure of a tandem pair of spectrin repeats of plectin reveals a modular organization of the plakin domain. *Journal of Molecular Biology*, *368*(5), 1379–1391. http://dx.doi.org/10.1016/j.jmb.2007.02.090.

Speicher, D. W., & Marchesi, V. T. (1984). Erythrocyte spectrin is comprised of many homologous triple helical segments. *Nature*, *311*(5982), 177–180.

Stradal, T., Kranewitter, W., Winder, S. J., & Gimona, M. (1998). CH domains revisited. *FEBS Letters*, *431*(2), 134–137.

Subramanian, A., Prokop, A., Yamamoto, M., Sugimura, K., Uemura, T., Betschinger, J., et al. (2003). Shortstop recruits EB1/APC1 and promotes microtubule assembly at the muscle-tendon junction. *Current Biology*, *13*(13), 1086–1095.

Sun, D., Leung, C. L., & Liem, R. K. (2001). Characterization of the microtubule binding domain of microtubule actin crosslinking factor (MACF): Identification of a novel group of microtubule associated proteins. *Journal of Cell Science*, *114*(Pt. 1), 161–172.

Sun, Y., Zhang, J., Kraeft, S. K., Auclair, D., Chang, M. S., Liu, Y., et al. (1999). Molecular cloning and characterization of human trabeculin-alpha, a giant protein defining a new family of actin-binding proteins. *Journal of Biological Chemistry*, *274*(47), 33522–33530.

Suozzi, K. C., Wu, X., & Fuchs, E. (2012). Spectraplakins: Master orchestrators of cytoskeletal dynamics. *Journal of Cell Biology*, *197*(4), 465–475. http://dx.doi.org/10.1083/jcb.201112034.

Svitkina, T. M., Verkhovsky, A. B., & Borisy, G. G. (1996). Plectin sidearms mediate interaction of intermediate filaments with microtubules and other components of the cytoskeleton. *Journal of Cell Biology*, *135*(4), 991–1007.

Tomita, K., Kubo, K., Ishii, K., & Nakajima, K. (2011). Disrupted-in-Schizophrenia-1 (Disc1) is necessary for migration of the pyramidal neurons during mouse hippocampal development. *Human Molecular Genetics*, *20*(14), 2834–2845. http://dx.doi.org/10.1093/hmg/ddr194.

Wiche, G., Gromov, D., Donovan, A., Castanon, M. J., & Fuchs, E. (1993). Expression of plectin mutant cDNA in cultured cells indicates a role of COOH-terminal domain in intermediate filament association. *Journal of Cell Biology*, *121*(3), 607–619.

Wu, X., Kodama, A., & Fuchs, E. (2008). ACF7 regulates cytoskeletal-focal adhesion dynamics and migration and has ATPase activity. *Cell*, *135*(1), 137–148. http://dx.doi.org/10.1016/j.cell.2008.07.045, S0092-8674(08)01011-8 [pii].

Wu, X., Shen, Q. T., Oristian, D. S., Lu, C. P., Zheng, Q., Wang, H. W., et al. (2011). Skin stem cells orchestrate directional migration by regulating microtubule-ACF7 connections through GSK3beta. *Cell*, *144*(3), 341–352. http://dx.doi.org/10.1016/j.cell.2010.12.033.

Yang, Y., Dowling, J., Yu, Q. C., Kouklis, P., Cleveland, D. W., & Fuchs, E. (1996). An essential cytoskeletal linker protein connecting actin microfilaments to intermediate filaments. *Cell*, *86*(4), 655–665.

CHAPTER EIGHTEEN

Functional and Genetic Analysis of Neuronal Isoforms of BPAG1

Anisha Lynch-Godrei[*,†], **Rashmi Kothary**[*,†,‡,§,1]
[*]Regenerative Medicine Program, Ottawa Hospital Research Institute, Ottawa, Ontario, Canada
[†]Department of Cellular and Molecular Medicine, University of Ottawa, Ottawa, Ontario, Canada
[‡]Department of Medicine, University of Ottawa, Ottawa, Ontario, Canada
[§]University of Ottawa Center for Neuromuscular Disease, Ottawa, Ontario, Canada
[1]Corresponding author: e-mail address: rkothary@ohri.ca

Contents

Abstract

The neuronal isoforms of bullous pemphigoid antigen 1 (BPAG1, and also known as dystonin) are a group of large cytoskeletal linker proteins predominantly expressed in sensory neurons. The major neuronal isoforms consist of the spectraplakins (BPAG1/dystonin-a1, -a2, -a3), which have an N-terminus actin-binding domain and a C-terminus microtubule-binding domain. These proteins have crucial roles in cytoskeletal organization and stability, organelle integrity, and intracellular transport. BPAG1 loss-of-function in mice results in a lethal movement disorder known as *dystonia musculorum* (*dt*), which is likely caused by rapid sensory neuron degeneration. A human

Methods in Enzymology, Volume 569
ISSN 0076-6879
http://dx.doi.org/10.1016/bs.mie.2015.05.004

disease termed hereditary and sensory autonomic neuropathy type VI was also identified to be associated with mutations in the *BPAG1* gene (*DST*). This chapter provides an overview of the type of experiments used for analysis of the different isoforms of BPAG1.

1. INTRODUCTION

1.1 BPAG1 Isoform Diversity and Structure

The bullous pemphigoid antigen 1 (BPAG1) protein was first identified in epithelia as the autoantigen in the human skin blistering disease known as bullous pemphigoid (Mueller, Klaus-Kovtun, & Stanley, 1989). This protein now referred to as BPAG1-e/dystonin-e, functions by tethering keratin intermediate filaments to hemidesmosomes within keratinocytes, thus playing an important role in maintaining structural integrity of the cell (Hopkinson & Jones, 2000; Mutasim et al., 1985). Since the discovery of BPAG1-e, several other isoforms with differential tissue expression and alternative splicing patterns have emerged (Leung, Zheng, Prater, & Liem, 2001). These include the muscle isoforms (BPAG1-b/dystonin-b) and neuronal isoforms (BPAG1-a/dystonin-a).

Historically, initial neuronal isoform variants had been described (BPAG1n1, BPAG1n2, BPAG1n3). However, whether these are true prominent isoforms and their importance is not entirely clear. Subsequently, the spectraplakins were described (BPAG1n4/-a1, -a2, -a3 also known as dystonin-a1, -a2, and -a3—hereon referred to as dystonin to easily distinguish between neuronal isoforms). These latter isoforms have been shown to be the predominant forms. The dystonin neuronal isoforms are cytoskeletal linker proteins with crucial roles in maintaining cellular shape and function. Each of the BPAG1n isoforms are believed to be derived from the same N-termini exons as the dystonin isoforms (Fig. 1; Suozzi, Wu, & Fuchs, 2012). BPAG1n1 and dystonin-a1 have an N-terminus actin-binding domain composed of two calponin homology repeats (Jefferson, Leung, & Liem, 2006). BPAG1n2 and dystonin-a2 also possess this actin-binding domain, but dystonin-a2's N-terminus is unique in that it is preceded by a transmembrane domain (Young, Pinheiro, & Kothary, 2006). Finally, the N-terminus of BPAG1n3 and dystonin-a3 isoforms has a myristoylation motif that lies upstream from a single calponin homology repeat (Fig. 1; Jefferson et al., 2006). Although all isoforms have a plakin domain just downstream from the calponin homology repeats, the main difference between the BPAG1n

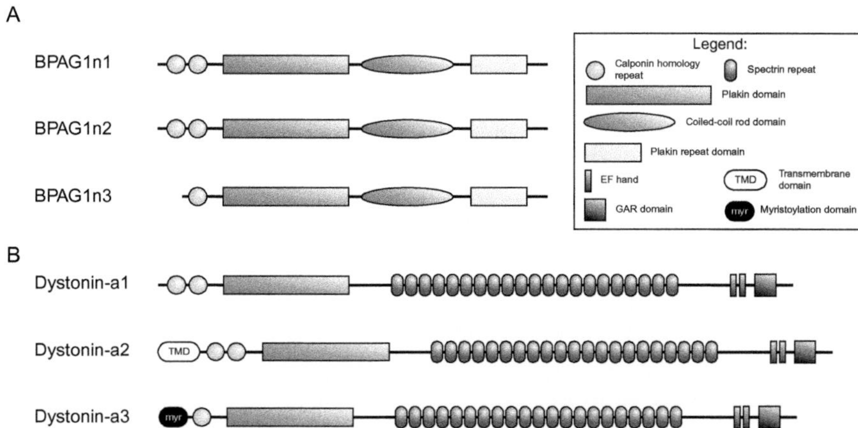

Figure 1 Schematic representation of the BPAG1 neuronal isoforms. (A) The putative neuronal BPAG1 plakin proteins consist of BPAG1n1, BPAG1n2, and BPAG1n3. BPAG1n1 and BPAG1n2 were identified as having actin-binding domains consisting of two calponin homology repeats at the N-terminus. All three isoforms were suggested to have a plakin repeat domain at their C-terminus, which permits intermediate filament binding. (B) The neuronal BPAG1 spectraplakin proteins are the more prominent isoforms. They are much larger in size to the BPAG1n isoforms and consist of dystonin-a1, dystonin-a2, and dystonin-a3. The neuronal dystonin proteins also differ from the BPAG1n isoforms in that they possess a functional microtubule-binding domain at their C-termini. Like BPAG1n1, dystonin-a1 has an actin-binding domain consisting of two calponin homology repeats. Dystonin-a2, however, has a transmembrane domain (TMD) that precedes its actin-binding domain, which allows it to localize to the perinuclear membranes. Unique to dystonin-a3 is an N-terminus myristoylation motif (myr), which is believed to permit localization to the plasma membrane.

and the dystonin isoforms lies in the composition of their C-terminus. BPAG1n isoforms have a coiled-coil rod domain (also known as the plectin repeat domain), followed by a plakin repeat domain (Fig. 1A). It is this plakin repeat domain that confers the ability of BPAG1n isoforms to interact with intermediate filaments (Yang et al., 1996). Dystonin isoforms on the other hand have a large spectrin repeat domain downstream from the plakin domain (Fig. 1B). This domain is believed to act as a spacer to separate the different functional domains of the N- and C-termini (Suozzi et al., 2012). The C-terminus of dystonin has a microtubule-binding domain made up of two EF-hand motifs and a growth-arrest specific 2 protein-related region (GAR) (Fig. 1B; Leung et al., 2001).

Evidently, the BPAG1n isoforms interact with intermediate filaments and have roles in cytoskeletal organization, such as neurofilament organization within sensory neurons (Yang et al., 1999). The dystonin isoforms are predominantly expressed in sensory neurons and have important roles in cytoskeletal organization, organelle integrity, and intracellular transport (Dalpé et al., 1998; Liu et al., 2003; Ryan, Ferrier, & Kothary, 2012; Ryan, Ferrier, Sato, et al., 2012). BPAG1/dystonin loss-of-function in mice results in a mutant phenotype known as *dystonia musculorum* (*dt*). It is a movement disorder characterized by rapid sensory neuron degeneration and death occurring around 3 weeks of age (Duchen, Strich, & Falconer, 1964). In 2012, a frameshift mutation in the human *BPAG1* gene (*DST*) was discovered in four infants (Edvardson et al., 2012). These individuals presented with a disease similar to murine *dt*, having symptoms such as joint contractures, autonomic irregularities, and a shortened lifespan (all succumbed to the disease before the age of 2). The human disease was termed hereditary sensory and autonomic neuropathy type VI (HSAN-VI), making it the most severe disease within this class of peripheral nervous system disorders (Rotthier, Baets, Timmerman, & Janssens, 2012). The recent discovery of HSAN-VI has increased the drive to elucidate the unique roles of the BPAG1 isoforms, which should provide novel insight into the etiology of the human disease.

1.2 Challenges of Studying BPAG1

In mice, the BPAG1 isoforms are encoded by a ~400 kb gene on chromosome 1, designated *Dst*. As the gene is composed of 107 exons (Pool, Larivière, Bernier, Young, & Kothary, 2005), the neuronal isoforms that it yields are substantially large (344–615 kD) (Leung et al., 2001). The large size of these proteins along with the high degree of similarity in amino acid sequence between isoforms makes them incredibly challenging to study. Isoform-specific antibodies are presently not available, which means that we had to develop alternate methods for investigating the unique BPAG1 isoforms. Much of the knowledge we have gained about the individual isoforms has come from isoform loss-of-function experiments, and through the use of fusion protein constructs containing different functional domains from various isoforms. Here, we describe some of the experimental techniques we have developed to study the neuronal BPAG1 isoforms.

2. MOUSE MODELS

2.1 *dt* Mice

The murine disease *dystonia musculorum* (*dt*) is a result of autosomal recessive mutations in the *Dst* gene. Most of the *Dst* mutations have arisen by spontaneous events, which is not surprising given the large size of the *Dst* gene. Others have come about by transgene insertion, chemical induction, and gene knockout (Table 1). Of these, only three have been characterized at the molecular level (Dst^{Tg4}, $Dst^{dt\text{-}Alb}$, and Dst^{tm1EFu}). In our laboratory, the primary mutant alleles used to study neuronal BPAG1 isoforms are Dst^{Tg4} and $Dst^{dt\text{-}27J}$. Dst^{Tg4} (also referred to as dt^{Tg4}) arose due to the insertion of a multicopy hsp68 promoter-*LacZ* transgene at the 5′ end of *Dst*, interrupting expression of dystonin-a1 and -a2 (Kothary et al., 1988). In contrast, the exact genetic mutation underlying the $Dst^{dt\text{-}27J}$ (also referred to as dt^{27J}) allele is unknown. However, evidence suggests that this mutation results in the disruption of all neuronal BPAG1 isoforms in this mouse (Bernier et al., 1995; Pool et al., 2005).

Initially, *dt* mice do not show any overt phenotype and are indistinguishable from their wild-type and heterozygous littermates. Between postnatal days 7–10, *dt* mice begin to display symptoms such as hindlimb clasping when picked up by the tail. Phenotype rapidly becomes more apparent as *dt* mice typically remain smaller in size, and exhibit symptoms such as uncoordinated movements, writhing and twisting of the trunk, as well as hyperflexion and pronation of the paws. Average lifespan of *dt* mice is about 3 weeks, however some differences in lifespan are observed between the different alleles. $Dst^{dt\text{-}27J}$ mice display a more severe phenotype and have a shorter lifespan (live to about P18–19) than other strains of *dt* mice. It is uncertain as to whether these differences are related to the mutation, or from the different genetic backgrounds of the *dt* mice ($Dst^{dt\text{-}27J}$ are on a C57BL/6 background, while Dst^{Tg4} mice have a CD1 background).

Table 1 The various *Dst* alleles

Mutation Type	Allele
Spontaneous mutation	$Dst^{dt\text{-}22J}$, $Dst^{dt\text{-}23J}$, $Dst^{dt\text{-}24J}$, $Dst^{dt\text{-}25J}$, $Dst^{dt\text{-}26J}$, $Dst^{dt\text{-}27J}$ (dt^{27J}), $Dst^{dt\text{-}29J}$, $Dst^{dt\text{-}30J}$, $Dst^{dt\text{-}31J}$, $Dst^{dt\text{-}32J}$, $Dst^{dt\text{-}35J}$, $Dst^{dt\text{-}39J}$, $Dst^{dt\text{-}Alb}$ (dt^{Alb}), $Dst^{dt\text{-}J}$, and Dst^{dt}
Chemically-induced mutation	$Dst^{dt\text{-}36J}$, $Dst^{dt\text{-}37J}$, and $Dst^{dt\text{-}38J}$
Targeted mutation	Dst^{tm1EFu} (dt^{tm1EFu}, *BPAG1* null)
Transgene insertion	Dst^{Tg4} (dt^{Tg4})

Information obtained from Mouse Genome Database (2015).

2.2 Genotyping

Polymerase chain reaction (PCR) can be used to differentiate mutant *Dst*^{Tg4} and mutant *Dst*$^{dt\text{-}27J}$ mice from their wild-type and heterozygous littermates.

1. Collect tail biopsies of about 3–5 mm and place in labeled microcentrifuge tubes.
2. Place 460 μl of tissue lysis buffer (0.1 *M* Tris–HCl at pH 8.5, 5 m*M* EDTA at pH 8.5, 0.2 *M* NaCl, and 2 mg/ml SDS dissolved in water) and 40 μl of 20.8 mg/ml Proteinase K to each tube. Briefly vortex and incubate tubes in a 54 °C heating block overnight.
3. The next morning, briefly vortex and transfer tubes to a ventilated fume hood.
4. Add 150 μl of chloroform, then 150 μl of UltraPure™ Buffer-Saturated Phenol (Life Technologies, Canada).
5. Shake tubes, then microfuge at 10,000 rpm for 10 min.
6. Using a wide bore tip, carefully remove the soluble upper fraction (~500 μl) and transfer to a new labeled centrifuge tube.
7. Add 500 μl of isopropanol and invert tubes until DNA precipitates out.
8. Pellet the DNA by centrifuging at 10,000 rpm for 10 min.
9. Discard the supernatant. Keep the tubes inverted and let air-dry for 15–20 min.
10. Add 100 μl of TE buffer (10 m*M* Tris–HCl and 1 m*M* EDTA) to each tube. Let it sit overnight to dissolve the DNA.

For *Dst*^{Tg4}, PCR must be performed in two separate tubes containing different primer sets, one to amplify the wild-type allele (~400 bp product) and another set to amplify the mutant allele containing the hsp68 promoter-*LacZ* transgene (~200 bp product).

Wild-type primers:

Forward: 5′-GGTGGTTCAGTGCCTTGATT-3′

Reverse: 5′-AGTAACAGCTCCCCCAACT-3′

***Dst*^{Tg4} primers:**

Forward: 5′-TTGGCCTGAACTGCCAGCTGGCGCAGG-3′

Reverse: 5′-TCCCGCAGCGCAGACCGTTTTCGCTCG-3′

To run the PCR for *Dst*^{Tg4}, each tube must contain: 1.5 μl of the DNA sample, 36.9 μl distilled water, 5 μl of 10 × PCR buffer, 0.2 m*M* dNTPs, 1.5 μl 50 m*M* $MgCl_2$, 0.2 μl Taq DNA Polymerase, and 0.2 μl of each 50 μ*M* wild-type primer in the first tube, and 0.2 μl of the 50 μ*M* *Dst*^{Tg4} primers in the second tube. The final volume before beginning the reaction should be 50.5 μl per tube. The reaction set up is as follows: (1) denature at 94 °C for 3 min, (2) denature at 94 °C for 1 min, (3) anneal at 50 °C for 1 min, (4) extend at 72 °C for 1 min, (5) Repeat steps 2–4 for 34 cycles, (6) finally extend at 72 °C for 10 min, and (7) hold at 4 °C until ready to run samples in a 2% agarose gel. By running each sample pair of PCR-amplified product side by side in the gel, the genotype of each mouse can be identified (Fig. 2A).

The $Dst^{dt\text{-}27J}$ PCR can be done in a single tube containing only one set of primers and is based on a polymorphism within the *Dst* gene. The wild-type allele will yield a product ~400 bp, while the $Dst^{dt\text{-}27J}$ allele will yield a product ~450 bp.

***Dst*$^{dt\text{-}27J}$ primers:**

Forward: 5′-GGATCTGCCCGACTTTCTGGG-3′

Reverse: 5′-CCAAGGTTCATTGCCTCCGTC-3′

To run the PCR for $Dst^{dt\text{-}27J}$, each tube must contain: 1.5 μl of the DNA sample, 16.8 μl distilled water, 2.5 μl of 10 × PCR buffer, 0.2 m*M* dNTPs, 1.5 μl 50 m*M* $MgCl_2$, 0.5 μl Taq DNA Polymerase, and 0.1 μl of each 50 μ*M* primer. The final volume before beginning the reaction should be 23 μl per tube. The reaction set up is as follows: (1) denature at 94 °C for 3 min, (2) denature at 94 °C for 45 s, (3) anneal at 58 °C for 45 s, (4) extend at 72 °C for 45 s, (5) repeat steps 2–4 for 29 cycles, (6) finally extend at 72 °C for 5 min, and (7) hold at 4 °C until ready to run samples in a 1.5% agarose gel. Genotype of each sample can be determined by analyzing the amplification products within each lane (Fig. 2B).

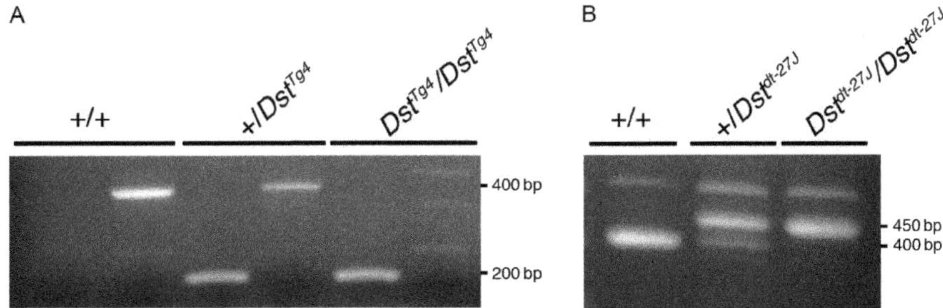

Figure 2 PCR genotyping of *Dst* mice. (A) Zygosity determination of Dst^{Tg4} mice by genomic PCR. Wild-type band is approximately 400 bp, while the Dst^{Tg4} band is approximately 200 bp. For each of these samples the left lane contains the Dst^{Tg4} PCR products, while the right lane contains the wild-type PCR products. (B) Zygosity determination of $Dst^{dt\text{-}27J}$ mice by genomic PCR. Wild-type band is approximately 400 bp, while the $Dst^{dt\text{-}27J}$ band is about 450 bp. Heterozygous mice will have both of these bands within the same lane. Since these bands are quite close together we emphasize adequate separation time to distinguish the different bands.

2.3 Breeding

Both Dst^{Tg4} and $Dst^{dt\text{-}27J}$ mouse lines are maintained in a barrier facility by the University of Ottawa Animal Care and Veterinary Service. New litters are genotyped at age P10, since this is the earliest stage at which an ear punching system can be employed to identify mice. To maintain the lines, male and female mice identified as heterozygous for the wild-type and mutant *Dst* alleles are bred according to standard protocol.

2.4 BPAG1 Rescue Mice

Recently, our laboratory generated a Dst^{Tg4} mouse model that has restored expression of dystonin-a2 in neurons. The transgenic mice exogenously express myc-his tagged dystonin-a2 under the control of the neuronal prion protein promoter (PrP) (Ferrier et al., 2014). Designated *PrP-dystonin-a2/PrP-dystonin-a2;*$dt^{Tg4/Tg4}$, these mice live much longer than Dst^{Tg4} mice with an average lifespan of 55 days and develop a milder phenotype. They also exhibit ameliorations in many of the pathological features known to be associated with *dt,* such as reduced sensory neuron death, increased number of myelinated sensory neuron axons, reduced muscle spindle degeneration, less fragmentation/dilation of Golgi and endoplasmic reticulum, as well as improved protein turnover, and autophagic flux (Ferrier et al., 2014). The lack of a complete rescue might be explained by insufficient expression of the transgene, as reverse transcription-qPCR analysis revealed much lower dystonin-a2 mRNA expression levels in *PrP-dystonin-a2/PrP-dystonin-a2;*$dt^{Tg4/Tg4}$ dorsal root ganglia (DRGs) tissue compared to wild type. It is also possible that the absence of other BPAG1 isoforms could be contributing to the lack of a full rescue (namely, dystonin-a1, BPAG1n1, or BPAG1n2).

The line is maintained as heterozygous *PrP-dystonin-a2/PrP-dystonin-a2;* $dt^{Tg4/+}$ mice in a barrier facility by the University of Ottawa Animal Care and Veterinary Service. Genotyping is only necessary to identify pups homozygous for the Dst^{Tg4} allele, as the parents are all homozygous for the *Prp-dystonin-a2* transgene (method described in Section 2.2).

3. CELL CULTURE SYSTEM

For the study of neuronal dystonin in primary cells, we recommend using sensory neurons derived from DRGs. This is due to the consistently high expression of the neuronal BPAG1 isoforms in sensory neurons

(Bernier et al., 1995; Dowling, Yang, Wollmann, Reichardt, & Fuchs, 1997), as well as its major involvement in the *dt* disease. It also seems that BPAG1 loss-of-function in sensory neurons is not compensated for by other members of the spectraplakin family (Leung et al., 2001), which makes this cell type ideal for investigating the specific cellular roles of BPAG1 isoforms.

3.1 Dorsal Root Ganglia Dissection

Typically, experiments on *dt* DRGs are conducted on prephenotype stage (P4–P5) or phenotype stage (P15–P18) mice. The steps to isolate DRGs are as follows (adapted from O'Meara, Ryan, Colognato, & Kothary, 2011):

1. Sacrifice mice according to institutional guidelines.
2. Extract the spine and place onto a clean Petri dish.
3. Remove as much of the tissue and bone surrounding the spine as possible, as this will later make extracting the DRGs easier.
4. Once complete, transfer the spine to a new Petri dish with its ventral side facing up and move the dish to a Leica MZ9.5 (Houston, Texas) dissecting microscope.
5. Using a scalpel, gently cut along the midline of the spine moving rostral to caudal (Fig. 3A). Be careful not to cut all the way through the spine; keeping the dorsal spine intact will facilitate DRG identification and extraction.
6. Using two sets of forceps, carefully open the spinal column to expose the spinal cord.
7. Before extracting DRGs from the left side of the spinal column, gently push the spinal cord over to the right side using closed forceps.

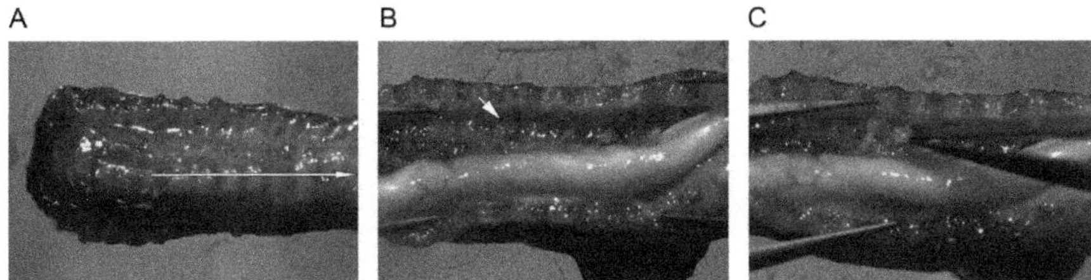

Figure 3 Dissection of dorsal root ganglia (DRG) in mice. (A) Spinal column, arrow indicates the location (midline) and direction (rostral to caudal) of incision. (B) Opened spinal column with the spinal cord pushed to the right side (bottom of the image) to expose the DRG of left side. Arrow points to a single DRG. (C) Removal of a single dorsal root ganglion. To preserve the shape of the DRG and keep cells intact, remove the DRG from behind by holding onto the roots as shown. (See the color plate.)

This will keep it out of the way, and prevent drying of DRGs on the opposite side.

8. DRGs can be identified as small translucent sphere-like structures that are found in small pockets between vertebrae on the dorsal side of the spinal column (Fig. 3B).
9. Using fine tipped forceps, extract the DRGs along the length of the left spinal column (Fig. 3C). Extracted DRGs can be placed in a pool of ice cold Hank's Balanced Salt Solution (HBSS; Gibco, Canada) located at the opposite end of the Petri dish used for dissecting. Note: cervical DRGs are very small and are surrounded by a dense fibrous network. Oftentimes removal of one cervical DRG will pull up others caught in this network.
10. Once DRGs are extracted from the left spinal column, push the spinal cord away from the right side of the spinal column and begin DRG extraction. Approximately 40 DRGs should be extracted in total.
11. If DRGs are not intended for culture of primary sensory neurons, the tissue can be collected from the HBSS and placed in a freezer safe 1.5 ml centrifuge tube, and subsequently flash frozen in liquid nitrogen. However, if the DRGs are intended for primary sensory neuron cultures, any excessively long roots must first be trimmed from the DRGs. This will help to reduce the number of contaminating cells in the culture (e.g., fibroblasts, glial cells). To trim the roots, lightly hold the root near the ganglion with a pair of forceps and cut adjacent to it using a scalpel.
12. Transfer all trimmed DRGs to a 1.5 ml centrifuge tube containing 1 ml of ice cold HBSS.

3.2 Culturing Primary Sensory Neurons

Once all dissections are complete, primary sensory neuron cultures can be established from the isolated DRGs by following these steps (adapted from O'Meara et al., 2011):

1. Centrifuge tubes at 4 °C for 5 min at 1200 rpm.
2. Transfer tubes to a sterile tissue culture hood.
3. Remove the HBSS and replace with 300 μl of 1.7 mg/ml papain (prewarmed to 37 °C). Incubate tubes in a 37 °C water bath for 10 min, shaking every 2 min to prevent aggregation.
4. Centrifuge at 4 °C for 5 min at 1200 rpm.
5. Transfer tubes to tissue culture hood and remove papain. Replace with 300 μl of 2 mg/ml Collagenase A (prewarmed to 37 °C). Incubate

tubes in a 37 °C water bath for 10 min, shaking every 2 min to prevent aggregation.

6. Centrifuge at 4 °C for 5 min at 1200 rpm.
7. In the tissue culture hood, remove the Collagenase A and replace with 1 ml DRG media (10% fetal bovine serum, 1% Penicillin/Streptomycin, in a Dulbecco's modified eagle medium (DMEM) base).
8. Centrifuge at 4 °C for 5 min at 1200 rpm.
9. Transfer tubes back into the tissue culture hood, removed the DRG media and replace with another 1300 μl of fresh DRG media. Invert the tube a few times.
10. Coat a flame-polished glass pasture pipette with 0.25% BSA dissolved in phosphate buffered saline (PBS). Using the BSA-coated pasture pipette, slowly triturate the digested DRGs, accelerating as the tissue starts dissociating.
11. Centrifuge at 4 °C for 5 min at 1200 rpm.
12. Carefully remove the supernatant off of the pellet, and resuspend in 1 ml fresh DRG media.
13. Using a hemocytometer, calculate the DRG neuron yield. DRG neurons can be identified as the large-bodied spherical cells.
14. For a 24-well plate, seed the DRG neurons at a density of 30,000–50,000 cells/well. Note: Plates and coverslips are to be coated with 4% LN2 (Human Merosin; Millipore, Canada) in 1 × PBS for at least 1 h, then wash three times with 1 × PBS. Fill wells with 1 ml DRG media and let equilibrate at 37 °C in an 8.5% CO_2 incubator.
15. After seeding, transfer plates to a 37 °C tissue culture incubator at 8.5% CO_2.
16. Early the next day, perform a full media change by replacing the DRG media with DRG differentiation media (1% Glutamax, 1% Penicillin/Streptomycin, 0.5% insulin, 0.5% fetal bovine serum, 0.15% holo-transferrin, 2% B27 supplement, 1% of 100 × Sato's supplement [40 μg/ml 3,3′,5-Triiodo-L-thyronine], 10.2 mg/ml bovine serum albumin, 6 μg/ml progesterone, 1.61 mg/ml putrescine, and 0.5 μg/ml sodium selenite in DMEM base) and 0.1% FuDR in a DMEM base.
17. On day 3, perform a 2/3 media change with the DRG differentiation media.
18. By day 5 the neurite bed should be quite extensive, cells can either be fixed for immunofluorescence analysis, or protein can be collected for immunoblot analysis at this stage. Cultures can be left for longer periods

of time to establish a denser neurite bed, with media changes occurring every other day (with the DRG differentiation media).

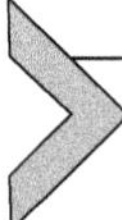

4. BPAG1 ISOFORM KNOCKDOWN IN IMMORTALIZED CELL LINES

One of the best tools we have to study the individual isoforms of dystonin is gene knockdown by means of siRNA molecules. We have previously been successful in knockdown of dystonin-a1 and dystonin-a2 in immortalized cell lines such as human embryonic kidney cells (HEK 293) and F11 cells (a hybrid between a mouse neuroblastoma cell line and rat embryonic DRG neurons). Although it would be ideal to carry out knockdown experiments on primary sensory neurons, this typically is not feasible, or is very challenging, as primary cells do not respond well to transfection agents such as Lipofectamine® 2000 (Life Technologies, Canada). The aforementioned cell lines are acceptable for studying how BPAG1 loss-of-function affects the cell, as they both express isoforms of BPAG1 and respond similar to *dt* primary sensory neurons under BPAG1 knockdown conditions. To perform knockdown of BPAG1 isoforms in HEK 293 cell lines, the following steps should be performed.

1. The day before transfection, make sure that cells are 85–90% confluent and replace media with an antibiotics-free HEK 293 T cell growth media (10% fetal bovine serum and 1% Glutamax in a DMEM base).
2. On the day of transfection, begin by making a 4% Lipofectamine® 2000 solution in OptiMEM® I Reduced Serum Medium (e.g., 10 μl Lipofectamine® 2000 for every 240 μl OptiMEM® I Reduced Serum Medium). Let this solution sit at room temperature for 5 min.
3. In a new tube, make a 60–80 n*M* siRNA solution in OptiMEM® I Reduced Serum Medium and let it sit for 5 min. Listed below are the siRNA sequences previously found to be effective for knockdown:
   Dystonin-a1 (GenBank accession number AF396878):
      Forward: 5′-ACAUGUACGUGGAGGAGCATT-3′
      Reverse: 5′-UGCUCCUCCACGUACAUGUAG-3′
   Dystonin-a2 (GenBank accession number DQ023311.2):
      Forward: 5′-CAAGCAUGAGAGAUCCAAATT-3′
      Reverse: 5′-UUUGGAUCUCUCAUGCUUGGG-3′
4. After the initial incubation period, combine equal parts of both solutions and let it sit at room temperature for 25 min to allow

siRNA–Lipofectamine 2000 complexes to form (final siRNA concentration should be 30–40 n*M*).

5. Add 100 μl of this solution for every 300 μl of media in the well. Let cells sit at 37 °C in a 5% CO_2 incubator for 5–6 h.
6. After the incubation period, remove the siRNA containing media from the cells and replace with the antibiotics-free HEK 293 cell growth media. Let cells continue to grow in the 37 °C/5% CO_2 incubator for 48 h before beginning gene knockdown assays.

5. BPAG1 EXPRESSION ANALYSIS

5.1 Reverse Transcription-qPCR

Reverse transcription-qPCR (RT-qPCR) is a great tool for studying tissue and developmental expression patterns of BPAG1 isoforms. Previously, isoform expression levels were assessed by RT-PCR but this lacked the necessary sensitivity and accuracy required for making biological comparisons.

1. Extract RNA from tissue/cells and reverse transcribe to obtain cDNA (we suggest cDNA synthesis using RT^2 First Strand Kit (Qiagen, USA)).
2. Initially, primers are dissolved in RNase-free water at a concentration of 100 μ*M*. Make up primer mixes for each isoform by combining 10 μl forward primer, 10 μl reverse primer, and 80 μl RNase-free water.
3. Reactions should be set up in triplicate for the BPAG1 isoform of interest, as well a reference gene such as actin. For the three different dystonin isoforms, the reaction mixes are the same with the exception of the different primer sets used. Each 25 μl reaction will be made up of 2× SsoFast™ EvaGreen® Supermix (Bio-Rad, Canada), 0.2 μl of the primer mix (0.8 μ*M* of each primer), RNase-free water, and a final concentration of cDNA between a range of 10–70 ng (amount of cDNA to be used is dependent on the expected BPAG1 expression level for the tissue/cell type in question). Note: we stress the importance of using SsoFast™ EvaGreen® Supermix for RT-qPCR experiments involving large amplicons such as the dystonin isoforms.

   The different primer sets are as follows:

   Dystonin-a1 primers (NCBI accession number NM_134448):

   Forward: 5′-CTACATGTACGTGGAGGAGCA-3′

   Reverse: 5′-CATCGTTTGCACCAATGCC-3′

   Dystonin-a2 primers (NCBI accession number NM_001276764):

   Forward: 5′-GAGGGCTGTGCTTCGGATAG-3′

   Reverse: 5′-CATCGTTTGCACCAATGCC-3′

Dystonin-a3 primers (NCBI accession number XM_006495669.1):
Forward: 5′-GTCTCCAAGGATGCACCTAGGGAT-3′
Reverse: 5′-CATCGTTTGCACCAATGCC-3′

4. Samples should first be spun down before loading into a real-time PCR machine. We use the CFX Connect™ Real-Time PCR detection system (Bio-Rad, Canada). The cycling parameters are also the same for each isoform: (1) denature at 95 °C for 10 min, (2) denature at 95 °C for 10 s, (3) anneal at 60 °C for 30 s, and (4) repeat steps 2–3 for 40 cycles.
5. Separation of the RT-qPCR products can be done by electrophoresis on a 2% agarose gel with RedSafe™ Nucleic Acid Staining Solution (FroggaBio, Canada). Products are expected to be of the following sizes: dystonin-a1—779 bp, dystonin-a2—741 bp, dystonin-a3—427 bp.

5.2 Immunofluorescence Staining for the myc-Tagged BPAG1a2

For the first time, the *PrP-dystonin-a2/PrP-dystonin-a2;dt*$^{Tg4/Tg4}$ mouse allowed for the intracellular detection of full-length dystonin-a2 protein. As predicted from previous work using fusion protein constructs, the dystonin-a2 isoform displays a perinuclear expression profile, which corresponds to its roles in the nuclear envelope, endoplasmic reticulum, and Golgi (Ferrier et al., 2014; Ryan, Bhanot, et al., 2012; Young & Kothary, 2008; Young et al., 2006). Here, we describe the staining protocol used for detection of the c-myc tag on dystonin-a2 in primary cultures of *PrP-dystonin-a2/PrP-dystonin-a2;dt*$^{Tg4/Tg4}$ sensory neurons.

1. In a ventilated fume hood, fix cells by removing half the volume of media in the well and replacing it with the same volume of 8% paraformaldehyde (PFA; final concentration 4% PFA).
2. PFA should remain in each well for no longer than 10 min. After that time period, wells should be washed with 1 × PBS three times for 5 min each.
3. Permeate the cells by placing a 0.4% Triton-X100 solution in the wells for 10 min, then wash with 1 × PBS three times for 5 min each.
4. Use 10% goat serum in 0.4% Triton-X100 in 1 × PBS to block for 30 min at room temperature.
5. In a 0.5% BSA and 0.4% Triton-X100 solution dissolved in 1X PBS, make up c-myc primary antibody dilutions according to the manufacturer's protocol. Let it sit in wells overnight in a hydrated chamber at 4 °C.
6. The following morning, wash wells 4 × with 1 × PBS for 5 min each.

7. Allow secondary antibodies diluted to 1:500 in 1 × PBS to sit in wells for 45 min to 1 h at room temperature. Cover plate to prevent exposure to light.
8. Remove secondary antibody solution from wells and replace with a 1:5000 Hoechst dilution in 1 × PBS. Let it sit for 5 min.
9. Wash twice more with 1 × PBS for 5 min each.
10. Coverslips can now be mounted onto slides using a fluorescent mounting media.

6. BPAG1 FUSION PROTEIN CONSTRUCTS

Due to the lack of isoform-specific antibodies, assessing the different intracellular localizations of the various BPAG1 isoforms is not feasible by immunofluorescence analysis. Therefore, fusion protein constructs have become an invaluable tool for assessing BPAG1 isoform distribution within cells. Through the use of various tagged constructs we have begun to elucidate the unique localization patterns of the neuronal BPAG1 isoforms and understand which domains are necessary and/or responsible for these expression patterns. Since neuronal BPAG1 isoforms vary most at their N-terminus, it is these N-termini domains that are most often used in fusion protein construct development. For example, our laboratory used a number of constructs containing an N-terminus FLAG tag to determine how the transmembrane domain of dystonin-a2 influences intracellular localization (Young et al., 2006). The vectors used for cloning the BPAG1 cDNA were chosen based on the type and location of the tag, and the cell type to be used for the experiment. In this case, the Stratagene (California, USA) pCMV-TAG2 vector was appropriate for transfection into the COS-1 cell line.

Although fusion proteins can be very useful, there are some disadvantages that come with their use. The main issue, especially in the case of studying BPAG1, is that vectors can only take up cDNA of a limited size. This can make it challenging to study the full-length BPAG1 protein. Also, the addition of a tag can interfere with the protein's function and intracellular localization.

It is important to keep this in mind and to use appropriate controls when carrying out this type of experiment.

7. CONCLUDING REMARKS

Over the past decade, we have made great advances in uncovering the isoform-specific roles of dystonin. Dystonin-a2 has been the center of attention as it seems to have the most unique role (not compensated for in sensory neurons), and since its absence results in the most persistent cellular defects (Ferrier et al., 2014; Ryan, Bhanot, et al., 2012; Ryan, Ferrier, Sato, et al., 2012). However, a better understanding of the other BPAG1 isoforms is necessary, especially since the discovery of the human disease HSAN-VI in which all BPAG1 isoforms are likely affected. One of the first steps toward this would be for the development of isoform-specific antibodies. It would allow for the confirmation of intracellular localization patterns and would permit such previously unfeasible experiments as immunoprecipitation, and colocalization assays. This would be especially beneficial for the identification of novel interacting partners of BPAG1.

Another major tool still missing in this area of research is the availability of a BPAG1 conditional knockout. However, there was a recent report of a reversible gene trapped *Dst* allele (named Dst^{Gt}) in mice of the C57BL/6 background (Horie et al., 2014). The gene trap targets the N-terminus actin-binding domain, thus effectively silencing BPAG1n-1 and -n2, as well as dystonin-a1 and -a2 isoforms. As we have previously observed in the dt^{Tg4} mice, loss of these isoforms is sufficient to cause the *dt* disease. The novelty of this Dst^{Gt} allele is the reversibility from mutant back to a fully functional gene based on treatment with either Cre or FLP recombinase. This new allele will be especially important for studying the role of BPAG1 isoforms during development.

Although we have been making progress toward uncovering the individual roles of the BPAG1 isoforms, there are still major hurdles to overcome. The work described in this review outlines the essential experiments used for studying the BPAG1 isoforms.

ACKNOWLEDGMENTS

This work was supported by grants awarded to R. K. from the Canadian Institutes of Health Research (MOP-126085). R. K. is also a recipient of a University of Ottawa Health Research Chair. A. L. G. is a recipient of a Frederick Banting and Charles Best Canada Graduate Scholarship at the Master's level.

REFERENCES

Bernier, G., Brown, A., Dalpé, G., De Repentigny, Y., Mathieu, M., & Kothary, R. (1995). Dystonin expression in the developing nervous system predominates in the neurons that degenerate in dystonia musculorum mutant mice. *Molecular and Cellular Neurosciences*, *6*(6), 509–520.

Dalpé, G., Leclerc, N., Vallée, A., Messer, A., Mathieu, M., De Repentigny, Y., et al. (1998). Dystonin is essential for maintaining neuronal cytoskeleton organization. *Molecular and Cellular Neurosciences*, *10*(5–6), 243–257.

Dowling, J., Yang, Y., Wollmann, R., Reichardt, L. F., & Fuchs, E. (1997). Developmental expression of BPAG1-n: Insights into the spastic ataxia and gross neurologic degeneration in dystonia musculorum mice. *Developmental Biology, 187*, 131–142. http://dx.doi.org/10.1006/dbio.1997.8567.

Duchen, L. W., Strich, S. J., & Falconer, D. S. (1964). Clinical and pathological studies of an hereditary neuropathy in mice (dystonia musculorum). *Brain, 87*, 367–378. http://dx.doi.org/10.1093/brain/87.2.367.

Edvardson, S., Cinnamon, Y., Jalas, C., Shaag, A., Maayan, C., Axelrod, F. B., et al. (2012). Hereditary sensory autonomic neuropathy caused by a mutation in dystonin. *Annals of Neurology, 71*(4), 569–572.

Ferrier, A., Sato, T., De Repentigny, Y., Gibeault, S., Bhanot, K., O'Meara, R. W., et al. (2014). Transgenic expression of neuronal dystonin isoform 2 partially rescues the disease phenotype of the Dystonia musculorum mouse model of hereditary sensory autonomic neuropathy VI. *Human Molecular Genetics, 23*(10), 2694–2710.

Hopkinson, S. B., & Jones, J. C. (2000). The N terminus of the transmembrane protein BP180 interacts with the N-terminal domain of BP230, thereby mediating keratin cytoskeleton anchorage to the cell surface at the site of the hemidesmosome. *Molecular Biology of the Cell, 11*, 277–286.

Horie, M., Watanabe, K., Bepari, A. K., Nashimoto, J.-I., Araki, K., Sano, H., et al. (2014). Disruption of actin-binding domain-containing Dystonin protein causes dystonia musculorum in mice. *The European Journal of Neuroscience, 40*(10), 3458–3471. http://dx.doi.org/10.1111/ejn.12711.

Jefferson, J. J., Leung, C. L., & Liem, R. K. H. (2006). Dissecting the sequence specific functions of alternative N-terminal isoforms of mouse bullous pemphigoid antigen 1. *Experimental Cell Research, 312*(15), 2712–2725. http://dx.doi.org/10.1016/j.yexcr.2006.04.025.

Kothary, R., Clapoff, S., Brown, A., Campbell, R., Peterson, A., & Rossant, J. (1988). A transgene containing lacZ inserted into the dystonia locus is expressed in neural tube. *Nature, 335*(6189), 435–437.

Leung, C. L., Zheng, M., Prater, S. M., & Liem, R. K. H. (2001). The BPAG1 locus: Alternative splicing produces multiple isoforms with distinct cytoskeletal linker domains, including predominant isoforms in neurons and muscles. *Journal of Cell Biology, 154*(4), 691–697.

Liu, J. J., Ding, J., Kowal, A. S., Nardine, T., Allen, E., Delcroix, J. D., et al. (2003). BPAG1n4 is essential for retrograde axonal transport in sensory neurons. *Journal of Cell Biology, 163*(2), 223–229.

Mouse Genome Database (MGD) at the Mouse Genome Informatics Website. (2015). http://www.informatics.jax.org/marker/MGI:104627 Accessed 16 February 2015.

Mueller, S., Klaus-Kovtun, V., & Stanley, J. R. (1989). A 230-kD basic protein is the major bullous pemphigoid antigen. *The Journal of Investigative Dermatology, 92*(1), 33–38.

Mutasim, D. F., Takahashi, Y., Labib, R. S., Anhalt, G. J., Patel, H. P., & Diaz, L. A. (1985). A pool of bullous pemphigoid antigen(s) is intracellular and associated with the basal cell cytoskeleton-hemidesmosome complex. *The Journal of Investigative Dermatology, 84*(1), 47–53.

O'Meara, R. W., Ryan, S. D., Colognato, H., & Kothary, R. (2011). Derivation of enriched oligodendrocyte cultures and oligodendrocyte/neuron myelinating co-cultures from post-natal murine tissues. *Journal of Visualized Experiments, 54*, http://dx.doi.org/10.3791/3324.

Pool, M., Larivière, C. B., Bernier, G., Young, K. G., & Kothary, R. (2005). Genetic alterations at the Bpag1 locus in dt mice and their impact on transcript expression. *Mammalian Genome, 16*(12), 909–917.

Rotthier, A., Baets, J., Timmerman, V., & Janssens, K. (2012). Mechanisms of disease in hereditary sensory and autonomic neuropathies. *Nature Reviews Neurology, 8*(2), 73–85. http://dx.doi.org/10.1038/nrneurol.2011.227.

Ryan, S. D., Bhanot, K., Ferrier, A., De Repentigny, Y., Chu, A., Blais, A., et al. (2012). Microtubule stability, Golgi organization, and transport flux require dystonin-a2-MAP1B interaction. *Journal of Cell Biology, 196*(6), 727–742.

Ryan, S. D., Ferrier, A., & Kothary, R. (2012). A novel role for the cytoskeletal linker protein dystonin in the maintenance of microtubule stability and the regulation of ER-Golgi transport. *BioArchitecture, 2,* 2–5.

Ryan, S. D., Ferrier, A., Sato, T., O'Meara, R. W., De Repentigny, Y., Jiang, S. X., et al. (2012). Neuronal dystonin isoform 2 is a mediator of endoplasmic reticulum structure and function. *Molecular Biology of the Cell, 23,* 553–566.

Suozzi, K. C., Wu, X., & Fuchs, E. (2012). Spectraplakins: Master orchestrators of cytoskeletal dynamics. *Journal of Cell Biology, 197,* 465–475.

Yang, Y., Bauer, C., Strasser, G., Wollman, R., Julien, J. P., & Fuchs, E. (1999). Integrators of the cytoskeleton that stabilize microtubules. *Cell, 98*(2), 229–238.

Yang, Y., Dowling, J., Yu, Q. C., Kouklis, P., Cleveland, D. W., & Fuchs, E. (1996). An essential cytoskeletal linker protein connecting actin microfilaments to intermediate filaments. *Cell, 86,* 655–665. http://dx.doi.org/10.1016/S0092-8674(00)80138-5.

Young, K. G., & Kothary, R. (2008). Dystonin/Bpag1 is a necessary endoplasmic reticulum/nuclear envelope protein in sensory neurons. *Experimental Cell Research, 314*(15), 2750–2761.

Young, K. G., Pinheiro, B., & Kothary, R. (2006). A Bpag1 isoform involved in cytoskeletal organization surrounding the nucleus. *Experimental Cell Research, 312*(2), 121–134.

CHAPTER NINETEEN

Functional and Genetic Analysis of Spectraplakins in *Drosophila*

Ines Hahn*, Matthew Ronshaugen*, Natalia Sánchez-Soriano†, Andreas Prokop*,[1]

*Faculty of Life Sciences, Michael Smith Building, Manchester, United Kingdom
†Cellular and Molecular Physiology, Institute of Translational Medicine, University of Liverpool, Liverpool, United Kingdom
[1]Corresponding author: e-mail address: andreas.prokop@manchester.ac.uk

Contents

Abstract

The cytoskeleton is a dynamic network of filamentous protein polymers required for virtually all cellular processes. It consists of three major classes, filamentous actin (F-actin), intermediate filaments, and microtubules, all displaying characteristic structural properties, functions, cellular distributions, and sets of interacting regulatory proteins. One unique class of proteins, the spectraplakins, bind, regulate, and integrate the functions of all three classes of cytoskeleton proteins. Spectraplakins are giant, evolutionary conserved multidomain proteins (spanning up to 9000 aa) that are true

Methods in Enzymology, Volume 569
ISSN 0076-6879
http://dx.doi.org/10.1016/bs.mie.2015.06.022

members of the plakin, spectrin, and Gas2-like protein families. They have OMIM-listed disease links to epidermolysis bullosa and hereditary sensory and autonomic neuropathy. Their role in disease is likely underrepresented since studies in model animal systems have revealed critical roles in polarity, morphogenesis, differentiation and maintenance, migration, signaling, and intracellular trafficking in a variety of tissues. This enormous diversity of spectraplakin function is consistent with the numerous isoforms produced from single genomic loci that combine different sets of functional domains in distinct cellular contexts. To study the broad range of functions and complexity of these proteins, *Drosophila* is a powerful model. Thus, the fly spectraplakin Short stop (Shot) acts as an actin–microtubule linker and plays important roles in many developmental processes, which provide experimentally amenable and relevant contexts in which to study spectraplakin functions. For these studies, a versatile range of relevant experimental resources that facilitate genetics and transgenic approaches, highly refined genomics tools, and an impressive set of spectraplakin-specific genetic and molecular tools are readily available. Here, we use the example of Shot to illustrate how the various tools and strategies available for *Drosophila* can be employed to decipher and dissect cellular roles and molecular mechanisms of spectraplakins.

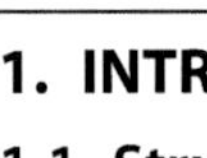

1. INTRODUCTION

1.1 Structure, Roles, and Disease Links of Spectraplakins

Spectraplakins are large proteins (up to 9000 aa, >400 nm) that are able to interact with all three classes of cytoskeletal filaments and link them either to each other or to other cellular structures, such as adhesion complexes. The best studied spectraplakins to date are Short stop (Shot) in *Drosophila*, Bullous Pemphigoid Antigen 1 (BPAG1/dystonin/BP320), and Microtubule–Actin Cross-linking Factor 1 (MACF1/Actin Cross-linking Family 7/ACF7) in mammals, as well as VAB-10 in *Caenorhabditis elegans*. As illustrated for Shot in Fig. 1A, the genomic loci encoding spectraplakins generate numerous isoforms with up to seven recognized functional domains that establish them as true members of three protein families, the plakins, spectrins, and Gas2-like proteins (Brown, 2008; Röper et al., 2002; Suozzi, Wu, & Fuchs, 2012): (1) The plakins (typical other members: plectin, desmoplakin, envoplakin, periplakin, and epiplakin) are cytoskeleton-associated scaffold proteins primarily at cell junctions crucial for maintaining tissues under mechanical stress (e.g., in the skin or heart; Sonnenberg & Liem, 2007). (2) The spectrins (α-/β-spectrin, α-actinin, dystrophin, and utrophin) are highly conserved actin-linked scaffolding proteins primarily at the cell cortex which form essential links between membrane and cytoplasmic proteins (Broderick & Winder,

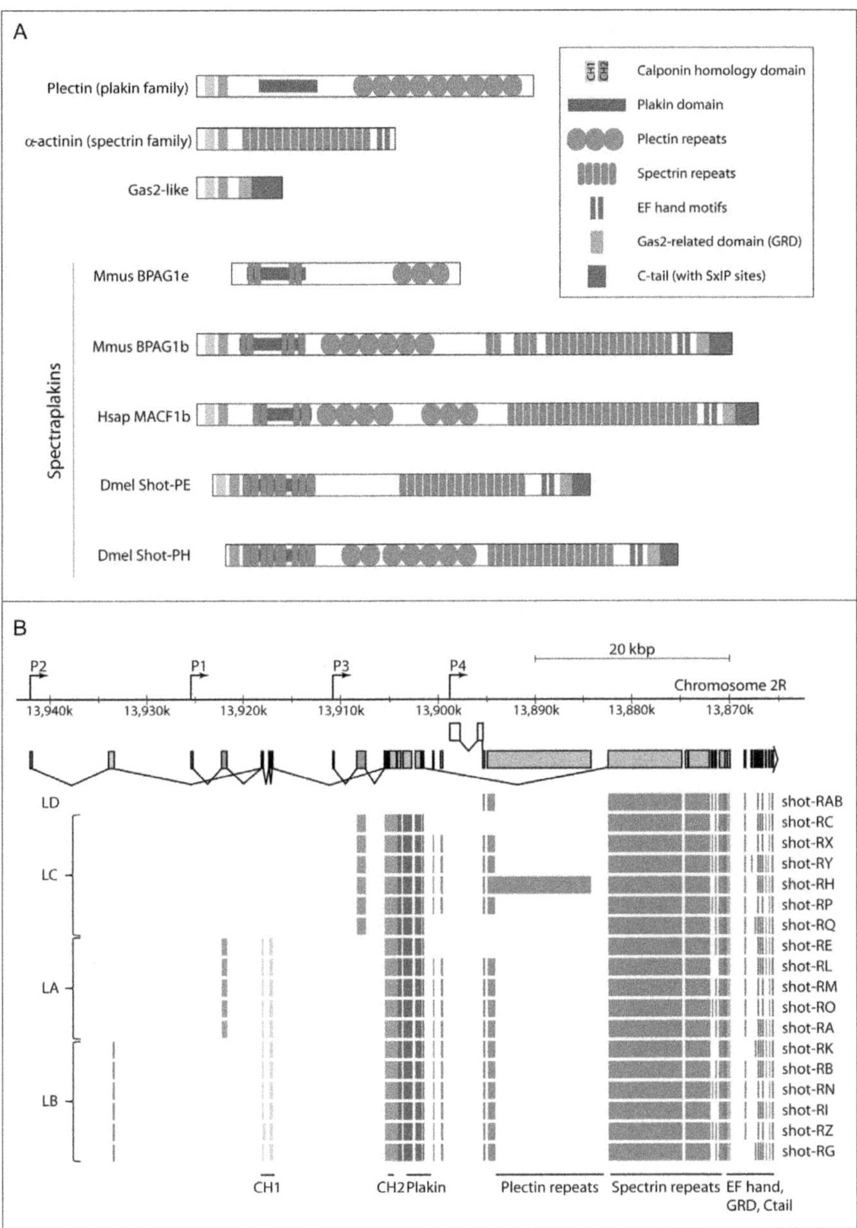

Figure 1 Family associations, homologues, functional domains, and isoforms of Shot. (A) Images show three typical representatives of the plakin, spectrin, and Gas2-like protein families, compared to specific isoforms of spectraplakins. Here shown are the shortest e- and longest b-isoform of mouse BPAG1 (Young, Pool, & Kothary, 2003), the longest b-isoform of human MACF1, and the long Shot-RE and Shot-RH isoforms

(Continued)

2005; Machnicka et al., 2014). (3) The Gas2-like proteins (Gas2, Gas2-like 1–3) act as linkers between MTs, end-binding (EB) proteins, and F-actin, important for cytoskeletal dynamics in cell division and development (Sharaby et al., 2014; Stroud et al., 2014; Wolter et al., 2012).

Notably, spectraplakins have maintained fundamental functions associated with all three protein families and, accordingly, the range of their functions is huge, and their loss often leads to early lethality. For example, the epidermal isoform of BPAG1/dystonin is an important constituent of hemidesmosomes in basal epithelial layers, particularly in the skin where autoimmune or hereditary functional aberration causes skin blistering of

Figure 1—Cont'd (compare B). The key (emboxed) shows the different functional domains: The N-terminus of Dystonin-a2 promotes localization to the nuclear envelope (Young & Kothary, 2007; Young et al., 2003). Two calponin homology domains (CH1 and CH2) form a classical actin-binding domain (ABD) essential for neuronal morphogenesis (Alves-Silva et al., 2012; Jefferson, Leung, & Liem, 2004; Kodama, Lechler, & Fuchs, 2004; Lee & Kolodziej, 2002b; Leung, Sun, Zheng, Knowles, & Liem, 1999). The plakin-homology domain (six to nine spectrin-like repeats and an SH3 domain embedded in the 5th repeat) links to transmembrane collagens and β4 integrin in BPAG1e (Aumailley, Has, Tunggal, & Bruckner-Tuderman, 2006; Koster, Geerts, Favre, Borradori, & Sonnenberg, 2003) and mediates compartmentalization of transmembrane proteins in fly neurons (Bottenberg et al., 2009). The spectrin/dystrophin repeat rod domain likely acts as spacer or in dimerization (Bottenberg et al., 2009; Leung et al., 1999; Machnicka et al., 2014; Röper, Gregory, & Brown, 2002). Plectin-repeat domains (PRD) typically bind intermediate filaments (Aumailley et al., 2006; Jefferson et al., 2004). The two helix–loop–helix EF-hand motifs bind calcium to regulate MT association (Kapur et al., 2012), the putative translational regulator Krasavietz/eIF5C (Kra) important for actin regulation during axonal pathfinding (Lee et al., 2007; Sánchez-Soriano et al., 2009), and intramolecularly to the CH domains to autoinhibit Shot functions (Applewhite, Grode, Duncan, & Rogers, 2013). The Gas2-related domain (GRD) binds MTs and protects them against destabilization. The flexible, positively charged Ctail region associates with MTs and contains one or two SxIP sites required for binding to end-binding proteins (EB1) which is essential for axon growth (Alves-Silva et al., 2012; Applewhite et al., 2010; Honnappa et al., 2009; Sun, Leung, & Liem, 2001). The C-terminus can be phosphorylated to regulate MT association (Wu et al., 2011) and is the region specifically deleted in human BPAG1/dystonin mutations linked to neurodegeneration (Edvardson et al., 2012). (B) According to FlyBase, the *shot* locus spans ~78 kb (genomic position 13,864,237–13,942,110) in cytogenetic map position 50C6–11 of chromosome 2R. The different isoforms Shot-R/P A–Z (R refers to transcript and P to protein isoforms) are generated from 4 transcriptional start sites (P1–4) giving rise to 4 N-terminal isoforms (termed Shot-LA-D as indicated on the left) and from the differential splicing of 39 different coding exons and 4 noncoding exons. *Modified from Röper and Brown (2003), Röper et al. (2002), and FlyBase (McQuilton, St Pierre, & Thurmond, 2012).* (See the color plate.)

the epidermolysis bullosa simplex type (Aumailley et al., 2006). Functional loss of the longer isoforms in humans as well as mice causes severe postnatal neurodegeneration classified as hereditary sensory and autonomic neuropathy type VI (HSAN6; Edvardson et al., 2012: #6627; Ferrier, Boyer, & Kothary, 2013; Young & Kothary, 2007). Furthermore, patient studies suggest potential roles in neural development (Giorda et al., 2004; Vincent et al., 2008), human melanoma (Shimbo et al., 2010), and the infection process of Herpes virus (Pasdeloup, McElwee, Beilstein, Labetoulle, & Rixon, 2013). Finally, studies of *dystonin* mutant mice reveal further defects in glial cells (Bernier, De Repentigny, Mathieu, David, & Kothary, 1998; Saulnier, De Repentigny, Yong, & Kothary, 2002) potentially linking to multiple sclerosis (Laffitte et al., 2005), and neuromuscular junction defects associated with intrinsic muscle weakness (Boyer, Bernstein, & Boudreau-Larivière, 2010; Dalpe et al., 1999; Poliakova et al., 2014).

Surprisingly, no human diseases or conditions have been associated with MACF1/ACF7, but mouse and zebrafish models suggest its involvement in a wide range of developmental processes, ranging from early embryogenesis (Chen et al., 2006; Gupta et al., 2010), cell migration of fibroblasts, or skin stem cells (Wu, Kodama, & Fuchs, 2008; Wu et al., 2011), as well as heart and brain development (Fassett et al., 2013; Goryunov, He, Lin, Leung, & Liem, 2010; Jorgensen et al., 2014; Sánchez-Soriano et al., 2009).

1.2 *Drosophila*: A Powerful Model for the Study of Spectraplakin Functions

To assess the biological relevance of molecules and mechanisms, *in vivo* analyses are essential and, for this, genetic invertebrate model organisms such as *Drosophila* are particularly well suited.

First, *in vivo* work is relatively fast in flies, and it is not unusual that new experimental ideas yield experimental data within a few weeks. Notably, there are no legal requirements, meaning that experiments can be performed straight off the drawing board with little or no need for an extensive ethical approval process. Using simple classical genetic crosses allows the fast generation of animals combining multiple genetic aberrations or transgenic constructs which can then be applied for studies of spectraplakin functions in development and disease. The basic knowledge required for this work is readily accessible and has recently been put together in a comprehensive manual for newcomers to *Drosophila* research (Prokop, 2013b; Roote & Prokop, 2013). Also the generation of transgenic animals takes just 2 months from injection to first experimentation and can be outsourced at low cost.

Second, 100 years of intense research has led to excellent conceptual understanding of fundamental biology in *Drosophila* (Bate & Martínez-Arias, 1993; Bellen, Tong, & Tsuda, 2010; Demerec, 1950; Kohler, 1994). This provides a prolific context in which to perform *in vivo* experiments, further fueled by countless genetic tools, usually readily available (all listed in FlyBase and many obtainable from resource centers; Cherbas & Gong, 2014; Matthews, Kaufman, & Gelbart, 2005; McQuilton et al., 2012; St. Pierre, Ponting, Stefancsik, McQuilton, & FlyBase Consortium, 2014).

Finally, *Drosophila* is small enough that whole embryos, adult brains, or ovaries can be visualized under the microscope, even at high resolution and/or using live imaging techniques (Tomer, Khairy, & Keller, 2011). Embryos are small enough so that four specimens can be mounted in parallel in one block for electron microscopic analysis, while being large enough to be dissected via simple techniques to carry out detailed microscopy studies of specific tissues, physiological measurements, or the collection of samples that allow tissue-specific transcriptomic or proteomic approaches (Budnik, Gorczyca, & Prokop, 2006). Consequently, highly refined descriptions of virtually all tissues and organs of the fly are available (Hartenstein, 1993), providing a rich resource of *in vivo* readouts.

This article will explain resources and key strategies to the non-drosophilist, illustrating how flies can be used as an efficient and cost-effective model to address fundamental mechanisms of spectraplakins, generating knowledge that can then be used to instruct research in higher animals.

2. STRATEGIES TO MANIPULATE SHOT FUNCTION

2.1 Available Gene Information and Classical Mutant Alleles for the *Drosophila shot* Gene

When searching for "shot" in FlyBase, a wealth of information is listed in ~17 categories in a logic, stringent, and well-linked-out manner (http://flybase.org/reports/FBgn0013733.html; McQuilton et al., 2012; St. Pierre et al., 2014). This information includes identifiers and location data (Prokop, 2013b), genomic maps, functional domains, gene ontology, expression and interaction data, orthologues, classical and transgenic alleles, external data, established fly models of human disease, available stocks and reagents, and a comprehensive list of over 200 publications in which *shot* is mentioned.

In the "Genomic Location/GBrowse" section of FlyBase, currently lists 22 different RNA/protein isoforms of *shot* (distinguished by an R/P suffix code; Fig. 1B), and these data are in line with the ensembl.org data base. The isoforms are generated through the alternative splicing of 43 coding exons (Fig. 1B), as well as the presence of four alternative transcription start sites (Fig. 1B) which generate four different N-termini (often classified as Shot-LA, -LB, -LC, -LD, indicated on the left side in Fig. 1B; Bottenberg et al., 2009; Lee, Harris, Whitington, & Kolodziej, 2000; Lee & Kolodziej, 2002b). According to the information available, most isoforms share the plakin domain, the second CH domain, spectrin repeats, EF-hand domains, GRD, and Ctail, whereas the first CH domain (LA and LB isoforms) is present in only 60% of the isoforms, and a huge plectin repeat domain (~3000 aa encoded by one exon) is restricted to the *shot-RH*/Shot-PH isoform only (Fig. 1B).

The "Alleles and Phenotypes" section of FlyBase, currently lists 80 different classical mutant alleles for *shot*, many of which have been molecularly characterized. They are mostly *shot*-specific mutations but also include 6 deficiencies, 1 duplication, and 24 transgenic lines for the targeted expression of RNAi constructs, full-length or deletion constructs (Fig. 2), or carrying Shot isoforms genomically tagged with GFP (see end of Section 2.3). With one click on any of these entries, a wealth of allele- or transgene-specific information will be listed in a new window including publicly available fly stocks and literature where these alleles or transgenic lines were used. Stock availability can also be retrieved for all *shot*-related stocks at once in the "Stocks and Reagents" section of FlyBase, with currently 50 entries. Other stocks are often still kept by individual fly groups who used them in published work and tend to share them freely.

Of all mutant alleles described, the most frequently used for functional studies (Sections 3 and 4) are $shot^{3}$ and $shot^{sf20}$. They are both considered null alleles and, accordingly, generate phenotypes comparable to smaller deficiencies uncovering the entire *shot* locus. Suitable deficiencies used in the literature are *Df(2R)MK1* [50B3–50B5; 50D4–50D7] (Sánchez-Soriano et al., 2009; Strumpf & Volk, 1998) or *Df(2R)BSC383* [50C6–50C6; 50D2–50D2] (Valakh, Walker, Skeath, & Diantonio, 2013), for which data in square brackets indicate the cytogenetic map position of their break points (i.e., the chromosomal area deleted). The genomic duplication *Dp(3R)6r35* has been used successfully to increase *shot* expression for immunohistochemical studies, thus complementing the analysis of loss-of-function (LOF) mutant alleles (Prokop, Uhler, Roote, & Bate, 1998). Another good example of how to make use of specific mutant alleles is the $shot^{kakP2}$ mutation. This allele affects the first two start sites of *shot* and generates only isoforms

UAS construct	Length [deleted region]	Isoform	Scheme	References
UAS-Shot-L(A)-GFP	5201 aa	shot-RE		1
UAS-Shot-L(A)-N'6myc	5201 aa	shot-RE		2
UAS-Shot-L(A)-N'6myc-C'GFP	5201 aa	shot-RE		2
UAS-Shot-L(C)-GFP	5160 aa	shot-RC		1
UAS-Shot-L(B)-GFP	5390 aa	shot-RB		1
UAS-Shot-L(A)-ΔPlakin-GFP	4364 aa [488–1325]	shot-RE		2
UAS-Shot-L(A)-Δrod1-GFP	1805 aa [1204–4600]	shot-RE		1
UAS-Shot-L(A)-Δrod2-GFP	4560 aa [1023–1664]	shot-RE		1
UAS-Shot-L(A)-Δrod3-GFP	4191aa [3589–4599]	shot-RE		1
UAS-Shot-L(A)-ΔEFhand-GFP	5063 aa [4711–4849]	shot-RE		1
UAS-Shot-L(A)-ΔGRD-GFP	5155 aa [4859–4905]	shot-RE		1
UAS-Shot-L(A)-ΔEB1-GFP	5165 aa [5154–5191]	shot-RE		2
UAS-Shot-L(C)-ΔGRD-GFP	4859–4905 aa of RE	shot-RC		2
UAS-Shot-RE-3MTLS*-GFP	5201 aa [4966-9, 5077-80, and 5141-4 replaced by AAAA]	shot-RE		4
UAS-Shot-RE-ΔCtail-GFP	4918 aa [4919–5201]	shot-RE		4
UAS-APT	1–1202 aa	shot-RE		5
UAS-PT	379–1202 aa	shot-RE		5
UAS-EF-GRD-Ctail-GFP	4780–5201 aa	shot-RE		5
UAS-GRD-GFP	4841–4923 aa	shot-RE		1
UAS-EB1-2HA	5140–5383 aa	shot-RE		2
UAS-Ctail	4918–5201 aa	shot-RE		4
UAS-Ctail-3MTLS*-GFP	4966-9, 5077–80, and 5141-4 replaced by AAAA	shot-RE		4

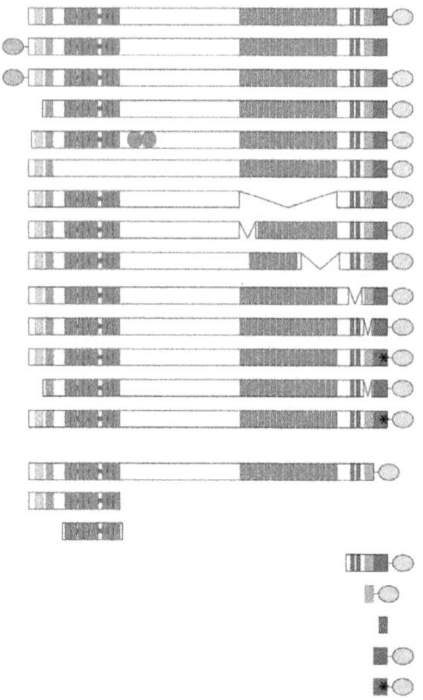

Figure 2 Published transgenic fly lines carrying UAS-linked *shot* constructs. Published name of the construct (1st column), length in amino acids (2nd column), the respective mother construct (3rd column), and schematic representations of the constructs (4th column). The domain key is given in Fig. 1; asterisks indicate mutated sequences; blue circles, c-myc; green/red circles, GFP/RFP. References: (1) Lee and Kolodziej (2002a), (2) Bottenberg (2006), (3) Bottenberg et al. (2009), (4) Alves-Silva et al. (2012), (5) Subramanian et al. (2003). (See the color plate.)

LC and LD lacking the first CH domain (Fig. 1B); $shot^{kakP2}$ was therefore used for structure–function analyses, complementary to transgenic rescues strategies (see Section 3) with the Shot-LC construct similarly lacking the first CH domain (Bottenberg et al., 2009; Röper & Brown, 2003). Another important application of *shot* mutant alleles or transgenic tools is to capitalize on the ease of *Drosophila* mating schemes (Roote & Prokop, 2013) and combine them with mutant alleles of other genes. The key principle is to use partial LOF conditions for *shot* (e.g., a heterozygous condition for a null allele, such as $shot^{sf20/+}$) and for another gene, where neither of the two alone causes a phenotype. If these two conditions combined in the same animal generate a phenotype, this suggests they are likely to operate in the same pathway or process. The nature of this interaction (e.g., sequential, parallel, in one complex) needs to be further investigated, for example, validated through biochemical studies. This strategy has worked well to illustrate functional and physical interactions of Shot with Eb1 and Krasavietz during cytoskeletal regulation required for axon growth (Alves-Silva et al., 2012; Lee et al., 2007; Sánchez-Soriano et al., 2009).

2.2 Genomic Engineering Technologies to Manipulate the *shot* Locus

In the past, mutant alleles in *Drosophila* had to be generated through fairly random procedures using X-ray, chemical or transposon mutagenesis, all well suited for forward genetics approaches (Prokop, 2013b). Now, also reverse genetics has become highly efficient through the advent of genomic engineering strategies.

TALEN (transcription activator-like effector nuclease)-induced strategies have been successfully applied in *Drosophila* in a couple of cases to modify genes endogenously (Kondo et al., 2014; Liu et al., 2012). This strategy is based on tailor-made site-specific DNA endonucleases which introduce double-strand breaks at specific loci causing errors or deletions leading to LOF alleles, or the brakes are used to introduce any DNA of choice via gap repair.

Before TALEN could be fully developed in flies, *CRISPR* (clustered regularly interspaced palindromic repeats) strategies were introduced which have clearly taken over as the key method for future genomic engineering approaches. Similar to TALEN, CRISPR allows the targeted deletion and/or replacement of genomic sequences but through a different mechanism. In CRISPR, the *Streptococcus*-derived RNA-guided DNA endonuclease Cas9 (CRISPR-associated protein 9) induces DNA double-strand breaks at sites for which complementary RNA sequences of 20 nucleotide length are provided in the form of small guide RNA (sgRNA) constructs. If two Cas9-dependent double-strand breaks are induced, this can be used to generate targeted deletions (Bassett, Tibbit, Ponting, & Liu, 2013; Gratz et al., 2013). If a construct for homology-directed repair (containing homology regions of around 1 kb) is provided in parallel, the cutout genomic section can be replaced by any sequence of choice. This strategy provides a powerful means to insert designed mutations, protein tags, or attP sites (see next paragraph) into the gene locus of choice (Bassett & Liu, 2014a, 2014b; Gratz et al., 2014). Through CRISPR technology, gene loci can be rapidly modified within ~2 months (Gokcezade, Sienski, & Duchek,

2014). The simplest method is to use standard injection procedures for early embryos (Prokop & Technau, 1993) to bring *cas9* plasmid DNA, the sgRNA(s) and, if homology-directed repair is intended, also the repair sequence into germline cells. If suitable markers were inserted during the process, successful targeting events can be screened ~10 days after injection when the individual flies eclose. Further improvements of CRISPR technology in flies include transgenic *nanos-Cas9* fly lines (stably expressing Cas9 in the germ line, so that Cas9 RNA or plasmid no longer needs to be coinjected; Ren et al., 2013), optimized sgRNA expression vectors (Port, Chen, Lee, & Bullock, 2014), and the development of online resources to facilitate construct design and prediction of potential off-target sites (Bassett et al., 2013; Gratz et al., 2013; Hsu et al., 2013; Ren et al., 2013; Yu et al., 2013). To keep up to date with CRISPR technology in *Drosophila*, a number of dedicated Web pages are available (http://www.flycrispr.molbio.wisc.edu, http://crispr.genome-engineering.org/, http://www.crisprflydesign.org).

φC31 recombinase-mediated strategies make use of targeted recombination involving two specific target sequences, attP and attB. When attB-bearing vectors are injected into *φ*C31-expressing fly strains carrying attP sites at a chosen genomic location, >90% of integration events occur site-directed at the attP sites (Bischof, Maeda, Hediger, Karch, & Basler, 2007). This strategy can be used to insert transgenes always in the same genomic position and avoid uncontrollable position effects, so that all transgenes will show comparable expression properties as a constant parameter during their functional comparison. Currently, ca. 130 different fly lines with defined attP landing sites are readily available from stock centers.

Recombinase-mediated cassette exchange (*RMCE*) provides a powerful strategy to replace chosen endogenous gene sequences in a targeted approach (Huang, Zhou, Dong, Watson, & Hong, 2009; Nagarkar-Jaiswal et al., 2015). In RMCE, first two attP sites flanking a gene or part of it are introduced, for example, using CRISPR/Cas9 technology (see previous paragraph). This flanked segment can then be substituted through *φ*C31-mediated cassette exchange, providing vectors carrying attB-flanked sequences of mutated, tagged, or otherwise nonhomologous DNA, and in this way whole series of endogenous gene variants can be produced (Gratz et al., 2014). Clearly, the advent of CRISPR/*φ*C31 in combination with RMCE has generated a step change in the precision of genetic manipulation and analysis possible in the fly.

2.3 Transgenic Ways of Depleting *shot*

A number of targeted approaches to deplete intrinsic gene products in a time-, tissue-, or even cell-specific manner can be used in *Drosophila*, thus providing a powerful complementary means to classical genetic methods. All these methods are based on well-established systems of targeted gene expression in the fly, in particular the GAL4/UAS system (Section 2.4.1).

Gene silencing by double-stranded RNA (RNAi, RNA interference) allows specific knockdown of any transcripts or their individual isoforms (Dykxhoorn, Novina, & Sharp, 2003). RNAi-mediated gene silencing is very effective in flies and has been used in a huge number of systematic genetic screens identifying regulators of various processes (Ejsmont & Hassan, 2014). In *Drosophila*, there are stable transgenic lines for the targeted knockdown of virtually every gene, most of which are readily available from public stock collections in Bloomington (USA; http://flystocks.bio.indiana.edu), Vienna (Austria; http://stockcenter.vdrc.at/control/main), and Mishima (Japan; http://www.shigen.nig.ac.jp/fly/nigfly). RNAi technology in flies has undergone gradual developments and improvements (Dietzl et al., 2007; Schmid, Schindelholz, & Zinn, 2002), and the newest versions of The Transgenic RNAi project (TRiP) combine a couple of important features: (1) All constructs are inserted into a defined attP landing site ensuring even expression levels across the collection (Section 2.2). (2) Insulator sequences were added to make expression levels independent of influences from the surrounding genome. (3) Highly efficient small shRNAs (22 bp length) are employed to deliver siRNAs which use the endogenous microRNA pathway (Ni et al., 2008, 2009, 2011). In particular, the newly introduced use of small shRNAs, as compared to long double-stranded RNA, makes it now possible to generate rescue constructs with codon modifications within the targeted sequence that are immune to the knockdown.

There are various sources of *shot* RNAi lines: Subramanian et al. successfully used the *UAS-dskakRNA* line which reproduced *shot* mutant phenotypes in tendon cells and the nervous system (Alves-Silva et al., 2008, 2012; Subramanian et al., 2003). A second-generation *shot* TRiP line (*GL01286*; Bloomington stock #41858) was shown to reduce *shot* transcript levels in the nervous system when expressed via *elav-GAL4* (Valakh et al., 2013). A first-generation TRiP line (*JF02971*; Bloomington stock #28336) exists, but seems not to have been used yet in published work. The efficiency of RNAi knockdown can be further enhanced by coexpression of the enzyme Dicer (a rate-limiting component of the RNAi machinery; Dietzl et al., 2007) or by rearing flies at higher temperatures, usually 29 °C, thus capitalizing on the temperature sensitivity of the UAS/GAL4 system (Section 2.4.1; Ni et al., 2009).

In recent years, additional strategies emerged to deplete gene function also at the protein level which, unlike RNAi, can overcome the problem of slow protein turnover rates. These strategies make use of camelid single-chain variable fragments (ScFv or nanobodies) which can be cloned and expressed in cell lines or used to generate stable transgenic fly lines. Nanobodies generated against a protein of interest retain their immune specificity in *Drosophila* cells and can functionally deplete the (cytoplasmic) proteins they were raised against (Layalle et al., 2011). The degrade Green Fluorescent Protein (deGradFP) system takes this strategy a step further by using an anti-GFP nanobody fused to a proteasome targeting signal. This fusion protein, when expressed in cells, rapidly depletes cytoplasmic pools of GFP- as well as YFP-tagged proteins (Caussinus, Kanca, & Affolter, 2012). Furthermore, deGradFP is temperature sensitive and, by simply shifting flies from 18 to 28 °C, protein pools can be depleted, and then restored again by reverting to 18 °C (Nagarkar-Jaiswal et al., 2015). This system can easily be combined with a huge variety of readily available protein trap lines where endogenous gene loci are genomically tagged with GFP or YFP. Such lines were systematically raised by the Flytrap, the MiMIC (Minos-Mediated Integration Cassette) and the Cambridge Protein Trap Insertion project (Kelso et al., 2004; Lowe et al., 2014; Nagarkar-Jaiswal et al., 2015; Venken et al., 2011). For Shot, a number of GFP trap lines exist [$shot^{CPTI001962}$, $shot^{CPTI\text{-}003653}$ (Lowe et al., 2014), $Mi\{MIC\}shot^{MI01617}$, $Mi\{MIC\}shot^{MI03583}$ (Venken et al., 2011)] which are all readily available for deGradFP knockdown experiments but also for live imaging studies (Section 3.2).

2.4 Transgenic *shot* Constructs

2.4.1 *Various* Drosophila *Expression Systems Enable Spatiotemporal Expression of Transgenes*

The GAL4/UAS system is considered the "Swiss Army knife" of *Drosophila* geneticists. It is a highly versatile means for the targeted spatiotemporal expression of genes or genetically encoded tools (Brand & Perrimon, 1993; del Valle Rodriguez, Didiano, & Desplan, 2012; Duffy, 2002). There is hardly a *Drosophila* publication where this system has not been applied and, in the case of *shot*, it has been used widely to drive deletion constructs (Section 2.4.2), to label specific cells or perform live imaging in *shot* mutant backgrounds (Sections 3 and 4), to knock down *shot* (Section 2.3), or to perform MARCM analysis (Section 3.1). GAL4/UAS is a binary system with two main components: The yeast transcription factor GAL4 (expressed in a tissue and/or time-specific pattern) activates transgenes under the control of a UAS (upstream activating sequence) promoter, thus enabling spatiotemporal specific transgene expression. Thousands of different *GAL4* fly lines have been raised, so that virtually any tissue and developmental stage can be targeted. To refine this approach, a number of tricks can be used to achieve even higher spatial or temporal resolution (del Valle Rodriguez et al., 2012; Pfeiffer et al., 2010): (1) The temperature-sensitive GAL4 repressor $Gal80^{ts}$ can be coexpressed, so that GAL4-induced expression is active only during periods when the experimental animals are shifted to 30 °C (McGuire, Mao, & Davis, 2004). (2) The GeneSwitch system uses tissue-specific GAL4 lines which are inactive under normal conditions but can be activated by supplementing the food with the steroid hormone mifepristone (Nicholson et al., 2008). (3) In the SplitGAL4 system, GAL4 is split into two halves which are expressed by two independent enhancers, so that only the very refined set of cells which has coincident expression of both halves shows GAL4 activity. A huge number of such lines are available from the Fly Light Split-GAL4 Driver Collection (http://splitgal4.janelia.org/cgi-bin/splitgal4.cgi; Aso et al., 2014; del Valle Rodriguez et al., 2012).

Finally, other binary expression systems exist, so that gene expression via the GAL4/UAS system can be combined with the expression of different genes or tools under a completely independent control (del Valle Rodriguez et al., 2012). First, in the colE1-derived LexA/LexAop system, the transcriptional activator LexA binds and activates the LexA operator (LexAop; Szuts & Bienz, 2000), and respective transgenic fly lines are readily available at the Bloomington stock center (http://flystocks.bio.indiana.edu/Browse/lexA/lexA_main.htm). Second, in the *Neurospora crassa*-derived QF/QUAS system, QF is the activator and can be repressed when coexpressing its antagonist QS (Riabinina et al., 2015), and respective fly lines are also available at the Bloomington stock center (http://flystocks.bio.indiana.edu/Browse/Qsystem/Qintro.htm). Since QS can be inhibited by quinic acid supplemented to the fly food, it can be used as a conditional system just like $Gal80^{ts}$ or GeneSwitch (Potter, Tasic, Russler, Liang, & Luo, 2010).

2.4.2 Generation of shot *Transgenes for Structure–Function Analysis*

A huge variety of *shot* constructs and transgenes have been generated (Fig. 2) and used in a variety of contexts to carry out structure–function analyses and unravel the mechanistic basis of the various Shot functions (see Sections 3 and 4). For most of them, the Shot-PE isoform (Lee et al., 2000; Fig. 1B) has been used as the template. Shot-PE has become the gold standard for domain deletion studies because it is able to fully or partially rescue all *shot* mutant phenotypes tested so far (Section 3). Apart from Shot-PE, the Shot-PH isoform might need future consideration because it contains an additional ~3000 aa plectin repeat domain encoded by one exon (Fig. 1B). Shot-PH is strongly expressed in the nervous system (Section 2.1), expected to play roles in epithelial maintenance (Roper & Brown, 2003), and it provides a potential path to understand the still unknown roles played by plectin/plakin-repeat-containing isoforms of the mammalian Shot homologue BPAG1/dystonin (Fig. 1A). One feasible path to generate *shot-RH* constructs is the use of BAC clones containing the entire genomic *shot* region (*P[acman] BAC CH321-44 M03*; listed at pacmanfly.org and distributed via bacpac.chori.org) which are designed for the use of recombineering technology.

Through the advent of BAC recombineering technology, in combination with site-directed attP/attB mutagenesis, the handling of the enormously large *shot* coding region (~16,000 bp for *shot-RE*) has become an efficient procedure when generating *shot* constructs or transgenic animals. For work with *shot* constructs, we normally use the *P[acman]* vector derivative *M-6-attB-UAS-1-3-4* which displays a number of helpful features (Alves-Silva et al., 2012; Venken, He, Hoskins, & Bellen, 2006): (1) It is generated on a BAC backbone suitable for the cloning and transgenesis of ≥100 kb DNA fragments. (2) The MCS-2 multiple cloning site contains rare restriction sites (we use *Asc*I as 5′ and *Pac*I as 3′ restriction sites) allowing digestion/ligation transfers of full-length *shot-RE* in one step. (3) Cloning of two homology arms via the MCS allows versatile construct mutagenesis or tagging through recombineering. (4) A 5xUAS fragment drives GAL4-induced expression in transgenic animals (Section 2.4.1). (5) Via an attB site, transgenic animals can be generated through *φ*C31-mediated position-specific integration at defined genomic attP sites (Section 2.2).

The pool of *shot* constructs (Fig. 2) can be employed for time- and tissue-specific overexpression experiments using the GAL4/UAS system (Section 2.4.1). Such experiments can be performed as gain-of-function analyses where constructs are overexpressed to assess their localization or potential dominant phenotypes (e.g., Alves-Silva et al., 2012; Bottenberg et al., 2009). Alternatively, in LOF analyses, expression of deletion constructs can be used to assess potential failure to rescue mutant phenotypes, thus determining the requirements of the various functional domains (Lee & Kolodziej, 2002b; Sánchez-Soriano et al., 2009). The very flexible use of these constructs for the study of *shot* mutations will keep them a powerful tool for the future, even in times of CRISPR technology (Section 2.2).

3. FUNCTIONAL ANALYSES OF *SHOT IN VIVO*

One important strength of the fly model is the ease with which mechanisms and roles of spectraplakins can be analyzed in biological contexts *in vivo*, revealing relevant mechanistic concepts that can then be used for studies in higher organisms (Prokop, Beaven, Qu, & Sánchez-Soriano, 2013). To illustrate these points, we will focus here on a number of *in vivo* readouts used for functional studies of *shot*.

3.1 Analysis of *shot* Function in the Nervous System

The first *shot* mutant phenotype reported consisted of reduced sensory axon growth in the embryo (hence the name *short stop*; Kolodziej, Jan, & Jan, 1995) which was later extended to motor axons, dendrites, and synaptic terminals (Gao, Brenman, Jan, & Jan, 1999; Lee et al., 2000; Löhr, Godenschwege, Buchner, & Prokop, 2002; Prokop, Uhler, et al., 1998). A key prerequisite for detecting and working with these phenotypes is the existence of enormously detailed descriptions of all embryonic motor and sensory neurons including their axonal and dendritic morphology (Landgraf, Bossing, Technau, & Bate, 1997; Merritt & Whitington, 1995; Rickert, Kunz, Harris, Whitington, & Technau, 2011). For functional analyses, they can either be stained with antibodies [all neurons (anti-HRP), (subsets of) sensory (22C10, Mab49C4) or motor axons (anti-FasII, anti-FasIII)] or using GAL4/UAS-based (Chiang et al., 2011) as well as manual single-cell labeling strategies (Hoang & Chiba, 2001; Landgraf et al., 1997). Simple strategies based on anti-HRP-labeled histological landmarks were developed to measure the extent of axon/nerve growth toward their respective target areas, thus generating quantitative data for mutant or experimental analyses (Bottenberg et al., 2009). Studies of motornerves in parallel to work with primary neurons (Section 4.2) have been used for very detailed structure–function analyses, making neurons the cellular context in which Shot's evolutionary conserved role as an actin–microtubule linker is currently best understood (Alves-Silva et al., 2012; Bottenberg et al., 2009; Lee & Kolodziej, 2002b; Sánchez-Soriano et al., 2009). The understanding of this role is further enhanced through additional work unraveling the functional interfaces of Shot with other actin- and MT-regulating proteins (Prokop et al., 2013).

Importantly, analysis of *shot* LOF can be extended into larval or adult stages, although this condition is embryonic lethal. For this, the expression of RNAi constructs can be targeted exclusively to specific neurons to knock down *shot* in these cells (Valakh et al., 2013), and this knockdown can be spatially and temporally refined as explained earlier (Section 2.4.1). Complementary to this, *shot* LOF mutant alleles can be used to the same end via the MARCM technique (mosaic analysis with a repressible cell marker). MARCM uses genetically induced Flippase/FRT-mediated homologous recombination between chromosome arms to generate clones of homozygous mutant neurons marked by a reporter gene, whereas the rest of the brain remains heterozygous mutant and is not marked (Wu & Luo, 2006). In this way, only the mutant neurons are labeled and their morphology can be analyzed. Notably, the *shot*^{3} mutant allele was the first ever mutant allele used for MARCM analyses, which have revealed Shot requirements during axon growth and neuronal polarity at pupal stages (Lee & Luo, 1999; Reuter, 2003: #1933).

3.2 Functional Studies of *shot* in Tendon Cells

Shot has also been extensively analyzed in specialized epidermal cells, called tendon cells, the nature of which is briefly outlined here. *Drosophila* has an exoskeleton in the form of an extracellular matrix (ECM) called cuticle which is secreted from the apical surface of the epidermis (Broadie, Baumgartner, & Prokop, 2011), whereas muscle tips anchor to the opposite basal surface of the epidermis via integrin- and ECM-dependent junctions (Prokop, Martín-Bermudo, Bate, & Brown, 1998). To withstand the enormous muscular pulling forces, the epidermis forms specialized tendon cells at these anchor points which contain dense arrays of actin and MT filaments linking the basally attached muscles to the apical cuticle on the other side (Alves-Silva et al., 2008). This organization is functionally analogous to basal keratinocytes where keratin filament networks link basal integrin-dependent hemidesmosomes to desmosomes on lateral and apical surfaces of the same cells (Aumailley et al., 2006). In both cell types, Shot and BPAG1/dystonin are required for linking the basal, integrin-dependent junctional complexes to the cytoskeleton, and their functional loss causes intracellular rupture at these junctions, leading to epidermolysis bullosa simplex in the case of BPAG1 (Aumailley et al., 2006; Prokop, Uhler, et al., 1998). Notably, epidermal isoforms of BPAG1/dystonin are short containing Plectin repeats (Fig. 1), whereas *Drosophila* Shot isoforms in tendon cells are long and bind MTs.

Roles of Shot in tendon cells become apparent not before the last third of embryogenesis (Campos-Ortega & Hartenstein, 1997). At this stage, functional studies are facilitated by the easy visualization and highly characteristic segmental pattern of tendon cells in the epidermis. Visualization is further facilitated by markers that can be used to specifically label tendon cells. These markers include endogenous proteins, such as integrins and their associated adhesion complex components; ECM proteins such as Tiggrin and Thrombospondin; the transcription factors Stripe, F-actin which is highly enriched in tendon cells, or the β1-tubulin isoform which is exclusive to the nervous system and tendon cells (Alves-Silva et al., 2008; Buttgereit, Leiss, Michiels, & Renkawitz-Pohl, 1991; Fogerty et al., 1994; Subramanian, Wayburn, Bunch, & Volk, 2007; Volohonsky, Edenfeld, Klambt, & Volk, 2007). For the detection of these proteins, antibodies or genomically GFP-tagged gene versions (Section 2.3) can be used, and also phalloidin works extremely well (Alves-Silva et al., 2008). In addition, *stripe-GAL4* can be used to drive any gene constructs of choice specifically in tendon cells, for example, to label them or to perform structure–function analyses with *shot* deletion constructs in *shot* mutant embryos (Alves-Silva et al., 2008; Bottenberg et al., 2009). Specimens can be either analyzed live, cut open, and dissected flat or injected with a fixation/labeling solution containing paraformaldehyde, phalloidin, and Triton-X (Alves-Silva et al., 2008; Budnik et al., 2006; Prokop & Technau, 1993). In flat dissected animals, collagenase or trypsin treatments can be used to dissolve ECM and detach muscles from tendon cells, thus providing an unhindered view at tendon cells in the absence of major mechanical forces (Alves-Silva et al., 2008). EM analyses at embryonic or larval stages are best performed in the sagittal plane close to the ventral midline which gives the best perspective of tendon cell architecture (Alves-Silva et al., 2008; Budnik et al., 2006; Prokop, Martín-Bermudo, et al., 1998).

Studies in tendon cells so far revealed that Shot is required for the integrity of MT arrays but not actin arrays (Alves-Silva et al., 2008). This function is absolutely dependent on the MT-binding C-terminal GRD and Ctail, partly the spectrin-repeat rod, but not at all the ABD or plakin-like domain (Fig. 1A; Alves-Silva et al., 2012; Bottenberg et al., 2009). Unlike BPAG1 at hemidesmosomes which binds keratin through its C-terminus and β4 integrin and collagen XVII through its plakin-like domain (Aumailley et al., 2006), a role of Shot as a physical linker in tendon cells remains debatable. Its primary roles may lie in the development and maintenance of these arrays, consistent also with findings that Shot promotes tubulin expression in tendon cells downstream of neuregulin signaling (Strumpf & Volk, 1998).

A number of other *in vivo* contexts have so far been less well studied but provide powerful readouts for studies of Shot. These include development of the foregut (where Shot is involved in Notch signaling; Fuss, Josten, Feix, & Hoch, 2004), the tubular transpiration system (called tracheae; where Shot is involved in cellular fusion; Lee & Kolodziej, 2002a), salivary glands (where Shot anchors minus ends of MTs; Booth, Blanchard, Adams, & Röper, 2014), epithelial junctions (Röper & Brown, 2003), the compound eye (where Shot is involved in junctional morphogenesis; Mui, Lubczyk, & Nam, 2011), and oocytes (where Shot is required for MT linkage to an internal membrane structure, the fusome, and subsequent oocyte specification; Röper & Brown, 2004), as well as the structural maintenance of muscle nuclei (where Shot, EB1 and nesprin protect nuclei from intrinsic or extrinsic forces; Wang, Reuveny, & Volk, 2015).

4. FUNCTIONAL ANALYSES OF SHOT IN CELL CULTURE

Complementary to *in vivo* analyses of Shot, cell culture studies offer efficient strategies to refine and speed up the investigation or provide alternative experimental means. Such studies can make use of ~200 existing *Drosophila* cell lines readily available from a dedicated stock center (https://dgrc.cgb.indiana.edu/cells/Catalog; Cherbas & Gong, 2014), primary cell cultures (cells directly harvested from the organism) in particular neurons and hemocytes (immune cells of the body fluid; Prokop, Küppers-Munther, & Sánchez-Soriano, 2012; Sampson & Williams, 2012), and also heterologous studies in mammalian cell lines are a feasible and often helpful strategy (Alves-Silva et al., 2012).

4.1 Studying Shot in S2 Cells

Studies in S2 cells have a long history in *Drosophila* with major contributions to genetic mechanisms of cell division (Moutinho-Pereira, Matos, & Maiato, 2010), and detailed protocols are readily available (www.flyrnai.org/DRSC-PRC.html; Ceriani, 2007; Rogers & Rogers, 2008). S2 cells are extremely easy to keep in Schneider's Drosophila Medium (Life Technologies), with passages being required every 3–4 days. With an average diameter of 10 μm, S2 cells are far smaller than fibroblasts but they present many helpful cytoskeletal readouts that can be used for studies of Shot: they contain acentrosomal microtubule arrays (Nye, Buster, & Rogers, 2014) and have actin-rich lamellipodia in the periphery but no stress fibers (although these can be artificially induced when manipulating formins; Lammel et al., 2014). Transfection of S2 cells is easy (Armknecht et al., 2005). Usually genes to be expressed are cloned into the *pAc*5.1 vector (Life Technologies) containing the constitutively active actin promoter. Alternatively, any constructs readily available on *pUAST*-plasmids (the most commonly used in the *Drosophila* field) can be cotransfected with a *pMT-GAL4* plasmid (which induces GAL4 expression upon 100 m*M* $CuSO_4$ supplementation to the medium) or *actin-GAL4* (driving constitutively). Over recent years, cell lines with defined genomic attP landing sites for *φ*C31-mediated construct integration have been generated, ideal to establish stable S2 cell lines (Groth, Fish, Nusse, & Calos, 2004), and CRISPR technology is now being used to manipulate S2 cells through genomic engineering (Bassett, Tibbit, Ponting, & Liu, 2014). However, S2 cells cannot be combined with classical genetics approaches, and knockdown strategies have to be used instead. S2 cells were originally generated from primary cultures of late-stage (20–24 h old) embryos and display a number of properties typical of hemocytes (Ribeiro, D'Ambrosio, & Vale, 2014). However, they do not adhere to ECM and are not very motile in the culture dish, so that the biological meaning of changes in cytoskeletal organization and dynamics is often not clear. Nevertheless, S2 cells are a powerful tool to unravel new molecular mechanisms which can then be studied in other contexts. In this way, S2 cells have provided us with first descriptions of mechanisms underlying the recruitment of Shot to MT plus ends (Applewhite et al., 2010; Slep et al., 2005) and a novel intramolecular autoinhibition mechanism (Applewhite et al., 2013).

4.2 Using Primary Cultures Combines Detailed Readouts with Powerful Genetics

Primary cell cultures have to be generated for each experiment anew, but, in contrast to cell lines, they offer full access to versatile *Drosophila* genetics and can be directly compared to *in vivo* contexts using animals with the identical genetic constellation. For example, hemocytes are a potential alternative to S2 cells. Their harvest from embryos or larvae is effortless (Sampson & Williams, 2012), and their study in culture can be complemented by work *in vivo*, including live imaging of cytoskeletal dynamics in developmental or wound healing contexts (Beaven et al., 2015; Evans, Zanet, Wood, & Stramer, 2010; Stramer et al., 2010). Surprisingly, Shot has not yet been investigated in these cells.

Instead, extensive studies of Shot have been performed in primary neurons. As explained and discussed in detail elsewhere (Prokop et al., 2012; Sánchez-Soriano et al., 2010), these neuron cultures can be generated from animals at all stages of development and a variety of different culture media can be used. Work on Shot has mostly been performed in primary neurons cultured in Schneider's medium and derived from mid-stage embryos (when most neurons are born and are about to grow axons; Campos-Ortega & Hartenstein, 1997). Neurons cultured in this way reproduce an impressive range of neuronal *in vivo-like* properties, including characteristic cytoskeletal organizations and dynamics or physiologically active presynaptic structures, both closely resembling descriptions from vertebrate neurons (Küppers, Sánchez-Soriano, Letzkus, Technau, & Prokop, 2003; Küppers-Munther et al., 2004; Sánchez-Soriano et al., 2010). Since the same combinatorial genetic strategies to combine mutations and/or transgenic elements can be applied in primary neurons just like *in vivo* (Beaven et al., 2015; Prokop et al., 2013), cultures offer important additional experimental opportunities: First, some of the readouts available in primary neurons are far more detailed than can be achieved *in vivo*, increasing the resolution of mechanistic studies. Second, genetic manipulations especially of the cytoskeleton often cause *in vivo* phenotypes so severe that they cannot be interpreted, whereas neurons cultures from such embryos still allow sensible experimentation. The only prerequisite is that embryos carrying these constellations develop neuronal precursor cells, so that primary neurons can be harvested from them and grown in culture. In addition to using transgenic manipulations, cells can be transfected, as has been demonstrated for RNAi constructs (Bai, Sepp, & Perrimon, 2009) as well as for larger constructs (our unpublished results).

To generate embryonic primary neuron cultures (Prokop et al., 2012), whole stage 11 embryos are selected for the right genotypes using standard fluorescent dissecting microscopes in combination with green balancer chromosomes (Casso, Ramirez-Weber, & Kornberg, 2000; Halfon et al., 2002). Embryos are cleaned and cells are mechanically extracted using a pistil, followed by chemical dispersion for 4 min, washes in fresh medium (involving centrifugation), their final dilution (ca. 4 embryos per 15 μl), and subsequent transfer of 30 μl drops into special culture chambers made from a slide with a drilled hole of 15 mm diameter glued onto another intact slide (both made from lead-free glass). Chambers are sealed with a cover slip and a bit of Vaseline and turned upside-down after 1.5 h (hanging drop culture) to avoid debris around the growing neurons. Mechanisms of axon growth are best analyzed in 6 to 8 h cultures, and synapses can be analyzed after ~15 h. Furthermore, neurons can be kept for up to a month in order to assess neurodegenerative phenotypes. Analyses of these neurons can be performed with live imaging using suitable fluorescently tagged constructs (Alves-Silva et al., 2012), or cells are fixed and stained using standard immunocytochemistry.

As summarized elsewhere (Prokop et al., 2013), work on *shot* performed in this system so far included detailed localization studies, LOF analyses, and structure–function approaches with *shot* deletion constructs. Furthermore, genetic interaction studies with other factors (e.g., *eb1* or *kra*; Section 2.1) were used complementary to biochemical binding studies to unravel the functional networks in which Shot operates (Alves-Silva et al., 2012; Sánchez-Soriano et al., 2009). These studies yielded understanding of novel spectraplakin roles in actin regulation, and molecular and cellular mechanisms underpinning roles of Shot in MT stabilization and guidance, which also provide simple conceptual explanations for neurodegenerative phenotypes observed in dystonin mutant mice and patients (Edvardson et al., 2012; Prokop, 2013a).

4.3 Analysing Shot Domain Requirements in NIH3T3 Fibroblasts

Finally, for the analysis of Shot constructs, NIH3T3 mouse fibroblasts provide a well-established and convenient system, offering highly refined cell biological descriptions while being easy to maintain and transfect (Blair, 2001; Jainchill, Aaronson, & Todaro, 1969; Todaro & Green, 1963). For example, GFP-tagged Shot constructs of ~16.5 kb length transfect with a ~3% success rate (Alves-Silva et al., 2012). For this, fibroblasts are cultured in DMEM (Sigma-Aldrich; supplemented with 1% L-glutamine, 1% penicillin/streptomycin, and 10% FCS in a humidified incubator at 5% CO_2) and passaged in a 1:10 dilution every 3 days. For transient DNA transfections, Lipofectamine and Plus reagent (Invitrogen) can be used according to the manufacturer's instructions, and cells are replated 5 h after transfection at 40% confluence in glass-bottom dishes (MatTek Corporation) coated with 10 g/ml bovine plasma fibronectin (Sigma-Aldrich). Immunohistochemistry follows standard protocols (Alves-Silva et al., 2012). For live imaging, cells are maintained in Ham's F-10 medium (Sigma-Aldrich) supplemented with 4% FCS. As explained in Section 2.3, we have most of our *shot* constructs flanked by *Pac*I and *Asc*I restriction sites. Using a modified version of the *pcDNA3.1* vector (Invitrogen) containing these restrictions sites, constructs can therefore be cloned over in a one-step process, making fibroblast studies an easily accessible part of the analysis process. Importantly, *shot* constructs in fibroblasts revealed localization and functional features reminiscent of comparable constructs in neurons and of ACF7/MACF1 or BPAG1/dystonin constructs in mammalian cells (Alves-Silva et al., 2012), but distinct from observations in S2 cells (Applewhite et al., 2010). They are therefore an efficient and valuable addition for work on Shot.

5. INFORMATICS APPROACHES TO SHOT PROTEIN FUNCTION

Whole-genome and large-scale "omics" approaches have generated a wealth of data that are often used to address specific biological questions. These extensive datasets constitute a resource that can be mined through the use of informatics tools to gain insight into regulatory processes. Much of this information has been aggregated and integrated into Web-based resources such as FlyBase that facilitate its use (Mohr, Hu, Kim, Housden, & Perrimon, 2014). One of the greatest explosions in big data has come from the development of next-generation RNA-sequencing approaches and the subsequent generation of transcriptomic profiles and profiles for DNA-binding and chromatin-interacting proteins.

As explained in greater detail elsewhere (McQuilton et al., 2012), FlyBase aggregates many available genome wide datasets (>60 transcriptomic datasets in the current release FB2015_02). These enable the examination and dissection of overall *shot* expression and isoform-specific expression levels throughout development, in specific tissues, and in a diverse array of *Drosophila*-derived cell lines. Much of these data are derived from the modENCODE consortia and is graphically accessible through the implementation of a generic genome browser (GBrowse), allowing diverse datasets to be examined simultaneously (Boley, Wan, Bickel, & Celniker, 2014). This transcriptomic data as well as a vast array of additional genomic annotations (more than 800 individual ChIP-chip and ChIP-seq experiments) can be displayed and customized through the selection of specific data under the GBrowse "Select Tracks" tab on FlyBase (datasets, descriptions and usage are reviewed elsewhere; Boley et al., 2014; Ejsmont & Hassan, 2014). For example, the inspection of *shot* transcription in whole embryos during development reveals that *shot* isoforms containing the large plectin repeat-encoding exon (Fig. 1B) are virtually absent in the early embryo prior to the initiation of zygotic transcription, but become predominant within the pool of zygotic transcript isoforms. Furthermore, tissue-specific transcriptomic datasets find the inclusion of this exon to be low in neuronal stem cells (neuroblasts) during larval stages, but greatly enriched in differentiated neurons at the same stage. The "Expression data/External data" section of FlyBase, also aggregates tissue-specific expression data revealed through *in situ* analysis of embryos and microarray-based profiles.

Beyond these diverse whole-genome ChIP-Seq and RNA-seq resources, FlyBase integrates links to additional external gene and protein interaction databases that aggregate physical and genetic interactions to help identify the network of components involved in a biological process. These resources are linked out in the "Interactions and Pathways/External data" section of FlyBase. For example, the BioGRID database, which collects physical and genetic interaction datasets for a number of diverse organisms (http://thebiogrid.org), suggests a number of physical contacts for *shot* that have not been explored experimentally yet. They include p115, an armadillo domain-containing protein that plays a role in vesicle transport and exocytosis (Kondylis & Rabouille, 2003), CG7379 a relatively uncharacterized zinc finger protein, and Kat60 an AAA-type ATPase component of the microtubule severing Katanin complex (Mao et al., 2014).

6. CONCLUSION

Given the enormous complexity of spectraplakin functions, work on the *Drosophila* spectraplakin Shot offers powerful opportunities to decipher their roles in cellular and tissue contexts. Such studies can capitalize on efficient combinatorial genetics available in the fly, an excellent infrastructure providing easy access to information and a rich pool of molecular tools, as well as the ease of *in vivo* work complementary to culture studies. Notably, understanding gained form studies on Shot can instruct work in vertebrates or mammals, thus accelerating research and often revealing a high degree of mechanistic conservation (Bellen et al., 2010).

ACKNOWLEDGMENTS

All authors were supported by the Biotechnology and Biological Sciences Research Council (BBSRC), in particular A.P., N.S.-S., and I.H. by the Project Grants BB/L000717/1 and BB/I002448/1, and M.R. through BB/H017801/1.

REFERENCES

Alves-Silva, J., Hahn, I., Huber, O., Mende, M., Reissaus, A., & Prokop, A. (2008). Prominent actin fibre arrays in *Drosophila* tendon cells represent architectural elements different from stress fibres. *Molecular Biology of the Cell, 19*(10), 4287–4297.

Alves-Silva, J., Sánchez-Soriano, N., Beaven, R., Klein, M., Parkin, J., Millard, T., et al. (2012). Spectraplakins promote microtubule-mediated axonal growth by functioning as structural microtubule-associated proteins and EB1-dependent +TIPs (Tip interacting proteins). *The Journal of Neuroscience, 32*(27), 9143–9158. http://dx.doi.org/10.1523/JNEUROSCI.0416-12.2012.

Applewhite, D. A., Grode, K. D., Duncan, M. C., & Rogers, S. L. (2013). The actin-microtubule cross-linking activity of *Drosophila* Short stop is regulated by intramolecular inhibition. *Molecular Biology of the Cell, 24*(18), 2885–2893. http://dx.doi.org/10.1091/mbc.E12-11-0798 (pii).

Applewhite, D. A., Grode, K. D., Keller, D., Zadeh, A., Slep, K. C., & Rogers, S. L. (2010). The Spectraplakin Short stop is an actin-microtubule crosslinker that contributes to organization of the microtubule network. *Molecular Biology of the Cell, 21*(10), 1714–1724.

Armknecht, S., Boutros, M., Kiger, A., Nybakken, K., Mathey-Prevot, B., & Perrimon, N. (2005). High-throughput RNA interference screens in Drosophila tissue culture cells. *Methods in Enzymology, 392*, 55–73. http://dx.doi.org/10.1016/S0076-6879(04)92004-6.

Aso, Y., Hattori, D., Yu, Y., Johnston, R. M., Iyer, N. A., Ngo, T. T., et al. (2014). The neuronal architecture of the mushroom body provides a logic for associative learning. *eLife, 3*, e04577. http://dx.doi.org/10.7554/eLife.04577.

Aumailley, M., Has, C., Tunggal, L., & Bruckner-Tuderman, L. (2006). Molecular basis of inherited skin-blistering disorders, and therapeutic implications. *Expert Reviews in Molecular Medicine, 8*(24), 1–21.

Bai, J., Sepp, K. J., & Perrimon, N. (2009). Culture of *Drosophila* primary cells dissociated from gastrula embryos and their use in RNAi screening. *Nature Protocols, 4*(10), 1502–1512.

Bassett, A., & Liu, J.-L. (2014a). CRISPR/Cas9 mediated genome engineering in Drosophila. *Methods, 69*(2), 128–136. http://dx.doi.org/10.1016/j.ymeth.2014.02.019.

Bassett, A. R., & Liu, J.-L. (2014b). CRISPR/Cas9 and genome editing in Drosophila. *Journal of Genetics and Genomics, 41*(1), 7–19. http://dx.doi.org/10.1016/j.jgg.2013.12.004.

Bassett, A. R., Tibbit, C., Ponting, Chris P., & Liu, J.-L. (2013). Highly efficient targeted mutagenesis of *Drosophila* with the CRISPR/Cas9 system. *Cell Reports, 4*, 1–9.

Bassett, A. R., Tibbit, C., Ponting, C. P., & Liu, J. L. (2014). Mutagenesis and homologous recombination in Drosophila cell lines using CRISPR/Cas9. *Biology Open, 3*(1), 42–49. http://dx.doi.org/10.1242/bio.20137120.

Bate, M., & Martínez-Arias, A. (1993). *The development of Drosophila melanogaster*, (Vols. 1 and 2). Cold Spring Harbor, New York: Cold Spring Harbor Laboratory Press.

Beaven, R., Dzhindzhev, N. S., Qu, Y., Hahn, I., Dajas-Bailador, F., Ohkura, H., et al. (2015). *Drosophila* CLIP-190 and mammalian CLIP-170 display reduced microtubule plus end association in the nervous system. *Molecular Biology of the Cell, 26*(8), 1491–1508. http://dx.doi.org/10.1091/mbc.E14-06-1083.

Bellen, H. J., Tong, C., & Tsuda, H. (2010). 100 years of *Drosophila* research and its impact on vertebrate neuroscience: A history lesson for the future. *Nature Reviews. Neuroscience, 11*(7), 514–522. http://dx.doi.org/10.1038/nrn2839. (pii).

Bernier, G., De Repentigny, Y., Mathieu, M., David, S., & Kothary, R. (1998). Dystonin is an essential component of the Schwann cell cytoskeleton at the time of myelination. *Development, 125*(11), 2135–2148.

Bischof, J., Maeda, R. K., Hediger, M., Karch, F., & Basler, K. (2007). An optimized transgenesis system for *Drosophila* using germ-line-specific varphiC31 integrases. *Proceedings of the National Academy of Sciences of the United States of America, 104*(9), 3312–3317.

Blair, D. G. (2001). The transforming activity of tumour cell DNA. *Developmental Biology (Basel), 106*, 283–288, discussion 288–290, 317–229.

Boley, N., Wan, K. H., Bickel, P. J., & Celniker, S. E. (2014). Navigating and mining modENCODE data. *Methods, 68*(1), 38–47. http://dx.doi.org/10.1016/j.ymeth.2014.03.007.

Booth, A. J. R., Blanchard, G. B., Adams, R. J., & Röper, K. (2014). A dynamic microtubule cytoskeleton directs medial actomyosin function during tube formation. *Developmental Cell, 29*(5), 562–576. http://dx.doi.org/10.1016/j.devcel.2014.03.023.

Bottenberg, W. (2006). *Neuronal differentiation and epithelial integrity: The role of Drosophila Short Stop*. Mainz: Johannes Gutenberg University.

Bottenberg, W., Sánchez-Soriano, N., Alves-Silva, J., Hahn, I., Mende, M., & Prokop, A. (2009). Context-specific requirements of functional domains of the spectraplakin short stop *in vivo*. *Mechanisms of Development, 126*(7), 489–502.

Boyer, J. G., Bernstein, M. A., & Boudreau-Larivière, C. (2010). Plakins in striated muscle. *Muscle & Nerve, 41*(3), 299–308. http://dx.doi.org/10.1002/mus.21472.

Brand, A., & Perrimon, N. (1993). Targeted gene expression as a means of altering cell fates and generating dominant phenotypes. *Development, 118*, 401–415.

Broadie, K., Baumgartner, S., & Prokop, A. (2011). Extracellular matrix and its receptors in *Drosophila* neural development. *Developmental Neurobiology, 71*(11), 1102–1130. http://dx.doi.org/10.1002/dneu.20935.

Broderick, M. J., & Winder, S. J. (2005). Spectrin, alpha-actinin, and dystrophin. *Advances in Protein Chemistry, 70*, 203–246.

Brown, N. H. (2008). Spectraplakins: The cytoskeleton's Swiss army knife. *Cell, 135*, 16–18.

Budnik, V., Gorczyca, M., & Prokop, A. (2006). Selected methods for the anatomical study of *Drosophila* embryonic and larval neuromuscular junctions. In V. Budnik & C. Ruiz-Cañada (Eds.), *International review of neurobiology: Vol. 75. The fly neuromuscular junction: Structure and function* (pp. 323–374). San Diego: Elsevier Academic Press.

Buttgereit, D., Leiss, D., Michiels, F., & Renkawitz-Pohl, R. (1991). During *Drosophila* embryogenesis the beta 1 tubulin gene is specifically expressed in the nervous system and the apodemes. *Mechanisms of Development, 33*(2), 107–118.

Campos-Ortega, J. A., & Hartenstein, V. (1997). *The embryonic development of* Drosophila melanogaster. Berlin: Springer Verlag.

Casso, D., Ramirez-Weber, F., & Kornberg, T. B. (2000). GFP-tagged balancer chromosomes for Drosophila melanogaster. *Mechanisms of Development, 91*(1–2), 451–454, S0925477300002483 [pii].

Caussinus, E., Kanca, O., & Affolter, M. (2012). Fluorescent fusion protein knockout mediated by anti-GFP nanobody. *Nature Structural & Molecular Biology, 19*(1), 117–121. http://dx.doi.org/10.1038/nsmb.2180.

Ceriani, M. F. (2007). Basic protocols for *Drosophila* S2 cell line. In E. Rosato (Ed.), *Circadian rhythms: Vol. 362* (pp. 415–422): New York: Humana Press.

Chen, H. J., Lin, C. M., Lin, C. S., Perez-Olle, R., Leung, C. L., & Liem, R. K. (2006). The role of microtubule actin cross-linking factor 1 (MACF1) in the Wnt signaling pathway. *Genes & Development, 20*(14), 1933–1945.

Cherbas, L., & Gong, L. (2014). Cell lines. *Methods, 68*(1), 74–81. http://dx.doi.org/10.1016/j.ymeth.2014.01.006.

Chiang, A. S., Lin, C. Y., Chuang, C. C., Chang, H. M., Hsieh, C. H., Yeh, C. W., et al. (2011). Three-dimensional reconstruction of brain-wide wiring networks in *Drosophila* at single-cell resolution. *Current Biology, 21*(1), 1–11. http://dx.doi.org/10.1016/j.cub.2010.11.056.

Dalpe, G., Mathieu, M., Comtois, A., Zhu, E., Wasiak, S., De Repentigny, Y., et al. (1999). Dystonin-deficient mice exhibit an intrinsic muscle weakness and an instability of skeletal muscle cytoarchitecture. *Developmental Biology, 210*(2), 367–380.

del Valle Rodriguez, A., Didiano, D., & Desplan, C. (2012). Power tools for gene expression and clonal analysis in *Drosophila*. *Nature Methods*, *9*(1), 47–55. http://dx.doi.org/10.1038/nmeth.1800.

Demerec, M. (1950). *Biology of Drosophila*. New York, London: Wiley, Chapman & Hall.

Dietzl, G., Chen, D., Schnorrer, F., Su, K. C., Barinova, Y., Fellner, M., et al. (2007). A genome-wide transgenic RNAi library for conditional gene inactivation in *Drosophila*. *Nature*, *448*(7150), 151–156.

Duffy, J. B. (2002). GAL4 system in *Drosophila*: A fly geneticist's Swiss army knife. *Genesis*, *34*(1–2), 1–15.

Dykxhoorn, D. M., Novina, C. D., & Sharp, P. A. (2003). Killing the messenger: Short RNAs that silence gene expression. *Nature Reviews. Molecular Cell Biology*, *4*(6), 457–467.

Edvardson, S., Cinnamon, Y., Jalas, C., Shaag, A., Maayan, C., Axelrod, F. B., et al. (2012). Hereditary sensory autonomic neuropathy caused by a mutation in dystonin. *Annals of Neurology*, *71*(4), 569–572. http://dx.doi.org/10.1002/ana.23524.

Ejsmont, R., & Hassan, B. (2014). The little fly that could: Wizardry and artistry of *Drosophila* genomics. *Genes*, *5*(2), 385–414.

Evans, I. R., Zanet, J., Wood, W., & Stramer, B. M. (2010). Live imaging of Drosophila melanogaster embryonic hemocyte migrations. *Journal of Visualized Experiments*, *12*(36). pii: 1696. http://dx.doi.org/10.3791/1696.

Fassett, J. T., Xu, X., Kwak, D., Wang, H., Liu, X., Hu, X., et al. (2013). Microtubule actin cross-linking factor 1 regulates cardiomyocyte microtubule distribution and adaptation to hemodynamic overload. *PLoS One*, *8*(9), e73887. http://dx.doi.org/10.1371/journal.pone.0073887.

Ferrier, A., Boyer, J. G., & Kothary, R. (2013). Cellular and molecular biology of neuronal Dystonin. *International Review of Cell and Molecular Biology*, *300*, 85–120. http://dx.doi.org/10.1016/B978-0-12-405210-9.00003-5.

Fogerty, F. J., Fessler, L. I., Bunch, T. A., Yaron, Y., Parker, C. G., Nelson, R. E., et al. (1994). Tiggrin, a novel *Drosophila* extracellular matrix protein that functions as a ligand for *Drosophila* alpha PS2 beta PS integrins. *Development*, *120*(7), 1747–1758.

Fuss, B., Josten, F., Feix, M., & Hoch, M. (2004). Cell movements controlled by the notch signalling cascade during foregut development in *Drosophila*. *Development*, *131*(7), 1587–1595.

Gao, F.-B., Brenman, J. E., Jan, L. Y., & Jan, Y. N. (1999). Genes regulating dendritic outgrowth, branching, and routing in *Drosophila*. *Genes & Development*, *13*, 2549–2561.

Giorda, R., Cerritello, A., Bonaglia, M. C., Bova, S., Lanzi, G., Repetti, E., et al. (2004). Selective disruption of muscle and brain-specific BPAG1 isoforms in a girl with a 6;15 translocation, cognitive and motor delay, and tracheo-oesophageal atresia. *Journal of Medical Genetics*, *41*(6), e71.

Gokcezade, J., Sienski, G., & Duchek, P. (2014). Efficient CRISPR/Cas9 plasmids for rapid and versatile genome editing in *Drosophila*. *G3 (Bethesda)*, *4*(11), 2279–2282. http://dx.doi.org/10.1534/g3.114.014126.

Goryunov, D., He, C. Z., Lin, C. S., Leung, C. L., & Liem, R. K. (2010). Nervous-tissue-specific elimination of microtubule-actin crosslinking factor 1a results in multiple developmental defects in the mouse brain. *Molecular and Cellular Neurosciences*, *44*(1), 1–14.

Gratz, S. J., Cummings, A. M., Nguyen, J. N., Hamm, D. C., Donohue, L. K., Harrison, M. M., et al. (2013). Genome engineering of *Drosophila* with the CRISPR RNA-guided Cas9 nuclease. *Genetics*, *194*(4), 1029–1035. http://dx.doi.org/10.1534/genetics.113.152710.

Gratz, S. J., Ukken, F. P., Rubinstein, C. D., Thiede, G., Donohue, L. K., Cummings, A. M., et al. (2014). Highly specific and efficient CRISPR/Cas9-catalyzed homology-directed repair in *Drosophila*. *Genetics*, *196*(4), 961–971. http://dx.doi.org/10.1534/genetics.113.160713.

Groth, A. C., Fish, M., Nusse, R., & Calos, M. P. (2004). Construction of transgenic *Drosophila* by using the site-specific integrase from phage phiC31. *Genetics*, *166*(4), 1775–1782.

Gupta, T., Marlow, F. L., Ferriola, D., Mackiewicz, K., Dapprich, J., Monos, D., et al. (2010). Microtubule actin crosslinking factor 1 regulates the Balbiani body and animal-vegetal polarity of the zebrafish oocyte. *PLoS Genetics*, *6*(8), e1001073.

Halfon, M. S., Gisselbrecht, S., Lu, J., Estrada, B., Keshishian, H., & Michelson, A. M. (2002). New fluorescent protein reporters for use with the *Drosophila* Gal4 expression system and for vital detection of balancer chromosomes. *Genesis*, *34*(1–2), 135–138.

Hartenstein, V. (1993). *Atlas of Drosophila development*. New York: Cold Spring Harbor Laboratory Press.

Hoang, B., & Chiba, A. (2001). Single-cell analysis of *Drosophila* larval neuromuscular synapses. *Developmental Biology*, *229*(1), 55–70.

Honnappa, S., Gouveia, S. M., Weisbrich, A., Damberger, F. F., Bhavesh, N. S., Jawhari, H., et al. (2009). An EB1-binding motif acts as a microtubule tip localization signal. *Cell*, *138*(2), 366–376.

Hsu, P. D., Scott, D. A., Weinstein, J. A., Ran, F. A., Konermann, S., Agarwala, V., et al. (2013). DNA targeting specificity of RNA-guided Cas9 nucleases. *Nature Biotechnology*, *31*(9), 827–832. http://dx.doi.org/10.1038/nbt.2647.

Huang, J., Zhou, W., Dong, W., Watson, A. M., & Hong, Y. (2009). From the cover: Directed, efficient, and versatile modifications of the *Drosophila* genome by genomic engineering. *Proceedings of the National Academy of Sciences of the United States of America*, *106*(20), 8284–8289. http://dx.doi.org/10.1073/pnas.0900641106, (pii).

Jainchill, J. L., Aaronson, S. A., & Todaro, G. J. (1969). Murine sarcoma and leukemia viruses: Assay using clonal lines of contact-inhibited mouse cells. *Journal of Virology*, *4*(5), 549–553.

Jefferson, J. J., Leung, C. L., & Liem, R. K. (2004). Plakins: Goliaths that link cell junctions and the cytoskeleton. *Nature Reviews. Molecular Cell Biology*, *5*(7), 542–553.

Jorgensen, L. H., Mosbech, M.-B., Faergeman, N. J., Graakjaer, J., Jacobsen, S. V., & Schroder, H. D. (2014). Duplication in the microtubule-actin cross-linking factor 1 gene causes a novel neuromuscular condition. *Scientific Reports*, *4*, 5180. http://dx.doi.org/10.1038/srep05180.

Kapur, M., Wang, W., Maloney, M. T., Millan, I., Lundin, V. F., Tran, T. A., et al. (2012). Calcium tips the balance: A microtubule plus end to lattice binding switch operates in the carboxyl terminus of BPAG1n4. *EMBO Reports*, *13*(11), 1021–1029. http://dx.doi.org/10.1038/embor.2012.140.

Kelso, R. J., Buszczak, M., Quinones, A. T., Castiblanco, C., Mazzalupo, S., & Cooley, L. (2004). Flytrap, a database documenting a GFP protein-trap insertion screen in Drosophila melanogaster. *Nucleic Acids Research*, *32*(Database issue), D418–D420. http://dx.doi.org/10.1093/nar/gkh014.

Kodama, A., Lechler, T., & Fuchs, E. (2004). Coordinating cytoskeletal tracks to polarize cellular movements. *The Journal of Cell Biology*, *167*(2), 203–207.

Kohler, R. E. (1994). *Lords of the fly. Drosophila genetics and the experimental life*. Chicago, London: The University of Chicago Press.

Kolodziej, P. A., Jan, L. Y., & Jan, Y. N. (1995). Mutations that affect the length, fasciculation, or ventral orientation of specific sensory axons in the *Drosophila* embryo. *Neuron*, *15*(2), 273–286.

Kondo, T., Sakuma, T., Wada, H., Akimoto-Kato, A., Yamamoto, T., & Hayashi, S. (2014). TALEN-induced gene knock out in *Drosophila*. *Development, Growth & Differentiation*, *56*(1), 86–91. http://dx.doi.org/10.1111/dgd.12097.

Kondylis, V., & Rabouille, C. (2003). A novel role for dp115 in the organization of tER sites in Drosophila. *The Journal of Cell Biology*, *162*(2), 185–198. http://dx.doi.org/10.1083/jcb.200301136.

Koster, J., Geerts, D., Favre, B., Borradori, L., & Sonnenberg, A. (2003). Analysis of the interactions between BP180, BP230, plectin and the integrin alpha6beta4 important for hemidesmosome assembly. *Journal of Cell Science*, *116*(Pt. 2), 387–399.

Küppers, B., Sánchez-Soriano, N., Letzkus, J., Technau, G. M., & Prokop, A. (2003). In developing *Drosophila* neurones the production of gamma-amino butyric acid is tightly regulated downstream of glutamate decarboxylase translation and can be influenced by calcium. *Journal of Neurochemistry*, *84*, 939–951.

Küppers-Munther, B., Letzkus, J., Lüer, K., Technau, G., Schmidt, H., & Prokop, A. (2004). A new culturing strategy optimises *Drosophila* primary cell cultures for structural and functional analyses. *Developmental Biology*, *269*, 459–478.

Laffitte, E., Burkhard, P. R., Fontao, L., Jaunin, F., Saurat, J. H., Chofflon, M., et al. (2005). Bullous pemphigoid antigen 1 isoforms: Potential new target autoantigens in multiple sclerosis? *The British Journal of Dermatology*, *152*(3), 537–540.

Lammel, U., Bechtold, M., Risse, B., Berh, D., Fleige, A., Bunse, I., et al. (2014). The *Drosophila* FHOD1-like formin Knittrig acts through Rok to promote stress fiber formation and directed macrophage migration during the cellular immune response. *Development*, *141*(6), 1366–1380. http://dx.doi.org/10.1242/dev.101352.

Landgraf, M., Bossing, T., Technau, G. M., & Bate, M. (1997). The origin, location, and projection of the embryonic abdominal motorneurons of *Drosophila*. *The Journal of Neuroscience*, *17*, 9642–9655.

Layalle, S., Volovitch, M., Mugat, B., Bonneaud, N., Parmentier, M. L., Prochiantz, A., et al. (2011). Engrailed homeoprotein acts as a signaling molecule in the developing fly. *Development*, *138*(11), 2315–2323, 138/11/2315 [pii].

Lee, S., Harris, K.-L., Whitington, P. M., & Kolodziej, P. A. (2000). *short stop* is allelic to *kakapo*, and encodes rod-like cytoskeletal-associated proteins required for axon extension. *The Journal of Neuroscience*, *20*(3), 1096–1108.

Lee, S., & Kolodziej, P. A. (2002a). The plakin short stop and the RhoA GTPase are required for E-cadherin-dependent apical surface remodeling during tracheal tube fusion. *Development*, *129*(6), 1509–1520.

Lee, S., & Kolodziej, P. A. (2002b). Short stop provides an essential link between F-actin and microtubules during axon extension. *Development*, *129*(5), 1195–1204.

Lee, T., & Luo, L. (1999). Mosaic analysis with a repressible neurotechnique cell marker for studies of gene function in neuronal morphogenesis. *Neuron*, *22*(3), 451–461.

Lee, S., Nahm, M., Lee, M., Kwon, M., Kim, E., Zadeh, A. D., et al. (2007). The F-actin-microtubule crosslinker Shot is a platform for Krasavietz-mediated translational regulation of midline axon repulsion. *Development*, *134*(9), 1767–1777.

Leung, C. L., Sun, D., Zheng, M., Knowles, D. R., & Liem, R. K. H. (1999). Microtubule actin cross-linking factor (MACF): A hybrid of dystonin and dystrophin that can interact with the actin and microtubule cytoskeleton. *The Journal of Cell Biology*, *147*(6), 1275–1285.

Liu, J., Li, C., Yu, Z., Huang, P., Wu, H., Wei, C., et al. (2012). Efficient and specific modifications of the Drosophila genome by means of an easy TALEN strategy. *Journal of Genetics and Genomics*, *39*(5), 209–215, S1673-8527(12)00077-X.

Löhr, R., Godenschwege, T., Buchner, E., & Prokop, A. (2002). Compartmentalisation of central neurons in *Drosophila*: A new strategy of mosaic analysis reveals localisation of pre-synaptic sites to specific segments of neurites. *The Journal of Neuroscience*, *22*(23), 10357–10367.

Lowe, N., Rees, J. S., Roote, J., Ryder, E., Armean, I. M., Johnson, G., et al. (2014). Analysis of the expression patterns, subcellular localisations and interaction partners of *Drosophila* proteins using a pigP protein trap library. *Development*, *141*(20), 3994–4005. http://dx.doi.org/10.1242/dev.111054.

Machnicka, B., Czogalla, A., Hryniewicz-Jankowska, A., Boguslawska, D. M., Grochowalska, R., Heger, E., et al. (2014). Spectrins: A structural platform for stabilization and activation of membrane channels, receptors and transporters. *Biochimica et Biophysica Acta*, *1838*(2), 620–634. http://dx.doi.org/10.1016/j.bbamem.2013.05.002.

Mao, C. X., Xiong, Y., Xiong, Z., Wang, Q., Zhang, Y. Q., & Jin, S. (2014). Microtubule-severing protein katanin regulates neuromuscular junction development and dendritic elaboration in Drosophila. *Development*, *141*(5), 1064–1074. http://dx.doi.org/10.1242/dev.097774.

Matthews, K. A., Kaufman, T. C., & Gelbart, W. M. (2005). Research resources for *Drosophila*: The expanding universe. *Nature Reviews. Genetics*, *6*(3), 179–193.

McGuire, S. E., Mao, Z., & Davis, R. L. (2004). Spatiotemporal gene expression targeting with the TARGET and gene-switch systems in Drosophila. *Science's STKE*, *2004*(220), pl6. http://dx.doi.org/10.1126/stke.2202004pl6.

McQuilton, P., St Pierre, S. E., & Thurmond, J. (2012). FlyBase 101—The basics of navigating FlyBase. *Nucleic Acids Research*, *40*, D706–D714. http://dx.doi.org/10.1093/nar/gkr1030.

Merritt, D. J., & Whitington, P. M. (1995). Central projections of sensory neurons in the *Drosophila* embryo correlate with sensory modality, soma position, and proneural gene function. *The Journal of Neuroscience*, *15*, 1755–1767.

Mohr, S. E., Hu, Y., Kim, K., Housden, B. E., & Perrimon, N. (2014). Resources for functional genomics studies in *Drosophila melanogaster*. *Genetics*, *197*(1), 1–18. http://dx.doi.org/10.1534/genetics.113.154344.

Moutinho-Pereira, S., Matos, I., & Maiato, H. (2010). Drosophila S2 cells as a model system to investigate mitotic spindle dynamics, architecture, and function. *Methods in Cell Biology*, *97*, 243–257. http://dx.doi.org/10.1016/S0091-679X(10)97014-3.

Mui, U. N., Lubczyk, C. M., & Nam, S. C. (2011). Role of spectraplakin in *Drosophila* photoreceptor morphogenesis. *PLoS One*, *6*(10), e25965. http://dx.doi.org/10.1371/journal.pone.0025965.

Nagarkar-Jaiswal, S., Lee, P.-T., Campbell, M. E., Chen, K., Anguiano-Zarate, S., Cantu Gutierrez, M., et al. (2015). A library of MiMICs allows tagging of genes and reversible, spatial and temporal knockdown of proteins in Drosophila. *eLife*, *4*. http://dx.doi.org/10.7554/eLife.08469.

Ni, J. Q., Liu, L. P., Binari, R., Hardy, R., Shim, H. S., Cavallaro, A., et al. (2009). A *Drosophila* resource of transgenic RNAi lines for neurogenetics. *Genetics*, *182*(4), 1089–1100.

Ni, J. Q., Markstein, M., Binari, R., Pfeiffer, B., Liu, L. P., Villalta, C., et al. (2008). Vector and parameters for targeted transgenic RNA interference in Drosophila melanogaster. *Nature Methods*, *5*(1), 49–51. http://dx.doi.org/10.1038/nmeth1146.

Ni, J. Q., Zhou, R., Czech, B., Liu, L. P., Holderbaum, L., Yang-Zhou, D., et al. (2011). A genome-scale shRNA resource for transgenic RNAi in Drosophila. *Nature Methods*, *8*(5), 405–407. http://dx.doi.org/10.1038/nmeth.1592.

Nicholson, L., Singh, G. K., Osterwalder, T., Roman, G. W., Davis, R. L., & Keshishian, H. (2008). Spatial and temporal control of gene expression in Drosophila using the inducible GeneSwitch GAL4 system. I. Screen for larval nervous system drivers. *Genetics*, *178*(1), 215–234. http://dx.doi.org/10.1534/genetics.107.081968.

Nye, J., Buster, D. W., & Rogers, G. C. (2014). The use of cultured *Drosophila* cells for studying the microtubule cytoskeleton. *Methods in Molecular Biology*, *1136*, 81–101. http://dx.doi.org/10.1007/978-1-4939-0329-0_6.

Pasdeloup, D., McElwee, M., Beilstein, F., Labetoulle, M., & Rixon, F. J. (2013). Herpesvirus tegument protein pUL37 interacts with dystonin/BPAG1 to promote capsid transport on microtubules during egress. *Journal of Virology*, *87*(5), 2857–2867. http://dx.doi.org/10.1128/JVI.02676-12.

Pfeiffer, B. D., Ngo, T.-T. B., Hibbard, K. L., Murphy, C., Jenett, A., Truman, J. W., et al. (2010). Refinement of tools for targeted gene expression in *Drosophila*. *Genetics*, *186*(2), 735–755. http://dx.doi.org/10.1534/genetics.110.119917.

Poliakova, K., Adebola, A., Leung, C. L., Favre, B., Liem, R. K. H., Schepens, I., et al. (2014). BPAG1a and b associate with EB1 and EB3 and modulate vesicular transport, Golgi apparatus structure, and cell migration in C2.7 myoblasts. *PLoS One*, *9*(9), e107535. http://dx.doi.org/10.1371/journal.pone.0107535.

Port, F., Chen, H. M., Lee, T., & Bullock, S. L. (2014). Optimized CRISPR/Cas tools for efficient germline and somatic genome engineering in Drosophila. *Proceedings of the National Academy of Sciences of the United States of America*, *111*(29), E2967–E2976. http://dx.doi.org/10.1073/pnas.1405500111.

Potter, C. J., Tasic, B., Russler, E. V., Liang, L., & Luo, L. (2010). The Q system: A repressible binary system for transgene expression, lineage tracing, and mosaic analysis. *Cell*, *141*(3), 536–548.

Prokop, A. (2013a). The intricate relationship between microtubules and their associated motor proteins during axon growth and maintenance. *Neural Development*, *8*, 17. http://dx.doi.org/10.1186/1749-8104-8-17.

Prokop, A. (2013b). A rough guide to *Drosophila* mating schemes. In: J. Roote & A. Prokop (Eds.), G3/Bethesda. How to design a genetic mating scheme: a basic training package for *Drosophila* genetics (pp. 353–358). *3*, http://dx.doi.org/10.6084/m6089.figshare.106631. Retrieved from http://figshare.com/articles/How_to_design_a_genetic_mating_scheme_a_basic_training_package_for_Drosophila_genetics/106631 doi:dx.doi.org/10.6084/m9.figshare.106631

Prokop, A., Beaven, R., Qu, Y., & Sánchez-Soriano, N. (2013). Using fly genetics to dissect the cytoskeletal machinery of neurons during axonal growth and maintenance. *Journal of Cell Science*, *126*(11), 2331–2341.

Prokop, A., Küppers-Munther, B., & Sánchez-Soriano, N. (2012). Using primary neuron cultures of *Drosophila* to analyse neuronal circuit formation and function. In B. A. Hassan (Ed.), *The making and un-making of neuronal circuits in Drosophila: Vol. 69*. (pp. 225–247). New York: Humana Press.

Prokop, A., Martín-Bermudo, M. D., Bate, M., & Brown, N. (1998). Absence of PS integrins or laminin A affects extracellular adhesion, but not intracellular assembly, of hemiadherens and neuromuscular junctions in *Drosophila* embryos. *Developmental Biology*, *196*, 58–76.

Prokop, A., & Technau, G. M. (1993). Cell transplantation. In D. Hartley (Ed.), *Cellular interactions in development: A practical approach* (pp. 33–57). London, New York: Oxford University Press.

Prokop, A., Uhler, J., Roote, J., & Bate, M. C. (1998). The *kakapo* mutation affects terminal arborisation and central dendritic sprouting of *Drosophila* motorneurons. *The Journal of Cell Biology*, *143*, 1283–1294.

Ren, X., Sun, J., Housden, B. E., Hu, Y., Roesel, C., Lin, S., et al. (2013). Optimized gene editing technology for Drosophila melanogaster using germ line-specific Cas9. *Proceedings of the National Academy of Sciences of the United States of America*, *110*(47), 19012–19017. http://dx.doi.org/10.1073/pnas.1318481110.

Reuter, J. E., Nardine, T. M., Penton, A., Billuart, P., Scott, E. K., Usui, T., et al. (2003). A mosaic genetic screen for genes necessary for Drosophila mushroom body neuronal morphogenesis. *Development (Cambridge, England)*, *130*, 1203–1213.

Riabinina, O., Luginbuhl, D., Marr, E., Liu, S., Wu, M. N., Luo, L., et al. (2015). Improved and expanded Q-system reagents for genetic manipulations. *Nature Methods*, *12*(3), 219–222. http://dx.doi.org/10.1038/nmeth.3250, 215 pp following 222.

Ribeiro, S. A., D'Ambrosio, M. V., & Vale, R. D. (2014). Induction of focal adhesions and motility in Drosophila S2 cells. *Molecular Biology of the Cell*, *25*(24), 3861–3869. http://dx.doi.org/10.1091/mbc.E14-04-0863.

Rickert, C., Kunz, T., Harris, K. L., Whitington, P. M., & Technau, G. M. (2011). Morphological characterization of the entire interneuron population reveals principles of neuromere organization in the ventral nerve cord of Drosophila. *The Journal of Neuroscience*, *31*(44), 15870–15883, 31/44/15870 [pii].

Rogers, S. L., & Rogers, G. C. (2008). Culture of *Drosophila* S2 cells and their use for RNAi-mediated loss-of-function studies and immunofluorescence microscopy. *Nature Protocols*, *3*(4), 606–611. http://dx.doi.org/10.1038/nprot.2008.18.

Roote, J., & Prokop, A. (2013). How to design a genetic mating scheme: A basic training package for *Drosophila* genetics. *G3 (Bethesda)*, *3*(2), 353–358. http://dx.doi.org/10.1534/g3.112.004820.

Röper, K., & Brown, N. H. (2003). Maintaining epithelial integrity: A function for gigantic spectraplakin isoforms in adherens junctions. *The Journal of Cell Biology*, *162*(7), 1305–1315. http://dx.doi.org/10.1083/jcb.200307089.

Röper, K., & Brown, N. H. (2004). A spectraplakin is enriched on the fusome and organizes microtubules during oocyte specification in *Drosophila*. *Current Biology*, *14*(2), 99–110.

Röper, K., Gregory, S. L., & Brown, N. H. (2002). The 'Spectraplakins': Cytoskeletal giants with characteristics of both spectrin and plakin families. *Journal of Cell Science*, *115*, 4215–4225.

Sampson, C., & Williams, M. (2012). Protocol for *ex vivo* incubation of *Drosophila* primary post-embryonic haemocytes for real-time analyses. In F. Rivero (Ed.), *Rho GTPases: Vol. 827* (pp. 359–367). New York: Springer.

Sánchez-Soriano, N., Gonçalves-Pimentel, C., Beaven, R., Haessler, U., Ofner, L., Ballestrem, C., et al. (2010). *Drosophila* growth cones: A genetically tractable platform for the analysis of axonal growth dynamics. *Developmental Neurobiology*, *70*(1), 58–71.

Sánchez-Soriano, N., Travis, M., Dajas-Bailador, F., Goncalves-Pimentel, C., Whitmarsh, A. J., & Prokop, A. (2009). Mouse ACF7 and *Drosophila* Short stop modulate filopodia formation and microtubule organisation during neuronal growth. *Journal of Cell Science*, *122*(Pt. 14), 2534–2542. http://dx.doi.org/10.1242/jcs.046268.

Saulnier, R., De Repentigny, Y., Yong, V. W., & Kothary, R. (2002). Alterations in myelination in the central nervous system of dystonia musculorum mice. *Journal of Neuroscience Research*, *69*(2), 233–242.

Schmid, A., Schindelholz, B., & Zinn, K. (2002). Combinatorial RNAi: A method for evaluating the functions of gene families in Drosophila. *Trends in Neurosciences*, *25*(2), 71–74.

Sharaby, Y., Lahmi, R., Amar, O., Elbaz, I., Lerer-Goldshtein, T., Weiss, A. M., et al. (2014). Gas2l3 is essential for brain morphogenesis and development. *Developmental Biology*, *394*(2), 305–313. http://dx.doi.org/10.1016/j.ydbio.2014.08.006.

Shimbo, T., Tanemura, A., Yamazaki, T., Tamai, K., Katayama, I., & Kaneda, Y. (2010). Serum anti-BPAG1 auto-antibody is a novel marker for human melanoma. *PLoS One*, *5*(5), e10566.

Slep, K. C., Rogers, S. L., Elliott, S. L., Ohkura, H., Kolodziej, P. A., & Vale, R. D. (2005). Structural determinants for EB1-mediated recruitment of APC and spectraplakins to the microtubule plus end. *The Journal of Cell Biology*, *168*(4), 587–598.

Sonnenberg, A., & Liem, R. K. (2007). Plakins in development and disease. *Experimental Cell Research*, *313*(10), 2189–2203.

St. Pierre, S. E., Ponting, L., Stefancsik, R., McQuilton, P., & FlyBase Consortium. (2014). FlyBase 102—Advanced approaches to interrogating FlyBase. *Nucleic Acids Research*, *42*(D1), D780–D788. http://dx.doi.org/10.1093/nar/gkt1092.

Stramer, B., Moreira, S., Millard, T., Evans, I., Huang, C. Y., Sabet, O., et al. (2010). Clasp-mediated microtubule bundling regulates persistent motility and contact repulsion in *Drosophila* macrophages *in vivo*. *The Journal of Cell Biology*, *189*(4), 681–689. http://dx.doi.org/10.1083/jcb.200912134.

Stroud, M. J., Nazgiewicz, A., McKenzie, E. A., Wang, Y., Kammerer, R. A., & Ballestrem, C. (2014). GAS2-like proteins mediate communication between microtubules and actin through interactions with end-binding proteins. *Journal of Cell Science*, *127*(Pt. 12), 2672–2682. http://dx.doi.org/10.1242/jcs.140558.

Strumpf, D., & Volk, T. (1998). Kakapo, a novel *Drosophila* protein, is essential for the restricted localization of the neuregulin-like factor, Vein, at the muscle-tendon junctional site. *The Journal of Cell Biology*, *143*, 1259–1270.

Subramanian, A., Prokop, A., Yamamoto, M., Sugimura, K., Uemura, T., Betschinger, J., et al. (2003). Short stop recruits EB1/APC1 and promotes microtubule assembly at the muscle-tendon junction. *Current Biology*, *13*(13), 1086–1095.

Subramanian, A., Wayburn, B., Bunch, T., & Volk, T. (2007). Thrombospondin-mediated adhesion is essential for the formation of the myotendinous junction in *Drosophila*. *Development*, *134*(7), 1269–1278.

Sun, D., Leung, C. L., & Liem, R. K. (2001). Characterization of the microtubule binding domain of microtubule actin crosslinking factor (MACF): Identification of a novel group of microtubule associated proteins. *Journal of Cell Science*, *114*(Pt. 1), 161–172.

Suozzi, K. C., Wu, X., & Fuchs, E. (2012). Spectraplakins: Master orchestrators of cytoskeletal dynamics. *The Journal of Cell Biology*, *197*(4), 465–475. http://dx.doi.org/10.1083/jcb.201112034.

Szuts, D., & Bienz, M. (2000). LexA chimeras reveal the function of Drosophila Fos as a context-dependent transcriptional activator. *Proceedings of the National Academy of Sciences of the United States of America*, *97*(10), 5351–5356.

Todaro, G. J., & Green, H. (1963). Quantitative studies of the growth of mouse embryo cells in culture and their development into established lines. *The Journal of Cell Biology*, *17*, 299–313.

Tomer, R., Khairy, K., & Keller, P. J. (2011). Shedding light on the system: Studying embryonic development with light sheet microscopy. *Current Opinion in Genetics & Development*, *21*(5), 558–565. http://dx.doi.org/10.1016/j.gde.2011.07.003.

Valakh, V., Walker, L. J., Skeath, J. B., & Diantonio, A. (2013). Loss of the spectraplakin short stop activates the DLK injury response pathway in *Drosophila*. *The Journal of Neuroscience*, *33*(45), 17863–17873. http://dx.doi.org/10.1523/JNEUROSCI.2196-13.2013.

Venken, K. J., He, Y., Hoskins, R. A., & Bellen, H. J. (2006). P[acman]: A BAC transgenic platform for targeted insertion of large DNA fragments in *D. Melanogaster*. *Science*, *314*(5806), 1747–1751.

Venken, K. J., Schulze, K. L., Haelterman, N. A., Pan, H., He, Y., Evans-Holm, M., et al. (2011). MiMIC: A highly versatile transposon insertion resource for engineering *Drosophila melanogaster* genes. *Nature Methods*, *8*(9), 737–743.

Vincent, J. B., Choufani, S., Horike, S., Stachowiak, B., Li, M., Dill, F. J., et al. (2008). A translocation t(6;7p11-p12;q22) associated with autism and mental retardation: Localization and identification of candidate genes at the breakpoints. *Psychiatric Genetics*, *18*(3), 101–109.

Volohonsky, G., Edenfeld, G., Klambt, C., & Volk, T. (2007). Muscle-dependent maturation of tendon cells is induced by post-transcriptional regulation of stripeA. *Development*, *134*(2), 347–356.

Wang, S., Reuveny, A., & Volk, T. (2015). Nesprin provides elastic properties to muscle nuclei by cooperating with spectraplakin and EB1. *The Journal of Cell Biology*, *209*(4), 529–538. http://dx.doi.org/10.1083/jcb.201408098.

Wolter, P., Schmitt, K., Fackler, M., Kremling, H., Probst, L., Hauser, S., et al. (2012). GAS2L3, a novel target gene of the dream complex, is required for proper cytokinesis and genomic stability. *Journal of Cell Science*, *125*(Pt. 10), 2393–2406. http://dx.doi.org/10.1242/jcs.097253, jcs.097253 [pii].

Wu, X., Kodama, A., & Fuchs, E. (2008). ACF7 regulates cytoskeletal-focal adhesion dynamics and migration and has ATPase activity. *Cell*, *135*, 137–148.

Wu, J. S., & Luo, L. (2006). A protocol for mosaic analysis with a repressible cell marker (MARCM) in Drosophila. *Nature Protocols*, *1*(6), 2583–2589. http://dx.doi.org/10.1038/nprot.2006.320, nprot.2006.320 [pii].

Wu, X., Shen, Q. T., Oristian, D. S., Lu, C. P., Zheng, Q., Wang, H. W., et al. (2011). Skin stem cells orchestrate directional migration by regulating microtubule-ACF7 connections through GSK3beta. *Cell*, *144*(3), 341–352.

Young, K. G., & Kothary, R. (2007). Dystonin/Bpag1—A link to what? *Cell Motility and the Cytoskeleton*, *64*(12), 897–905.

Young, K. G., Pool, M., & Kothary, R. (2003). Bpag1 localization to actin filaments and to the nucleus is regulated by its N-terminus. *Journal of Cell Science*, *116*(Pt. 22), 4543–4555.

Yu, Z., Ren, M., Wang, Z., Zhang, B., Rong, Y. S., Jiao, R., et al. (2013). Highly efficient genome modifications mediated by CRISPR/Cas9 in Drosophila. *Genetics*, *195*(1), 289–291. http://dx.doi.org/10.1534/genetics.113.153825.

CHAPTER TWENTY

Functional and Genetic Analysis of VAB-10 Spectraplakin in *Caenorhabditis elegans*

Christelle Gally*,[1], **Huimin Zhang**†, **Michel Labouesse**‡,[1]

*IGBMC, Development and Stem Cells Program, CNRS (UMR 7104)/INSERM (U964)/Université de Strasbourg, Illkirch, France
†Jiangsu Key Laboratory of Infection and Immunity, Institutes of Biology and Medical Sciences, Soochow University, Suzhou, Jiangsu, PR China
‡IBPS, Paris, France
[1]Corresponding authors: e-mail address: gally@igbmc.fr; michel.labouesse@upmc.fr

Contents

Abstract

Intermediate filaments (IFs) are involved in multiple cellular processes that are essential for the maintenance of cell and tissue integrity. To achieve this crucial function, IFs have to be organized as long and resistant filaments across the cells and to be tightly anchored at the cell periphery. This anchoring takes place at the level desmosomes and hemidesmosomes through proteins belonging to the spectraplakin family. Here, we focus on the sole nematode *Caenorhabditis elegans* spectraplakin locus *vab-10* that

Methods in Enzymology, Volume 569
ISSN 0076-6879
http://dx.doi.org/10.1016/bs.mie.2015.05.005

is essential to connect the epidermis to the cuticle apically and to the muscles basally. After briefly reviewing the structure of the gene, we first present the genetic tools available to study this gene as well as the reagents to examine the distribution of its translation products. We discuss the functional assays that enable examining their function. Finally, we detail a genetic method to identify spectraplakin functional partners through RNAi screens, and a biochemical method to examine the phosphorylation status of IFs.

1. INTRODUCTION

Intermediate filaments (IFs) are involved in multiple cellular processes that are essential for the maintenance of cell and tissue integrity (Toivola, Strnad, Habtezion, & Omary, 2010). In particular, IF reorganization plays a central role in response to mechanical shear stress in endothelial cells (Helmke, Goldman, & Davies, 2000) and in conferring mechanical integrity to keratinocytes (Loschke, Seltmann, Bouameur, & Magin, 2015; Ramms et al., 2013). IFs are anchored to desmosomes and hemidesmosomes through proteins belonging to the spectraplakin family (Sonnenberg & Liem, 2007). Spectraplakins are large multidomain proteins that connect actin, IFs, and microtubules. Whereas vertebrates harbor seven loci encoding spectraplakins (BPAG1, MACF/ACF7, Plectin, Desmoplakin, Envoplakin, Periplakin, Epiplakin), the nematode *Caenorhabditis elegans* encodes a single spectraplakin locus named *vab-10*, which will be the focus of this chapter. *Drosophila* also has a single spectraplakin locus, *Shot*, but no cytoplasmic IFs.

Vertebrate spectraplakins are subdivided into two subfamilies: the IF-binding epithelial plakins and the complex microtubule-binding spectraplakins BPAG1 and MACF/ACF7, which undergo further alternative splicing events. Invertebrate spectraplakins are related to the latter. Among plakins, Plectin and BPAG1e are part of the hemidesmosome adhesion complex anchoring basal keratinocytes to the underlying basement membrane and play a key role in organizing keratins in epithelia (Osmani & Labouesse, 2015). Vertebrate type I hemidesmosomes include, in addition to Plectin and BPAG1e, the α6β4 integrin and type XVII collagen as transmembrane receptors to laminin-332. In *Plectin*, and to a lesser extent in *Bpag1e*, mutant mice, the basal layer of epidermal keratinocytes displays cytoplasmic disruption reminiscent of the epidermolysis bullosa simplex (EBS) disorder described in humans (Andra et al., 1997; Coulombe, Kerns, & Fuchs, 2009).

In this chapter, we describe tools to address VAB-10 cellular functions, as well as methods to identify its partners through genetic and molecular approaches. In particular, we describe methods to examine IF organization *in vivo* and phosphorylation state *in vitro* by a 2D gel approach.

2. TOOLS

2.1 Genomic Organization

The *vab-10* locus, which has been identified by positional cloning, contains 32 exons, and undergoes multiple alternative splicing events (Bosher et al., 2003). It produces two broad categories of alternative isoforms sharing a common 5′ region encoded by exons 1–15, with two distinct 3′ regions (Fig. 1A). These two sets of isoforms are reminiscent of the spectraplakin proteins Plectin/BPAG1e (VAB-10A) and MACF1a/BPAG1a (VAB-10B). Only the main two sets of isoforms are represented in Fig. 1, as several additional alternative splicing events, mostly altering the first six exons, exon 16, and exon 27 have been described.

2.2 Domain Composition

The common exons (from 1 to 15) code for a pair of calponin homology (CH) domains forming a functional actin-binding domain (ABD; exons 2–8) and six spectrin repeats (SRs; exons 9–15) with an SH3 domain embedded in the larger SR3 (SR5 in vertebrate Plectin). All Plakin family members share this particular domain composition (Sonnenberg & Liem, 2007). SRs contain around 100 residues forming three α helices connected by short loops (Pascual, Pfuhl, Rivas, Pastore, & Saraste, 1996). The crystal structure of the SH3-containing SR5 of human Plectin reveals that the SH3 domain is distorted and occluded by intramolecular interactions with the SR4, raising the possibility that it could be accessible only under certain conditions (Ortega, Buey, Sonnenberg, & de Pereda, 2011). For example, it might bind microtubule-associated proteins 1 and 2 (MAP1 and MAP2) leading to MT destabilization (Herrmann & Wiche, 1987; Valencia et al., 2013).

The *vab-10A*-specific exons (exons 16 and 17) code for a short coiled-coil domain followed by 16 plectin repeats. Among them, a cluster of five plectin repeats at the COOH end (Fig. 1B) could form an IF-binding domain by comparison with the tight arrangement of five plectin repeats that bind IFs in rat plectin and human desmoplakin (Choi, Park-Snyder, Pascoe, Green, & Weis, 2002; Nikolic, Mac Nulty, Mir, & Wiche, 1996).

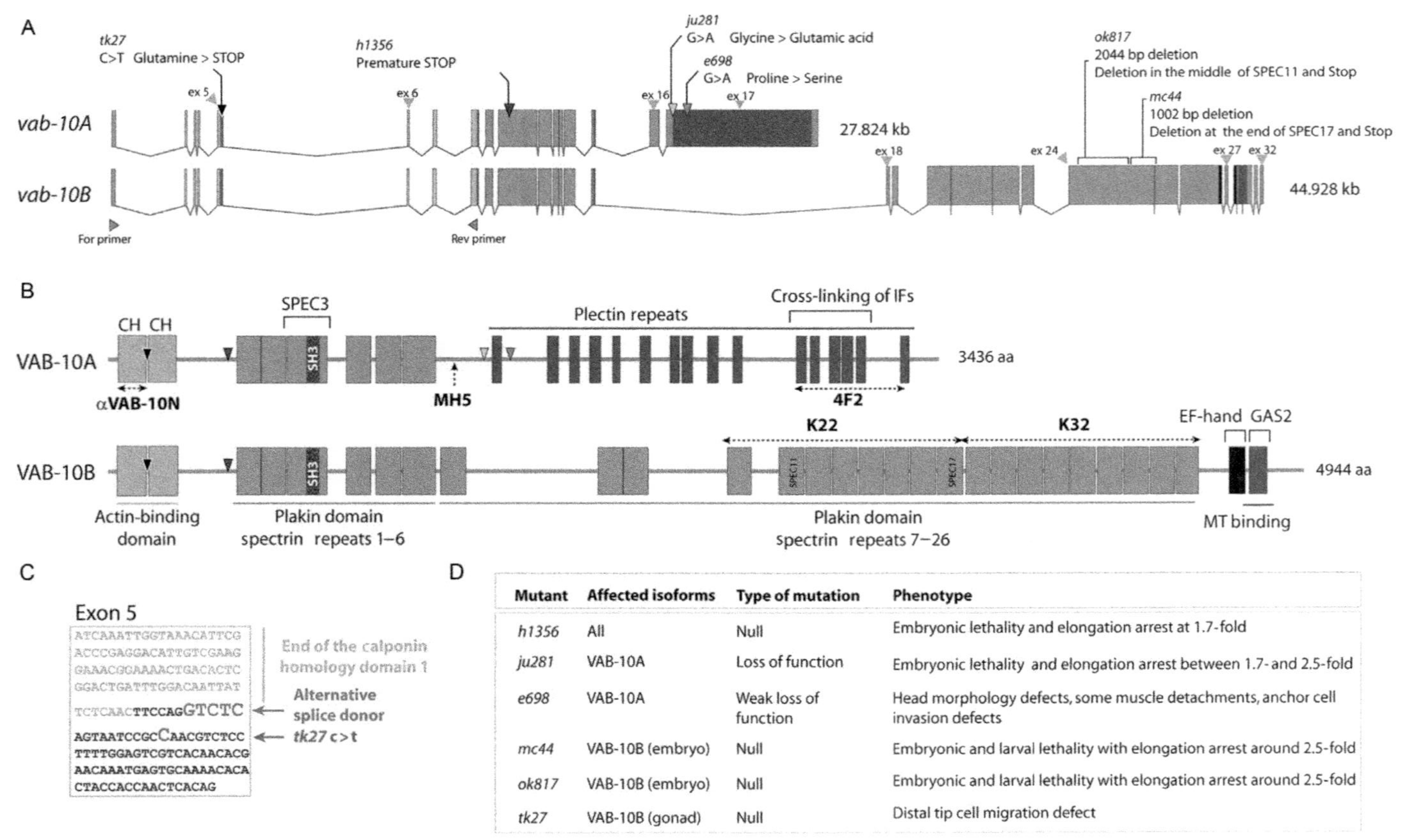

Mutant	Affected isoforms	Type of mutation	Phenotype
h1356	All	Null	Embryonic lethality and elongation arrest at 1.7-fold
ju281	VAB-10A	Loss of function	Embryonic lethality and elongation arrest between 1.7- and 2.5-fold
e698	VAB-10A	Weak loss of function	Head morphology defects, some muscle detachments, anchor cell invasion defects
mc44	VAB-10B (embryo)	Null	Embryonic and larval lethality with elongation arrest around 2.5-fold
ok817	VAB-10B (embryo)	Null	Embryonic and larval lethality with elongation arrest around 2.5-fold
tk27	VAB-10B (gonad)	Null	Distal tip cell migration defect

The *vab-10B*-specific exons (18–32) code for 20 additional SRs, an EF-Hand domain followed by a Growth-arrest-specific protein 2 (GAS2) domain. This latter is known to be a functional microtubule-binding domain (Fig. 1B; see next paragraph for further details).

2.3 VAB-10 Reagents and VAB-10-Associated Cellular Structures

Antibodies and GFP reagents specific for VAB-10 isoforms have identified cellular structures playing important development roles. Both VAB-10A and VAB-10B are gigantic proteins (predicted to be about 390 and 560 kDa, respectively) with a variety of isoforms, making very difficult the generation of GFP fusions. Until very recently, our knowledge of their distribution relied on specific polyclonal or monoclonal antibodies.

There are now two antibodies specific for the VAB-10A isoforms, MH5 (monoclonal; Francis & Waterston, 1991) and 4F2 (polyclonal; Bosher et al., 2003); two for VAB-10B isoforms, K22 and K32 (both polyclonal; Bosher et al., 2003); and one specific for the first CH domain of both VAB-10A and VAB-10B, VAB-10N (Kim et al., 2011). VAB-10A-specific antibodies detect a signal at the apical and basal sides of the ventral and dorsal epidermal

Figure 1 *vab-10* encodes two distinct plakins. (A) Schematic representation of the *vab-10* locus. It encodes by alternative splicing two major classes of isoforms designated *vab-10A* (with exons 16–17 as specific exons) and *vab-10B* (with all or a subset of exons 18–32 as specific exons); the first 15 exons are in common. The mutants described in the literature are listed above. The color-coded exons correspond to the color-coded domains described in (B). Red (gray in the print version) arrowheads are forward and reverse primers used to detect the alternative splice variant describe in (C). Some of the exons are shown with gray inverted triangle. (B) Schematic representation of the VAB-10A and VAB-10B major isoforms. The common region encodes two calponin homology (CH) domains forming a predicted actin-binding domain, and a Plakin domain including six spectrin repeats (SRs). The third SR is larger and hides an SH3 domain (dark rectangle). The specific part of VAB-10A contains 16 plectin repeats. The five but last plectin repeats are predicted to be IF cross-linkers based on homology with Plectin. The specific part of VAB-10B contains 20 additional SR, an EF-hand pair, and a microtubule-binding domain consisting of a growth-arrest-specific protein 2 (GAS2) domain. The position of the mutations (with the same color code as in A) and the regions recognized by the available antibodies (dashed lines) are also shown. (C) Exon 5 special alternative splicing. The exon 5 presents an unusual alternative splice donor (green (light gray in the print version) sequence) just before the *tk27* point mutation (red (gray in the print version)) that might explain the absence of embryonic lethality in *tk27* homozygous mutants. (D) Table listing the *vab-10* mutants and their associated phenotypes.

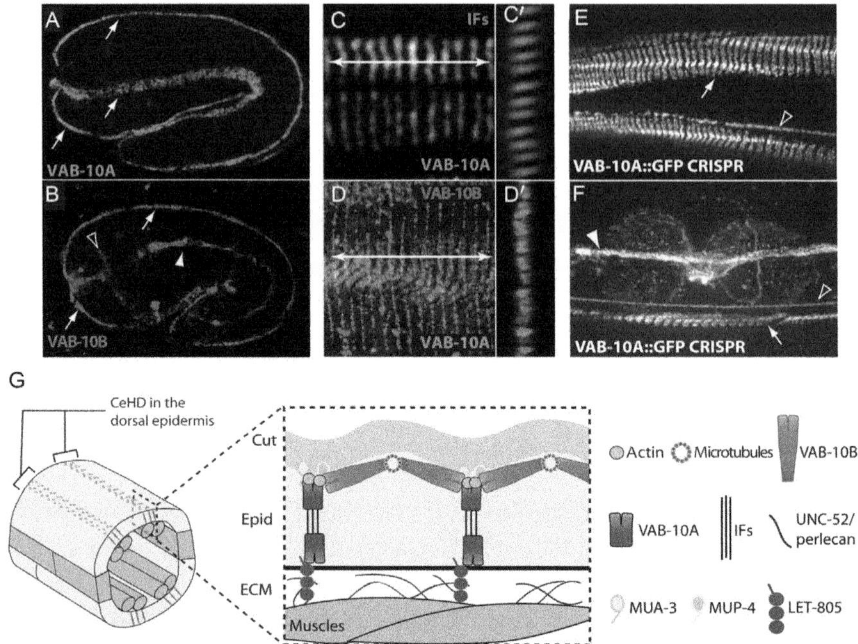

Figure 2 Subcellular localizations of VAB-10A and VAB-10B. (A) VAB-10A (MH5) and (B) VAB-10B (K22) immunostainings on wild-type elongating embryos. Arrows, visible hemidesmosomes; arrowhead, apical side of the intestine; open arrowhead, nerve ring. Anterior is to the left and dorsal is to the top. (C) Coimmunostaining of VAB-10A (4F2) and IFs (MH4) on L1 larvae showing that they perfectly colocalize. (D) Coimmunostaining of VAB-10A (MH5) and VAB-10B (K22) on L1 larvae showing that both proteins are intercalated. Panels (C′ and D′) show an optical section through the apico–basal axis along the area marked with a double arrow in C and D, respectively. (E and F) Confocal images of the VAB-10A::GFP CRISPR knock-in in adults. Hemidesmosomes (arrow) and mechanosensory neurons (open arrowhead) are clearly visible (E and F) as well as the utse (arrowhead in F). (G) Schematic representation of CeHD during embryonic elongation (left) and model for the distributions of VAB-10 isoforms with respect to the different cytoskeletal elements in the epidermis (right). Note that the lateral epidermal cells (dark blue in the left panel) are not connected to muscles and therefore do not present CeHDs. The VAB-10B localization is speculative and deduced from the respective distribution of VAB-10A, VAB-10B, actin filaments, microtubules, and intermediate filaments. (See the color plate.)

cells in the elongating embryos colocalizing with IFs (Fig. 2A, C, and C′), which corresponds to *C. elegans* hemidesmosomes (CeHDs; Fig. 2G). The basal and apical CeHDs, which are bridged by the IFs, act as a transepithelial structure connecting the epidermis basally to muscles and apically to the exoskeletal cuticle. Of note, the *C. elegans* cytoplasmic IFs do not resemble any of their vertebrate counterparts and are related to nuclear lamins (Geisler, Schunemann, Weber, Haner, & Aebi, 1998). However, they have the same general structure with a central rod domain and form obligate heterodimers, which in the epidermis correspond to IFB-1 and IFA-3 in embryos, or IFB-1 and MUA-6/IFA-2 in larvae (Karabinos, Schulze, Schunemann, Parry, & Weber, 2003). As in vertebrates, depletion of IFB-1/IFA-3 or IFB-1/MUA-6 heterodimers mimics the loss of VAB-10A isoforms (Hapiak et al., 2003; Karabinos, Schmidt, Harborth, Schnabel, & Weber, 2001; Woo, Goncharov, Jin, & Chisholm, 2004; Zhang & Labouesse, 2010). The CeHD composition differs from vertebrate HDs, as they do not contain integrin receptors; instead the presumptive receptors are large nematode-specific transmembrane proteins: LET-805 basally; MUP-4 and MUA-3 apically (Fig. 3G; Hapiak et al., 2003; Hong et al., 2001; Hresko, Schriefer, Shrimankar, & Waterston, 1999). A key extracellular matrix protein in the basement membrane separating muscles from the epidermis is UNC-52/perlecan (Rogalski, Williams, Mullen, & Moerman, 1993).

Recently, the lab of David Sherwood has generated a CRISPR knock-in strain for *vab-10A* by introducing the GFP at the end of exon 17. In addition to the hemidesmosomes and the axons of the six mechanosensory neurons already reported with the antibodies (Fig. 2E), this strain labels a novel structure called B-LINK, which links the basement membranes made by the somatic gonad and the ventral epidermis (Morrissey et al., 2014). The B-LINK involves VAB-10A, the integrin dimer INA-1/PAT-3, and the extracellular component hemicentin. In the adult, VAB-10A::GFP highlights the uterine-seam cell syncytium or utse (Fig. 2F).

VAB-10B-specific antibodies detect a signal in the epidermis and the intestine (Fig. 2B). Since the signal is clearly apical in the intestine, we assume it is also the case in the epidermis (Fig. 2G). The VAB-10B protein is also present in the nerve ring (Fig. 2B), as well as in muscles (Hahn BS and Labouesse M, unpublished). Both VAB-10A and VAB-10B epidermal signals form four longitudinal rows (two dorsally and two ventrally) in the embryo (Fig. 2A and B) that evolve into four circumferentially oriented and alternating bands above muscle quadrants (Fig. 2C, D′, and G). Antibodies against VAB-10B, as well as against the first CH domain (VAB-10N) also detect a signal in the distal tip cells (DTCs), which drive the extension of the somatic gonad (Fig. 3B).

GFP fusions to specific VAB-10 domains have been generated and mark the cytoskeleton. Specifically, expression of the two CH domains fused to

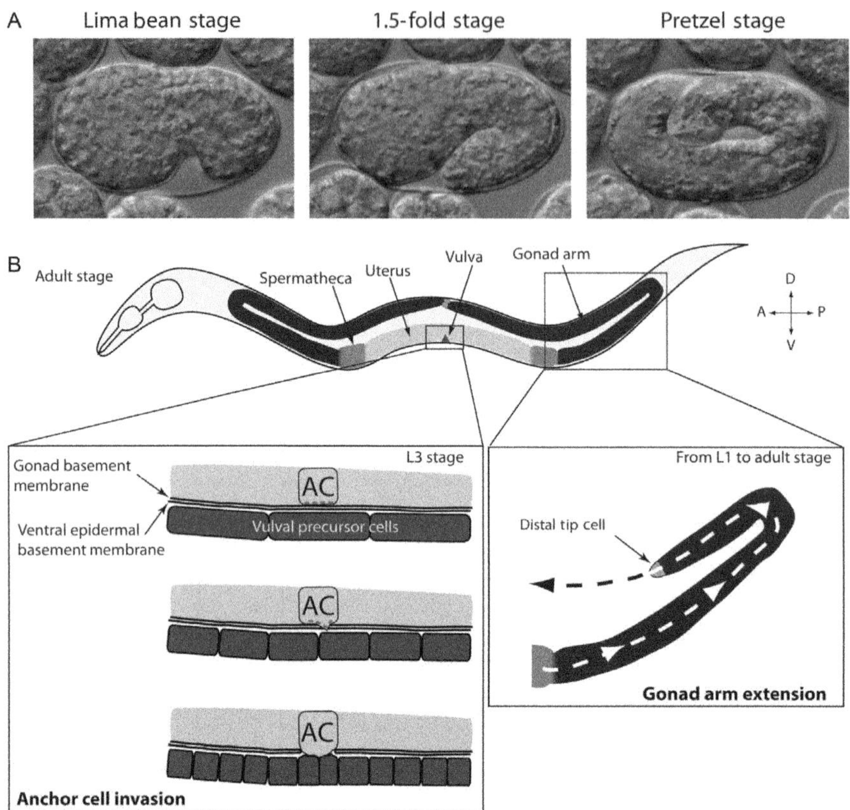

Figure 3 Cellular processes affected by *vab-10* mutants. (A) Embryonic elongation is a process that transforms a ball of cells into worm-shaped larvae ready to hatch. Embryonic elongation stages are described based on embryo shape from lima bean (1.2-fold stage, where the embryo length is 1.2 × that of the eggshell) to pretzel (or 4-fold stage, the final length of the embryo). (B) Simplified adult representation showing the reproductive system affected in *vab-10* mutants (upper panel). During the third larval stage, the anchor cell (AC) breaches the two layers of the basement membranes separating the uterine from the vulval precursors (left panel). This process requires the presence of the VAB-10A protein at the basal side of the AC (red dashed line). The AC invasion subsequently allows the formation of a passageway for egg laying between the ventral uterine tissue and the vulva, after the fusion of the AC with the uterine-seam cell (utse). From the L1 to the adult stages, the left and right arms of the somatic gonad extend laterally and reflex after migrating dorsally to form a U-shaped extension (right panel). This process is mainly driven by one cell at the tip of each arm, the distal tip cell (DTC). The DTCs undergo a rearrangement of their microtubule network that is dependent of the VAB-10B gonad isoform when they undergo dorsal migration. (See the color plate.)

GFP in the epidermis (known as VAB-10AABD::GFP) highlights actin filaments (Gally et al., 2009) and behaves as moesin-GFP in *Drosophila*. The expression of the VAB-10B GAS2 domain fused to GFP in the epidermis displays a microtubule pattern reminiscent of the TBA-2::GFP line developed in our lab (unpublished results). The homologous domains from MACF/ACF7 and *Drosophila* Shortstop behave in the same way, decorating actin filaments and microtubules (Applewhite et al., 2010; Lee & Kolodziej, 2002; Leung, Sun, Zheng, Knowles, & Liem, 1999; Sun, Leung, & Liem, 2001).

2.4 Mutant Description

Several *vab-10* alleles have been isolated over the years, some of which have been well characterized genetically (Fig. 1D). These mutants affect embryonic elongation (Fig. 3A), the uterus–vulva connection and gonad arm migration (Fig. 3B). For general reviews on embryonic elongation and gonad development, see Chisholm and Hardin (2005), Gupta, Hanna-Rose, and Sternberg (2012), Pasti and Labouesse (2014), and Wong and Schwarzbauer (2012).

The first *vab-10* allele, *vab-10A(e698)*, was isolated based on its head morphology defects and poor mating efficiency; it was later shown to partially affect muscle attachment to the cuticle (Hodgkin, 1983; Plenefisch, Zhu, & Hedgecock, 2000). The *h1356* mutation was identified in a two-step genetic screen for morphogenetic defects (Bosher et al., 2003; Chanal & Labouesse, 1997) and proved to be a *vab-10* null allele failing to complement *vab-10A(e698)*. The *ju281* and *mc44* alleles were identified in separate screens and proved to also be *vab-10* alleles (Bosher et al., 2003). Recently, other *vab-10* alleles, *ok817* and *tk27*, which affect DTC migration, have been described (Kim et al., 2011).

The *h1356* mutation induces an arrest during embryonic elongation at the 1.7-fold stage associated with irregular body morphology (for a description of stages, see Fig. 3A). Homozygous *vab-10(h1356)* mutants show an enlarged epidermis reminiscent of the phenotype associated with plectin mutations in humans EBS disorders.

ju281 and *e698* are point mutations in the last exon of VAB-10A (Fig. 1A). Homozygous *vab-10A(ju281)* embryos have strong morphogenetic defects, although in general less severe than *vab-10A(h1356)* (Bosher, 2003). Homozygous *vab-10A(e698)* animals have minor head morphology defects and some muscle detachments but are viable (Zahreddine, Zhang, Diogon, Nagamatsu, & Labouesse, 2010). At the L3 stage, 35% *vab-10A (e698)* animals present no or partial invasion of the uterine anchor cell (AC) during the initiation of the uterine–vulval contact (Fig. 3B; Morrissey et al., 2014).

mc44 is a 1-kb deletion at the end of the 24th exon specific for VAB-10B. Hundred percent homozygous *mc44* embryos display severe body morphology defects and partial elongation; upon hatching, they show an abnormal organization of muscle fibers and defects in the intestinal lumen (Bosher et al., 2003; Legouis R and Labouesse M, unpublished data). *ok817* is a 2-kb deletion in exon 24 behaving like *mc44*.

Genetically, *vab-10A* and *vab-10B* alleles behave as two independent transcription units, since *vab-10A(ju281)* and *vab-10A(e698)* alleles can complement the *vab-10B(mc44)* allele. Based on complementation tests with *vab-10(h1356)*, *vab-10A(ju281)* and *vab-10A(e698)* are not *vab-10A* null alleles, whereas *vab-10B(mc44)* represents a null allele for *vab-10B* (Bosher et al., 2003).

tk27 is a nonsense mutation in the 5th exon that prevents normal DTC migration (Kim et al., 2011). Although it should truncate all isoforms, *tk27* animals are fully viable; furthermore, *tk27/h1356* and *tk27/mc44* transheterozygous animals also only display a DTC migration defect. Therefore, *tk27* affects a subclass of VAB-10B isoforms involved in gonad development termed VAB-10B1. In a recent molecular characterization of the fifth *vab-10* exon, we identified an unusual splice donor in the middle of this exon that splices out the *tk27* mutation from the embryonic VAB-10A and B isoforms by introducing a space shortening between the two CH domains (Fig. 1C; Gally C and Labouesse M, unpublished data). We suggest that the *vab-10* isoforms expressed in embryos and in the somatic gonad may be different.

2.5 Functional Tests

VAB-10 proteins are involved in a large number of cellular processes, and over the years, we and others have developed a series of functional tests to address their implications at different steps of development.

2.5.1 Differential Interference Contrast-Based Assays

Differential interference contrast (DIC) microscopy, which allows the visualization of nuclei and basement membranes, remains a very simple yet powerful approach in *C. elegans* because animals are naturally transparent at all development stages.

Time-lapse DIC microscopy provides information about the rate of embryonic elongation and about the general shape of the embryo. As both processes depend on normal CeHDs, it indirectly assays how well these have been assembled (Zhang & Labouesse, 2010). A detailed protocol for mounting embryos for DIC microscopy and making images can be found elsewhere (Hardin, 2011).

DIC microscopy also provides information about the shape of the somatic gonad, and on the successful connection between the uterus and the vulva, which depends on a specialized uterine cell breaching the basement membrane separating the somatic gonad from the ventral epidermis (Matus et al., 2014). A detailed protocol for mounting larvae for DIC microscopy can be found elsewhere (Lee & Cram, 2009).

2.5.2 Immunofluorescence Assays

A more direct way to assess CeHD integrity is to stain embryos with antibodies recognizing specific CeHD components, such as those recognizing VAB-10A (Figs. 1 and 2), epidermal IFs (monoclonal antibody MH4), or the basal transmembrane CeHD receptor LET-805 (monoclonal antibody MH46). The monoclonal antibodies MH4, MH5, and MH46 (Francis & Waterston, 1991) are available from the Developmental Studies Hybridoma Bank at University of Iowa (dshb.biology.uiowa.edu). Additionally, embryos can be stained with antibodies recognizing body wall muscle components, such as vinculin (monoclonal antibody MH24, available from the DSHB) or the unknown antigen recognized by the monoclonal antibody NE8/4C6 (available from the antibody collection of the MRC-LMB in Cambridge, UK; Schnabel, 1994). In normal mid-stage embryos, CeHDs and body wall muscle form four thick lines running along the anterior–posterior axis adopting the curved shape of the embryo. In CeHD-defective embryos, muscles detach from the cuticle and fail to adopt the curved line of the embryo, yielding to a straight line that is easily spotted in areas of maximum embryo curvature.

The state of IF anchoring can be addressed using the MH4 antibody. When IFs are well anchored, they make a thick line (see above). When their anchorage becomes defective, they can assemble long squiggles in the epidermis outside of the area where CeHDs are normally found, as was first described when the epidermal mechanotransduction pathway responding to muscle tension is defective (Zhang et al., 2011).

2.5.3 Live Fluorescence Imaging of CeHDs and Actin

Several markers are available to visualize CeHD formation and function: VAB-10A::GFP (see below) and VAB-19::GFP for CeHD (Ding,

Goncharov, Jin, & Chisholm, 2003) and IFB-1::GFP for IFs (Kaminsky et al., 2009; Woo et al., 2004). The IFB-1::GFP marker produced a stronger GFP signal than the other two. In addition, we have used the actin marker VAB-10A$_{ABD}$::GFP that we developed to indirectly visualize the extent of muscle contractions; their displacement is compromised when VAB-10 function is defective (Zhang et al., 2011). A detailed protocol for mounting embryos for time-lapse spinning disk microscopy can be found elsewhere (Hardin, 2011). Visualizing actin displacement requires restricting the number of focal planes to a minimum (less than three) and taking images at high frequency (ideally a new stack every 0.5 s or faster), as muscle contractions occur every half-second to a few seconds.

2.5.4 Live-Imaging Assay for AC Invasion of the Basement Membrane

The Sherwood laboratory has created several fluorescent markers including Laminin::GFP and VAB-10A::GFP fusion to study how the AC breaches the basement membrane (Morrissey et al., 2014).

3. HIGH-THROUGHPUT RNAi SCREEN IN L4 WORMS

To better understand how VAB-10A functions during morphogenesis, we have performed various screens to identify VAB-10A physical and genetic interactors. In the first strategy, our lab and coworkers have performed a yeast two-hybrid screen using the spectrin repeats 1–6 as bait (Hetherington et al., 2011). From this work, we identified a new hemidesmosome protein in *C. elegans*, PAT-12, as a physical interactor of VAB-10.

In parallel, we have undertaken genome-wide RNAi screens in the weak *vab-10(e698)* viable mutant to identify genetic enhancers inducing embryonic elongation arrest. From this screen, we have identified a number of hemidesmosome regulators and among them the ubiquitin ligase EEL-1 and the Calreticulin/CRT-1, which fine-tunes the abundance of the myotactin/LET-805 and perlecan/UNC-52, respectively (Zahreddine et al., 2010).

Genome-wide RNAi screens should ideally be performed in liquid. For this purpose, we have adapted the protocol developed by the lab of Andrew Fraser (Fig. 4; Lehner, Tischler, & Fraser, 2006), starting the RNAi feeding step with L4 larvae mothers. This allowed us to successfully recover genes playing important roles in early embryonic development, which otherwise

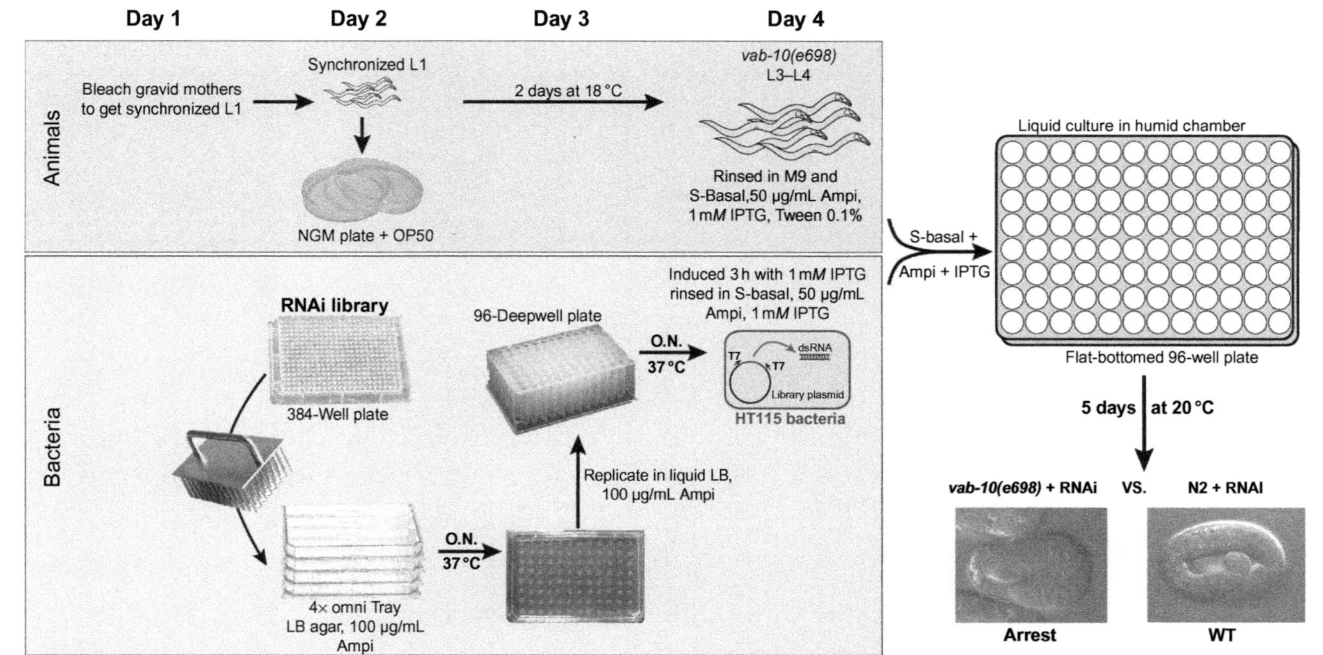

Figure 4 Flowchart of the protocol to perform genome-wide RNAi screens on L4 animals. See text for details.

would have been missed if RNAi feeding had been started at the L1 larval stage. This strategy is described in detail below.

3.1 Reagents

3.1.1 Library

This screen is performed using the Ahringer RNAi library (Fraser et al., 2000; Kamath et al., 2003), commercially available from Geneservice, Ltd. The library consists in HT115 bacteria clones (RNase III-deficient *Escherichia coli* strain, with IPTG-inducible T7 polymerase activity). Upon IPTG induction, each bacteria clone is capable of expressing double-stranded RNA (dsRNA) corresponding to a single gene. This library consists in 52 individual 384-well plates.

3.1.2 Solutions

Prepare the following solutions and keep at room temperature (RT):

S-Basal: 5.85 g NaCl, 1 g K_2HPO_4, 6 g KH_2PO_4, 1 mL cholesterol (5 mg/mL in ethanol), H_2O to 1 L. Sterilize by autoclaving.

Bleach solution: 1.5 mL bleach, 0.5 mL 10 N NaOH, 5 mL sterile H_2O prepared immediately prior to use.

M9 solution: 3 g KH_2PO_4, 6 g Na_2HPO_4, 5 g NaCl, 1 mL 1 *M* $MgSO_4$, H_2O to 1 L. Sterilize by autoclaving.

3.2 Procedure and Equipment

Day 1

- *Animals*: Bleach gravid adults 5 min in bleach solution and rinse four times in M9. Let embryos hatch in M9 over night (O.N.) at 20 °C. Without food (OP50 *E. coli* strain; Brenner, 1974), animals hatch and stop their development at the L1 stage.

Day 2

- *Bacteria*: Use a 96 pin replicator to replicate the bacteria library on Omnitray (Nunc™) LB agar plate supplemented with 50 μg/mL ampicillin. Note that the Ahringer library is available on 384-well plates, so prepare four 96 replicates for one 384-well library.
- *Animals*: Centrifuge the newly hatched L1 larvae to concentrate them and put a drop containing around 50 animals on 10 OP50 containing plates. Grow the animals at 18 °C for 2 days.

Day 3

- *Bacteria*: Prepare four 96-deep well plates (1000 μL, Eppendorf), containing 150 μL LB, ampicillin 50 μg/mL in each well. Inoculate these deep well plates with the bacteria culture obtained in the Omnitray plate using the 96 pin replicator. Cover the deep well plate with breatheable sealing films (Sealer BREATHseal™ from Greiner Bio-One) and grow O.N. at 37 °C in a shaking incubator at 300 rpm.

Day 4

- *Bacteria*: Induce the transcription of dsRNA by adding 5 μL of 25 m*M* filtrated IPTG to each well of bacterial cultures using a multichannel pipet and incubate at 37 °C for 3 h. Pellet bacteria by centrifugation at 2500 × *g* for 20 min. Discard the supernatant by rapid inversion to avoid cross-contamination and resuspend in 200 μL of S-Basal supplemented with 50 μg/mL ampicillin and 1 m*M* IPTG. Aliquot 30 μL of these cultures to flat-bottomed 96-well plates (96-well microplate, PS, F-bottom from Greiner Bio-One) using a multichannel pipet. One well represents one RNAi clone. Make as many flat-bottomed 96-well plates as you have genetic backgrounds.
- *Animals*: Rinse the synchronized L3–L4 worms in S-Basal. Centrifuge 2 min at 600 × *g* and repeat the rinsing twice. Resuspend the animals in S-Basal supplemented with 50 μg/mL ampicillin, 1 m*M* IPTG, and 0.01% Tween 20. Adjust the concentration to 7–10 animals in 20 μL. Aliquot 20 μL of this animal preparation into the aforementioned flat-bottomed plates containing the bacteria.

Incubate the animal-RNAi cultures into a humid chamber (Varibox, Dominique Dutscher) without shaking, at 20 °C for 4–5 days depending on the genotype.

Days 8–9

Look for embryonic elongation arrest under a regular dissecting binocular. Positive hits can be treated with 5 μL levamisole at 12.5 m*M* for 5 min and photographed on an inverted Zeiss AxioPlan with a 20 × objective.

3.3 Tips

- This protocol has been successfully used in five different viable mutant backgrounds in our lab.
- We established that one experimentator can easily handle a screen of 2–4 different 384-well plates of the Ahringer library (four times as many 96-well plates) in two different genotypes per week. Therefore, a genome-wide screen performed in two genetic backgrounds takes about 3 months, yet allows spotting partial phenotypes.

- The manual observation of embryonic lethality/elongation arrest in a 96-well flatted-bottom plate takes approximately 15 min.

4. 2D GEL ON INTERMEDIATE FILAMENTS

Phosphorylation is one of the most important posttranslational modifications regulating the functions of IFs, and therefore, one of the most extensively studied (Snider & Omary, 2014). Comprehensive views have been provided in several recent review articles about the triggering factors, enzymes, sites, and functions of IF phosphorylation (Hyder, Pallari, Kochin, & Eriksson, 2008; Omary, Ku, Tao, Toivola, & Liao, 2006; Snider & Omary, 2014). In summary, IF proteins are often phosphorylated at multiple Ser/Thr sites, and at a higher frequency in their head and tail regions. Phosphorylation of IF proteins affects their subcellular localization, organization, transport, turnover, as well as interaction with IF-associated proteins. As such, IF proteins can regulate diverse physiological events such as cell shape change, growth, migration, and stress response. Perhaps not surprisingly, the mechanotransduction pathways relaying the input of muscle contractions, in which VAB-10A plays an important initiating role, lead to IF phosphorylation (Zhang et al., 2011).

Two-dimensional gel electrophoresis (2-DE) coupled with immunoblotting is a frequently used method on protein phosphorylation in standard laboratory settings (Rabilloud, 2002). This method is based on the fact that phosphorylation induces an acidic shift on the isoelectric point of the protein by replacing the neutral hydroxyl groups on Ser/Thr/Tyr residues with negatively charged phosphate group(s). Proteins of different phosphorylation states can be separated by electrophoreses in a pH gradient based on their isoelectric points, followed by standard SDS-PAGE and immunoblotting. Here, we describe a detailed protocol for phosphorylation analysis of *C. elegans* IFs using 2-DE coupled with immunoblotting (Fig. 5).

4.1 Reagents

Prepare the following solutions and keep at −20 °C: 9.8 *M* urea (2 mL aliquots), 40% CHAPS (500 μL aliquots), 1 *M* DTT (1 mL aliquot), Endo Nuclease (Sigma), cOmplete Protease inhibitor 100× (Roche), PhosSTOP Phosphatase inhibitor 100× (Roche).

Prepare the equilibration base buffer: urea 18 g, 5 mL of 20% SDS, 1.65 mL of 1.5 *M* Tris/HCl, pH 8.8, and 11.5 mL of 87% glycerol. Adjust to 50 mL with H_2O.

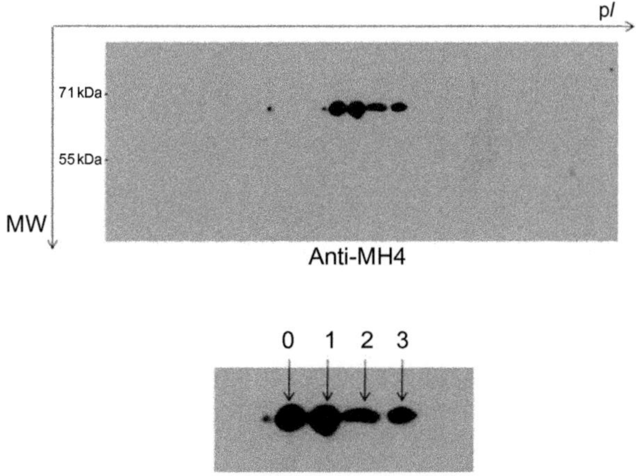

Figure 5 Two-dimensional immunoblotting analysis showing the phosphorylation status of *C. elegans* IFA proteins. Arrows point to distinct species separated by isoelectric focusing using, pH 5–8, IPG strips, followed by SDS-PAGE and immunoblotting of MH4 antibody recognizing *C. elegans* intermediate filament A. The experiment was performed as described in Section 4 of this chapter. The predicted pI value of each phosphorylated form is calculated by Scansite program (http://scansite.mit.edu). pI, isoelectric point; MW, molecular weight.

For the DTT equilibration buffer, add 1 g DTT to 50 mL base buffer.

Prepare 40% Bio-lyte ampholyte, pH 3–10 (BioRad, suitable for IPG strips at all pH) Iodoacetamide, powder and keep at 4 °C.

Prepare the sealing agarose: 0.5% low-melting agarose, 0.01% bromophenol blue and adjust to 10 mL with H_2O. Melt in a 75 °C water bath and store at RT.

Prepare 1 mL of rehydration buffer (freshly made before each experiment) as follow: 9.8 *M* urea 816 μL, 40% CHAPS 75 μL, 1 *M* DTT 50 μL, 40% Bio-lyte ampholyte 5 μL, Endo Nuclease (Sigma) 1 μL, Protease inhibitor 10 μL, Phosphatase inhibitor 10 μL and adjust to 1 mL with H_2O.

4.2 Equipment

BioRad PROTEAN® i12™ IEF System, 11 cm IPG strips, pH 5–8, Mineral oil (Bio-rad).

4.3 Procedure

Step One: Rehydration

- Homogenize samples in rehydration buffer on ice slush using a tissue grinder. Check under a dissecting microscope to ensure that tissues are homogenized. Centrifuge to remove debris.
- Measure the protein concentration (blank must contain the same volume of rehydration buffer). Adjust protein concentration to 0.5 μg/μL with rehydration buffer.
- Take out the desired number of IPG strips that have been stored at −20 °C and allow them to equilibrate to RT by incubation for at least 5 min.
- Apply the sample carefully onto IEF cells (200 μL/well for 11 cm strips), so that it covers approximately half of the well near the "+" electrode. Double-check to make sure that no bubbles have been introduced. If working with multiple samples, never add samples in wells adjacent to each other. Always leave at least one empty well between two samples.
- Using two forceps designated for handling IPG strips, carefully remove the plastic cover from the strip. Save the plastic covers and count afterward to make sure that the cover of all IPG strips had been properly removed.
- Hold the strip on the "−" end with one forceps, gel side down, and carefully put the strip in the IEF cell. Make sure the "+" end touch the "+" electrode first. Slowly lower the "−" end onto the "−" electrode to allow sample solution spreading over the entire strip. After sample solution is evenly spread, double-check for bubbles and remove any bubbles under the strip by gently tapping the strip using clean pipette tips.
- After all IPG strips have been put safely in the IEF cells, the cell tray is sealed with Saran Wrap to prevent evaporation of the sample solutions. Put a few layers of tissue paper on top of the Saran Wrap for better sealing. Cover the rehydration tray with the plastic cover and firmly press by adding a heavy object (e.g., a 500-mL bottle full of solution). Incubate the whole apparatus at 20 °C for 11–16 h.

Step Two: Isoelectric focusing

- Remove the Saran Wrap from the IEF cell. Double-check cells to make sure that most solution has been absorbed by the strips and no excessive sample solution remains in the cell.
- Wet a paper wick by dipping it into an Eppendorf tube filled with distilled H_2O. Allow water to spread evenly over the paper wick. Gently lift one end of the IPG strip from the electrode using forceps. Insert the

paper wick between the IPG strip and the electrode (the wire should be positioned at the center of the wick). Then gently place the strip back on top of the paper wick. After applying paper wicks on both ends of the IPG strip, double-check and remove any bubbles introduced during this process.

- After applying paper wicks on all IPG strips, install the IEF cell tray onto the IEF focusing machine, make sure that the electrodes attach to the corresponding "+" and "−" slots correctly.
- Apply mineral oil onto each IEF well with IPG strips (1.6 mL for 11 cm strip) slowly to allow even spreading of the oil. Double-check and remove any bubbles introduced during this process.
- Apply the cover of the cell tray, making sure that the wires are positioned at the center of the protruding blocks.
- Close and secure the cover of the machine. Program the machine according to BioRad instructions. Monitor the program during the first 1–2 h for potential arcing accidents and make sure the voltage increase is normal. Connect the machine to the printer and print reports every 5 min if necessary.
- After the program is finished, remove the IEF cell tray from the machine. IPG strips together with the tray can be wrapped in foil and stored at −20 °C for 1 week before the second dimension.
- Clean IEF cell tray using hot water and soap. Rinse with deionized water before storage.

Step Three: SDS-PAGE

- Cast the gel in a cassette designated for IPG strips. The surface of the gel should be about 10–15 mm from the top edge of the cassette.
- Take out the equilibration base buffer and DTT equilibration buffer stock from −20 °C and allow them to thaw out completely. Prepare a 75 °C water bath.
- Take out the IPG strips from −20 °C and allow them to equilibrate to RT. Remove excessive mineral oil by tapping strips on a filter paper. Transfer IPG strips gel side up to an equilibration tray designated for SDS buffers.
- Add DTT equilibration buffer to each well (1.6 mL for 11 cm strip) in the ventilation hood, cover equilibration tray with the plastic lid. Gently rock for 15 min at RT.
- During this period, dissolve iodoacetamide in equilibration base buffer (25 mg/mL).

- Carefully decant DTT buffer. Blot away excessive buffer using a tissue paper. Add iodoacetamide equilibration buffer (1.6 mL for 11 cm strip). Gently rock for 15 min at RT.
- During this period, put a tube of 0.5% low-melting agarose in 75 °C water bath. Agarose should be cooled down at RT for 5 min before use.
- Fill a 15-mL conical tube with 1 × SDS-PAGE running buffer. Hold the IPG strip on one end with forceps and briefly rinse in running buffer to remove residue equilibration buffer on the strip.
- Immerse the top of the polyacrylamide gel in running buffer. Carefully place the IPG strip on top of the gel surface, immersed in buffer. The plastic side of the strip should be in contact with the longer plate of the cassette. Make sure that no gaps or bubbles form between the IPG strip and the gel surface.
- Remove most running buffer by pipetting. Soak a piece of $5 \times 10\ mm^2$ filter paper with protein MW marker then insert the paper wick next to the "−" end of the IPG strip.
- Overlay the IPG strip and the filter paper with lukewarm molten low-melting agarose to secure them in place. Allow the agarose to solidify for 10 min.
- Mark the positions of the "+" and "−"end on the cassette. Assemble the apparatus and start the second dimension running. Do not move the cassette during the process.
- At the end of gel running, double-check the gel temperature (do not overheat). Proceed with standard western procedure (mark the position of the "+" and "−" end on the membrane after transfer).
- To detect epidermal IFs, use the MH4 monoclonal antibody (see above).

4.4 Tips

- Rehydration buffer should never be heated above 30 °C at any step during the experiment until proteins are safely separated by SDS-PAGE.
- Homogenized samples in rehydration buffer cannot sustain extensive storage, even in liquid nitrogen. Protein samples should always be made fresh and applied to the IEF cells as soon as possible.
- If the isoelectric focusing process does not run properly (arcing, super slow voltage increase or failure to reach maximum voltage), it is unlikely to get successful results.
- Recommended IEF program for IFA proteins is: 250 V (linear) 1 H, 1000 V (rapid) 1 H, 8000 V (linear) 4 H, 8000 V (rapid) to 40,000 V total, and 500 V hold.

5. CONCLUSION

Spectraplakins play important roles in multilayered tissues to organize cellular architecture and connections between cells. The large size and complex genomic organization of spectraplakin loci make very difficult the study of their function *in vivo*. During the past 10 years, *C. elegans* studies on VAB-10 proteins have brought important information about their roles during embryonic elongation, mechanotransduction, tissue invasion, and cellular migration. The implementation of novel genome-editing technologies by CRISPR to probe the role of specific domains coupled to high-resolution microscopy techniques is very promising to potentially unravel other roles of spectraplakins.

ACKNOWLEDGMENTS

Work in the M.L. laboratory is supported by Agence Nationale de la Recherche and European Research Council grants to M.L.; by institutional funds from the CNRS, INSERM, University of Strasbourg; and by the grant ANR-10-LABX-0030-INRT, a French State fund managed by the Agence Nationale de la Recherche under the frame programme Investissements d'Avenir labeled ANR-10-IDEX-0002-02 to the IGBMC.

REFERENCES

Andra, K., Lassmann, H., Bittner, R., Shorny, S., Fassler, R., Propst, F., et al. (1997). Targeted inactivation of plectin reveals essential function in maintaining the integrity of skin, muscle, and heart cytoarchitecture. *Genes & Development*, *11*(23), 3143–3156.

Applewhite, D. A., Grode, K. D., Keller, D., Zadeh, A. D., Slep, K. C., & Rogers, S. L. (2010). The spectraplakin short stop is an actin-microtubule cross-linker that contributes to organization of the microtubule network. *Molecular Biology of the Cell*, *21*(10), 1714–1724. http://dx.doi.org/10.1091/mbc.E10-01-0011.

Bosher, J. M., Hahn, B. S., Legouis, R., Sookhareea, S., Weimer, R. M., Gansmuller, A., et al. (2003). The Caenorhabditis elegans vab-10 spectraplakin isoforms protect the epidermis against internal and external forces. *The Journal of Cell Biology*, *161*(4), 757–768. http://dx.doi.org/10.1083/jcb.200302151.

Brenner, S. (1974). The genetics of Caenorhabditis elegans. *Genetics*, 77(1), 71–94.

Chanal, P., & Labouesse, M. (1997). A screen for genetic loci required for hypodermal cell and glial-like cell development during Caenorhabditis elegans embryogenesis. *Genetics*, *146*(1), 207–226.

Chisholm, A. D., & Hardin, J. (2005). Epidermal morphogenesis. *WormBook*, *1*, 1–22. http://dx.doi.org/10.1895/wormbook.1.35.1.

Choi, H. J., Park-Snyder, S., Pascoe, L. T., Green, K. J., & Weis, W. I. (2002). Structures of two intermediate filament-binding fragments of desmoplakin reveal a unique repeat motif structure. *Nature Structural Biology*, *9*(8), 612–620. http://dx.doi.org/10.1038/nsb818.

Coulombe, P. A., Kerns, M. L., & Fuchs, E. (2009). Epidermolysis bullosa simplex: A paradigm for disorders of tissue fragility. *The Journal of Clinical Investigation*, *119*(7), 1784–1793. http://dx.doi.org/10.1172/JCI38177.

Ding, M., Goncharov, A., Jin, Y., & Chisholm, A. D. (2003). C. elegans ankyrin repeat protein VAB-19 is a component of epidermal attachment structures and is essential for epidermal morphogenesis. *Development*, *130*(23), 5791–5801. http://dx.doi.org/10.1242/dev.00791.

Francis, R., & Waterston, R. H. (1991). Muscle cell attachment in Caenorhabditis elegans. *The Journal of Cell Biology*, *114*(3), 465–479.

Fraser, A. G., Kamath, R. S., Zipperlen, P., Martinez-Campos, M., Sohrmann, M., & Ahringer, J. (2000). Functional genomic analysis of C. elegans chromosome I by systematic RNA interference. *Nature*, *408*(6810), 325–330. http://dx.doi.org/10.1038/35042517.

Gally, C., Wissler, F., Zahreddine, H., Quintin, S., Landmann, F., & Labouesse, M. (2009). Myosin II regulation during C. elegans embryonic elongation: LET-502/ROCK, MRCK-1 and PAK-1, three kinases with different roles. *Development*, *136*(18), 3109–3119. http://dx.doi.org/10.1242/dev.039412.

Geisler, N., Schunemann, J., Weber, K., Haner, M., & Aebi, U. (1998). Assembly and architecture of invertebrate cytoplasmic intermediate filaments reconcile features of vertebrate cytoplasmic and nuclear lamin-type intermediate filaments. *Journal of Molecular Biology*, *282*(3), 601–617. http://dx.doi.org/10.1006/jmbi.1998.1995.

Gupta, B. P., Hanna-Rose, W., & Sternberg, P. W. (2012). Morphogenesis of the vulva and the vulval-uterine connection. *WormBook*, *30*, 1–20. http://dx.doi.org/10.1895/wormbook.1.152.1.

Hapiak, V., Hresko, M. C., Schriefer, L. A., Saiyasisongkhram, K., Bercher, M., & Plenefisch, J. (2003). mua-6, a gene required for tissue integrity in Caenorhabditis elegans, encodes a cytoplasmic intermediate filament. *Developmental Biology*, *263*(2), 330–342.

Hardin, J. (2011). Imaging embryonic morphogenesis in C. elegans. *Methods in Cell Biology*, *106*, 377–412. http://dx.doi.org/10.1016/B978-0-12-544172-8.00014-1.

Helmke, B. P., Goldman, R. D., & Davies, P. F. (2000). Rapid displacement of vimentin intermediate filaments in living endothelial cells exposed to flow. *Circulation Research*, *86*(7), 745–752.

Herrmann, H., & Wiche, G. (1987). Plectin and IFAP-300K are homologous proteins binding to microtubule-associated proteins 1 and 2 and to the 240-kilodalton subunit of spectrin. *The Journal of Biological Chemistry*, *262*(3), 1320–1325.

Hetherington, S., Gally, C., Fritz, J. A., Polanowska, J., Reboul, J., Schwab, Y., et al. (2011). PAT-12, a potential anti-nematode target, is a new spectraplakin partner essential for Caenorhabditis elegans hemidesmosome integrity and embryonic morphogenesis. *Developmental Biology*, *350*(2), 267–278. http://dx.doi.org/10.1016/j.ydbio.2010.11.025.

Hodgkin, J. (1983). Male phenotypes and mating efficiency in CAENORHABDITIS ELEGANS. *Genetics*, *103*(1), 43–64.

Hong, L., Elbl, T., Ward, J., Franzini-Armstrong, C., Rybicka, K. K., Gatewood, B. K., et al. (2001). MUP-4 is a novel transmembrane protein with functions in epithelial cell adhesion in Caenorhabditis elegans. *The Journal of Cell Biology*, *154*(2), 403–414.

Hresko, M. C., Schriefer, L. A., Shrimankar, P., & Waterston, R. H. (1999). Myotactin, a novel hypodermal protein involved in muscle-cell adhesion in Caenorhabditis elegans. *The Journal of Cell Biology*, *146*(3), 659–672.

Hyder, C. L., Pallari, H. M., Kochin, V., & Eriksson, J. E. (2008). Providing cellular signposts—Post-translational modifications of intermediate filaments. *FEBS Letters*, *582*(14), 2140–2148. http://dx.doi.org/10.1016/j.febslet.2008.04.064.

Kamath, R. S., Fraser, A. G., Dong, Y., Poulin, G., Durbin, R., Gotta, M., et al. (2003). Systematic functional analysis of the Caenorhabditis elegans genome using RNAi. *Nature*, *421*(6920), 231–237. http://dx.doi.org/10.1038/nature01278.

Kaminsky, R., Denison, C., Bening-Abu-Shach, U., Chisholm, A. D., Gygi, S. P., & Broday, L. (2009). SUMO regulates the assembly and function of a cytoplasmic intermediate filament protein in C. elegans. *Developmental Cell*, *17*(5), 724–735. http://dx.doi.org/10.1016/j.devcel.2009.10.005.

Karabinos, A., Schmidt, H., Harborth, J., Schnabel, R., & Weber, K. (2001). Essential roles for four cytoplasmic intermediate filament proteins in Caenorhabditis elegans development. *Proceedings of the National Academy of Sciences of the United States of America*, *98*(14), 7863–7868. http://dx.doi.org/10.1073/pnas.121169998.

Karabinos, A., Schulze, E., Schunemann, J., Parry, D. A., & Weber, K. (2003). In vivo and in vitro evidence that the four essential intermediate filament (IF) proteins A1, A2, A3 and B1 of the nematode Caenorhabditis elegans form an obligate heteropolymeric IF system. *Journal of Molecular Biology*, *333*(2), 307–319.

Kim, H. S., Murakami, R., Quintin, S., Mori, M., Ohkura, K., Tamai, K. K., et al. (2011). VAB-10 spectraplakin acts in cell and nuclear migration in Caenorhabditis elegans. *Development*, *138*(18), 4013–4023. http://dx.doi.org/10.1242/dev.059568.

Lee, M., & Cram, E. J. (2009). Quantitative analysis of distal tip cell migration in C. elegans. *Methods in Molecular Biology*, *571*, 125–136. http://dx.doi.org/10.1007/978-1-60761-198-1_8.

Lee, S., & Kolodziej, P. A. (2002). Short Stop provides an essential link between F-actin and microtubules during axon extension. *Development*, *129*(5), 1195–1204.

Lehner, B., Tischler, J., & Fraser, A. G. (2006). RNAi screens in Caenorhabditis elegans in a 96-well liquid format and their application to the systematic identification of genetic interactions. *Nature Protocols*, *1*(3), 1617–1620. http://dx.doi.org/10.1038/nprot.2006.245.

Leung, C. L., Sun, D., Zheng, M., Knowles, D. R., & Liem, R. K. (1999). Microtubule actin cross-linking factor (MACF): A hybrid of dystonin and dystrophin that can interact with the actin and microtubule cytoskeletons. *The Journal of Cell Biology*, *147*(6), 1275–1286.

Loschke, F., Seltmann, K., Bouameur, J. E., & Magin, T. M. (2015). Regulation of keratin network organization. *Current Opinion in Cell Biology*, *32C*, 56–64. http://dx.doi.org/10.1016/j.ceb.2014.12.006.

Matus, D. Q., Chang, E., Makohon-Moore, S. C., Hagedorn, M. A., Chi, Q., & Sherwood, D. R. (2014). Cell division and targeted cell cycle arrest opens and stabilizes basement membrane gaps. *Nature Communications*, *5*. http://dx.doi.org/10.1038/ncomms5184, 4184.

Morrissey, M. A., Keeley, D. P., Hagedorn, E. J., McClatchey, S. T., Chi, Q., Hall, D. H., et al. (2014). B-LINK: A hemicentin, plakin, and integrin-dependent adhesion system that links tissues by connecting adjacent basement membranes. *Developmental Cell*, *31*(3), 319–331. http://dx.doi.org/10.1016/j.devcel.2014.08.024.

Nikolic, B., Mac Nulty, E., Mir, B., & Wiche, G. (1996). Basic amino acid residue cluster within nuclear targeting sequence motif is essential for cytoplasmic plectin-vimentin network junctions. *The Journal of Cell Biology*, *134*(6), 1455–1467.

Omary, M. B., Ku, N. O., Tao, G. Z., Toivola, D. M., & Liao, J. (2006). "Heads and tails" of intermediate filament phosphorylation: Multiple sites and functional insights. *Trends in Biochemical Sciences*, *31*(7), 383–394. http://dx.doi.org/10.1016/j.tibs.2006.05.008.

Ortega, E., Buey, R. M., Sonnenberg, A., & de Pereda, J. M. (2011). The structure of the plakin domain of plectin reveals a non-canonical SH3 domain interacting with its fourth spectrin repeat. *The Journal of Biological Chemistry*, *286*(14), 12429–12438. http://dx.doi.org/10.1074/jbc.M110.197467.

Osmani, N., & Labouesse, M. (2015). Remodeling of keratin-coupled cell adhesion complexes. *Current Opinion in Cell Biology*, *32C*, 30–38. http://dx.doi.org/10.1016/j.ceb.2014.10.004.

Pascual, J., Pfuhl, M., Rivas, G., Pastore, A., & Saraste, M. (1996). The spectrin repeat folds into a three-helix bundle in solution. *FEBS Letters*, *383*(3), 201–207.

Pasti, G., & Labouesse, M. (2014). Epithelial junctions, cytoskeleton, and polarity. *WormBook*, *4*, 1–35. http://dx.doi.org/10.1895/wormbook.1.56.2.

Plenefisch, J. D., Zhu, X., & Hedgecock, E. M. (2000). Fragile skeletal muscle attachments in dystrophic mutants of Caenorhabditis elegans: Isolation and characterization of the mua genes. *Development*, *127*(6), 1197–1207.

Rabilloud, T. (2002). Two-dimensional gel electrophoresis in proteomics: Old, old fashioned, but it still climbs up the mountains. *Proteomics*, *2*(1), 3–10.

Ramms, L., Fabris, G., Windoffer, R., Schwarz, N., Springer, R., Zhou, C., et al. (2013). Keratins as the main component for the mechanical integrity of keratinocytes. *Proceedings of the National Academy of Sciences of the United States of America*, *110*(46), 18513–18518. http://dx.doi.org/10.1073/pnas.1313491110.

Rogalski, T. M., Williams, B. D., Mullen, G. P., & Moerman, D. G. (1993). Products of the unc-52 gene in Caenorhabditis elegans are homologous to the core protein of the mammalian basement membrane heparan sulfate proteoglycan. *Genes & Development*, 7(8), 1471–1484.

Schnabel, R. (1994). Autonomy and nonautonomy in cell fate specification of muscle in the Caenorhabditis elegans embryo: A reciprocal induction. *Science*, *263*(5152), 1449–1452.

Snider, N. T., & Omary, M. B. (2014). Post-translational modifications of intermediate filament proteins: Mechanisms and functions. *Nature Reviews. Molecular Cell Biology*, *15*(3), 163–177. http://dx.doi.org/10.1038/nrm3753.

Sonnenberg, A., & Liem, R. K. (2007). Plakins in development and disease. *Experimental Cell Research*, *313*(10), 2189–2203. http://dx.doi.org/10.1016/j.yexcr.2007.03.039.

Sun, D., Leung, C. L., & Liem, R. K. (2001). Characterization of the microtubule binding domain of microtubule actin crosslinking factor (MACF): Identification of a novel group of microtubule associated proteins. *Journal of Cell Science*, *114*(Pt. 1), 161–172.

Toivola, D. M., Strnad, P., Habtezion, A., & Omary, M. B. (2010). Intermediate filaments take the heat as stress proteins. *Trends in Cell Biology*, *20*(2), 79–91. http://dx.doi.org/10.1016/j.tcb.2009.11.004.

Valencia, R. G., Walko, G., Janda, L., Novacek, J., Mihailovska, E., Reipert, S., et al. (2013). Intermediate filament-associated cytolinker plectin 1c destabilizes microtubules in keratinocytes. *Molecular Biology of the Cell*, *24*(6), 768–784. http://dx.doi.org/10.1091/mbc.E12-06-0488.

Wong, M. C., & Schwarzbauer, J. E. (2012). Gonad morphogenesis and distal tip cell migration in the Caenorhabditis elegans hermaphrodite. *Wiley Interdisciplinary Reviews. Developmental Biology*, *1*(4), 519–531. http://dx.doi.org/10.1002/wdev.45.

Woo, W. M., Goncharov, A., Jin, Y., & Chisholm, A. D. (2004). Intermediate filaments are required for C. elegans epidermal elongation. *Developmental Biology*, *267*(1), 216–229. http://dx.doi.org/10.1016/j.ydbio.2003.11.007.

Zahreddine, H., Zhang, H., Diogon, M., Nagamatsu, Y., & Labouesse, M. (2010). CRT-1/calreticulin and the E3 ligase EEL-1/HUWE1 control hemidesmosome maturation in C. elegans development. *Current Biology*, *20*(4), 322–327. http://dx.doi.org/10.1016/j.cub.2009.12.061.

Zhang, H., & Labouesse, M. (2010). The making of hemidesmosome structures in vivo. *Developmental Dynamics*, *239*(5), 1465–1476. http://dx.doi.org/10.1002/dvdy.22255.

Zhang, H., Landmann, F., Zahreddine, H., Rodriguez, D., Koch, M., & Labouesse, M. (2011). A tension-induced mechanotransduction pathway promotes epithelial morphogenesis. *Nature*, *471*(7336), 99–103. http://dx.doi.org/10.1038/nature09765.

PART IV

Functional and Genetic Analysis of Lamin-Associated Proteins

CHAPTER TWENTY-ONE

Tagged Chromosomal Insertion Site System: A Method to Study Lamina-Associated Chromatin

Jennifer C. Harr, Karen L. Reddy[1]
Department of Biological Chemistry and Center for Epigenetics, Johns Hopkins University, Baltimore, MD, USA
[1]Corresponding author: e-mail address: kreddy4@jhmi.edu

Contents

Abstract

The three-dimensional (3D) organization of the genome is important for chromatin regulation. This organization is nonrandom and appears to be tightly correlated with or regulated by chromatin state and scaffolding proteins. To understand how specific DNA and chromatin elements contribute to the functional organization of the genome, we developed a new tool—the tagged chromosomal insertion site (TCIS) system—to identify and study minimal DNA sequences that drive nuclear compartmentalization and applied this system to specifically study the role of *cis* elements in targeting DNA to the nuclear lamina. The TCIS system allows Cre-recombinase-mediated site-directed integration of any DNA fragment into a locus tagged with *lacO* arrays, thus enabling both functional molecular studies and positional analysis of the altered locus. This system can be used to study the minimal DNA sequences that target the nuclear periphery (or other nuclear compartments), allowing researchers to understand how genome-wide results obtained, for example, by DNA adenine methyltransferase identification, chromosome conformation capture (HiC), or related methods, connect to the

Methods in Enzymology, Volume 569
ISSN 0076-6879
http://dx.doi.org/10.1016/bs.mie.2015.09.028

actual organization of DNA and chromosomes at the single-cell level. Finally, TCIS allows one to test roles for specific proteins in chromatin reorganization and to determine how changes in nuclear environment affect chromatin state and gene regulation at a single locus.

1. INTRODUCTION

The genome is organized on multiple levels such that the interaction of genes/genic elements with particular nuclear compartments influences their regulation and, conversely, specific chromatin states appear to be sequestered into specific regions of the nucleus (Harr et al., 2015; Luperchio, Wong, & Reddy, 2014; Reddy, Zullo, Bertolino, & Singh, 2008; Wong, Luperchio, & Reddy, 2014; Zullo et al., 2012). For example, the region near the inner nuclear membrane (INM) is enriched in heterochromatin, and many developmentally regulated and cell type-specific genes interact dynamically with this compartment, dependent on cell state and gene activity (Finlan et al., 2008; Harr et al., 2015; Kosak et al., 2002; Kumaran & Spector, 2008; Meister, Towbin, Pike, Ponti, & Gasser, 2010; Reddy et al., 2008; Szczerbal, Foster, & Bridger, 2009; Williams et al., 2006; Zullo et al., 2012). In mammalian cells, the INM includes a multitude of resident INM proteins that interact closely with nuclear intermediate filament proteins (B-type and A-type lamins) to form peripheral nuclear "lamina" networks. Lamins have diverse roles including nuclear structure and function, genome organization, and tissue-specific signaling and gene regulation (Wilson & Berk, 2010). The nuclear lamina may play a role in maintaining large 0.1–10 Mb regions of DNA, known as "lamina-associated domains" (LADs), at the nuclear periphery (Guelen et al., 2008; Peric-Hupkes et al., 2010). These LADs are correlated with the presence of specific histone modifications that silence genes (Bian, Khanna, Alvikas, & Belmont, 2013; Harr et al., 2015; Reddy et al., 2008; Towbin et al., 2012). How these large DNA regions interact and are maintained at the nuclear periphery, and the contributions of the nuclear lamina, resident INM proteins, transcription factors, and chromatin to their dynamic organization are important open questions in biology.

This chapter provides protocols for the tagged chromosomal insertion site (TCIS) system, which is designed to study how specific regions of DNA are targeted to a particular nuclear environment, such as the nuclear periphery. The TCIS system construct consists of a recombination cassette (to allow site-specific Cre-mediated heterologous recombination) and *lacO* arrays (for direct visualization in single cells) and is integrated as a single copy into the genome, either by random integration or via CRISPR-mediated insertion as diagrammed in Fig. 1 (Belmont, Li, Sudlow, & Robinett, 1999; Esperet et al., 2000; Feng et al., 1999). The inverted LoxP sites allow specific DNA sequences of interest to be introduced from a "switch" cassette, which consists of matching inverted LoxP sites interrupted by the DNA sequence of interest (Fig. 1). Thus, this system allows you to integrate a single copy of any DNA fragment into the genome by directed recombination into the TCIS site. Importantly, the adjacent *lacO* sites allow for direct visualization by microscopy of the integrated insert via bound fluorescent (EGFP-LacI) proteins. In addition, the *lacO* sites can be used to direct recruitment of any protein of interest (POI) through binding of a POI-LacI fusion protein. This system successfully identified specific DNA elements from LADs that are sufficient to target the nuclear periphery (Harr et al., 2015). TCIS also helped identify DNA-binding proteins involved in targeting chromatin to the nuclear periphery. Furthermore, a variety of genome-wide studies (e.g., ChIP, DamID, chromatin conformation capture, and other methods) are providing data about genome organization and specific DNA elements, chromatin states, or protein complexes; these relationships can be directly tested using the TCIS system (e.g., specific DNA element targeting to the nuclear periphery; Fig. 2). Most importantly, the TCIS system allows one to ask questions about 3D chromatin organization and dynamics at the single-cell level, and in living cells—questions that cannot be answered by genome-wide sequencing-based strategies. To enable both the immediate use and future novel adaptations of the TCIS system, the protocols below describe each stage in making and using TCIS constructs, and strategies for testing and manipulating the system.

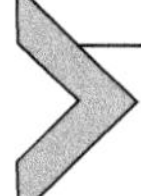

2. DESIGN AND PREPARATION OF THE TCIS SYSTEM CONSTRUCT

TCIS combines and slightly alters two previous technologies: recombination-mediated cassette exchange (RMCE) and the *lacO*/LacI system. RMCE is used to integrate any DNA sequence of interest, and the *lacO*/LacI system is used to visualize where this DNA localizes within the nucleus (Belmont et al., 1999; Feng et al., 1999). The RMCE element consists of inverted LoxP sites that surround cDNA encoding both hygromycin phosphotransferase and thymidine kinase (HYTK), allowing both positive

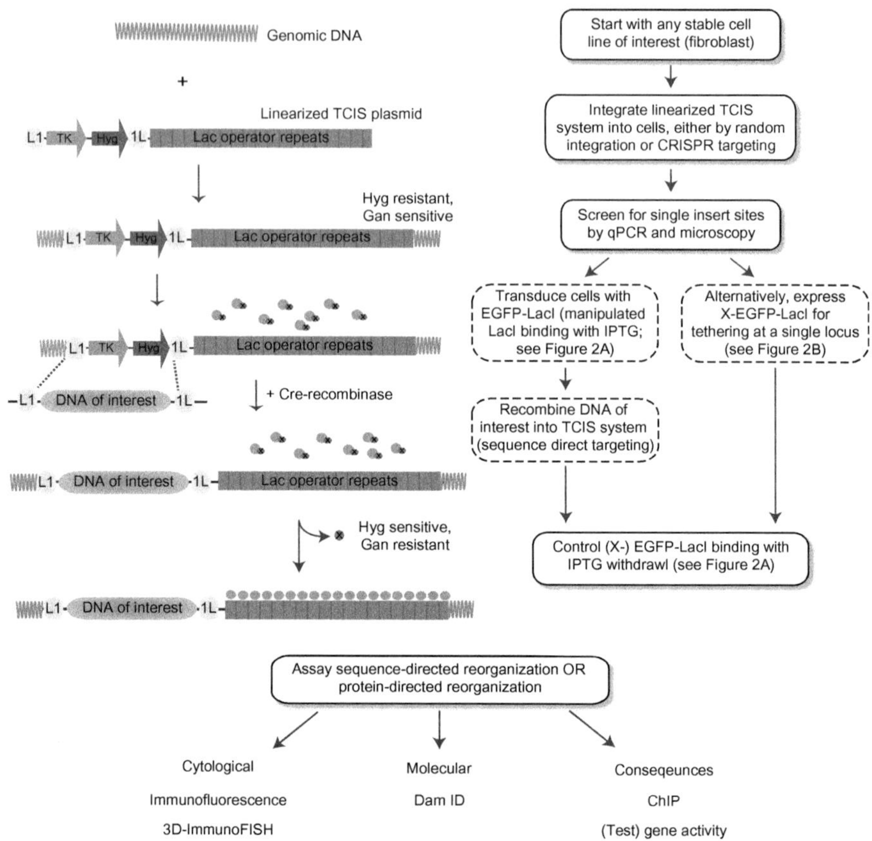

Figure 1 The TCIS system and a flow chart showing the main steps for creating TCIS cell lines and integrating a DNA fragment of interest. The TCIS system is designed to allow for site-specific integration of any "DNA of interest" into the genome. This is accomplished by integrating the TCIS construct into the genome of a cell line of choice (e.g., murine fibroblasts). The linearized TCIS construct (derived from two previous technologies, the *lacO*/LacI system and recombination-mediated cassette exchange [RMCE]) is integrated into the cell line. The TCIS construct includes a thymidine kinase (TK)/hygromycin gene, and successful integration renders cells hygromycin-resistant. Isolated individual clones are screened by qPCR and microscopy (visualized via EGFP-LacI protein accumulation at *lacO* arrays in the TCIS site) to identify those with single integrations. These cell lines can be used to study recruitment of specific proteins to the *lacO* array (protein of interest fused to EGFP-LacI) or to study the effects of specific DNA fragments inserted into the TCIS site. Binding of EGFP-LacI proteins is controlled by adding or removing IPTG. Without IPTG, EGFP-LacI proteins will bind the *lacO* arrays in the TCIS system. (See the color plate.)

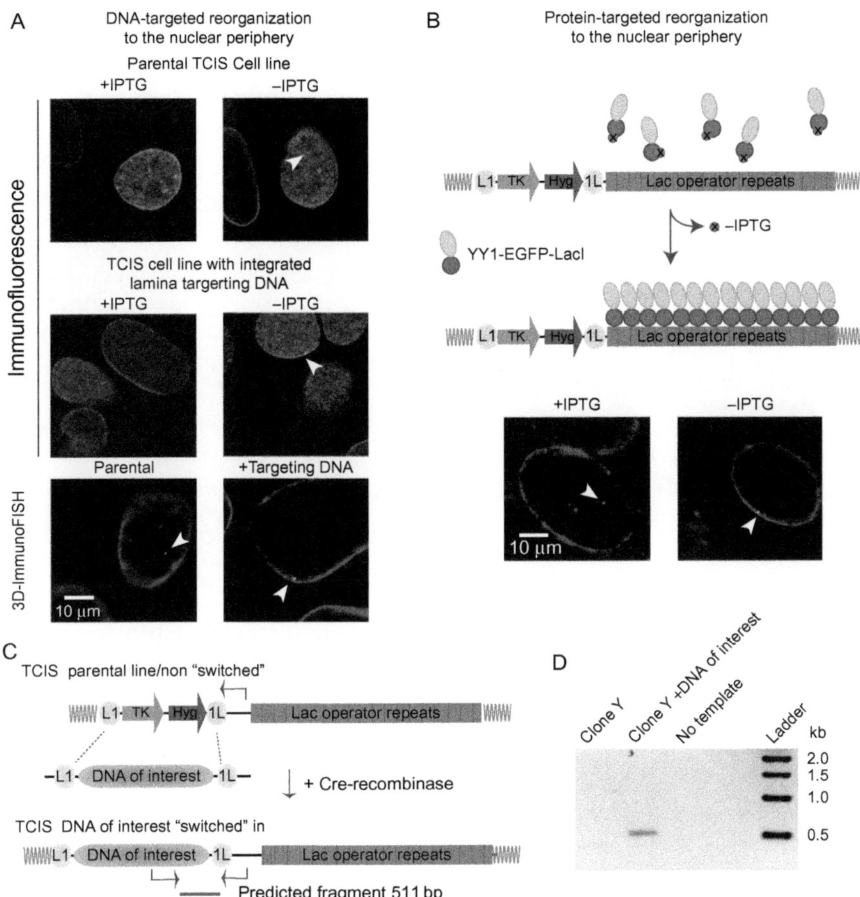

Figure 2 TCIS system and recruitment. (A) The TCIS system has been used to investigate the mechanisms by which chromatin regions relocalize to the nuclear periphery. Analysis by different methods—live cell imaging, immunofluorescence (IF), and 3D-immunoFISH—yields similar results. Interaction with a subcompartment of interest (e.g., the nuclear periphery) is determined by overlap of IF signals (e.g., EGFP-LacI and anti-Lamin B1). (B) The TCIS system is used to study protein-targeted reorganization of chromatin to a subcompartment of interest (e.g., the nuclear periphery) by tethering a protein of interest (POI) to the *lacO* arrays through EGFP-LacI (POI-EGFP-LacI). Top: schematic showing that IPTG inhibition of LacI binding to *lacO* is reversed by removal of IPTG. Bottom, YY1-EGFP-LacI tethering to the TCIS locus causes this locus to target the nuclear periphery. (C) The TCIS system contains a RMCE region that harbors a thymidine kinase hygromycin gene. This allows for negative selection against cells that did not recombine the DNA of interest. PCR can be used to verify the integration of your DNA of interest, using primers that flank the integrated DNA. Successful integration yields a gancyclovir-resistant culture and a PCR band of predicted size (D). (See the color plate.)

and negative selection by drugs (Fig. 1). The 256-repeat *lacO* array is integrated upstream of the LoxP site (Harr et al., 2015). The original TCIS constructs were linearized before transfection into mammalian cells to facilitate integration into the genome. However, we now have new constructs suitable for CRISPR-directed targeting (Cong et al., 2013; Jinek et al., 2012, 2013; Mali et al., 2013) and integration into the genome (Fig. 3). The following protocol will allow you to produce usable amounts of TCIS constructs (adapted from Reddy & Singh, 2008) and includes specific steps needed to ensure that the *lacO* arrays do not recombine during amplification in bacteria (Reddy et al., 2008).

TCIS constructs are available upon request to the Reddy Lab (kreddy4@jhmi.edu).

Recombination-deficient bacteria: We use Stbl3 competent *E. coli* (Life Technologies #C7373-03) whose genotype is: F^- *mcr*B *mrrhsd*S20 (r_B^-, m_B^-) *rec*A13 *sup*E44 *ara*-14 *gal*K2 *lac*Y1 *pro*A2 *rps*L20(Str^R)*xyl*-5 λ^-*leumtl*-1.

Generate TCIS plasmid DNA

2.1. Transform electro- or chemically competent recombination-deficient bacterial strain, such as Stbl3 (Life Technologies) with the TCIS construct. Recombination-deficient bacteria are essential when propagating repetitive or unstable DNA elements (e.g., the *lacO* array).

2.2. Plate transformed bacterial onto an LB agar plate under appropriate antibiotic selection (e.g., Ampicillin).

2.3. Grow at 30 °C overnight (~16 h). This lower incubation temperature helps reduce recombination events and loss of the full 256-repeat *lacO* array.

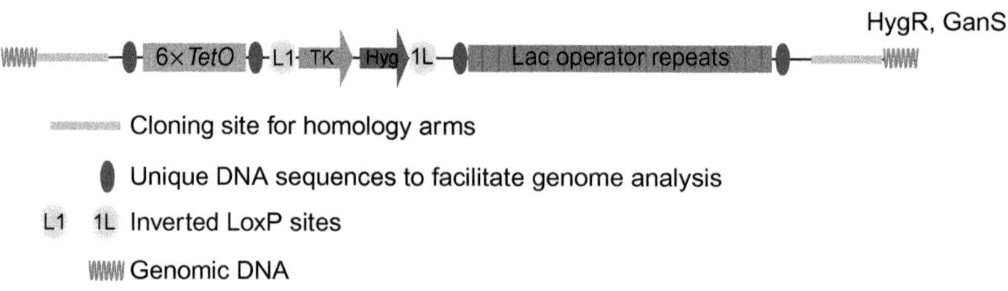

Figure 3 Second-generation tagged chromosomal insertion site (TCIS-2) for testing interrelationships between epigenetic changes and 3D localization. Improvements to TCIS include unique primer docking sites (orange (gray in the print version) ovals), *TetO*-binding sites to recruit proteins (e.g., histone modifying proteins, "HMP") and homology arms for CRISPR-mediated integration.

2.4. Early the next morning, pick 24–48 colonies. Inoculate each into 3 ml LB medium with appropriate selection and culture 10–12 h (same day) at 30 °C. Plan to prepare and screen DNA that same evening for best outcomes; this is a long day in the lab. The more colonies you screen, the better your chance of getting nonrecombined constructs. (Alternatively, you can culture overnight at 30 °C, but this involves refrigeration the next day and longer incubation times, increasing the likelihood of recombination.)

2.5. Prepare DNA from 2 ml of each culture to screen in step 2.6. Save the remaining ~1 ml on ice or at 4 °C; positives identified by screening will be used to inoculate large cultures in step 2.7.

2.6. Screen plasmids by digesting DNA with restriction enzymes that remove the *lacO* cassette, and electrophoresis in 1% agarose containing ethidium bromide or SYBR Safe. Our original TCIS construct can be cut using *Sal*I and *Xho*I (Belmont et al., 1999), which yields two bands at 6 and 10 kb (includes *lacO* repeats). Alternatively, you can digest with *Pvu*II, which yields two bands at 2.4 and 13.5 kb (includes *lacO* repeats). For the TCIS-2 construct, which enables CRISPR-mediated integration, restriction sites will have to be chosen carefully by the end user since the homology arms will harbor additional restriction sites. Different restriction conditions may apply to future modified TCIS constructs. The point here is to identify an intact plasmid that did not recombine during amplification in bacteria. Recombination yields multiple *lacO* array-containing bands, so it is most ideal to identify a restriction site that will yield a single band harboring the *lacO* array, while not cutting too frequently in the remaining plasmid. Recombined *lacO* arrays appear as smaller-than-expected band(s) or "ladders".

2.7. Choose at least six clones that appear intact (expected number of appropriately sized bands in step 2.6) for large-scale culture. Inoculate 250 ml LB medium containing the selection agent (ampicillin) with each reserved culture and grow for a "short" overnight at 30 °C—no more than 15–16 h.

2.8. The next morning, pellet ~2 ml of each culture, resuspend in LB in 50% glycerol, flash freeze on dry ice, and store at −80 °C for future use (for future experiments, you will need to streak from this frozen stock onto LB plates and prepare multiple cultures as described in steps 2.1–2.6, to insure full length *lacO* repeats). Use the rest of each culture to prepare DNA by a standard method (e.g., alkaline lysis). Digest a small aliquot of each (see step 2.6) to verify constructs did not recombine. You now have a large stock(s) of purified TCIS plasmid to integrate into your cells of choice in Section 3.

3. GENERATION OF STABLE MAMMALIAN TCIS CLONAL LINES VIA RANDOM INTEGRATION

This protocol (adapted from Reddy & Singh, 2008) takes 2–4 weeks and was used to create clonal TCIS-bearing lines of murine fibroblasts and pro-B cells. This protocol can be adapted to any cultured cells of interest. The TCIS construct can also be used as-is or in a recently adapted version (CRISPR targeted integration, TCIS-2), to create mouse lines where it could be present in every cell. The protocol below is based on the original TCIS construct but can be modified as needed for other (e.g., CRISPR-compatible) constructs. Cells are typically cultured under typical conditions (37 °C, high humidity, 5% CO_2).

3.1. To isolate and linearize the TCIS fragment: (a) cut 1–5 μg DNA with *Pvu*II, (b) resolve bands by electrophoresis in 1% agarose with SYBR Safe, (c) excise the desired DNA fragment using blue light illumination, avoiding UV illumination, and (d) purify using a gel extraction cleanup column for large DNA fragments (e.g., Geneaid DF100/DF300 or other silica-based columns designed for fragments >10 kb).

3.2. Transfect the isolated linearized TCIS fragment containing the *lac*O array and HYTK cassette (*Pvu*II digest, 13.5 kb fragment) into your cell line of choice. Transfection via electroporation or reagents such as Fugene6 (Promega #E2691) each work well. We recommend transfection at low DNA concentrations to increase the probability of single integrations. Perform pilot studies using the amount of DNA recommended for transient transfection protocols and also test 10-, 100- and 1000-fold higher levels. Ideally, a 150-mm plate will show very few colonies growing after selection (see below).

3.3. At 48 h posttransfection, place cultures under Hygromycin B (positive) selection. The concentration of hygromycin will vary depending on cell type; 500 μg/ml works well for TCIS integration into murine fibroblasts and up to1 mg/ml works well for selecting pro-B cells. Once colonies are visible, pick and seed each colony in individual wells of a 24-well plate. (For suspension cultures, grow cells under selection for 1 week, and then plate [via dilution or single-cell sorting] single cells in a 96-well plate and wait for them to grow up.) Maintain continuous growth and hygromycin selection until cells are confluent in a 12- or 6-well dish. Freeze at least half of each cell line for long-term storage in liquid nitrogen in culture media (without selection agents) plus 5–10% DMSO.

3.4. Split cells into two dishes: one dish for maintenance and one dish for analysis. Verify integration of the TCIS construct by quantitative PCR (qPCR) using the following primers: TCIS insert "C", 5′-GAT GTT GGC GAC CTC GTA TT-3′, and TCIS insert "W": 5′-ATT TCG GCT CCA ACA ATG TC-3′.

3.5. To determine the number of inserted TCIS elements, prepare genomic DNA (gDNA) from each cell line. Since single integrations are very difficult to obtain by random integration, this is the limiting step that will be ameliorated by future CRISPR-mediated targeting. Using qPCR and TCIS-specific primers (see step 3.4), compare the TCIS copy number to any single-copy gene. A single integration of TCIS will yield half as many copies as a single-copy gene in a diploid cell line. Since cultured cells can be aneuploid, we recommend testing several different single-copy genes; the TCIS element should be present at half (or less) of their average frequency. Including standard curves in the qPCR scheme allows a more direct comparison of DNA amount using Ct (threshold cycle) values. Our newer TCIS construct (TCIS-2) includes computer-generated sequences that are not found in the human or mouse genome to serve as docking sites for specific primers. Other options include primers specific for the HYTK cassette (see step 3.4) or regions flanking this cassette (TCIS_ChIP_F1, 5′-AGC TTG GCG TAA TCA TGG TC-3′, and ChIP_R5, 5′-ATT AGG CAC CCC AGG CTT TA-3′). Amplification of the *lacO* array is not recommended. Do not design primers that fall outside the linearized fragment. Caution is warranted using primers designed to the primer backbone as these are quite easily contaminated in a typical molecular or cell biology lab environment.

3.6. Grow up and freeze stocks (in culture media containing 5–10% DMSO) of each clonal line that contains a single copy of the TCIS integrant. These are your parental cell lines. Discard or segregate vials frozen in step 3.3 that do not contain single integrations.

3.7. The next step is to visually localize TCIS insertion sites within the nucleus of each clonal line. Your goal is to identify clones in which the (PCR-verified) single TCIS site is both visible by microscopy and positioned appropriately within the nuclear volume. This is accomplished by transiently transfecting or stably transducing an

EGFP-LacI construct into the cells to detect the integrated *lacO* arrays (Sections 5 and 6); hundreds of LacI proteins will bind the *lacO* array, creating a distinct focus of EGFP fluorescence. You are screening for cells that harbor a clear single integration of TCIS. These foci should be visible in 10–30% of cells (see Section 6), either by live microscopy or after fixation. If the background GFP fluorescence is high (e.g., if the EGFP-LacI construct was introduced by transient transfection), then visualize your TCIS insertions using 3D-ImmunoFISH with a *lacO* array DNA probe (see Section 6). The TCIS focus should localize similarly whether imaged by immunofluorescence, 3D-ImmunoFISH or in living cells. When EGFP-LacI is present in the cell line, you must always culture in the presence of 1 m*M* IPTG (see Section 6 for further considerations) unless an experiment is being performed. Withdraw IPTG *only* when conducting an experiment. Choose TCIS clones that are clearly single copy and have an integration site location compatible with your studies. Our studies focused on DNA sequences that are sufficient to target to the nuclear periphery, so we chose clones in which TCIS sites were located centrally in the nucleus (i.e., not in contact with the nuclear periphery, marked by lamin B1 staining; Fig. 2). We used TCIS to integrate DNA elements that might direct association with the nuclear periphery. Adapt these criteria to your needs.

These parental cell lines now contain a single TCIS integration, with or without EGPF-LacI. You should store liquid nitrogen aliquots of both parental cell lines—one with the EGFP-LacI protein and one without. Maintain these parental cell lines at 70–80% confluence when in culture (37 °C, high humidity, 5% CO_2) under appropriate selection at all times: hygromycin (500 μg/ml) for all TCIS lines, and both 1 μg/ml puromycin and 1 m*M* IPTG when EGFP-LacI is being expressed.

4. INSERT YOUR "DNA OF INTEREST" INTO A GENOMIC TCIS SITE

Any DNA of interest (DOI) can be recombined into a stably integrated TCIS site. For example, we chose to integrate large fragments of DNA from bacterial artificial chromosomes (BACs, BACPAC Children's Hospital Oakland Research Institute in Oakland, California, USA). To do this, you must first insert your DOI into a "switch" vector, flanked by inverted LoxP sites, to enable integration into the genomic TCIS site, between its inverted LoxP sites. Our switch vector was created by *Bam*HI digestion of the original RMCE vector L1HYTK1L (Feng et al., 1999) to remove the CMV promoter and hygromycin/thymidine kinase cassette, leaving behind the inverted LoxP sites. We then inserted the multiple cloning site (MCS) from pBlueScript between the LoxP sites (Harr et al., 2015). The MCS makes it easy to integrate your DOI into the "switch" vector, using standard protocols. (Recombination in bacteria during propagation is not an issue since this vector has no *lacO* repeats.) Before starting the protocol below, clone your DOI into the MCS of the switch vector. Note that your prepared switch vector with DOI and Cre-recombinase must be expressed *simultaneously* in the stable TCIS clone line. In addition, highly efficient transfection rates are required (90–100%) to facilitate successful integration of your DOI. Note that gancyclovir is used to select against cells that did not recombine successfully. Use typical cell culture conditions (37 °C, high humidity, 5% CO_2).

4.1. Defrost parental TCIS clone lines and grow to 70–80% confluence, under hygromycin selection (500 μg/ml).

4.2. Once cells reach desired confluence, remove media and wash cells once with PBS. Add 0.05% trypsin/EDTA to release adherent cells from the dish, quench cells with an equal amount of fresh (no drugs) media, and transfer to a sterile tube. Determine the concentration of cells. Each electroporation reaction will be for 10^6 cells. For non-adherent cells, spin cells down and wash once with fresh media.

4.3. Centrifuge 10^6 cells for 10 min at 90–100 $\times g$ at room temperature (25 °C) to pellet cells. Remove supernatant and resuspend each cell pellet in 110 μl of room-temperature electroporation solution (Mirus Bio LLC, MIR 50111 or Lonza electroporation products). Please note: we use the Amaxa electroporation system (Lonza, Nucleofector 4) for electroporation to ensure high efficiency of transfection. Actual electroporation conditions need to be determined empirically for the cell line of interest.

4.4. Transfer 110 μl cells (in electroporation solution) into a 1.5-ml tube preloaded with 1–5 μg highly purified plasmid DNA, in a maximum volume of 5 μl. For recombination, use a 1:1 mix containing 2.5 μg Cre-recombinase (Plasmid 24593, AAV-pgk-cre from Addgene, GenbankID: AY056050) and 2.5 μg "switch" vector harboring your DOI. Transfer the cells and DNA mixture to a cuvette, making sure the sample is at the bottom of the cuvette and no bubbles are present.

4.5. Electroporate cells using the appropriate Amaxa program (U-30 for murine fibroblasts). Immediately transfer cells from the cuvette to a

1.5-ml tube containing 500 μl prewarmed (37 °C) RPMI media and incubate cells 15 min at 37 °C. This incubation in calcium-free media allows cells to recover after electroporation and improves survival. Process no more than six samples at a time, to minimize the time cells spend in electroporation solution (<15 min).

4.6. Transfer cells to one well of a 6-well dish containing prewarmed (37 °C) growth media (e.g., DMEM high, 10% FBS, pen/strep and glutamine for fibroblasts) *lacking* hygromycin and place in tissue culture incubator. Do not disturb cells for 24 h. *Important*: Starting now, cells are not exposed to hygromycin, because successful recombination events remove the hygromycin phosphotransferase–thymidine kinase gene; hygromycin will kill the cells you want. This Amaxa protocol transfects nearly 100% of cells, with minimal cell death. Within 7 days posttransfection, all transient DNA is undetectable, reducing the likelihood of random integration events that would complicate analysis.

4.7. At 24–48 h posttransfection, seed 10,000 cells per well of a 6-well dish in DMEM high media (10% FBS, pen/strep, and glutamine) containing 1 μ*M* gancyclovir. Incubate 24–48 h. Cells that successfully recombined (inserted the DOI) will no longer contain the hygromycin phosphotransferase (Hyg) or thymidine kinase (TK) genes and will be therefore unaffected by gancyclovir. However, if recombination was not successful and the Hyg-TK cassette is still present, the TK protein will phosphorylate gancyclovir. Phosphorylated gancyclovir is toxic and is also released into the media, where it can potentially kill neighboring cells, even those that are gancyclovir resistant. To avoid this "toxic neighbor" effect, cells must be (a) transfected with very high efficiency and (b) maintained at low confluence during gancyclovir selection. Inadequate transfection rates or overconfluence will kill your cultures.

4.8. After 24–48 in gancylclovir-containing medium, replace the culture medium with DMEM high (10% FBS, pen/strep, and glutamine) *without gancyclovir*. Maintain cells at 10–40% confluence for the next 7 days, changing media every 24 h. The short (24–48 h) incubation with gancyclovir in step 4.7 represents a balance between selection and survival: nearly all TCIS parental cells (with or without EGFP-LacI) will be killed (you may see some stragglers), whereas the Cre-recombined cells should grow relatively robustly.

4.9. Expand each surviving TCIS line and store aliquots in liquid nitrogen for future use. These cells should be resistant to gancyclovir and sensitive to hygromycin, but are *no longer* cultured under selection. If desired, subclones of this cell line can be generated.

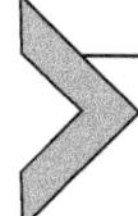

5. DESIGN AND USE A CUSTOMIZED (OR STANDARD) EGFP-LacI CONSTRUCT FOR TCIS VISUALIZATION

Inducible recruitment of EGFP-LacI to the *lacO* arrays of the TCIS system is used to visualize or manipulate the stable integration site. Accumulation of EGFP-LacI (after IPTG withdrawal) at the 256-copy *lacO* array enables visualization by live cell imaging or immunofluorescence. FISH to the *lacO* repeats themselves (not dependent on LacI binding) can also be employed for detection. The EGFP-LacI retroviral vector was made by ligating a *Dra*I fragment containing GFP-LacI from p3′ssEGFP-LacI into an *Hpa*I site in pMSCV-puro (Clonetech) (Reddy & Singh, 2008). *Customization*: The *lacO* array in TCIS can recruit any POI to the site of integration, in an IPTG-regulated manner, if your POI is fused to either EGFP-LacI or LacI alone. This allows you to test the functional consequences of POI recruitment to specific loci. Novel POI-EGFP-LacI fusion proteins are generated by ligating the POI cDNA (lacking a stop codon) upstream and in frame with EGFP-LacI in a retroviral vector (pMSCV-puro-EGFP-LacI; Reddy et al., 2008). Using this system, we discovered that the transcription factor Ying-Yang 1 targets the TCIS locus to the nuclear periphery (Fig. 2), as assayed by live cell imaging and FISH, and confirmed by chromatin immunoprecipitation (ChIP) and DamID (DNA Adenine Methyltransferase Identification) analysis (Harr et al., 2015). The following protocol describes how to make stocks of the EGFP-LacI construct (pMSCV-puro-EGFP-LacI), with or without customization, and use them to visualize TCIS insertion sites in the genome.

5.1. Transiently transfect pMSCV-puro-EGFP-LacI into a viral packaging line (PlatE, Platinum-E, retroviral packaging Cell line, Cell Biolabs, Inc.). After 12–16 h, remove media and replace with fresh medium (DMEM high, 10% FBS, pen/strep, L-glutamine).

5.2. After an additional 8–12 h (24 h after transient transfection), collect the medium (this is your first virus-containing supernatant) and provide fresh medium. Virus-containing supernatant is collected every 12 h thereafter (i.e., at 36, 48, and 60 h after transfection). Do not collect viral supernatant prior to the 24 h timepoint.

5.3. Centrifuge the collected medium (containing EGFP-LacI packaged retrovirus) for 5 min at 500 × *g* (4 °C) to remove cell debris. The recovered supernatant can be used immediately or snap-frozen in liquid nitrogen and stored at −80 °C for later use. *Note*: viral infectivity decreases after freeze/thaw.

5.4. Infect the parental TCIS cell lines with EGFP-LacI virus. E.g., we culture murine fibroblast lines in viral supernatant with 0.5 μg/ml polybrene and 1 m*M* IPTG for 24 h. Follow specific transduction/infection protocols suitable for your cells. *Note*: it is important to introduce the virus by transduction (not transient transfection) so that the entire population of cells will express EGFP-LacI at consistent, persistent, and moderate levels. It is also important to infect cells in the presence of IPTG.

5.5. Remove the virus-containing medium and replace with fresh medium such as DMEM high (10% FBS, pen/strep, and glutamine) containing 1 μg/ml puromycin (to select for EGFP-LacI) and 1 m*M* IPTG (to induce expression of your POI). Adjust medium for your cell type, if needed.

5.6. Image cells as described in Section 6.

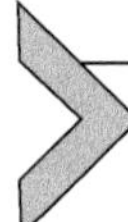

6. DETECT AND LOCALIZE TCIS SITES BY IMMUNOFLUORESCENCE

Grow adherent cells on coverslips (for suspension cells, attach to poly-L-lysine-coated coverslips), e.g., 25-mm round coverslips in 6-well plates or 12-mm coverslips in 24-well plates. Cells must not be allowed to dry out during the following procedures. For *DNA targeting experiments*, you should have (a) control cells containing the parental TCIS insert, (b) cells containing the integrated DOI (experimental line), and possibly (c) cells containing an integrated control DNA fragment. All of these are grown without IPTG and without selection.

For *tethered fusion protein experiments*, you should have two sets (duplicate coverslips) for cells harboring either EGFP-LacI (control fusion) or POI-fused GFP-LacI (experimental fusion). One set of coverslips is cultured with IPTG and the other without IPTG. The POI-GFP-LacI cells minus IPTG are the experimental cells.

6.1. Plate cells on irradiated coverslips (in a 6- or 12-well culture dish) with selection agents and 1 m*M* IPTG. Once cells reach 50% confluence, remove media and replace with media containing 1 m*M* IPTG (control) or media alone (allowing visualization). Do not include selection reagents after this point.

6.2. Grow cells 24–48 h until they reach 75–90% confluence. Do not overgrow; cells will start to peel off or overlap too much for imaging.

6.3. Fix cells on coverslips by incubating 10–15 min in PBS containing 4% formaldehyde. Fixation time is determined empirically for each cell type; we fix murine fibroblasts for 15 min, pro-B cells for 12 min.

6.4. Rinse three times (2–3 min each) in PBS to wash out all fixative.

6.5. Permeabilize cells by incubating 10–20 min in PBS containing 0.5% Triton X-100. This also helps "clear" the nucleus of unbound EGFP-LacI proteins, making the GFP focus more visible.

6.6. Rinse three times (5 min each) in PBS to remove all detergent.

6.7. Block in PBS containing 4% BSA for 1–2 h at room temperature in a humid chamber (e.g., soak a paper towel in water, place in a covered container with your coverslips), as follows. (a) Prepare a piece of parafilm by placing a little "pond" of blocking solution for each coverslip: ~100 μl for each 25-mm round coverslip or ~50 μl per 12-mm coverslip. (b) Place each coverslip, cells facing down, onto the blocking solution. If your antibodies give high background, block with PBS containing 4% BSA and 0.05% Triton X-100. Do not block with serum; serum does not work as well, or as consistently, as BSA (or BSA plus Triton).

6.8. Dilute your primary antibodies in blocking solution. Start with the recommended concentration for immunofluorescence. Concentrations recommended for Western blotting or ChIP are not necessarily ideal for immunofluorescence; test several dilutions to determine which gives the best signal to noise ratio. Use a final volume consistent with coverslip size (e.g., 80–100 μl antibody solution per 25-mm round coverslip).

6.9. On a new piece of parafilm, place "drops" comprising the diluted antibodies and transfer each coverslip from the blocking solution onto the antibody solution (cells down, as above). If a coverslip is difficult to remove, pipette ~100 μl blocking solution under its edge to help loosen it.

6.10. Incubate in a humid chamber at room temperature at least 1 h at room temperature or overnight at 4 °C.

6.11. Wash coverslips three times (10 min each) in PBS. If more stringent conditions are needed, wash in PBS containing 0.1% Triton-X100. Gently lift coverslips from parafilm and place (cells facing up) in 6-well plates and add wash solution to the wells. Aspirate wash solution and then replace with fresh wash buffer. The coverslips never leave the dish during these washes. The coverslips should never dry out.

6.12. Dilute secondary antibodies in the blocking solution. In a humid chamber, place 100-μl "puddles" onto parafilm and place coverslips (cells-down) as described above. Incubate 1–2 h at room temperature. Avoid overnight incubation in secondary antibodies.

6.13. Gently return coverslips to 6-well plates (cells facing up) and wash three times (10 min each) in PBS as above.

6.14. If needed, stain cells in DAPI or Hoechst (1/10,000 dilution in PBS; 10 μg/ml stock solution) for 30–60 s (in the same 6-well dish after aspirating the last wash). This short incubation time is essential to avoid bright Hoechst/DAPI signals that block visualization of fine chromatin organization details. Rinse once with PBS. Do not use DAPI-containing mounting medium.

6.15. Prepare slides by adding 25 μl VECTASHIELD (H-1000, Vector laboratories) or Slow-Fade Gold (s36937, Life Sciences) mounting media. Gently dry each coverslip by touching its edge to a dry paper towel and invert the coverslip (cells down) onto the mounting media. Blot excess mounting media by very gently blotting with a paper towel. Seal the coverslip with nail polish. Allow slides to dry (a few hours or overnight) in the dark before imaging. This allows the mounting media, which contains an antifade solution, to fully penetrate the sample. Slides can be stored at 4 °C for a few weeks.

Imaging notes: TCIS foci can be difficult to see. They are easier to see via immunofluorescence than by live cell imaging (likely because the permeabilization step "clears" unbound EGFP-LacI), but they do take time to find. We find that foci are best located using the computer interface on a digital microscope camera system. We typically pan from field to field, and adjust Z positions (focus up and down) while viewing on the computer screen. In a typical experiment, 20–40% of cells will display obvious foci. To obtain statistically significant results, you will need to quantify subnuclear localization of the TCIS site with respect to a reference nuclear landmark (e.g., nuclear lamina) in two independent experiments with a total of 50–100 nuclei for each condition. Scoring is based on the extent of overlap or distance from your selected landmark.

7. APPLICATIONS

This section outlines experiments you can do to test the activity of a specific DNA fragment (within the TCIS), or a tethered POI, in terms of targeting to a specific subnuclear compartment.

To investigate disposition of TCIS-integrated DNA fragments

Thaw three cell lines containing (a) the DOI integrated into TCIS (experimental), (b) a control DNA fragment integrated into TCIS, and (c) the parental TCIS cell line (control to determine original disposition of locus). All three lines should also harbor EGFP-LacI. Grow cells in appropriate selection media (1 m*M* IPTG, and/or puromycin, and/or hygromycin), until the experiment begins—not allowing cells to reach confluence. Experiments ideally begin when cultures reach ~80–90% confluence, at which point cells are plated onto coverslips or slides for live cell, immunofluorescence, or fluorescent *in situ* hybridization protocols. For example, TCIS parental lines with EGFP-LacI are grown in hygromycin, puromycin, and IPTG. "Switched" TCIS clones are grown in puromycin and IPTG. These slides/coverslips are then processed for imaging analyses, and the EGFP-LacI foci are scored (percent of cells displaying focus and disposition of focus relative to a nuclear landmark) for each condition. The difference between parental disposition and clones harboring a specific DOI can be determined by imaging 50–100 nuclei.

To investigate POI-mediated targeting

Thaw two TCIS cell lines containing (a) EGFP-LacI (control cells), or (b) POI-EGFP-LacI (experimental cells), and culture in the presence of 1 m*M* IPTG (to prevent precocious recruitment of the POI). If the TCIS integration is in the parental configuration (i.e., no DOI switched in), the experimental (POI-EGFP-LacI) cells should also be kept under hygromycin (500 μg/ml). When cultures reach 70–80% confluence, plate onto coverslips (immunofluorescence), slides (ImmunoFISH) or a 6-well dish (gDNA or protein lysates) depending on the experiment (see Section 6). Cells are cultured either with IPTG (control) or without IPTG (experimental/recruitment) for 24–48 h before harvesting for imaging, gDNA extraction, or making protein lysates. Hygromycin can be omitted at this point.

Detect DNA reorganization by immunofluorescence: Which assay to use?

The TCIS system allows you to visualize changes in chromatin localization in living cells or by immunofluorescence or FISH analysis in fixed cells. For quantitative studies, we recommend immunofluorescence (protocol provided in Section 6), because TCIS sites are easier to see in fixed and permeabilized cells due to lower GFP backgrounds. For live cell imaging, we recommend either epifluorescence microscope or spinning disk confocal microscopy. To visualize TCIS sites and endogenous proteins (regardless of LacI binding), we recommend 3D-immunoFISH or immunofluorescence. Some POI-fused EGFP-LacI proteins have punctate localization patterns in the nucleus, making it difficult or impossible to detect specific LacI/*lacO* signals at the TCIS site (e.g., YY1-EGFP-LacI in Fig. 2). To image these cells, and parental control cells that lack EGFP-LacI, localization via 3D-immunoFISH is essential. Finally, 3D-ImmunoFISH is very efficient and detects TCIS sites in virtually 100% of cells, unlike live cell or immunofluorescence techniques in which only 20–40% of cells have foci visible above the diffuse nucleoplasmic background. 3D-immunoFISH protocols are not described here; see Solovei et al. (2002) and Reddy et al. (2008) for detailed information.

Further applications of the TCIS system

TCIS system was designed to visually track the targeting of a single locus within the nucleus. TCIS site visualization via binding of EGFP-LacI/*lacO* is easily assayed by microscopy but yields only cytological data. For molecular analysis, the DamID method has been adapted to assay molecular contact between DNA and the nuclear lamina or any nuclear proteins of interest (INM proteins, such as Lap2β.) This protocol was originally published by Greil, Moorman, and van Steensel (2006) and adapted for mammalian cells as described in Vogel, Peric-Hupkes, and van Steensel (2007), Guelen et al. (2008), Reddy et al. (2008), and Zullo et al. (2012). We find DamID is very useful for detecting *de novo* TCIS contact with the nuclear lamina and may be adaptable to other subnuclear compartments. A detailed protocol can be found in Greil et al. (2006).

Altered subnuclear localization can also manifest as changes in chromatin state. Indeed, we find that DNA targeted to the nuclear lamina adopts a more facultative heterochromatin state, including enrichment of histone H3 methylated at lysine 9 and 27 (H3K9me2/3 and H3K9me3) (Harr et al., 2015). These types of chromatin changes can be assayed by ChIP using established protocols (e.g., Carey, Peterson, & Smale, 2009) or commercial ChIP kits (e.g., Abcam, Millpore, Invitrogen). Finally, enrichment of DNA segments around binding sites of interacting proteins/complexes or even the LacI fusions can be detected by ChIP—all you need is a specific antibody.

For DamID and ChIP analysis, you must specifically detect the engineered TCIS locus. For the current TCIS system (TCIS-1), the following primer pairs are used: TCIS_ChIP_F1, AGC TTG GCG TAA TCA TGG TC, and ChIP_R5, ATT AGG CAC CCC AGG CTT TA. These primer pairs lie outside the switched region, but do not contain the *lacO* repeats. You will also need to design primers specific for other regions of the genome as positive and negative controls. For ChIP experiments, the ActB (beta-actin gene) promoter provides an excellent negative control when assaying for repressive (H3k9me2/3 and H3K27me3) modifications. In fibroblasts, the muscle-specific MyoD promoter serves as a positive control for silent modifications (Harr et al., 2015).

8. CONCLUDING REMARKS

These protocols provide an assay for the directed reorganization of chromatin to any nuclear subcompartment of interest. This system has been used to successfully identify DNA sequences (zip codes) from LAD border regions that mediate repositioning of a normally centrally disposed ectopic site to the nuclear periphery (Harr et al., 2015). These studies used the TCIS system to identify YY1 binding, and subsequent H3K27me3 modification of the locus, as important steps in relocating to the nuclear lamina (Harr et al., 2015). A second generation (TCIS-2) system, available in early 2016, will include several useful modifications that allow direct integration of TCIS-2 constructs via CRISPR-mediated homologous recombination (Fig. 3), and more facile detection and manipulation of TCIS integration sites via unique primer and protein (*TetO*) docking sites. Flexibility could also be extended in future by incorporating other inducible protein recruitment strategies, such as rapamycin-inducible dimerization of fusions to FKBP and FK1012 (DeRose, Miyamoto, & Inoue, 2013). In summary, the TCIS system can be used to examine any nuclear compartment of interest and enables the discovery of new targeting DNA elements and gene control mechanisms.

ACKNOWLEDGMENTS

We thank the Taverna and Feinberg laboratories for valuable scientific input. J.H. and K. L.R. acknowledge support from NIH (R01 GM106024).

REFERENCES

Belmont, A. S., Li, G., Sudlow, G., & Robinett, C. (1999). Visualization of large-scale chromatin structure and dynamics using the lac operator/lac repressor reporter system. *Methods in Cell Biology*, *58*, 203–222.

Bian, Q., Khanna, N., Alvikas, J., & Belmont, A. S. (2013). β-Globin cis-elements determine differential nuclear targeting through epigenetic modifications. *The Journal of Cell Biology*, *203*, 767–783.

Carey, M. F., Peterson, C. L., & Smale, S. T. (2009). Chromatin immunoprecipitation (ChIP). *Cold Spring Harbor Protocols*, *1*, 2. pdb.prot5279.

Cong, L., Ran, F. A., Cox, D., Lin, S., Barretto, R., Habib, N., et al. (2013). Multiplex genome engineering using CRISPR/Cas systems. *Science*, *339*, 819–823.

DeRose, R., Miyamoto, T., & Inoue, T. (2013). Manipulating signaling at will: Chemically-inducible dimerization (CID) techniques resolve problems in cell biology. *Pflügers Archiv: European Journal of Physiology*, *465*, 409–417.

Esperet, C., Sabatier, S., Deville, M. A., Ouazana, R., Bouhassira, E. E., Godet, J., et al. (2000). Non-erythroid genes inserted on either side of human HS-40 impair the activation of its natural alpha -globin gene targets without being themselves preferentially activated. *The Journal of Biological Chemistry*, *275*, 25831–25839.

Feng, Y. Q., Seibler, J., Alami, R., Eisen, A., Westerman, K. A., Leboulch, P., et al. (1999). Site-specific chromosomal integration in mammalian cells: Highly efficient CRE recombinase-mediated cassette exchange. *Journal of Molecular Biology*, *292*, 779–785.

Finlan, L. E., Sproul, D., Thomson, I., Boyle, S., Kerr, E., Perry, P., et al. (2008). Recruitment to the nuclear periphery can alter expression of genes in human cells. *PLoS Genetics*, *4*, e1000039.

Greil, F., Moorman, C., & van Steensel, B. (2006). DamID: Mapping of in vivo protein-genome interactions using tethered DNA adenine methyltransferase. *Methods in Enzymology*, *410*, 342–359.

Guelen, L., Pagie, L., Brasset, E., Meuleman, W., Faza, M. B., Talhout, W., et al. (2008). Domain organization of human chromosomes revealed by mapping of nuclear lamina interactions. *Nature*, *453*, 948–951.

Harr, J. C., Luperchio, T. R., Wong, X., Cohen, E., Wheelan, S. J., & Reddy, K. L. (2015). Directed targeting of chromatin to the nuclear lamina is mediated by chromatin state and A-type lamins. *Journal of Cell Biology*, *208*, 33–52.

Jinek, M., Chylinski, K., Fonfara, I., Hauer, M., Doudna, J. A., & Charpentier, E. (2012). A programmable dual-RNA-guided DNA endonuclease in adaptive bacterial immunity. *Science*, *337*, 816–821.

Jinek, M., East, A., Cheng, A., Lin, S., Ma, E., & Doudna, J. (2013). RNA-programmed genome editing in human cells. *eLife*, *2*, e00471.

Kosak, S. T., Skok, J. A., Medina, K. L., Riblet, R., Le Beau, M. M., Fisher, A. G., et al. (2002). Subnuclear compartmentalization of immunoglobulin loci during lymphocyte development. *Science*, *296*, 158–162.

Kumaran, R. I., & Spector, D. L. (2008). A genetic locus targeted to the nuclear periphery in living cells maintains its transcriptional competence. *Journal of Cell Biology*, *180*, 51–65.

Luperchio, T. R., Wong, X., & Reddy, K. L. (2014). Genome regulation at the peripheral zone: Lamina associated domains in development and disease. *Current Opinion in Genetics & Development*, *25C*, 50–61.

Mali, P., Yang, L., Esvelt, K. M., Aach, J., Guell, M., DiCarlo, J. E., et al. (2013). RNA-guided human genome engineering via Cas9. *Science*, *339*, 823–826.

Meister, P., Towbin, B. D., Pike, B. L., Ponti, A., & Gasser, S. M. (2010). The spatial dynamics of tissue-specific promoters during *C. elegans* development. *Genes & Development*, *24*, 766–782.

Peric-Hupkes, D., Meuleman, W., Pagie, L., Bruggeman, S. W., Solovei, I., Brugman, W., et al. (2010). Molecular maps of the reorganization of genome-nuclear lamina interactions during differentiation. *Molecular Cell, 38*, 603–613.

Reddy, K. L., & Singh, H. (2008). Using molecular tethering to analyze the role of nuclear compartmentalization in the regulation of mammalian gene activity. *Methods, 45*, 242–251.

Reddy, K. L., Zullo, J. M., Bertolino, E., & Singh, H. (2008). Transcriptional repression mediated by repositioning of genes to the nuclear lamina. *Nature, 452*, 243–247.

Solovei, I., Cavallo, A., Schermelleh, L., Jaunin, F., Scasselati, C., Cmarko, D., et al. (2002). *Experimental Cell Research, 276*, 10–23.

Szczerbal, I., Foster, H. A., & Bridger, J. M. (2009). The spatial repositioning of adipogenesis genes is correlated with their expression status in a porcine mesenchymal stem cell adipogenesis model system. *Chromosoma, 118*, 647–663.

Towbin, B. D., González-Aguilera, C., Sack, R., Gaidatzis, D., Kalck, V., Meister, P., et al. (2012). Step-wise methylation of histone H3K9 positions heterochromatin at the nuclear periphery. *Cell, 150*, 934–947.

Vogel, M. J., Peric-Hupkes, D., & van Steensel, B. (2007). Detection of in vivo protein-DNA interactions using DamID in mammalian cells. *Nature Protocols, 2*, 1467–1478.

Williams, R. R., Azuara, V., Perry, P., Sauer, S., Dvorkina, M., Jorgensen, H., et al. (2006). Neural induction promotes large-scale chromatin reorganisation of the Mash1 locus. *Journal of Cell Science, 119*, 132–140.

Wilson, K. L., & Berk, J. M. (2010). The nuclear envelope at a glance. *Journal of Cell Science, 123*, 1973–1978.

Wong, X., Luperchio, T. R., & Reddy, K. L. (2014). NET gains and losses: The role of changing nuclear envelope proteomes in genome regulation. *Current Opinion in Cell Biology, 28C*, 105–120.

Zullo, J. M., Demarco, I. A., Pique-Regi, R., Gaffney, D. J., Epstein, C. B., Spooner, C. J., et al. (2012). DNA sequence-dependent compartmentalization and silencing of chromatin at the nuclear lamina. *Cell, 149*, 1474–1487.

CHAPTER TWENTY-TWO

Lamin-Binding Proteins in *Caenorhabditis elegans*

Agnieszka Dobrzynska*, Peter Askjaer*,[1], Yosef Gruenbaum†,[1]
*Andalusian Center for Developmental Biology, CSIC-Junta de Andalucia-Universidad Pablo de Olavide, Carretera de Utrera, Seville, Spain
†Department of Genetics, The Alexander Silberman Institute of Life Sciences, The Hebrew University of Jerusalem, Jerusalem, Israel
[1]Corresponding authors: e-mail address: pask@upo.es; gru@vms.huji.ac.il

Contents

Methods in Enzymology, Volume 569
ISSN 0076-6879
http://dx.doi.org/10.1016/bs.mie.2015.08.036

Abstract

The nuclear lamina, composed of lamins and numerous lamin-associated proteins, is required for mechanical stability, mechanosensing, chromatin organization, developmental gene regulation, mRNA transcription, DNA replication, nuclear assembly, and nuclear positioning. Mutations in lamins or lamin-binding proteins cause at least 18 distinct human diseases that affect specific tissues such as muscle, adipose, bone, nerve, or skin, and range from muscular dystrophies to lipodystrophy, peripheral neuropathy, or accelerated aging. *Caenorhabditis elegans* has unique advantages in studying lamin-binding proteins. These advantages include the low complexity of genes encoding lamin and lamin-binding proteins, advanced transgenic techniques, simple application of RNA interference, sophisticated genetic strategies, and a large collection of mutant lines. This chapter provides detailed and comprehensive protocols for the genetic and phenotypic analysis of lamin-binding proteins in *C. elegans*.

1. INTRODUCTION

1.1 The Nuclear Lamina

The nuclear envelope (NE) includes an outer nuclear membrane (ONM), an inner nuclear membrane (INM), nuclear pore complexes (NPCs), and nuclear lamina. The nuclear lamina is a complex structure positioned between the INM and peripheral chromatin. Lamins, as nuclear intermediate filaments, are the building blocks of the nuclear lamina. There are numerous lamin-associated proteins, many of which bind lamins either directly or indirectly (Wilson & Foisner, 2010). The nuclear lamina has many roles in cell and nuclear structure, mechanotransduction, higher order chromatin structure and provides a dynamic platform for protein complexes that regulate cell signaling. A small fraction of lamins resides in the nuclear interior and binds specific proteins involved in cell cycle regulation, DNA replication, and cell signaling (Dechat, Gesson, & Foisner, 2010). Mutations in lamins or lamin-binding proteins (LBPs) cause numerous heritable diseases, termed nuclear envelopathies, that show phenotypes in neurons, muscle, bone, fat, or skin cells, as well as premature aging (Worman, Ostlund, & Wang, 2010).

The nuclear lamina globally organizes the chromosomes whereby gene-poor chromosomes tend to localize closer to the NE and gene-rich chromosomes tend to localize in the interior (Croft et al., 1999). Chromatin associates with the nuclear lamina through lamina-associated domains (LADs) that are enriched in transcriptionally silent genes (Guelen et al., 2008).

Chromatin association with the nuclear lamina is dynamic and functionally nuanced (Zheng, Kim, & Zheng, 2015). For example, the nuclear lamina regulates the spatial positions of developmentally regulated genes within the nucleus (Mattout et al., 2011; Towbin et al., 2012).

1.2 Lamin-Binding Proteins

Proteomic analyses of isolated nuclear membranes revealed hundreds of NE transmembrane proteins (NETs) (Korfali et al., 2012; Schirmer, Florens, Guan, Yates, & Gerace, 2003). While some NETs are expressed ubiquitously, many others show tissue-specific expression (Korfali et al., 2012; Malik et al., 2010), raising the hypothesis that NETs define the tissue-specific functions of the nuclear lamina. Certain soluble proteins also enrich near the NE; these include chromatin modifiers, transcriptional regulators, and proteins that contribute to cell mechanics (Bank & Gruenbaum, 2011). Only a small fraction of LBPs have been experimentally shown to bind lamins directly (Simon & Wilson, 2013). Some LBPs can be grouped into families based on the presence of evolutionarily conserved *L*AP2, *e*merin, and *M*AN1 (LEM) or *S*ad1 and *UN*C-84 homology (SUN) domains. Other characterized mammalian LBPs include the lamin B receptor (Olins, Rhodes, Welch, Zwerger, & Olins, 2010) and lamin-associated protein 1 (LAP1) (Shin, Dauer, & Worman, 2014).

1.3 LBPs in *C. elegans*: LEM Proteins, BAF-1, SUN Proteins

The nematode *Caenorhabditis elegans* is an excellent model organism to study LBPs and decipher their biological functions. *C. elegans* has several known LBPs including the LEM-domain proteins emerin and LEM-2 (also called Ce-MAN1) (Liu et al., 2003), the SUN-domain protein UNC-84 (Lee et al., 2002), barrier-to-autointegration factor 1 (BAF-1) (Margalit, Brachner, Gotzmann, Foisner, & Gruenbaum, 2007), in addition to core histones, which also bind lamin (Mattout, Goldberg, Tzur, Margalit, & Gruenbaum, 2007). Thanks to the combined efforts of the *C. elegans* research community, an enormous variety of genetic tools and strains are freely available. Detailed information about *C. elegans* biology and breeding is accessible at www.wormbase.org, www.wormbook.org, and www.wormatlas.org. For *C. elegans* maintenance, see Stiernagle (2006). Relevant mutant alleles of LBPs and related reporters are listed in Tables 1 and 2.

Table 1 *C. elegans* LBPs

LBP[a]	Mutant Allele[b]	Specific Phenotypes	Reference
ANC-1 (KASH)	*e1873*	Nuclear anchorage defective	Starr and Han (2002)
BAF-1	*gk324*	Premature fusion of seam cells; larval arrest; sterile	Margalit, Neufeld, et al. (2007)
	t1639(ts)	Embryonic lethal and nuclear assembly defective at 25 °C	Gorjánácz et al. (2007)
EMR-1 (LEM)	*gk119(0)*	Decreased neuromuscular junction activity	González-Aguilera et al. (2014)
LEM-2 (LEM)	*tm1582(0)*	Reduced muscle activity and lifespan; nuclear positioning defective	Barkan et al. (2012) and Morales-Martínez, Dobrzynska, and Askjaer (2015)
LEM-3 (LEM)	*op444*	Irradiation-induced embryonic lethality	Dittrich et al. (2012)
LEM-4	*ax475(ts)*	Embryonic lethal at 25 °C; hyperphosphorylation of BAF-1	Asencio et al. (2012)
SUN-1 (SUN)	*jf18*	Defective in SUN-1 mobility during meiotic pairing of homologous chromosomes	Baudrimont et al. (2010)
	gk199	Sterile; reduced number of germ cells	Fridkin et al. (2004)
UNC-83 (KASH)	*e1408(0)*	Nuclear positioning defective; lethargic	Starr et al. (2001) and Sulston and Horvitz (1981)
UNC-84 (SUN)	*e1410(ts)*	Nuclear positioning defective	Malone, Fixsen, Horvitz, and Han (1999) and Sulston and Horvitz (1981)
ZYG-12 (KASH)	*or577(ts)*	Embryonic lethal at 25 °C; defective in centrosomes and nuclei attachment	Malone et al. (2003)

[a]Presence of KASH, LEM, or SUN domains is indicated in parentheses.
[b]Only selected mutant alleles are listed: (0) = null allele, (ts) = thermosensitive allele.

Table 2 *C. elegans* LBP Fluorescent and DamID Strains

LBP	Reporter	Reference
ANC-1	$P_{hsp\text{-}16.2}$::*NH2-anc-1::gfp*	Starr and Han (2002)
BAF-1	$P_{baf\text{-}1}$::*gfp::baf-1*	Margalit, Neufeld, et al. (2007)
	$P_{pie\text{-}1}$::*gfp::baf-1*	Gorjánácz et al. (2007)
EMR-1	$P_{emr\text{-}1}$::*emr-1::mCherry*	Morales-Martínez et al. (2015)
	$P_{emr\text{-}1}$::*emr-1::dendra2*	Morales-Martínez et al. (2015)
	$P_{emr\text{-}1}$::*emr-1::gfp*	Morales-Martínez et al. (2015)
	$P_{hsp\text{-}16.41}$::*Dam::Myc::emr-1*	González-Aguilera et al. (2014)
LEM-2	$P_{lem\text{-}2}$::*lem-2::gfp*	Morales-Martínez et al. (2015)
	$P_{lem\text{-}2}$::*lem-2::mCherry*	González-Aguilera et al. (2014)
LEM-3	$P_{npp\text{-}1}$::*yfp::lem-3*	Dittrich et al. (2012)
LEM-4	$P_{pie\text{-}1}$::*lem-4::gfp*	Asencio et al. (2012)
LMN-1	$P_{lmn\text{-}1}$::*lmn-1::gfp*	Liu et al. (2000)
	$P_{hsp\text{-}16.41}$::*Dam::Myc::lmn-1*	González-Aguilera et al. (2014)
SUN-1	$P_{sun\text{-}1}$::*sun-1::gfp*	Penkner et al. (2009)
UNC-84	$P_{unc\text{-}84}$::*unc-84A::gfp*	Malone et al. (1999)
ZYG-12	$P_{pie\text{-}1}$:*gfp::zyg-12*	Malone et al. (2003)

Nonexhaustive list of strains expressing *C. elegans* LBPs fused to fluorescent proteins or Dam.

1.3.1 LEM-Domain Proteins

The LEM-domain family of proteins is characterized by a ~40 residue sequence, which mediates binding to BAF-1. Four LEM proteins are annotated in the *C. elegans* genome: emerin (EMR-1), LEM-2, LEM-3, and LEM-4, although LEM-4 lacks a recognizable LEM domain (Asencio et al., 2012). Three are anchored at the NE; the exception is LEM-3 which has no transmembrane domain (Dittrich et al., 2012). Single *emr-1* and *lem-2* null mutants are viable and fertile; however, simultaneous depletion of both proteins leads to defects in chromosome segregation and early embryonic lethality (Barkan et al., 2012; Liu et al., 2003), suggesting that they have overlapping functions. EMR-1 and LEM-2 are expressed in all tissues (Gruenbaum, Lee, Liu, Cohen, & Wilson, 2002; Liu et al., 2003) but their relative distributions differ across tissues and they have different kinetics during nuclear assembly (Morales-Martínez et al., 2015; Fig. 1), indicating

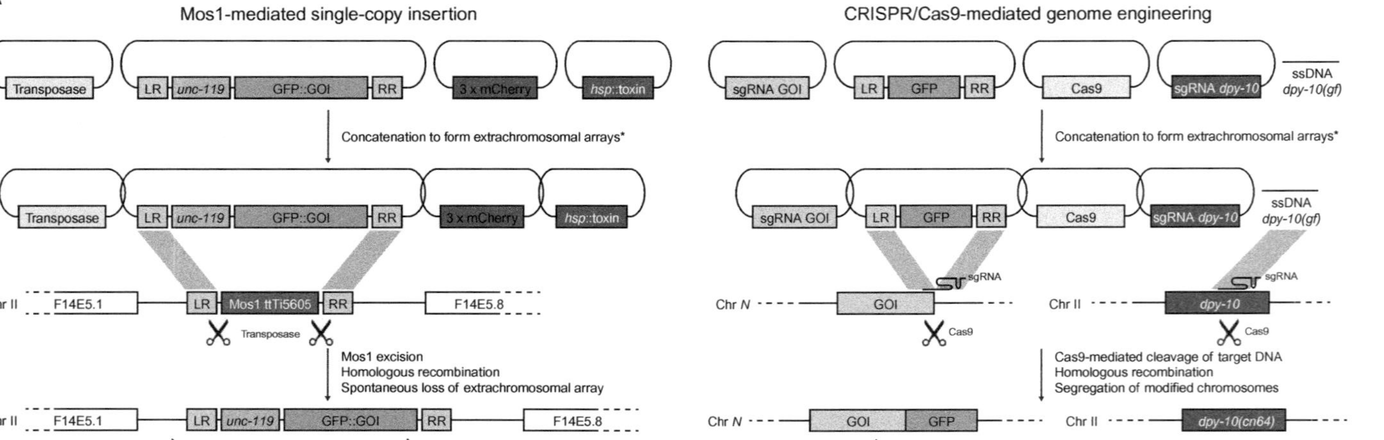

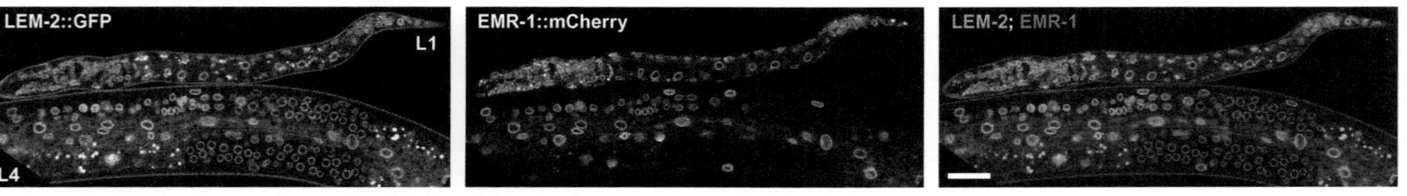

Figure 1 Transgenesis in *C. elegans*. (A) Mos1-mediated single-copy insertion (MosSCI; left) and CRISPR/Cas9-mediated genome engineering (right) are based on site-directed generation of double-stranded DNA breaks in the genome followed by repair via template-mediated homologous recombination. MosSCI is typically used to insert entire transgenes (e.g., gene of interest [GOI] fused to GFP) including transgenesis markers (e.g., *unc-119*), whereas CRISPR/Cas9 is mainly used to introduce protein tags into endogenous loci. However, both methods are versatile and offer many other options. The genomic position of an Mos1 transposon in the host strain determines the integration site in MosSCI, whereas sgRNA molecules are designed to guide the Cas9 nuclease to a genomic target site. Co-conversion to generate an easy tractable allele (e.g.,*dpy-10*(*cn64*)) may increase CRISPR/Cas9-mediated transgenesis. LR and RR denote homologous left and right recombination sequences, respectively. Asterisks indicate that genome modification may take place in the germ line of the injected animals or their progeny, in which case the injected DNA molecules presumably form extrachromosomal arrays. See text for details. (B) Confocal micrographs of an L1 larva and central part of an L4 larva expressing EMR-1::mCherry and LEM-2::GFP from single-copy transgenes (Morales-Martínez et al., 2015) generated by MosSCI.

distinct roles in nuclear organization and cell differentiation. For example, *emr-1* mutants have specific defects in neuromuscular junction activity (González-Aguilera et al., 2014), whereas *lem-2* mutants have reduced lifespan, muscle contraction defects (Barkan et al., 2012), and abnormal separation of daughter nuclei (Morales-Martínez et al., 2015). LEM-4 is specifically implicated in coordinating kinase(s) and phosphatase(s) that regulate BAF-1 during mitosis (Asencio et al., 2012).

1.3.2 BAF-1

BAF-1 is a highly conserved 10-kDa protein that directly binds double-stranded DNA, LEM-domain proteins, lamin, and histones. BAF-1 localizes predominantly to the NE but is also found in the nucleoplasm and cytoplasm (Margalit, Neufeld, et al., 2007; Margalit, Segura-Totten, Gruenbaum, & Wilson, 2005). The NE localization of BAF-1, lamin, and LEM proteins is interdependent, since depletion of EMR-1 and LEM-2 leads to mislocalization of BAF-1, and depletion of BAF-1 alters the distribution of lamin and LEM proteins (Liu et al., 2003; Margalit et al., 2005). Moreover, knockdown of BAF-1 by RNAi leads to embryonic lethality and defects in chromatin structure and chromosome segregation (Gorjánácz et al., 2007; Margalit et al., 2005). Many NE proteins are phosphorylated early in mitosis to induce NE disassembly and are later dephosphorylated as a prerequisite for nuclear reassembly. In particular, phosphorylation of BAF-1 by the vaccinia-related kinase 1 (VRK-1) is critical to release BAF-1 from chromatin in prophase (Gorjánácz et al., 2007). Interestingly VRK-1 is inhibited by LEM-4, which also recruits a specific PP2A phosphatase to facilitate NE reformation (Asencio et al., 2012). LEM-4 activity ensures rapid BAF-1 dephosphorylation (required for BAF-1 to bind chromatin) and the subsequent incorporation of other NE components (Asencio et al., 2012; Gorjánácz et al., 2007). During interphase, the association of BAF-1 with the nuclear lamina is highly dynamic (Bar et al., 2014; Margalit, Neufeld, et al., 2007). Interestingly, BAF-1 mobility is influenced by several types of stress, such as caloric restriction, food deprivation, and heat shock (Bar et al., 2014; Fig. 2).

1.3.3 SUN-Domain Proteins

The SUN domain is a highly conserved ~120-residue motif located at the C terminus of the *C. elegans* UNC-84 and matefin/SUN-1 proteins (Fridkin, Penkner, Jantsch, & Gruenbaum, 2009). The SUN domain is positioned in the lumenal space between the INM and ONM where it binds the

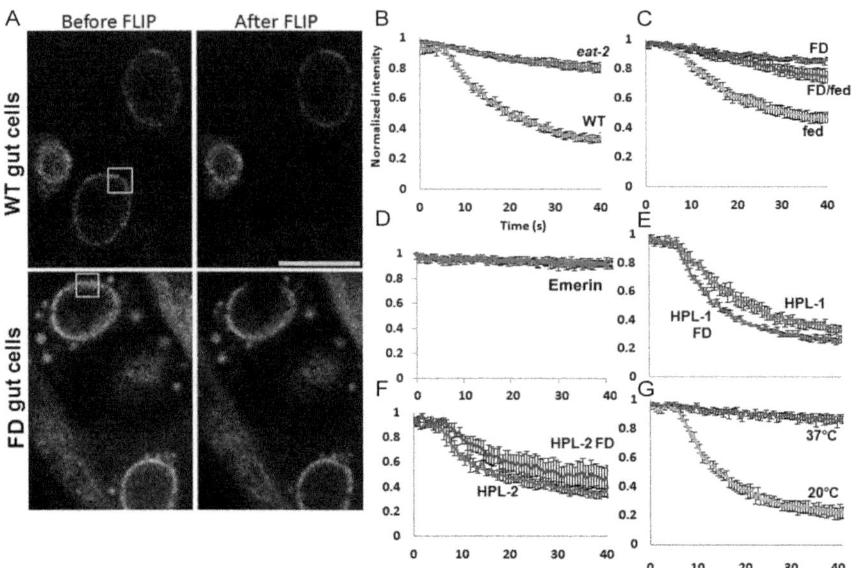

Figure 2 FLIP analysis of intestine cells in L1 larvae reveals reduced GFP::BAF-1 mobility in response to dietary restriction, food deprivation, or 1 h heat shock. (A) FLIP analysis of GFP::BAF-1 in intestine cells of L1 larvae in wild-type (upper panel) or *eat-2* animals (lower panel). Scale bar, 5 μm. (B) Mobility plot of the relative intensity of the GFP:: BAF-1 in wild-type (green line) and *eat-2* animals (purple). (C) Mobility plot of the relative intensity of the GFP::BAF-1 in *C. elegans* L1 larvae that were either well fed (blue line), animals that were food deprived (FD) overnight (red line), or following 2 h recovery from FD (green line). FLIP analysis of emerin (D), HPL-1 (E), and HPL-2 (F) fused to GFP did not reveal any difference in mobility after food deprivation. (G) Mobility plot of the relative intensity of GFP::BAF-1 in wild-type animals with time following 1 h heat shock at 37 °C. Error bars indicate SEM. For each experiment in (B), $n=7$; in (C–G), $n=6$. *X*-axis: time. *Y*-axis: normalized fluorescence intensity. Bleaching area: 2 μm^2. *Figure taken from Bar et al. (2014).* (See the color plate.)

KASH domain of ONM-localized KASH domain proteins (Sosa, Kutay, & Schwartz, 2013). Based on studies in *C. elegans*, Lee et al. (2002) were first to suggest the bridging model, in which the SUN–KASH protein complex, also known as LINC complexes, spans both nuclear membranes to connect the nuclear lamina to cytoskeletal structures.

The NE localization of UNC-84 depends on binding to lamin (Lee et al., 2002). UNC-84 is required to anchor the centrosome to the nuclear periphery in early embryonic stages. UNC-84 is also required to anchor nuclei within the hypodermal syncytium, and for nuclear migration in P cells during early embryogenesis, as well as for the migrations of the two distal gonadal tip cells (Starr & Fridolfsson, 2010). Migration failure of P cell nuclei causes P cell death, which leads to an uncoordinated (unc) phenotype.

The *C. elegans* Matefin/SUN-1 protein is expressed in the germline in adult hermaphrodites and is deposited to the embryos (Fridkin et al., 2004). The maternally deposited matefin/SUN-1 is required for embryogenesis because downregulation of *sun-1* causes embryonic death around the ~300-cell stage with defects in nuclear structure, DNA content, and chromatin morphology. Elimination of UNC-84 has no effect on larvae downregulated for *sun-1* (Fridkin et al., 2004). The LINC complex of matefin/SUN-1 and the KASH domain protein ZYG-12 is required for centrosome attachment in early embryonic divisions (Malone et al., 2003). This complex also plays an essential role in meiosis where it transmits cytoplasmic microtubule forces that allow chromosomal movement and homologous pairing while preventing nonhomologous synapsis (Penkner et al., 2009). In the meiotic transition zone, Matefin/SUN-1 aggregates at sites of chromosome attachment to the NE; the formation of these aggregates and dynamic chromosome movements require phosphorylation of matefin/SUN-1 (Penkner et al., 2009).

2. MOS1-MEDIATED SINGLE-COPY INSERTION

Two genome engineering methods are particularly relevant to study LBPs in *C. elegans*: Mos1-mediated single-copy integration (MosSCI) and CRISPR/Cas9. Both methods involve the induction and repair of DNA double-stranded breaks (DSBs) followed by homologous recombination. The major difference is that MosSCI is restricted to defined sites in the genome (except for the miniMos1 variant; Frøkjær-Jensen et al., 2014), whereas CRISPR/Cas9 can be directed to practically any site (Waaijers & Boxem, 2014). If the goal is to introduce exogenous reporters, or test a series of similar constructs (e.g., expression activity of promoter truncations or rescue capacity of different protein fragments), we recommend MosSCI because it integrates large transgenes with high efficiency. By contrast, if your goal is to study the behavior of endogenous genes by tagging and/or point mutations, CRISPR/Cas9 is preferable.

MosSCI involves replacing a Mos1 transposon in the *C. elegans* genome with a DNA sequence containing one or several transgenes (Frøkjær-Jensen, Davis, Ailion, & Jorgensen, 2012; Frøkjær-Jensen et al., 2008) (http://www.wormbuilder.org). Strains with single Mos1 insertions in intergenic loci on each of the six chromosomes (I–V, X) are available to facilitate genetic crosses to combine different transgenes and mutant alleles. Genomic sequences flanking the Mos1 transposon are included in the vector containing the transgene(s) to direct homologous recombination (Fig. 1). To ease the identification of transgenic animals, a selectable marker is integrated together with the transgene. Several markers are available for this purpose, including genes that rescue visible phenotypes in the host strain, and antibiotic resistance genes. In particular, a wild-type copy of the *unc-119* gene from sister species *Caenorhabditis briggsae* is widely used to complement the uncoordinated behavior of *C. elegans unc-119* mutants. The original MosSCI protocol can be used for transgenes up to 14–16 kb, whereas the miniMos system can handle constructs up to 45 kb. However, miniMos integration does not occur at defined sites and may therefore suffer position effects.

2.1 Molecular Cloning of Plasmids for Transgenesis

The most frequently used integration site for MosSCI is the *ttTi5605* locus on chromosome II. If downstream experiments involve other alleles on chromosome II, the *cxTi10882* locus on chromosome IV is a good alternative. New strains that carry insertions of the *ttTi5605* sequence in other chromosomes are also available, so the same vector can be used for transgene insertion into different loci (Frøkjær-Jensen et al., 2014). Targeting vectors for the *ttTi5605* locus include the Gateway compatible pCFJ150 and multiple cloning vectors pCFJ350 and pBN8. These vectors contain two ~1.5 kb recombination sequences and the *unc-119* gene from *C. briggsae*.

The strategy for cloning the transgene will naturally depend on your goals. Numerous promoters for expression in specific tissues are available, including pan-neuronal (*unc-33*, *unc-119*, *rgef-1*), muscle (*myo-3*), and intestinal (*elt-2*) promoters. Ubiquitous promoters that also produce robust expression in the germ line include *his-72*, *baf-1*, and *emr-1*. The choice of 3′-UTR can also greatly influence expression. We generally recommend using promoter and 3′-UTR sequences from the same gene.

2.2 Prepare the Injection Mixture

An injection mixture for MosSCI contains the plasmid carrying the transgene, a plasmid encoding the Mos1 transposase, a plasmid encoding a heat-shock inducible toxin for negative selection and a series of plasmids encoding either bright red or bright green fluorescent markers in the cytosol of pharyngeal and muscle cells and in most nuclei. The fluorescent markers are initially used to identify successful injection events, and later used to discriminate between animals carrying all the injected plasmids in extrachromosomal arrays and animals in which the transgene is integrated into the relevant locus. Both classes of animals have the *unc-119* rescue gene and will move like wild type, whereas only the latter (integrated transgene) will lose expression of the fluorescent markers. The heat-shock inducible toxin facilitates negative selection against animals carrying extrachromosomal arrays.

Prepare the 20 μl injection mixture in 1.5 ml tube:

Plasmid containing the transgene, *unc-119* rescue gene, and recombination sequences	50 ng/μl[a]
pCFJ601 (transposase)	50 ng/μl
pMA122 (toxin; optional)	10 ng/μl
pCFJ90 (red pharynx) or pBN41 (green pharynx)	2.5 ng/μl
pCFJ104 (red muscles) or pBN42 (green muscles)	5.0 ng/μl
pBN1 (red nuclei) or pBN40 (green nuclei)	10 ng/μl

[a]Tip: Certain transgenes may cause toxicity: if no transgenic animals are obtained try lower concentrations (10–30 ng/μl).

Centrifuge the mixture 15 min at 10,000–13,000 × *g* and aspirate 15 μl without touching the bottom. Transfer to a fresh tube.

2.3 Cultivate *C. elegans* for Microinjection

The host strain for MosSCI should carry a unique Mos1 insertion in the locus corresponding to the recombination sequences of the transgene plasmid. Moreover, the strain should be suitable for easy selection of transgenic animals. As mentioned above, *unc-119* loss of function alleles are frequently used. *unc-119* animals are shorter than wild type, with very restricted mobility. In practical terms, a single transgenic nematode with a wild-type *unc-119* rescue gene is easily detected on plates with hundreds of uncoordinated animals. Frøkjær-Jensen and coworkers created a series of *unc-119* host strains that cover all chromosomes:

Chromosome I: EG8078 (*ttTi5605*), EG6701 (*ttTi4348*), EG6702 (*ttTi4391*)
Chromosome II: EG6699 (*ttTi5605*), EG8079 (*ttTi5605*)
Chromosome III: EG8080 (*ttTi5605*)
Chromosome IV: EG8081 (*ttTi5605*), EG6703 (*cxTi10816*), EG6700 (*cxTi10882*)
Chromosome V: EG8082 (*ttTi5605*), EG8083 (*ttTi5605*)
Chromosome X: EG6705 (*ttTi14024*)

All strains are available from the *Caenorhabditis* Genetics Center (CGC; http://www.cbs.umn.edu/research/resources/cgc).

2.4 Microinjection

Microinjection is the most critical step of the MosSCI protocol. Good success rates require dedication and skill, but the technique is relatively simple and most researchers learn quickly after initial frustration.

Prepare injection pads in advance: Place a thin film of 2% agarose in the center of each 24 × 60 mm glass coverslip, and dry overnight at 37 °C. These pads can be stored several weeks at room temperature.

2.4.1 The day before injecting, transfer uncoordinated L4 larvae to fresh NGM plates. It is important to inject young adults with well-distinguishable gonads.

2.4.2 The day of injection, transfer worms to an NGM plate without food for at least 20 min to cleanse them of bacteria, which can interfere with immobilization of the worm and might clog the needle during the injection.

2.4.3 Load the injection needle. First centrifuge the injection mix for 15 min at 10,000–13,000 × *g*. Aspirate 2 μl without touching the bottom and gently load the injection needle. We recommend Femtotips II microinjection capillaries (Eppendorf #930000043), or prepare needles manually. Wait 5–10 min to allow the injection mix to fill the capillary.

2.4.4 Check the injection needle under a microscope to make sure the needle is not clogged, and the liquid flows easily.

2.4.5 Place the worm in a drop of microinjection oil (Halocarbon oil, Sigma, cat. #H8898) on an injection pad (described above) to prevent the worm from dehydrating. Use an eyelash fixed to a Pasteur pipette with parafilm or glued to a toothpick to immobilize the worm on the agarose pad.

2.4.6 Move the slide to the injection microscope and locate the worm using a 10 × objective. Change to a 40 × objective with DIC optics. Orient the slide so the injection needle aligns with the syncytial part of one of the gonads (germ nuclei should be clearly visible) at an angle of ~30–45°.

2.4.7 Gently insert the needle into the gonad and inject the mix until the gonad shows pronounced swelling. With experience, it should be possible to inject both gonads, but beginners should inject a single gonad. Injecting the mix during retraction of the injection needle prevents needle clogging, and helps keep the worm hydrated.

2.4.8 Transfer the worm to a drop of M9 buffer (KH_2PO_4 22 m*M*, Na_2HPO_4 34 m*M*, NaCl 86 m*M*, $MgSO_4$ 1 m*M*; see also Stiernagle, 2006) on an NGM plate with food. Incubate animals at 25 °C.

2.5 Identify Transgenic Animals

Plasmids that encode fluorescent proteins make it possible to evaluate injection success after 1 day. Ideally, 10–50% of the injected P0 animals should produce fluorescent embryos and larvae, visible by dissection microscope. Four days after injection, start screening for wild-type moving animals without fluorescence, i.e., animals in which the transgene and the *unc-119*(+) gene are integrated. We usually observe plates every other day for about 2 weeks after injection. Integration in later generations is very rare. If you included the pMA122 plasmid in the injection mixture, heat shock (2 h at 34 °C) after 7–8 days to kill animals that carry the injected plasmids as extrachromosomal arrays. Although this negative selection step is not always fully effective, it is particularly useful when screening a large number of plates. Candidate animals are transferred to fresh plates and homozygous animals are identified based on 100% wild-type moving offspring. Integrity of the inserted transgene should be evaluated using relevant methods (e.g., PCR; behavior of fluorescent proteins; rescue capacity). We routinely outcross MosSCI strains twice to a wild-type reference strain to reduce the potential risk of novel mutations induced when excising Mos1. Finally, we determine by PCR whether the transposon has been reinserted elsewhere in the genome, and select strains without Mos1.

Single-nematode PCR genotyping:

2.5.1 Add 1 μl proteinase K (10 mg/ml) to 99 μl worm lysis buffer (Tris–HCl [pH 8.3] 10 m*M*, KCl 50 m*M*, $MgCl_2$ 2.5 m*M*, NP-40 0.45%, Tween-20 0.45%, gelatin 0.01%).

2.5.2 Aliquot 2.5 μl worm lysis buffer with proteinase K in PCR tubes. Keep PCR tubes on ice.

2.5.3 Place worms in PCR tubes and overlay with a drop of mineral oil. Freeze at −80 °C (or on dry ice) for at least 15 min.

2.5.4 Incubate tubes 1 h at 60 °C in a thermocycler followed by heat inactivation of proteinase K for 15 min at 95 °C.

2.5.5 Add 22.5 μl of the following PCR mix to each tube:

ddH_2O	9.375 μl
Promega GoTaq 5 × PCR green buffer	5 μl
$MgCl_2$ 25 m*M*	2 μl
dNTP (each 2.5 m*M*)	2 μl
Forward primer (10 μ*M*)	2 μl
Reverse primer (10 μ*M*)	2 μl
Promega GoTaq polymerase (#M780B)	0.125 μl

2.5.6 PCR parameters:

95 °C	1 min
95 °C	30 s
55–58 °C	30 s
72 °C	2 min
Go to step 2 and repeat 34 times	

To detect Mos1, we use primers 5′-CAACCTTGACTGTCGAACCA CCATAG and 5′-TCTGCGAGTTGTTTTTGCGTTTGAG, which produce a ~350-bp PCR fragment. To detect integrated transgenes, you may need to design two pairs of PCR primers, in which 0.5 μl of the first PCR reaction is used as template for a second, nested PCR reaction.

3. CRISPR/CAS9-MEDIATED GENOME ENGINEERING

The use of RNA-guided Cas9 endonuclease to generate specific DSBs has revolutionized genome engineering, including mutagenesis and transgenesis (Waaijers & Boxem, 2014). Briefly, the specificity of the Cas9 enzyme is controlled by a small guide RNA (sgRNA) molecule with 20 nt complementary to a target sequence in the genome, whose only requirement is a 3′ flanking NGG trinucleotide. After recruitment, Cas9 cleaves both strands at the NGG sequence. When DSBs are repaired by non-homologous end joining, small insertions or deletions (InDels) are typically produced. However, exogenous repair templates push the equilibrium toward homologous recombination allowing the introduction of specific nucleotide substitutions or protein tags (Fig. 1). We focus here on CRISPR/Cas9 for transgenesis. Tremendous efforts are devoted to optimizing and modifying CRISPR/Cas9-mediated genome engineering protocols, hence novel variations are being reported. In particular, several co-conversion and antibiotics resistance approaches are proposed to ease identification of transgenic animals (Arribere et al., 2014; Dickinson, Pani, Heppert, Higgins, & Goldstein, 2015; Kim et al., 2014; Ward, 2015), and short ~30 bp homology arms may suffice when preparing repair templates (Paix et al., 2014; Zhao, Zhang, Ke, Yue, & Xue, 2014). The protocol below merges information from recent publications and our own experience.

3.1 Design the sgRNA Vector

The target sequence (also known as the protospacer) can conveniently be introduced into the sgRNA vector pDD162 (Dickinson, Ward, Reiner, & Goldstein, 2013) using the New England Biolabs Q5 Site-Directed Mutagenesis Kit (#E0554S). Alternatives include conventional cloning into pRB1017 (Arribere et al., 2014) or PCR stitching (Ward, 2015).

The sgRNA consists of a constant domain and a variable $GN_{19–20}$ sequence at the 5′ end. The first G is included to facilitate transcription initiation by RNA polymerase III and is present in most sgRNA cloning vectors. If the G at position +1 anneals to the target, usually an N_{19} sequence is included in the sgRNA, whereas an $N_{19–20}$ sequence is used for targets where the first G does not anneal. Several online tools are available to search for potential sgRNA sequences in given target genes (e.g., http://crispr.mit.edu and http://www.e-crisp.org). These services also identify putative off-targets with significant homology to the primary target sequence. However, Farboud and Meyer (2015) reported that selecting target sites with a GG dinucleotide immediately before the PAM sequence may be more relevant for cleavage efficiency than the scores provided by online tools. Farboud and Meyer also compared sgRNAs containing the original secondary structure with sgRNAs engineered with extended stem-loop structures and found no enhancement of Cas9 activity, whereas Ward (2015) reported increased efficiency.

3.2 Prepare the Repair Template

Initial studies used repair templates with 500–2000 bp homology arms flanking the cleavage site to knock-in protein tags or convert genes

(Chen, Fenk, & de Bono, 2013; Dickinson et al., 2013; Katic & Grosshans, 2013; Tzur et al., 2013). However, new studies with single-stranded DNA oligomers and repair templates with short homology arms showed that high efficiencies can be achieved with as little as 30 nt homology on each side of the DSB (Arribere et al., 2014; Paix et al., 2014; Zhao et al., 2014). When tagging endogenous proteins, this implies that repair templates can be prepared by a single PCR step with 50–70-mers, although an additional nested PCR with shorter primers is recommended (Paix et al., 2014). Importantly, the repair template can be designed to insert an exogenous sequence ~60 bp or even further from the cleavage site but homology arms must be extended equivalently. To avoid sgRNA-mediated cleavage of the repair template, silent mutations should be introduced in the protospacer or PAM sequences.

3.3 Co-Conversion

Simultaneous targeting of an unrelated locus to produce an easily detectable phenotype has been shown to facilitate isolation of transgenic events (Arribere et al., 2014; Kim et al., 2014; Ward, 2015). In particular, we find the *dpy-10* co-conversion marker very useful because it can be applied to most genetic backgrounds and allows detection of candidate animals in the F_1 progeny of injected hermaphrodites. Using a *dpy-10* sgRNA (protospacer sequence 5′-GCTACCATAGGCACCACGAG) and single-stranded repair oligomer (sequence 5′-CACTTGAACTTCAATACGGC AAGATGAGAATGACTGGAAACCGTACCGCATGCGGTGCCTA TGGTAGCGGAGCTTCACATGGCTTCAGACCAACAGCCTAT), a single amino acid substitution is introduced into the *dpy-10* locus, which causes a roller or dumpy phenotype in heterozygous or homozygous animals, respectively (Arribere et al., 2014).

3.4 Prepare the Injection Mixture

Prepare 20 μl injection mixture in 1.5 ml tube.

Cas9 expression plasmid (Peft-3::cas9-SV40_NLS::tbb-2 3′-UTR; Addgene #46168)	50 ng/μl
sgRNA designed to target gene of interest	25 ng/μl
Repair template	50 ng/μl
dpy-10 sgRNA	25 ng/μl
dpy-10 single-stranded oligo	500 n*M*

Centrifuge the mixture for 15 min at maximum speed and aspirate 15 μl without touching the bottom. Transfer to a fresh tube.

3.5 Microinjection

Microinjection follows the same protocol as MosSCI (Section 2.4) except that wild-type animals are usually used as hosts for CRISPR/Cas9. Inhibiting nonhomologous end-joining DNA repair via RNAi knockdown of *cku-80* prior to injection is proposed to increase repair template-mediated transgenesis (Ward, 2015).

3.6 Identify Transgenic Animals

Three to five days after injection, plates are screened for roller and dumpy F_1 animals, which are transferred to individual plates. Once the F_1s have produced embryos, they are sacrificed to test for modification of the locus of interest. Insertion of fluorescent tags can be tested directly by microscopy, whereas other modifications, or large numbers of F_1 candidates require PCR analysis. Design a PCR scheme with one primer inside the modification and another in the flanking sequence. A pair of nested primers increases sensitivity and enables PCR analysis of pools of 5–10 animals in each PCR tube. Construction of an artificial template as positive PCR control is strongly recommended.

Individual F_2s are then singled out from positive F_1 animals. Select animals with a wild-type appearance if the locus of interest is not tightly linked to the co-conversion marker (e.g., close to *dpy-10* on chromosome II). Repeat the PCR or microscopy test on F_2 animals after egg laying to ensure stable inheritance of the transgene. Positive F_1 animals may or may not carry the modification in their germ line. Finally, outcross offspring of positive F_2 animals twice with wild-type animals to reduce the risk of mutations in off-target genes, and isolate animals homozygous for the desired genome modification.

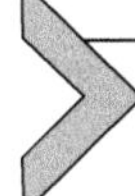

4. DamID ANALYSIS OF LBP INTERACTIONS WITH CHROMATIN

Extensive contacts exist between chromatin and the nuclear lamina and it is becoming increasingly clear that alterations in these contacts influence gene expression patterns. Mapping the contact points in chromatin is therefore relevant for understanding the function of individual LBPs and may provide insight into the mechanisms of laminopathies. Two methods have successfully been applied in *C. elegans*: chromatin immunoprecipitation (ChIP) with BAF-1 (Margalit, Neufeld, et al., 2007) and LEM-2 (Ikegami, Egelhofer, Strome, & Lieb, 2010) as well as Dam-mediated identification (DamID) with EMR-1 and LMN-1 (González-Aguilera et al., 2014). ChIP and DamID have advantages and limitations that were recently discussed (Askjaer, Ercan, & Meister, 2014). We provide detailed DamID protocols here and ChIP protocols in Section 5.

DamID technology relies on *in vivo* expression of the *E. coli* DNA adenine methyltransferase (Dam) fused to a chromatin-associated protein of interest (van Steensel & Dekker, 2010). DNA near the Dam-fused protein is methylated at GATC sites, which are subsequently isolated by enzymatic methods. To control for unspecific methylation, a strain expressing Dam fused to GFP is analyzed in parallel. To minimize experimental variability, we recommend using strains with integrated single-copy Dam fusion transgenes (see MosSCI protocol and Askjaer et al. (2014) for further details).

4.1 Nematode Culture

4.1.1 Collect embryos from asynchronous cultures by standard hypochlorite treatment (1 N NaOH, 30% household bleach solution). Check worms regularly in a dissection stereoscope. Proceed to the next step when half of the worms are broken up.

4.1.2 Wash embryos five times in 12 ml M9. Pellet embryos by centrifugation at 2000 rpm for 3 min. After the last wash, resuspend in 5 ml M9 with 0.01% Tween-20.

4.1.3 Count number of embryos in 2–10 μl aliquots.

4.1.4 Leave embryos to hatch overnight at 16–20 °C with gentle agitation.

4.1.5 Count hatched L1s in 2–10 μl aliquots.

4.1.6 Place 500–1000 L1s per 10 cm plate (or 150–250 L1s per 3 cm plate) containing a thick lawn of Dam− *E. coli* bacteria (e.g., strain GM119) as food source. Use a total of 4 large or 16 small plates per strain.

4.1.7 Incubate worms at 20 °C and collect when they are enriched for the life stage to be analyzed (e.g., 53 h for nongravid young adults; 66 h to accumulate young embryos, etc.). Harvest either embryos (Towbin et al., 2012) or—for better signal/noise ratio—adults (González-Aguilera et al., 2014). Harvest embryos by hypochlorite treatment; wash three times with M9. Harvest adults by centrifugation; wash five to seven times with M9 to remove bacteria.

4.1.8 Make aliquots containing ~30 μl embryo or adult material. Remove excess liquid, snap-freeze in liquid nitrogen and store at −80 °C until needed.

4.2 Purification and Amplification of Dam-Methylated DNA

4.2.1 Purify genomic DNA (gDNA) from 30 mg embryos or adults; we use the DNeasy Blood and Tissue Kit (QIAGEN #69504).

4.2.2 Elute the gDNA in 200 μl and measure its concentration (expected yield is 4–7 μg).

4.2.3 Digest 500 ng adult gDNA using 10 units *Dpn*I (NEB #R0176S; cuts methylated G^{m}ATC) in 10 μl for 6 hours at 37 °C in a thermocycler. Include an additional reaction with *Dpn*I for one of the biological samples to be used as ligation control (control A) and also a control reaction without *Dpn*I (control B). Inactivate *Dpn*I by heating at 80 °C for 20 min.

4.2.4 Prepare double-stranded adaptors. Mix 50 μl of primer AdRt (5′-CTAATACGACTCACTATAGGGCAGCGTGGTCGCG GCCGAGGA, 100 μ*M*) and 50 μl of AdRb (5′-TCCTCGGCCG; 100 μ*M*). Heat to 95 °C and let cool slowly to room temperature.

4.2.5 Prepare the adapter ligation reaction mixtures on ice, then incubate at 16 °C in a thermocycler overnight. Also include a reaction with *Dpn*I-digested DNA without T4 DNA ligase (control A).

ddH_2O	6.2 μl
*Dpn*I-digested gDNA	10 μl
10× ligation buffer	2 μl
Double-stranded adaptors	0.8 μl
Roche T4 DNA ligase (#10799009001, 5 U/μl)	1 μl
Inactivate the T4 DNA ligase by incubating 10 min at 65 °C	

4.2.6 Purify the reaction with Agencourt AMPure XP® (Beckman Coulter; #A63880) using a magnetic particle concentrator (e.g., Magnum™ EX Universal Magnet Plate; Alphaqua® #A000380) and elute in 20 μl.

4.2.7 Digest the ligation reactions with 10 units of *Dpn*II (NEB #R0543S; cuts unmethylated GATC) in a volume of 50 μl for 1 h at 37 °C. Inactivate *Dpn*II by incubating 20 min at 80 °C.

4.2.8 Purify the reaction with AMPure XP as above and elute in 25 μl.

4.2.9 Amplify methylated DNA using Advantage® cDNA Polymerase Mix (Clontech #639105). Include a control without DNA template (control C).

ddH_2O	13.75 μl
10 × PCR reaction buffer	5 μl
*Dpn*II digested DNA	25 μl
Primer Adr (5′-NNNNGGTCGCGGCCGAGGATC; 50 μ*M*)	1.25 μl
dNTP mix (each 2.5 m*M*)	4 μl
Advantage cDNA Polymerase Mix	1 μl

PCR parameters:

68 °C	10 min
94 °C	1 min
65 °C	5 min
68 °C	15 min
94 °C	1 min
65 °C	1 min
68 °C	10 min
Go to step 5 and repeat three times	
94 °C	1 min
65 °C	1 min
68 °C	2 min
Go to step 8 and repeat 23 times (embryos) or 20 times (adults)	

4.2.10 Check 5 μl of each PCR reaction on an agarose gel. Successful amplification of Dam-methylated DNA should yield a 400–1000 bp smear.

4.2.11 Purify the PCR products using QIAquick PCR Purification Kit (QIAGEN #28104). Perform an extra wash with 700 μl of 35% guanidine hydrochloride before PE washing. Elute in 30 μl. Measure DNA concentration.

4.2.12 Use End-It™ DNA End-Repair Kit (Epicentre #ER0720) to blunt the PCR products. Incubate the reaction 45 min at room temperature.

ddH_2O	To a final volume of 5 μl
PCR product	1 μg
10 × end-repair buffer	5 μl
dNTP mix	5 μl
ATP	5 μl
End-repair enzyme mix	1 μl

4.2.13 Purify DNA using the QIAquick PCR Purification Kit. Perform an extra wash with 700 μl guanidine hydrochloride 35% before PE washing. Elute DNA in 25 μl.

4.2.14 Add 3′-A overhang to blunt-ended DNA fragments with Klenow. Incubate 30 min at 37 °C, then put tubes on ice.

ddH_2O	18.5 μl
Blunt-ended DNA	25 μl
10 × NEB buffer 2	5 μl
dATP 10 m*M*	1 μl
Klenow fragment 50 U/μl (NEB #E2612)	0.5 μl

4.2.15 Purify DNA using QIAquick PCR Purification Kit. Perform an extra wash with 400 μl guanidine hydrochloride 35% before PE washing. Elute DNA in 20 μl.

4.2.16 Prepare Y-shaped Illumina adaptors: mix 1:1 of primers 5′-ACACTCTTTCCCTACACGACGCTCTTCCGATCT and phosphorylated 5′-P-GATCGGAAGAGCACACGTCT (100 μ*M* each). Heat to 95 °C and let cool slowly to room temperature.

4.2.17 Ligate Y-shaped Illumina adapters. Incubate 2 h at 25 °C, then inactivate T4 ligase by incubating 10 min at 65 °C.

DNA with 3′-A overhang	8 μl (~250 ng)
10 × ligation buffer	1 μl
T4 DNA ligase (5 U/μL)	0.5 μl
Illumina Y-Adapter (50 μ*M*)	0.5 μl

4.2.18 Purify DNA (QIAquick PCR Purification Kit). Elute DNA in 20 μl.

4.2.19 Amplify by Illumina Index PCR (MyTaq™ Mix; BIOLINE #BIO-25041).

DNA with Y-adapter	8 μl (~100 ng)
Illumina index primer (5 μ*M*)	1 μl
Fwd P5 primer (5 μ*M*)	1 μl
2× MyTaq enzyme mix	10 μl

PCR parameters:

94 °C	1 min
94 °C	30 s
58 °C	30 s
72 °C	30 s
Go to step 2 and repeat five to eight times	
72 °C	2 min

4.2.20 Purify the Index PCR material with AMPure XP as above.

4.2.21 Quantify libraries using QUBIT® Fluorometer (QuantiFluor™ dsDNA System; Promega #E2670) or NanoDrop spectrophotometer (Thermo Scientific), and submit for next-generation sequencing (Sharma et al., 2014).

5. CHROMATIN IMMUNOPRECIPITATION OF LBPs

5.1 *C. elegans* expressing GFP–LBPs are grown on six 9-cm NGM OP50 plates and collected either as mixed-staged animals or at a desired stage.

5.2 Wash worms twice with M9, fix 30 min with 2% formaldehyde at room temperature, wash once with 100 m*M* Tris, pH 7.5, wash twice with M9 buffer, and wash once with homogenization buffer (50 m*M* HEPES-KOH, pH 7.5, 1 m*M* EDTA, 140 m*M* KCl, 0.5% NP-40, 10% glycerol, 5 m*M* DTT, and protease inhibitors).

5.3 Freeze animals in liquid nitrogen, and sonicate on ice 10 times (30 s each; we use Sonic; Heat Systems Ultrasonic, Inc.). Centrifuge at 6500 rpm for 20 min at 4 °C. Collect the supernatant and sonicate five times (30 s each) on ice to shear DNA. Centrifuge at 14,000 rpm for 20 min. Collect the supernatant, and test an aliquot for GFP–LBP expression by immunoblotting, and freeze positives in liquid nitrogen and store at −80 or in liquid nitrogen.

5.4 Immunoprecipitate by incubating ~100 μl lysates with 5 μg anti-GFP antibody for 2 h at 4 °C. We have used anti-GFP antibodies from Roche (Margalit, Neufeld, et al., 2007) or GFP-Trap antibodies per manufacture protocol (Chromotek GFP-TRAP_M) (Bar et al., 2014).

5.5 Remove cellular debris by centrifugation (6500 rpm, 15 min, 4 °C). Recover the supernatant, and centrifuge 10 min at 14,000 rpm before adding 50 μl protein G-Sepharose (Roche) to the supernatant.

5.6 Wash immunoprecipitates twice with ChIP buffer (50 m*M* HEPES/KOH, pH 7.5, 1 m*M* EDTA, 0.5%, NP-40, and 5 m*M* DTT with protease inhibitors) containing 100 m*M* KCl.

5.7 Wash immunoprecipitates twice with ChIP buffer containing 1 *M* KCl and Tris–EDTA.

5.8 Elute complexes with ~100 μl elution buffer (1% SDS, 10 m*M* Tris–HCl, pH 8). Add 16 μl of 5 *M* NaCl to the eluate and heat overnight at 65 °C.

5.9 Isolate DNA using standard phenol–chloroform extraction, and resuspend in 20 μl Tris–EDTA, pH 8.

5.10 Quantitative PCR with locus-specific primers is used to assess the quality and quantity of the eluted DNA. To a 35 μl reaction add 250 n*M* of each primer, 1.25 units of *Taq* enzyme (Promega), 1× reaction buffer, 1.5 m*M* $MgCl_2$, 0.5 m*M* dNTP (Invitrogen), 0.2× SYBR green I dye (Molecular Probes), and an appropriate amount of DNA. Usually we use 3% of the eluted DNA.

6. PHYSIOLOGICAL ASSAYS FOR LBP FUNCTION

When LBP homozygous-null animals are viable, experiments are performed on mutant animals grown on NGM plates seeded with *E. coli* OP50 and compared to wild-type (N2) worms. Alternatively, wild-type worms are grown on RNAi feeding plates containing 25 μg/ml ampicillin and 1 m*M* isopropyl β-D-thiogalactoside (IPTG) seeded with *E. coli* HT115(DE3) expressing LBP dsRNA from the feeding vector L4440.

6.1 Motility

Adult worms are treated with a bleach solution (0.5 N NaOH, 0.5% sodium hypochlorite) for 10 min at 22 °C and then washed several times with M9 buffer. This procedure dissolves adult worms and larvae but does not affect the eggs. The isolated embryos are grown at 20 °C until they reach the desired stage of development. Normally, 30 animals are placed per feeding plate. Animals that are downregulated for the desired LBP are subsequently recorded on their growing plates for 1 min sessions (5 frames per second), with 1 min interval between sessions. At least 10 animals are used every time and for every treatment. Each experiment is repeated three times. Monitor head movement using N.J. Carter's freeware Retrac for image tracking. A recording is made using Dino-Eye (Dino-Lite Microscope, NY) on a Nikon SZM800 binocular. *P*-values are calculated using an unpaired equal distribution *t*-test.

6.2 Fertility

Single L4 animals, either mutants or animals grown from L1 on RNAi plates, are transferred each to a separate NGM plate. A typical experiment consists of at least 30 animals. During the first 3 days, each animal is transferred daily to a fresh plate, while keeping the older plates in the incubator. On the third day, the animals are given 24 h to lay eggs and then discarded. The progeny from all plates are scored as the mean number of laid eggs per animal. Progeny survival rate is calculated as the mean of the percentage of eggs that develop into larvae from the total eggs laid. Egg laying experiments are usually performed at 20 °C.

6.3 Lifespan

Ninety animals at the L4 stage, either mutants or animals grown from L1 on RNAi plates, are transferred to 3 NGM plates (30 worms per plate). Initially, the animals are transferred daily to new NGM plates until they no longer lay eggs. Animals that die due to the transfer procedure are discarded from the experiment. Animals that do not move when prodded with a platinum wire are considered dead. The experiments continue until all animals die. Lifespan experiments are normally performed at 20 °C.

6.4 Rate of Development

Synchronize animals by bleaching adult *C. elegans* and mildly shaking the isolated embryos overnight in a tube containing M9 buffer at 20 °C until they reach the L1 stage. Place the synchronized L1 stage animals on NGM plates. To obtain statistically significant results, each experiment should include at least 3 NGM plates each containing 30 animals. Assess developmental stage(s) daily under a high-quality binocular microscope until the worms start to lay eggs.

7. FRAP AND FLIP ANALYSIS

Fluorescence recovery after photobleaching (FRAP) is a fluorescence microscopy technique used to quantify two-dimensional lateral diffusion within a cell. Fluorescence loss in photobleaching (FLIP) is another fluorescence microscopy technique that examines molecular movements within cells (Fig. 2).

To perform FRAP and FLIP experiments, we normally acquire live fluorescence images with a Leica SP5 confocal microscope and a 63×/1.4 oil-immersion objective. Imaging is performed and initially analyzed with Leica 116 Application suite 2.3.1. Photobleaching is done on a single Z-plane, at 100% transmission. Imaging typically requires 1–6% of laser power. For pre- and postbleach images, four frames are extracted and averaged from corresponding movies using the ImageJ package (http://imagej.nih.gov/ij/). Graphs are generated after normalization and averaging from multiple regions of interest. During the experiment, animals are mounted on agarose pads (see Section 2.4), paralyzed by adding a drop of 1–10 m*M* levamisole and immediately taken for imaging. Typically, animals are imaged for 5–30 min after adding levamisole. Photobleaching of GFP-fused LBPs performed by a 488-nm laser in a defined region of each cell, and imaged with 488- or 496-nm lasers. For FRAP analysis, the fluorescence intensity in the bleached area, the background area, and the total cell area are measured as a function of time after bleaching and are normalized as described (Rabut & Ellenberg, 2005).

For FLIP experiments of GFP-fused LBPs, we use a 488-nm laser line to continuously bleach the GFP in a region of interest, while imaging the entire nucleus at low intensity with a 496-nm laser line. In FRAP experiments, a region of interest is bleached with the 488-nm laser line and imaged at high frame rate over ~1 min. The half-life (time for GFP intensity to return to half of its maximal value) is extracted from the plot/equation as described (Rabut & Ellenberg, 2005).

8. SUMMARY

The functions of LBPs in vertebrate cells are mostly unknown. These protocols, combined with powerful genetic tools in *C. elegans*, can provide important insight into the biological roles of any *C. elegans* gene, and are particularly valuable in understanding the functions of conserved LBPs whose roles may be more difficult to elucidate in mammals.

ACKNOWLEDGMENTS

We gratefully acknowledge Georgina Gómez-Saldivar for discussion on the DamID protocol and funding from the Muscular Dystrophy Association (MDA), the Binational Israel-USA Science Foundation (BSF 2007215), and the Niedersachsen-Israeli Research Cooperation program to Y.G. and funding from the Spanish Ministry of Economy and Competitiveness (BFU2013-42709P), the Autonomous Government of Andalusia (P08-CVI-3920), and the European Regional Development Fund to P.A.

REFERENCES

Arribere, J. A., Bell, R. T., Fu, B. X., Artiles, K. L., Hartman, P. S., & Fire, A. Z. (2014). Efficient marker-free recovery of custom genetic modifications with CRISPR/Cas9 in *Caenorhabditis elegans*. *Genetics*, *198*, 837–846.

Asencio, C., Davidson, I. F., Santarella-Mellwig, R., Ly-Hartig, T. B., Mall, M., Wallenfang, M. R., et al. (2012). Coordination of kinase and phosphatase activities by Lem4 enables nuclear envelope reassembly during mitosis. *Cell*, *150*, 122–135.

Askjaer, P., Ercan, S., & Meister, P. (2014). Modern techniques for the analysis of chromatin and nuclear organization in *C. elegans*. WormBook ed. The *C. elegans* Research Community, WormBook, doi:10.1895/wormbook.1.169.1.

Bank, E. M., & Gruenbaum, Y. (2011). The nuclear lamina and heterochromatin: A complex relationship. *Biochemical Society Transactions*, *39*, 1705–1709.

Bar, D. Z., Davidovich, M., Lamm, A. T., Zer, H., Wilson, K. L., & Gruenbaum, Y. (2014). BAF-1 mobility is regulated by environmental stresses. *Molecular Biology of the Cell*, *25*, 1127–1136.

Barkan, R., Zahand, A. J., Sharabi, K., Lamm, A. T., Feinstein, N., Haithcock, E., et al. (2012). Ce-emerin and LEM-2: Essential roles in *Caenorhabditis elegans* development, muscle function, and mitosis. *Molecular Biology of the Cell*, *23*, 543–552.

Baudrimont, A., Penkner, A., Woglar, A., Machacek, T., Wegrostek, C., Gloggnitzer, J., et al. (2010). Leptotene/zygotene chromosome movement via the SUN/KASH protein bridge in *Caenorhabditis elegans*. *PLoS Genetics*, *6*, e1001219.

Chen, C., Fenk, L. A., & de Bono, M. (2013). Efficient genome editing in *Caenorhabditis elegans* by CRISPR-targeted homologous recombination. *Nucleic Acids Research*, *41*, e193.

Croft, J. A., Bridger, J. M., Boyle, S., Perry, P., Teague, P., & Bickmore, W. A. (1999). Differences in the localization and morphology of chromosomes in the human nucleus. *The Journal of Cell Biology*, *145*, 1119–1131.

Dechat, T., Gesson, K., & Foisner, R. (2010). Lamina-independent lamins in the nuclear interior serve important functions. *Cold Spring Harbor Symposia on Quantitative Biology*, *75*, 533–543.

Dickinson, D. J., Ward, J. D., Reiner, D. J., & Goldstein, B. (2013). Engineering the *Caenorhabditis elegans* genome using Cas9-triggered homologous recombination. *Nature Methods*, *10*, 1028–1034.

Dickinson, D. J., Pani, A. M., Heppert, J. K., Higgins, C. D., & Goldstein, B. (2015). Streamlined genome engineering with a self-excising drug selection cassette. *Genetics*, *200*, 1035–1049.

Dittrich, C. M., Kratz, K., Sendoel, A., Gruenbaum, Y., Jiricny, J., & Hengartner, M. O. (2012). LEM-3—A LEM domain containing nuclease involved in the DNA damage response in *C. elegans*. *PLoS One*, *7*, e24555.

Farboud, B., & Meyer, B. J. (2015). Dramatic enhancement of genome editing by CRISPR/ Cas9 through improved guide RNA design. *Genetics*, *199*, 959–971.

Fridkin, A., Mills, E., Margalit, A., Neufeld, E., Lee, K. K., Feinstein, N., et al. (2004). Matefin, a *C. elegans* germ-line specific SUN-domain nuclear membrane protein, is essential for early embryonic and germ cell development. *Proceedings of the National Academy of Sciences of the United States of America*, *101*, 6987–6992.

Fridkin, A., Penkner, A., Jantsch, V., & Gruenbaum, Y. (2009). SUN-domain and KASH-domain proteins during development, meiosis and disease. *Cellular and Molecular Life Sciences*, *66*, 1518–1533.

Frøkjær-Jensen, C., Davis, M. W., Ailion, M., & Jorgensen, E. M. (2012). Improved Mos1-mediated transgenesis in *C. elegans*. *Nature Methods*, *9*, 117–118.

Frøkjær-Jensen, C., Davis, M. W., Hopkins, C. E., Newman, B. J., Thummel, J. M., Olesen, S. P., et al. (2008). Single-copy insertion of transgenes in *Caenorhabditis elegans*. *Nature Genetics*, *40*, 1375–1383.

Frøkjær-Jensen, C., Davis, M. W., Sarov, M., Taylor, J., Flibotte, S., LaBella, M., et al. (2014). Random and targeted transgene insertion in *Caenorhabditis elegans* using a modified Mos1 transposon. *Nature Methods*, *11*, 529–534.

González-Aguilera, C., Ikegami, K., Ayuso, C., de Luis, A., Íñiguez, M., Cabello, J., et al. (2014). Genome-wide analysis links emerin to neuromuscular junction activity in *Caenorhabditis elegans*. *Genome Biology*, *15*, R21.

Gorjánácz, M., Klerkx, E. P. F., Galy, V., Santarella, R., López-Iglesias, C., Askjaer, P., et al. (2007). *C. elegans* BAF-1 and its kinase VRK-1 participate directly in postmitotic nuclear envelope assembly. *The EMBO Journal*, *26*, 132–143.

Gruenbaum, Y., Lee, K. K., Liu, J., Cohen, M., & Wilson, K. L. (2002). The expression, lamin-dependent localization and RNAi depletion phenotype for emerin in *C. elegans*. *Journal of Cell Science*, *115*, 923–929.

Guelen, L., Pagie, L., Brasset, E., Meuleman, W., Faza, M. B., Talhout, W., et al. (2008). Domain organization of human chromosomes revealed by mapping of nuclear lamina interactions. *Nature*, *453*, 948–951.

Ikegami, K., Egelhofer, T. A., Strome, S., & Lieb, J. D. (2010). *Caenorhabditis elegans* chromosome arms are anchored to the nuclear membrane via discontinuous association with LEM-2. *Genome Biology*, *11*, R120.

Katic, I., & Grosshans, H. (2013). Targeted heritable mutation and gene conversion by Cas9-CRISPR in *Caenorhabditis elegans*. *Genetics*, *195*, 1173–1176.

Kim, H., Ishidate, T., Ghanta, K. S., Seth, M., Conte, D. J., Shirayama, M., et al. (2014). A co-CRISPR strategy for efficient genome editing in *Caenorhabditis elegans*. *Genetics*, *197*, 1069–1080.

Korfali, N., Wilkie, G. S., Swanson, S. K., Srsen, V., de Las Heras, J., Batrakou, D. G., et al. (2012). The nuclear envelope proteome differs notably between tissues. *Nucleus*, *3*, 552–564.

Lee, K. K., Starr, D., Cohen, M., Liu, J., Han, M., Wilson, K., et al. (2002). Lamin-dependent localization of UNC-84, a protein required for nuclear migration in *C. elegans*. *Molecular Biology of the Cell, 13*, 892–901.

Liu, J., Lee, K. K., Segura-Totten, M., Neufeld, E., Wilson, K. L., & Gruenbaum, Y. (2003). MAN1 and emerin have overlapping function(s) essential for chromosome segregation and cell division in *C. elegans*. *Proceedings of the National Academy of Sciences of the United States of America, 100*, 4598–4603.

Liu, J., Rolef Ben-Shahar, T., Riemer, D., Treinin, M., Spann, P., Weber, K., et al. (2000). Essential roles for *Caenorhabditis elegans* lamin gene in nuclear organization, cell cycle progression, and spatial organization of nuclear pore complexes. *Molecular Biology of the Cell, 11*, 3937–3947.

Malik, P., Korfali, N., Srsen, V., Lazou, V., Batrakou, D. G., Zuleger, N., et al. (2010). Cell-specific and lamin-dependent targeting of novel transmembrane proteins in the nuclear envelope. *Cellular and Molecular Life Sciences, 67*(8), 1353–1369.

Malone, C. J., Fixsen, W. D., Horvitz, H. R., & Han, M. (1999). UNC-84 localizes to the nuclear envelope and is required for nuclear migration and anchoring during *C. elegans* development. *Development, 126*, 3171–3181.

Malone, C. J., Misner, L., Le Bot, N., Tsai, M. C., Campbell, J. M., Ahringer, J., et al. (2003). The *C. elegans* Hook protein, ZYG-12, mediates the essential attachment between the centrosome and nucleus. *Cell, 115*, 825–836.

Margalit, A., Brachner, A., Gotzmann, J., Foisner, R., & Gruenbaum, Y. (2007). Barrier-to-autointegration factor—A BAFfling little protein. *Trends in Cell Biology, 17*, 202–208.

Margalit, A., Neufeld, E., Feinstein, N., Wilson, K. L., Podbilewicz, B., & Gruenbaum, Y. (2007). Barrier to autointegration factor blocks premature cell fusion and maintains adult muscle integrity in *C. elegans*. *The Journal of Cell Biology, 178*, 661–673.

Margalit, A., Segura-Totten, M., Gruenbaum, Y., & Wilson, K. L. (2005). Barrier-to-autointegration factor is required to segregate and enclose chromosomes within the nuclear envelope and assemble the nuclear lamina. *Proceedings of the National Academy of Sciences of the United States of America, 102*, 3290–3295.

Mattout, A., Goldberg, M., Tzur, Y., Margalit, A., & Gruenbaum, Y. (2007). Specific and conserved sequences in *D. melanogaster* and *C. elegans* lamins and histone H2A mediate the attachment of lamins to chromosomes. *Journal of Cell Science, 120*, 77–85.

Mattout, A., Pike, B. L., Towbin, B. D., Bank, E. M., Gonzalez-Sandoval, A., Stadler, M. B., et al. (2011). An EDMD mutation in *C. elegans* lamin blocks muscle-specific gene relocation and compromises muscle integrity. *Current Biology, 21*, 1603–1614.

Morales-Martínez, A., Dobrzynska, A., & Askjaer, P. (2015). Inner nuclear membrane protein LEM-2 is required for correct nuclear separation and morphology. *Journal of Cell Science, 128*, 1090–1096.

Olins, A. L., Rhodes, G., Welch, D. B., Zwerger, M., & Olins, D. E. (2010). Lamin B receptor: Multi-tasking at the nuclear envelope. *Nucleus, 1*, 53–70.

Paix, A., Wang, Y., Smith, H. E., Lee, C. Y., Calidas, D., Lu, T., et al. (2014). Scalable and versatile genome editing using linear DNAs with microhomology to Cas9 Sites in *Caenorhabditis elegans*. *Genetics, 198*, 1347–1356.

Penkner, A. M., Fridkin, A., Gloggnitzer, J., Baudrimont, A., Machacek, T., Woglar, A., et al. (2009). Meiotic chromosome homology search involves modifications of the nuclear envelope protein Matefin/SUN-1. *Cell, 139*, 920–933.

Rabut, G., & Ellenberg, J. (2005). Photobleaching techniques to study mobility and molecular dynamics of proteins in live cells: FRAP, iFRAP and FLIP (Vol. 1). In R. D. Goldman & D. L. Spector (Eds.), *Live cell imaging* (pp. 101–126). Cold Spring Harbor: Cold Spring Harbor Laboratory Press.

Schirmer, E. C., Florens, L., Guan, T., Yates, J. R., & Gerace, L. (2003). Nuclear membrane proteins with potential disease links found by subtractive proteomics. *Science, 531*, 1380–1382.

Sharma, R., Jost, D., Kind, J., Gómez-Saldivar, G., van Steensel, B., Askjaer, P., et al. (2014). Differential spatial and structural organization of the X chromosome underlies dosage compensation in *C. elegans*. *Genes & Development*, *28*, 2591–2596.

Shin, J. Y., Dauer, W. T., & Worman, H. J. (2014). Lamina-associated polypeptide 1: Protein interactions and tissue-selective functions. *Seminars in Cell & Developmental Biology*, *29*, 164–168.

Simon, D. N., & Wilson, K. L. (2013). Partners and post-translational modifications of nuclear lamins. *Chromosoma*, *122*, 13–31.

Sosa, B. A., Kutay, U., & Schwartz, T. U. (2013). Structural insights into LINC complexes. *Current Opinion in Structural Biology*, *23*, 285–291.

Starr, D. A., & Fridolfsson, H. N. (2010). Interactions between nuclei and the cytoskeleton are mediated by SUN-KASH nuclear-envelope bridges. *Annual Review of Cell and Developmental Biology*, *26*, 421–444.

Starr, D. A., & Han, M. (2002). Role of ANC-1 in tethering nuclei to the actin cytoskeleton. *Science*, *298*, 406–409.

Starr, D. A., Hermann, G. J., Malone, C. J., Fixsen, W., Priess, J. R., Horvitz, H. R., et al. (2001). *unc-83* encodes a novel component of the nuclear envelope and is essential for proper nuclear migration. *Development*, *128*, 5039–5050.

Stiernagle, T. (2006). Maintenance of *C. elegans*. WormBook ed. The *C. elegans* Research Community, WormBook, doi:10.1895/wormbook.1.101.1.

Sulston, J. E., & Horvitz, H. R. (1981). Abnormal cell lineages in mutants of the nematode *Caenorhabditis elegans*. *Developmental Biology*, *82*, 41–55.

Towbin, D. B., González-Aguilera, C., Sack, R., Gaidatzis, D., Kalck, V., Meister, P., et al. (2012). Step-wise methylation of histone H3K9 positions chromosome arms at the nuclear periphery in *C. elegans* embryos. *Cell*, *150*, 934–947.

Tzur, Y. B., Friedland, A. E., Nadarajan, S., Church, G. M., Calarco, J. A., & Colaiácovo, M. P. (2013). Heritable custom genomic modifications in *Caenorhabditis elegans* via a CRISPR-Cas9 system. *Genetics*, *195*, 1181–1185.

van Steensel, B., & Dekker, J. (2010). Genomics tools for unraveling chromosome architecture. *Nature Biotechnology*, *28*, 1089–1095.

Waaijers, S., & Boxem, M. (2014). Engineering the *Caenorhabditis elegans* genome with CRISPR/Cas9. *Methods*, *68*, 381–388.

Ward, J. D. (2015). Rapid and precise engineering of the *Caenorhabditis elegans* genome with lethal mutation co-conversion and inactivation of NHEJ repair. *Genetics*, *199*, 363–377.

Wilson, K. L., & Foisner, R. (2010). Lamin-binding proteins. *Cold Spring Harbor Perspectives in Biology*, *2*(4), a000554.

Worman, H. J., Ostlund, C., & Wang, Y. (2010). Diseases of the nuclear envelope. *Cold Spring Harbor Perspectives in Biology*, *2*, a000760.

Zhao, P., Zhang, Z., Ke, H., Yue, Y., & Xue, D. (2014). Oligonucleotide-based targeted gene editing in *C. elegans* via the CRISPR/Cas9 system. *Cell Research*, *24*, 247–250.

Zheng, X., Kim, Y., & Zheng, Y. (2015). Identification of lamin B-regulated chromatin regions based on chromatin landscapes. *Molecular Biology of the Cell*, *26*, 2685–2697.

CHAPTER TWENTY-THREE

Pathologically Relevant Prelamin A Interactions with Transcription Factors

Arantza Infante, Clara I. Rodríguez[1]
Stem Cells and Cell Therapy Laboratory, BioCruces Health Research Institute, Cruces University Hospital, Barakaldo, Spain
[1]Corresponding author: e-mail address: cirodriguez@osakidetza.eus

Contents

Abstract

LMNA-linked laminopathies are a group of rare human diseases caused by mutations in *LMNA* or by disrupted posttranslational processing of its largest encoded isoform, prelamin A. The accumulation of mutated or immature forms of farnesylated prelamin A, named progerin or prelamin A, respectively, dominantly disrupts nuclear lamina structure with toxic effects in cells. One hypothesis is that aberrant lamin filament networks disrupt or "trap" proteins such as transcription factors, thereby interfering with their normal activity. Since laminopathies mainly affect tissues of mesenchymal origin, we tested this hypothesis by generating an experimental model of laminopathy by inducing prelamin A accumulation in human mesenchymal stem cells (hMSCs). We provide detailed protocols for inducing and detecting prelamin A accumulation in hMSCs, and describe the bioinformatic analysis and *in vitro* assays of transcription factors potentially affected by prelamin A accumulation.

Methods in Enzymology, Volume 569
ISSN 0076-6879
http://dx.doi.org/10.1016/bs.mie.2015.08.032

1. INTRODUCTION

1.1 The Nuclear Lamina

The nuclear lamina is a meshwork of nuclear intermediate filaments formed by A- and B-type lamins, located primarily near the inner membrane of the nuclear envelope, but also found at low levels within the nucleoplasm (Dittmer & Misteli, 2011). In mammals, the *LMNA* gene encodes four "A-type" lamins (named A, C, AΔ10, C2) by alternative splicing, of which lamins A and C are the major isoforms in somatic cells. B-type lamins are encoded by *LMNB1* (lamin B1) and *LMNB2* (lamin B2 and B3) (Burke & Stewart, 2013). The *ratios* of A- and B-type lamins, important for nuclear stiffness and mechanical stability, are tightly regulated in different cell types and tissues (Swift et al., 2013). Lamins function as scaffolds for the binding of chromatin and diverse proteins related to the nucleoskeleton, nuclear pore complexes, and gene regulation (Wilson & Foisner, 2010). Through these and other interactions, lamins participate in essential cellular processes such as mitosis, cell signaling, proliferation, and differentiation.

1.2 Lamin A Maturation Process in Health and Disease (Genetic Versus Acquired Laminopathies)

Mutations in *LMNA* or genes required to posttranslationally modify prelamin A cause human diseases (laminopathies), including segmental progeroid (premature aging) syndromes. Under normal conditions, the largest A-type lamin is produced as a precursor, named prelamin A. Prelamin A has a C-terminal CaaX motif that triggers three sequential enzymatic modifications, starting with the addition of a 15-carbon farnesyl lipid to the cysteine residue by protein farnesyltransferase (FTase) (Zhang & Casey, 1996) (Fig. 1A). Farnesylation is a prerequisite for subsequent steps: removal of the last three residues ("aaXing") by either RCE1 or ZMPSTE24 (Fig. 1) (Bergo et al., 2002), and carboxymethylation of the exposed farnesylcysteine by ICMT (Dai et al., 1998). ZMPSTE24 performs the final cleavage that removes the last 15 residues (including the farnesylated/methylated Cys) to generate mature lamin A (Beck, Hosick, & Sinensky, 1990; Weber, Plessmann, & Traub, 1989). This final step does not occur in the absence of ZMPSTE24 (Bergo et al., 2002; Pendás et al., 2002). Farnesylated prelamin A is the only known substrate for ZMPSTE24 (Pendás et al., 2002).

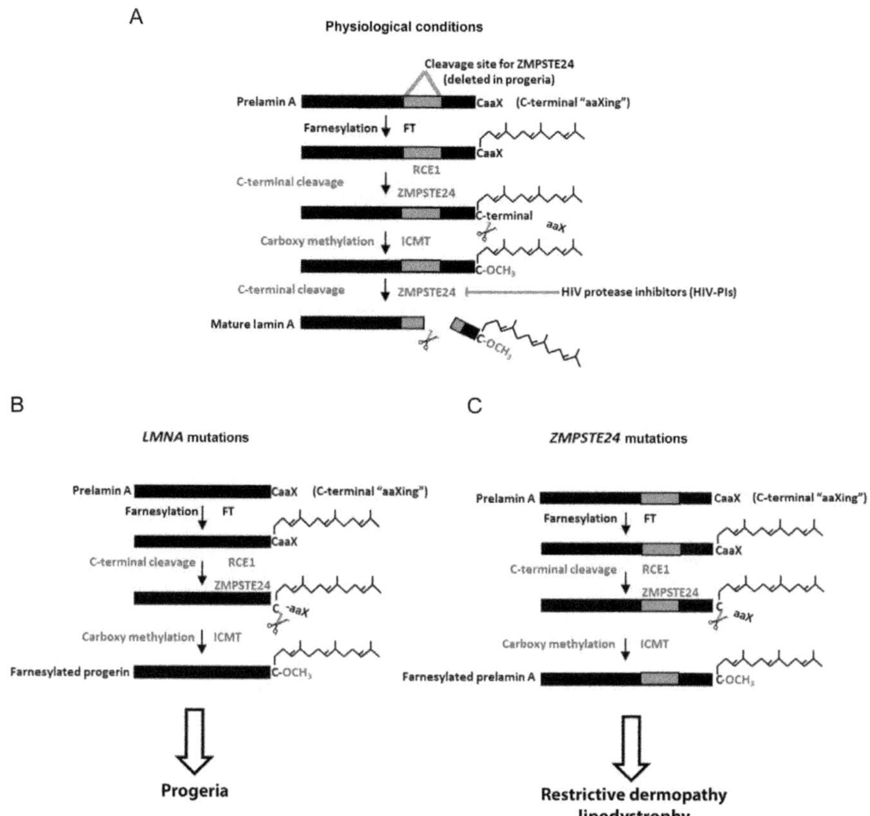

Figure 1 Maturation of lamin A and formation of farnesylated progerin and prelamin A. (A) Under physiological conditions, lamin A is synthesized as a precursor named prelamin A, which is sequentially posttranslationally modified to generate mature lamin A. Protease inhibitors used to treat HIV patients also inhibit ZMPSTE24 activity, blocking the final step of lamin A maturation. (B and C) Mutations in *LMNA* (B) or *ZMPSTE24* (C) can each interfere with processing, leading to the accumulation of farnesylated progerin or prelamin A, which act dominantly and cause human disease(s).

Certain mutations in *LMNA* or genes needed for prelamin A maturation, such as *ZMPSTE24*, cause premature aging syndromes (e.g., Hutchinson-Gilford progeria syndrome, "HGPS"), restrictive dermopathy or lipodystrophy (Fig. 1B and C). In the case of HGPS, specific mutations in the *LMNA* gene can activate a cryptic splice site resulting in a partial deletion in *LMNA* exon 11, where the cleavage site for ZMPSTE24 is located, leading to the accumulation of a permanently farnesylated, uncleaved protein named progerin (De Sandre-Giovannoli et al., 2003; Eriksson et al., 2003) (Fig. 1B). Mutations that inactivate *ZMPSTE24* cause the accumulation of full-length farnesylated prelamin A (Fig. 1C) (Bergo et al., 2002). The accumulation of progerin or prelamin A is toxic for cells, profoundly altering the architecture and function of the nuclear lamina. Multiple pathological mechanisms have been proposed, including the hypothesis that aberrant lamina architecture disrupts interactions with chromatin and transcription factors (TFs), perturbing gene regulation (Capanni et al., 2005; Infante, Ruiz de Eguino, Gago, & Rodriguez, 2013; Ruiz de Eguino et al., 2012).

1.2.1 Acquired Laminopathy

Prelamin A accumulation can also be induced pharmacologically, as seen in HIV patients undergoing "highly active antiretroviral therapy" (HAART). The HAART regimen includes HIV protease inhibitors (HIV-PIs) such as lopinavir and tipranavir (TPV). In addition to blocking the HIV protease (needed for virus replication), these drugs also inhibit ZMPSTE24, a metalloprotease, leading to the accumulation of farnesylated prelamin A (Fig. 1A) (Coffinier et al., 2007; Hudon et al., 2008). This discovery may explain some of the metabolic side effects of HAART, in which patients develop an "acquired laminopathy" that combines metabolic syndrome with partial lipodystrophy and premature aging (Vigouroux et al., 1999).

1.3 Human Cell Models of Laminopathies

Much of our knowledge about the molecular pathogenesis of laminopathies is based on mouse models carrying human mutations in endogenous *LMNA* or *ZMPSTE24* alleles, or mutated human or mouse *LMNA* transgenes (Zhang, Kieckhaefer, & Cao, 2013). However, no single mouse model recapitulates all human symptoms, underscoring the need for better models to unravel the pathological mechanisms and identify new therapeutic targets.

Most *LMNA*-linked syndromes are rare, limiting access to patient samples. Current knowledge comes largely from cultured fibroblasts derived from accessible tissue such as skin biopsies (Caron et al., 2007). To overcome the scarcity of primary cell sources in *LMNA*-linked diseases, several disease models have been developed based on induced pluripotent stem cells or human mesenchymal stem cells (hMSCs), which can be differentiated to specific cell types affected in laminopathies (Scaffidi & Misteli, 2008; Xiong, LaDana, Wu, & Cao, 2013; Zhang et al., 2011). Studying hMSC biology in the context of laminopathies is especially intriguing, since tissues derived from mesenchyme (adipose, bone, muscle, connective tissue) are most affected in *LMNA*-linked syndromes (Hennekam, 2006). To test this hypothesis, we generated the first human model of premature aging based on primary hMSCs that accumulate prelamin A in response to treatment with the protease inhibitor, TPV (Infante, Gago, de Eguino, et al., 2014; Ruiz de Eguino et al., 2012). These cell models are being used to characterize metabolic phenotypes and identify TFs that are dysregulated due to prelamin A accumulation (Infante & Rodríguez, 2015; Sánchez et al., 2015). Described below are protocols for inducing and detecting prelamin A in hMSCs. We also outline strategies for bioinformatic analysis of TFs whose activity is potentially impaired by prelamin A accumulation, and describe protocols to validate this alteration.

2. PRELAMIN A ACCUMULATION IN hMSCS

2.1 Induction of Prelamin A Accumulation in hMSCs

TPV-inducible accumulation of prelamin A in hMSCs was first reported in 2012 (Ruiz de Eguino et al., 2012). Since our goal was to simulate the effects of long-term TPV treatment (e.g., as seen in HIV patients), hMSCs were treated with TPV at clinically relevant concentrations of 20–30 μ*M* (Coffinier et al., 2007) for at least 35 days. Higher concentrations of TPV (>50 μ*M*) are toxic to hMSCs. Hence, if your experimental goal is to model laminopathy *per se* by acute induction of prelamin A accumulation, which requires higher TPV concentrations, treatment times must be shorter (48–72 h).

2.2 Detection of Prelamin A Accumulation in hMSCs

Prelamin A accumulation in cultured cells can be detected by Western blot, indirect immunofluorescence or "in-cell Westerns," described below.

2.2.1 Western Blot

Use Western blots to detect prelamin A in hMSCs only when prelamin A is present at high levels, for example, in cells treated with TPV at concentrations >50 μ*M* (Fig. 2A).

1. Culture cells in hMSC culture media (DMEM low glucose [Sigma #D5546], 2 m*M* glutamax [GIBCO #11574466], 10% FBS [Sigma #F7524], 100 U/ml penicillin/streptomycin [GIBCO #11548876]). As the positive control for prelamin A accumulation, treat parallel cultures for 24–48 h with 5 μ*M* FTI-277 (farnesyl transferase inhibitor; Sigma #F9803).

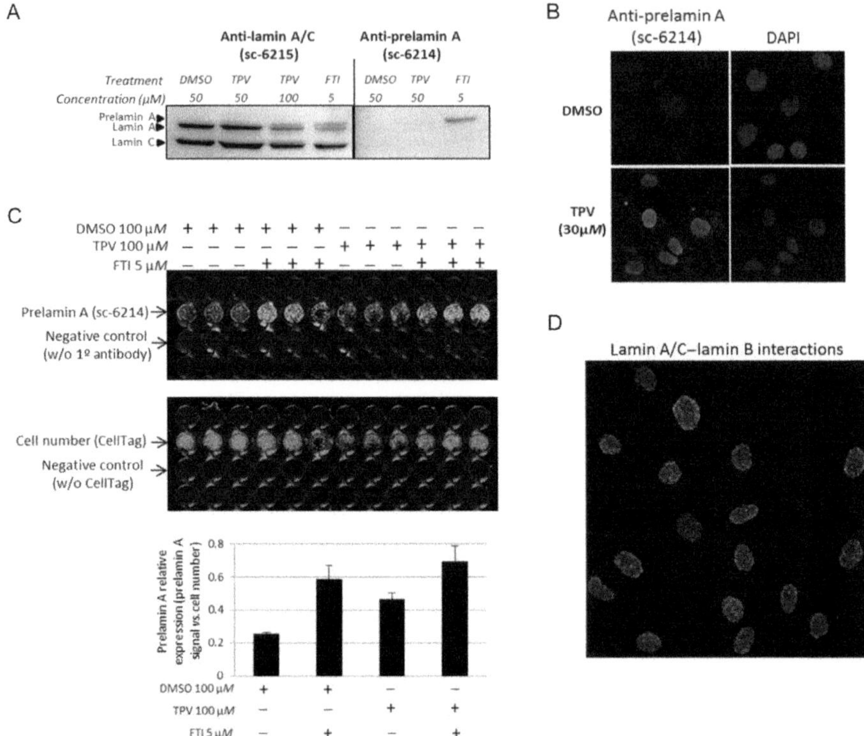

Figure 2 (A) Western blot showing prelamin A accumulation triggered by high concentrations of TPV, probed with antibodies that recognize either all A-type lamins (sc-6215), or only prelamin A (sc-6214). Negative control cells were treated with vehicle alone (DMSO). As the positive control for prelamin A accumulation, hMSCs were treated with a farnesyl transferase inhibitor (FTI). (B) Confocal immunofluorescence images showing prelamin A accumulation in hMSCs (red) treated with a low concentration of TPV. Cells were stained with the prelamin A-specific antibody (sc-6214). Nuclei are stained with DAPI (blue). Scale bar: 20 μm. (C) In-cell Western detection of prelamin A expression in treated cells in triplicate. This image shows a 96-well two-color in-cell Western with the 800 and 700 nm channels detecting prelamin A (sc-6214) and total cell numbers, respectively. Results are quantified in the graph below (mean ± standard deviation of triplicates). (D) PLA detection of lamin A/C interactions with lamin B in hMSCs. Each red dot reveals a physical protein–protein interaction. Nuclei are stained with DAPI (blue). (See the color plate.)

2. At selected timepoints, wash cells with PBS Dulbecco's w/o Ca Mg 1 × (PBS −/−) (Gibco #14190169).
3. Detach cells by incubating with TrypLE™ Select Enzyme (Gibco #12563011) in an incubator (5% CO_2, 37 °C) for 6 min.

4. Centrifuge cells (300 × *g*, 5 min). Wash the cell pellet in PBS Dulbecco's 1 × +Ca,+Mg (PBS +/+) (Gibco #14040174). Then, add 50–100 μl (per 1×10^6 cells) of ice-cold lysis buffer [1% NP-40 (Sigma #I8896), 0.25% sodium deoxycholate (Sigma #30970), 150 m*M* NaCl (Sigma #S3014), 1 m*M* EGTA (Sigma #E3889), 1 m*M* PMSF (Sigma #93482), 1 m*M* NaF (Sigma #S7920), and 1 × protease inhibitor cocktail (Thermo #78410)]. Incubate pellets in ice 30 min.
5. Clear lysates by centrifugation (12,000 × *g*, 4 °C, 15 min); recover the supernatant.
6. Measure lysate protein concentrations (e.g., BCA protein assay kit; Pierce #23227).
7. Prepare samples for gel electrophoresis with 10 μl per lane, consisting of 80 μg protein in NuPAGE LDS Sample Buffer 1 × (Invitrogen #NP0008) with NuPAGE Reducing Agent 1 × (Invitrogen #NP0004) in deionized water.
8. Heat samples 10 min at 70 °C, and load onto 4–12% Bis–Tris gels (Invitrogen #NP0321BOX). We use MOPS 1 × running buffer (NuPage 20 × MOPS stock from Invitrogen #NP0001) to separate prelamin A, lamin A, and lamin C which resolve at 74, 70, and 65 kDa, respectively. Since the samples were reduced in Step 7, also include 500 μl NuPAGE antioxidant (Invitrogen #NP0005) per 200 mL running buffer in the upper chamber.
9. Run the gel at 4 °C (to avoid over-heating) at 200 V for at least 2 h, to separate farnesylated prelamin A from lamin A (~4 kDa difference).
10. Transfer proteins to a PVDF membrane; this takes 8 min using the iBlot Dry Blotting system (Invitrogen #IB21001).
11. Block the PVDF membrane 1 h at room temperature with blocking buffer: TBS 1 × (50 m*M* Tris–HCl, pH 7.5, 150 m*M* NaCl) containing 0.1% Tween 20 (Sigma #P9416) and 5% powdered milk.
12. Incubate overnight at 4 °C with primary antibody diluted in blocking buffer. We use:
    a. Goat polyclonal anti-lamin A/C (Santa Cruz #sc-6215) at 1:100 dilution. This antibody recognizes the N-terminus of all A-type human lamins, whether modified or not.
    b. Goat polyclonal anti-prelamin A (Santa Cruz #sc-6214) at 1:100 dilution. This serum was raised against a C-terminal peptide of unprocessed human prelamin A, and thus recognizes only prelamin A (farnesylated or not). Note: this antibody shows batch-to-batch differences, and a new batch must be checked to ensure efficacy.

13. Wash the membrane three times (10 min each) in TBS containing 0.1% Tween 20.
14. Incubate membrane 1 h (RT) with secondary rabbit anti-goat IgG (H+L) HRP Conjugate (Bio-Rad #172-1034) at 1:6000 dilution in blocking buffer.
15. Wash the membrane three times (10 min each) in TBS containing 0.1% Tween 20.
16. Incubate membrane 5 min with ECL: SuperSignal West Femto Maximum Sensitivity Substrate (Pierce #34095).
17. Detect chemiluminescence; we use the ChemiDoc MP system (Bio-Rad).

2.2.2 Immunofluorescence

Low levels of prelamin A accumulation (e.g., hMSCs treated with clinical concentrations [20–30 μ*M*] of TPV) can be detected by immunofluorescence (Fig. 2B). All steps are performed at room temperature (RT) unless otherwise indicated.

1. Place sterile glass coverslips in 24-well plates (Corning #3524). Coat with 0.1 mg/ml poly-D-lysine hydrobromide (Sigma #P6407) for 1 h at 37 °C, or overnight at 4 °C. This treatment promotes hMSC adhesion to coverslips.
2. Wash coverslips twice with PBS (+/+) and let dry before seeding hMSCs.
3. At desired timepoints, seed hMSCs on prepared coverslips; the optimal density is 20,000–30,000 hMSCs per coverslip.
4. Wash cells with PBS (+/+), then fix 15 min in 4% paraformaldehyde (Sigma #HT5011-1CS).
5. Wash twice with PBS (+/+).
6. In a humid chamber, block coverslips 1 h with 125 μl blocking solution: PBS (+/+) containing 5% FBS to saturate nonspecific binding, and 0.1% Triton X-100 to permeabilize cells (Sigma #T8787).
7. Incubate 1 h with primary prelamin A antibody (Santa Cruz #sc-6214) diluted 1:100 in blocking solution.
8. Wash coverslips twice with PBS (+/+) containing 0.1% Triton X-100.
9. Incubate coverslips 1 h with secondary antibody: Cy3-AffiniPure Fab Fragment Donkey Anti-Goat IgG (Jackson InmunoResearch #705-166-147) diluted 1:300 in blocking solution.
10. Wash coverslips twice with PBS (+/+) containing 0.1% Triton X-100.

11. Invert each coverslip onto a slide containing a drop of mounting media: Prolong Gold antifade reagent with DAPI (Molecular Probes #P36931).
12. Seal covership edges with transparent nail polish and dry overnight.
13. Observe samples with a fluorescence microscope.

2.2.3 In-Cell Western

The in-cell Western, a microplate-based immunocytochemical assay, is very useful when hMSC cell numbers are low or when quantifying differences in prelamin A levels between samples (Fig. 2C). This assay uses primary antibodies highly specific for the target protein, infrared-conjugated secondary antibodies, and an infrared imaging system such as Odyssey CLx (Li-Cor).

1. Seed the hMSCs in 96-well plates (Corning #3360) at 2000 cells per well.
2. When appropriate, aspirate media and immediately fix cells 20 min in 4% paraformaldehyde.
3. Wash three times (5 min each) with PBS (+/+) containing 0.1% Triton X-100, to permeabilize.
4. Block cells by adding 100 μl Odyssey Blocking Buffer (Li-Cor #927-40100) to each well, and shake gently for 1.5 h at room temperature.
5. Shake cells gently overnight at 4 °C with primary antibody anti-prelamin A (Santa Cruz #sc-6214) diluted 1:100 in Odyssey Blocking Buffer in a final volume of 50 μl.
6. Wash each well three times (5 min each) with 200 μl PBS containing 0.1% Tween 20, with gentle shaking.
7. Dilute the fluorescent secondary antibody, donkey anti-goat IrDye800 (Li-Cor #926-32214), at 1:1000 in Odyssey Blocking Buffer. Then, add Tween 20 to a final concentration of 0.2% (to lower background), and add CellTag 700 Stain (Li-Cor #926-41090) to a dilution of 1:600 to normalize for the number of cells. Add 50 μl of this solution to each well, and incubate 1 h with gentle shaking.
8. Wash each well three times (5 min each) with 200 μl PBS containing 0.1% Tween 20, with gentle shaking. After the final wash, remove wash solution completely.
9. You then have two choices. For best results, scan the plate immediately per manufacturer's instructions (Li-Cor). Alternatively, add PBS to prevent drying, and store plates at 4 °C until scanning.

3. IDENTIFICATION OF TRANSCRIPTION FACTORS DYSREGULATED DUE TO PRELAMIN A ACCUMULATION

3.1 *In Silico* Analysis from Microarray Data

We used gene expression microarrays to identify genes potentially misregulated in hMSCs and hMSC-derived adipocytes that accumulated prelamin A (Infante, Gago, de Eguino, et al., 2014; Ruiz de Eguino et al., 2012). The transcriptome alteration, due to prelamin A accumulation, was further evaluated by a bioinformatics strategy that we outline here. We used the Database for Annotation, Visualization, and Integrated Discovery (DAVID), a bioinformatics program that predicts biological meaning from large gene lists (Huang, Sherman, & Lempicki, 2009). To identify specific dysregulated TFs that might govern the observed changes in gene expression, we used a free Web-based tool known as DiRE (Distant Regulatory Elements of coregulated genes) (Gotea & Ovcharenko, 2008).

3.1.1 Gene Ontology Classification by DAVID

The aim of gene ontology analysis by DAVID is to determine if dysregulated genes from the microarray data can be grouped into functional categories to obtain a global biological vision of the gene expression results. The DAVID program determines whether a specific annotation is "enriched" in the gene list when compared to background, and uses a hypergeometric test-based method to evaluate the potential significance of the results. The null hypothesis is that the enrichment of an annotation is by chance. Significance is measured by the *p*-value: the smaller the *p*-value, the more significant the finding.

1. Go to the link: http://david.abcc.ncifcrf.gov.
2. Click on the Functional annotation link. Follow the steps that appear at the left of the window.
3. *Step 1: Enter a gene list* → Upload the list of differentially expressed genes using the option "A" and paste a list of differentially expressed genes, or use option "B" to upload a previously saved file.
4. *Step 2: Select Identifier* → Gene lists can be uploaded using many identifiers. If official gene symbols have been uploaded, select the item "official gene symbol."
5. *Step 3: List Type* → Choose "Gene List."

6. *Step 4: Submit List* → Click the button "Submit List."
7. Now enter information about the gene list you uploaded. You must select the species (e.g., *Homo sapiens*) and in the List Manager below, you may rename your gene list.
8. Click "use" and enter the background you want to compare your gene list with. The most common backgrounds are the background of the same species (e.g., *Homo sapiens*) or the entire list of genes interrogated in microarrays (e.g., Illumina Human HT-12_v3).
9. Execute the analysis. Results appear to the right of the "Annotation Summary Results" window.
10. To start, go to the button "Functional Annotation Table." This table lists all terms associated with each gene.
11. Click on "Functional Annotation Chart" to obtain the enriched categories within the list of dysregulated genes.
12. Click on the button "Functional Annotation Clustering." Annotation terms are listed, along with a count of how many genes from your list are associated with each term.
13. To analyze more options, go back to the "Annotation Summary Results" window of the browser and select other options. To study pathways, find the annotation category "Pathways," and click on the "+" to the left of it to expand the category.

3.1.2 Computational Promoter Analysis by DiRE

DiRE performs statistical tests based on a hypergeometric distribution to identify TFs whose binding site signatures are significantly more enriched in the list of dysregulated genes, compared to the background set. DiRE retrieves a list of candidate TFs that could be responsible for the altered gene expression detected in microarrays. Two parameters are shown. "*Occurrence*" is the percentage of candidate regulatory elements that include a conserved binding site for a particular transcription factor. "*Importance*" is the product of TF occurrence and TF weight (DiRE optimization procedure calculates a weight for each TF as a measure of its association with the input gene set). Scores >0.1 generally indicate good confidence predictions.

1. Go to the link: http://dire.dcode.org.
2. Paste a target set (the list of dysregulated genes) in the "coregulated genes" option.
3. Enter the gene annotation (gene symbols) and the species (human).

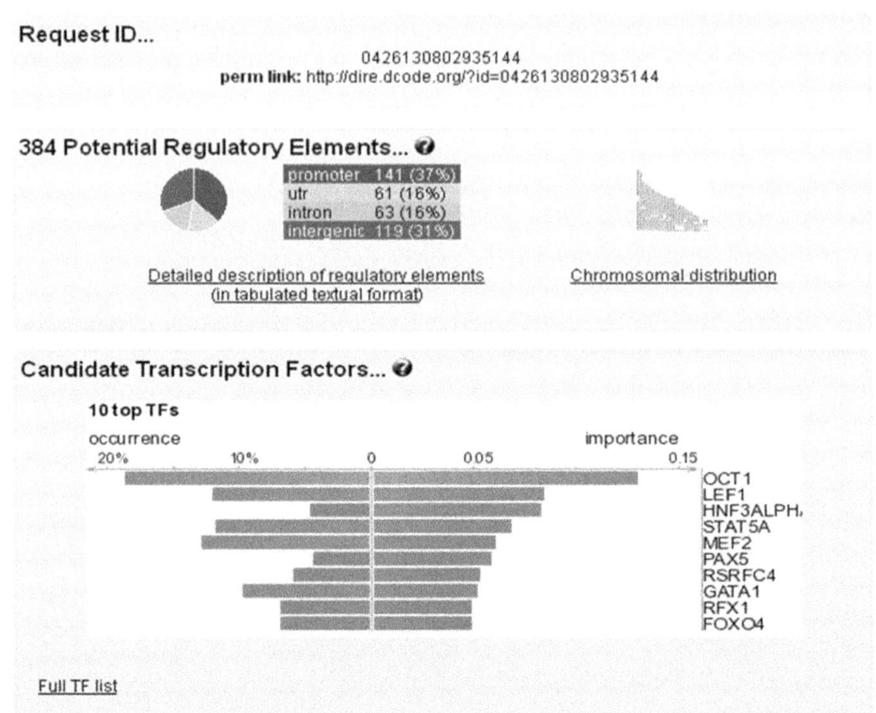

Figure 3 Sample results from DiRE bioinformatics program. The Request ID at the top is used for data retrieval. Detected regulatory elements (RE) in the input genes (categorized as promoter, intronic, intergenic or UTR elements) are graphically represented based on their chromosomal distribution. The bottom panel shows candidate transcription factor scores.

4. Enter a background set of promoters (random set of 5000 genes). The entire set of genes represented on the microarray can alternatively be used as the background set.
5. Click on the button "Submit." Results are illustrated in Fig. 3.

3.2 Validation of Transcription Factor Candidates

In silico results retrieved by DiRE are mathematical predictions that must be tested and validated biologically to determine if the candidate TF(s) are functionally impaired, in our case, due to prelamin A accumulation. For this validation, we use dual-luciferase reporter assays to measure the endogenous activity of specific TFs.

3.2.1 Dual-Luciferase Reporter Assays

Genetic reporter systems can be used to study the activation or inhibition of a gene promoter under specific circumstances. Dual reporter assays are commonly used to improve experimental accuracy: they combine the simultaneous expression and measurement of two individual reporter enzymes, namely firefly (*Photinus pyralis*) luciferase and *Renilla* (*Renilla reniformis*) luciferase, within a single system. The firefly luciferase plasmid is typically the experimental reporter, whereas the Renilla-luciferase construct is the internal control reporter that measures the baseline response. Normalizing the activity of the experimental reporter to the activity of the internal control minimizes experimental variability caused by differences in cell viability or transfection efficiency.

Approximately 60% of hMSCs become transfected, based on the percentage of GFP (green fluorescent protein)-positive cells in a parallel experiment, using the protocol below.

1. In a Nucleofector machine (Amaxa Biosystems), select the program optimized for hMSC nucleofection.
2. Prepare solutions following the instructions of the Human MSC Nucleofector® Kit (Amaxa #VPE-1001). Resuspend 400,000 hMSCs in 100 μl nucleofection buffer, and nucleofect with 2 μg of firefly plasmid DNA engineered to include the promoter of interest, and 0.005 μg of *Renilla* plasmid DNA.
3. Immediately after nucleofection, plate the hMSCs into one well of a 6-well plate filled with prewarmed medium.
4. Incubate cells 24 h, then wash gently with PBS (+/+) three times. Add the appropriate culture or induction medium and incubate 24 h longer.
5. Measure activities using the Dual-Luciferase® Reporter Assay System (Promega #E1980), following the manufacturer's instructions.

If the activity of the candidate TF is altered (e.g., inhibited or hyperactive) in response to prelamin A accumulation, the next step is to determine if the TF directly binds mature lamin A or prelamin A. For that, we use the Proximity Ligation Assay (PLA) technique.

3.2.2 In Situ *Proximity Ligation Assay*

In situ PLA detects protein–protein interactions with high specificity and sensitivity (Söderberg et al., 2006). Two primary antibodies (from two different host species) are used together with Duolink species-specific secondary antibodies that contain unique DNA strands that template the hybridization of added oligonucleotides. When in close proximity

(<40 nm), the oligonucleotides are ligated by a ligase to form a circular template. This template, still anchored to the antibody, is then amplified and detected using complementary labeled oligonucleotide probes. This technology generates discrete fluorescent dot signals that can be counted, showing the exact subcellular location of the protein–protein interaction. This technique has many advantages over immunoprecipitation, since it requires few cells and small amounts of antibody, and can be performed in hours.

1. Culture, fix, and block hMSCs on coverslips as indicated for immunofluorescence assays (steps 1–6, above).
2. Remove as much blocking solution as possible, and add 125 μl of primary antibody solution to each coverslip. As positive controls, we use antibodies specific for lamins A/C (Santa Cruz #sc-6215) and lamin B (Santa Cruz #sc-365962) as shown in Fig. 2D. Negative control cells are incubated without primary antibodies.
3. Wash coverslips twice (5 min each) with 125 μl of 0.1% Triton X-100 in PBS (+/+).
4. Dilute and mix together the two PLA probes (Duolink II anti-goat MINUS and Duolink II anti-Rabbit PLUS) 1:5 in Antibody Diluent, and add 95 μl per coverslip.
5. Incubate 1 h in a humid chamber.
6. Wash coverslips twice (5 min each) with 125 μl of 0.1% Triton X-100 in PBS (+/+).
7. Dilute Duolink II Ligation stock (5 ×) in high purity water to 1 × final, and mix. When calculating the volume of water, take into account the volume of ligase that will be added at a final dilution of 1:40 just before adding the mix to the coverslips. Add 40 μl of the final solution to each coverslip. Incubate slides 30 min in a preheated (37 °C) humid chamber at 37 °C.
8. Wash slides twice (5 min each, room temperature) in 1 × Wash Buffer A.
9. Dilute the Duolink II Amplification stock (5 ×) in high purity water to 1 × final, and mix. When calculating the volume of water, take into account the volume of polymerase that will be added at a final dilution of 1:80 just before adding this mix to the wells. Add the polymerase at the last possible minute at 1:80 dilution, and vortex. Add 40 μl of the final solution to each coverslip. Incubate slides in a humid chamber for 100 min at 37 °C.
10. Wash twice (10 min each) in 1 × Duolink II Wash Buffer B at room temperature.

11. Wash once (1 min) in 0.01% Duolink II Wash Buffer B (diluted in high purity water).
12. Let coverslips dry.
13. Seal coverslip edges with transparent nail polish, and dry overnight.
14. Observe samples with a fluorescence microscope.

4. CONCLUSIONS

Primary hMSCs that accumulate prelamin A are a unique experimental model to study the molecular pathology of *LMNA*-linked diseases. Although TPV treatment might affect other pathways independent of prelamin A, to our knowledge the only cellular protease inhibited by TPV is ZMPSTE24 (Coffinier et al., 2007). This experimental model also reliably summarizes the different phenotypes seen in cells from patients affected by *LMNA*-linked lipodystrophies (Ruiz de Eguino et al., 2012). This model can function as a platform to screen drugs against potential targets, including transcription factors whose activities are disrupted by laminopathy (Infante, Gago, & Rodriguez, 2014). The above protocols provide powerful platforms for functional analysis of laminopathy-relevant TFs, for example, by shRNA-dependent downregulation (Infante, Gago, de Eguino, et al., 2014) or chemical inhibitors (Ruiz de Eguino et al., 2012) to understand the pathological consequences of altered transcriptional activity.

ACKNOWLEDGMENTS

This work was supported by grants from the European Union, Marie Curie International Reintegration Grant MIRG-CT-2006-044914, Fondo de Investigación Sanitaria (Instituto de Salud Carlos III, Madrid, Spain) Grant PI06-1335, and the Department of Industry of the Basque Country Government (SAIOTEK Research Program 2013).

REFERENCES

Beck, L. A., Hosick, T. J., & Sinensky, M. (1990). Isoprenylation is required for the processing of the lamin A precursor. *The Journal of Cell Biology*, *110*, 1489–1499.

Bergo, M. O., Gavino, B., Ross, J., Schmidt, W. K., Hong, C., Kendall, L. V., et al. (2002). Zmpste24 deficiency in mice causes spontaneous bone fractures, muscle weakness, and a prelamin A processing defect. *Proceedings of the National Academy of Sciences of the United States of America*, *99*, 13049–13054.

Burke, B., & Stewart, C. L. (2013). The nuclear lamins: Flexibility in function. *Nature Reviews Molecular Cell Biology*, *14*, 13–24.

Capanni, C., Mattioli, E., Columbaro, M., Lucarelli, E., Parnaik, V. K., Novelli, G., et al. (2005). Altered pre-lamin A processing is a common mechanism leading to lipodystrophy. *Human Molecular Genetics*, *14*, 1489–1502.

Caron, M., Auclair, M., Donadille, B., Béréziat, V., Guerci, B., Laville, M., et al. (2007). Human lipodystrophies linked to mutations in A-type lamins and to HIV protease inhibitor therapy are both associated with prelamin A accumulation, oxidative stress and premature cellular senescence. *Cell Death and Differentiation, 14*, 1759–1767.

Coffinier, C., Hudon, S. E., Farber, E. A., Chang, S. Y., Hrycyna, C. A., Young, S. G., et al. (2007). HIV protease inhibitors block the zinc metalloproteinase ZMPSTE24 and lead to an accumulation of prelamin A in cells. *Proceedings of the National Academy of Sciences of the United States of America, 104*, 13432–13437.

Dai, Q., Choy, E., Chiu, V., Romano, J., Slivka, S. R., Steitz, S. A., et al. (1998). Mammalian prenylcysteine carboxyl methyltransferase is in the endoplasmic reticulum. *The Journal of Biological Chemistry, 273*, 15030–15034.

De Sandre-Giovannoli, A., Bernard, R., Cau, P., Navarro, C., Amiel, J., Boccaccio, I., et al. (2003). Lamin a truncation in Hutchinson-Gilford progeria. *Science, 300*, 2055.

Dittmer, T. A., & Misteli, T. (2011). The lamin protein family. *Genome Biology, 12*, 222.

Eriksson, M., Brown, W. T., Gordon, L. B., Glynn, M. W., Singer, J., Scott, L., et al. (2003). Recurrent *de novo* point mutations in lamin A cause Hutchinson–Gilford progeria syndrome. *Nature, 423*, 293–298.

Gotea, V., & Ovcharenko, I. (2008). DiRE: Identifying distant regulatory elements of co-expressed genes. *Nucleic Acids Research, 36*, W133–W139.

Hennekam, R. C. (2006). Hutchinson-Gilford progeria syndrome: Review of the phenotype. *American Journal of Medical Genetics Part A, 140*, 2603–2624.

Huang, D. W., Sherman, B. T., & Lempicki, R. A. (2009). Systematic and integrative analysis of large gene lists using DAVID bioinformatics resources. *Nature Protocols, 4*, 44–57.

Hudon, S. E., Coffinier, C., Michaelis, S., Fong, L. G., Young, S. G., & Hrycyna, C. A. (2008). HIV-protease inhibitors block the enzymatic activity of purified Ste24p. *Biochemical and Biophysical Research Communications, 374*, 365–368.

Infante, A., Gago, A., de Eguino, G. R., Calvo-Fernández, T., Gómez-Vallejo, V., Llop, J., et al. (2014). Prelamin A accumulation and stress conditions induce impaired Oct-1 activity and autophagy in prematurely aged human mesenchymal stem cell. *Aging (Albany NY), 6*, 264–280.

Infante, A., Gago, A., & Rodriguez, C. I. (2014). A human experimental model of laminopathy based on adult stem cells establishes the relevance of Oct1 transcription factor in the aging process. *Stem Cell and Translational Investigation, 1*, e276.

Infante, A., & Rodríguez, C. I. (2015). Prelamin A and Oct-1: A puzzle of aging. *Oncotarget, 6*, 3475–3476.

Infante, A., Ruiz de Eguino, G., Gago, A., & Rodriguez, C. I. (2013). *LMNA*-linked lipodystrophies: Experimental models to unravel the molecular mechanisms. In Y. Lin & X. Cai (Eds.), *Adipogenesis, signaling pathways, molecular regulation and impact on human disease* (pp. 35–93). New York: Nova Science Publishers, Inc.

Pendás, A. M., Zhou, Z., Cadiñanos, J., Freije, J. M., Wang, J., Hultenby, K., et al. (2002). Defective prelamin A processing and muscular and adipocyte alterations in Zmpste24 metalloproteinase-deficient mice. *Nature Genetics, 31*, 94–99.

Ruiz de Eguino, G., Infante, A., Schlangen, K., Aransay, A. M., Fullaondo, A., Soriano, M., et al. (2012). Sp1 transcription factor interaction with accumulated prelamin A impairs adipose lineage differentiation in human mesenchymal stem cells: Essential role of sp1 in the integrity of lipid vesicles. *Stem Cells Translational Medicine, 1*, 309–321.

Sánchez, P., Infante, A., Ruiz de Eguino, G., Fuentes-Maestre, J. A., García-Verdugo, J. M., & Rodriguez, C. I. (2015). Age-related lipid metabolic signature in human LMNA-lipodystrophic stem cell-derived adipocytes. *The Journal of Clinical Endocrinology and Metabolism, 100*, E964–E973.

Scaffidi, P., & Misteli, T. (2008). Lamin A-dependent misregulation of adult stem cells associated with accelerated ageing. *Nature Cell Biology, 10*, 452–459.

Söderberg, O., Gullberg, M., Jarvius, M., Ridderstrale, K., Leuchowius, K. J., Jarvius, J., et al. (2006). Direct observation of individual endogenous protein complexes in situ by proximity ligation. *Nature Methods*, *3*, 995–1000.

Swift, J., Ivanovska, I. L., Buxboim, A., Harada, T., Dingal, P. C., Pinter, J., et al. (2013). Nuclear lamin-A scales with tissue stiffness and enhances matrix-directed differentiation. *Science*, *341*, 1240104.

Vigouroux, C., Gharakhanian, S., Salhi, Y., Nguyen, T. H., Chevenne, D., Capeau, J., et al. (1999). Diabetes, insulin resistance and dyslipidaemia in lipodystrophic HIV-infected patients on highly active antiretroviral therapy (HAART). *Diabetes & Metabolism*, *25*, 225–232.

Weber, K., Plessmann, U., & Traub, P. (1989). Maturation of nuclear lamin A involves a specific carboxy-terminal trimming, which removes the polyisoprenylation site from the precursor; implications for the structure of the nuclear lamina. *FEBS Letters*, *257*, 411–414.

Wilson, K. L., & Foisner, R. (2010). Lamin-binding proteins. *Cold Spring Harbor Perspectives in Biology*, *2*, a000554.

Xiong, Z. M., LaDana, C., Wu, D., & Cao, K. (2013). An inhibitory role of progerin in the gene induction network of adipocyte differentiation from iPS cells. *Aging*, *5*, 288–303.

Zhang, F. L., & Casey, P. J. (1996). Protein prenylation: Molecular mechanisms and functional consequences. *Annual Review of Biochemistry*, *65*, 241–269.

Zhang, H., Kieckhaefer, J. E., & Cao, K. (2013). Mouse models of laminopathies. *Aging Cell*, *12*, 2–10.

Zhang, J., Lian, Q., Zhu, G., Zhou, F., Sui, L., Tan, C., et al. (2011). A human iPSC model of Hutchinson Gilford Progeria reveals vascular smooth muscle and mesenchymal stem cell defects. *Cell Stem Cell*, *8*, 31–45.

CHAPTER TWENTY-FOUR

MCLIP Detection of Novel Protein–Protein Interactions at the Nuclear Envelope

Mohammed Hakim Jafferali, Ricardo A. Figueroa, Einar Hallberg[1]
Department of Neurochemistry, Stockholm University, Stockholm, Sweden
[1]Corresponding author: e-mail address: einar.hallberg@neurochem.su.se

Contents

Abstract

The organization and function of the nuclear envelope (NE) involves hundreds of nuclear membrane proteins and myriad protein–protein interactions, most of which are still uncharacterized. Many NE proteins interact stably or dynamically with the nuclear lamina or chromosomes. This can make them difficult to extract under non-denaturing conditions, and greatly limits our ability to explore and identify functional protein interactions at the NE. This knowledge is needed to understand nuclear envelope structure and the mechanisms of human laminopathy diseases. This chapter provides detailed protocols for MCLIP (membrane cross-linking immunoprecipitation) identification of novel protein–protein interactions in mammalian cells.

1. INTRODUCTION

In eukaryotic cells, the nucleoplasm is separated from the cytoplasm by the nuclear envelope (NE). The outer nuclear membrane (ONM) is continuous with the ER; however, the ONM and to an even greater extent the

Methods in Enzymology, Volume 569
ISSN 0076-6879
http://dx.doi.org/10.1016/bs.mie.2015.08.022

inner nuclear membrane (INM) each have unique integral membrane proteins. Over 200 unique transmembrane proteins have been identified that specifically localize at the NE (Worman & Schirmer, 2015). Many novel NE transmembrane (NET) proteins are expressed only in specific cell types (Korfali et al., 2012; Schirmer, Florens, Guan, Yates, & Gerace, 2003; Wilkie et al., 2011). Most characterized NE membrane proteins bind lamins (Table 1), and mutations in many of these lamin-binding proteins are linked to human disease. Yet very few NE proteins have been characterized in terms of their binding to lamins or functional partners (Table 1).

To understand the functional organization of the NE, it is essential to explore new NE proteins and their interactions. However, NE proteins are especially challenging due to their poor solubility. Many different biochemical and genetic approaches are available to identify and characterize protein–protein interactions, which can be stable or transient, strong, or weak.

However, three approaches are specifically capable of detecting novel protein–protein interactions in mammalian cells, in combination with mass spectrometry (MS): coimmunoprecipitation (CoIP), biotinylation identification (BioID), and chemical cross-linking. We discuss the advantages and limitations of each strategy, and provide a detailed protocol for a method, membrane cross-linking immunoprecipitation (MCLIP), that combines chemical cross-linking with CoIP as a discovery tool.

1.1 Coimmunoprecipitation as a Discovery Tool

CoIP, a widely used method to study protein–protein interactions, relies on the specificity and binding affinity of antibodies (Harlow, Whyte, Franza, & Schley, 1986; Lane & Crawford, 1979; Masters, 2004). In CoIP, cells are lysed with mild buffers to retain protein–protein interactions; the protein of interest and associated proteins, are then captured and separated from unbound proteins using specific antibodies immobilized on beads. Precipitated proteins and coprecipitated endogenous partners or associated protein complexes are then typically resolved by SDS-PAGE and evaluated by Western blotting. The main advantage of CoIP is that one can study interactions among native endogenous proteins; no exogenous proteins or cell engineering are required. However, CoIP requires specific antibodies that can efficiently immunoprecipitate the protein of interest, which may not be available, or which may not recognize the target under all conditions (e.g., epitope may be blocked by posttranslational modifications or associated partners). To surmount this problem, the target protein can be affinity tagged, expressed exogenously, and captured using high-affinity commercial

Table 1 Strategies Used to Identify Partners for Selected NE Membrane Proteins and Lamins

NE Proteins	Interacting Partners	Protein–Protein Interactions Studies: CoIP	Pull-Down Experiments	Y2H[#]	Other Methods	Type of Interaction Direct: + Unknown: ?	References
A-type lamins	Sun1, 2		+			?	Crisp et al. (2006)
	Sun1	+	+			?	Haque et al. (2006)
	Lamin B				Affinity chromatography and solid phase binding assay	+	Georgatos, Stournaras, and Blobel (1988)
	Lap2α	+			Blot overlay	+	Dechat et al. (2000)
	Lap 1A, 1B		+			+	Foisner and Gerace (1993)
	Nup 153		+		Blot overlay assay	+	Al-Haboubi, Shumaker, Koser, Wehnert, and Fahrenkrog (2011)
B-type lamins	Lap 2		+			+	Foisner and Gerace (1993)
	Lap 1A, 1B		+				
	LBR				Ligand binding assay	?	Worman, Yuan, Blobel, and Georgatos (1988)
	Nup 153		+		Blot overlay assay	+	Al-Haboubi et al. (2011)
Emerin	BAF				FRET	+	Shimi et al. (2004)
	Lamin A, B, C	+				?	Vaughan et al. (2001)
	Lamin A, C			+		?	Sakaki et al. (2001)
	Sun1, 2	+	+			?	Haque et al. (2010)
	Man1	+			Blot overlay assay	+	Mansharamani and Wilson (2005)
	Myosin IIB	+				?	Chang, Folker, Worman, and Gundersen (2013)
	Tubulin		+			?	Salpingidou, Smertenko, Hausmanowa-Petrucewicz, Hussey, and Hutchison (2007)
MAN1	BAF		+		Microtiter assay	+	Mansharamani and Wilson (2005)
	Lamin A						
	Lamin B1						

Continued

Table 1 Strategies Used to Identify Partners for Selected NE Membrane Proteins and Lamins—cont'd

NE Proteins	Interacting Partners	Protein–Protein Interactions Studies: CoIP	Pull-Down Experiments	Y2H[#]	Other Methods	Type of Interaction Direct: + Unknown: ?	References
Sun1 and Sun2	Nesprin 1, 2		+	+		+	Padmakumar et al. (2005)
		+				?	Haque et al. (2006)
		+				?	Crisp et al. (2006)
	Nesprin 3α	+				?	Ketema et al. (2007)
POM 121	Nup 155		+			+	Mitchell, Mansfeld, Capitanio, Kutay, and Wozniak (2010)
	Nup 98, 160						
LBR	Histone H3/H4		+			?	Polioudaki et al. (2001)
	Histone H4		+			+	Hirano et al. (2012)
	HP1	+	+	+		+	Ye and Worman (1996)
Lamin A	Lap1, Lap 2α				BioID-LaA/mass s pectrometry	?	Roux, Kim, Raida, and Burke (2012)
	Emerin, Man1						
	Samp1						
	Nup 153, 50, ELYS						
Samp1	Emerin	+				?	Gudise, Figueroa, Lindberg, Larsson, and Hallberg (2011)
	Lamin B				MCLIP	?	Jafferali et al. (2014)
	Ran						
	Sun1						
	Emerin		+			+	
	Sun2	+				?	Borrego-Pinto et al. (2012)

antibodies against the engineered tag (e.g., GFP, FLAG, HA, and MYC). Another disadvantage is that CoIP is done after cell lysis, and cannot distinguish if interactions are direct or mediated by unknown components. CoIP may also fail to recover low-abundance, transient, or weaker-affinity partners (Berggard, Linse, & James, 2007; Masters, 2004).

1.2 BioID as a Discovery Tool

Biotinylation identification (BioID) is a unique and sensitive new method to screen for protein–protein interactions in living cells (Roux et al., 2012). BioID protocols are provided in this volume (see chapter "BioID identification of lamina-associated proteins" by A. A. Mehus, R. H. Anderson, & K. J. Roux), and have also been adapted for use in *Dictyostelium amoebae* (see chapter "Proximity-dependent biotin identification (BioID) in Dictyostelium amoebae" by P. Batsios, I. Meyer, & R. Gräf; this volume). In BioID, the protein of interest is fused to a "promiscuous" biotin ligase and stably expressed in cells; the fusion protein biotinylates proximal proteins only when excess biotin is added to the medium. High-affinity avidin/streptavidin-beads are used to capture biotinylated candidates in a single step. BioID has great advantages, particularly its sensitivity in recovering novel candidates. However, biotinylation of the prey or bait protein might disturb their function. Protein labeling in BioID occurs over time, and maximal labeling requires relatively long (16–18 h) incubation times (Mehus et al., this volume). Thus, there is less experience, so far, using BioID to study protein interactions that take place during short time intervals, e.g., during mitosis.

1.3 Chemical Cross-linking as a Discovery Tool

Chemical cross-linking has long been used to map large protein complexes, but can also capture dynamic, transient, or weak interactions that elude other methods (Percipalle et al., 2002; Smith, Friedman, Yu, Carnahan, & Reynolds, 2011). Membrane-permeable cross-linkers allow investigators to probe protein–protein interactions *in vivo*. The *N*-hydroxysuccinimide (NHS) ester of cross-linkers such as DSP (shown in Fig. 1) and DTSSP react

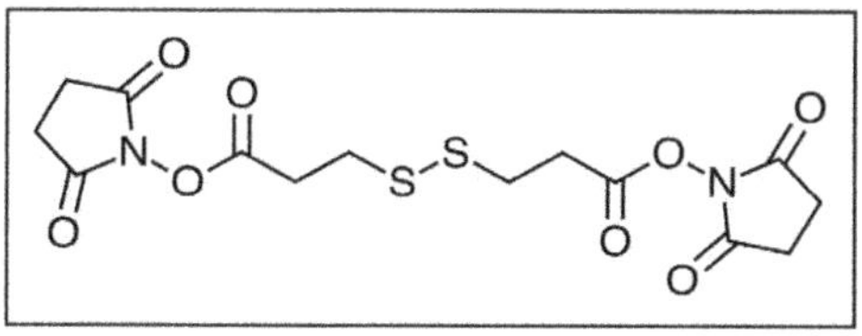

Figure 1 Structure of DSP.

efficiently with primary amines on proteins, forming stable amide bonds (Lomant & Fairbanks, 1976) that resist the harsh denaturing conditions required to solubilize NE proteins (Jafferali et al., 2014).

Cross-linked proteins are then identified by MS, theoretically allowing the discovery of novel and unexpected interaction candidates. However, several caveats must be considered. Cross-linking via DSP or other NHS esters covalently modifies primary amines, and therefore changes the mass of modified peptides. Standard algorithms might fail to identify these peptides, reducing coverage; in this case, protein identification is limited by the availability of unmodified peptides. Furthermore, cross-link-modified residues might not be recognized by enzymes, such as trypsin, typically used to digest proteins prior to MS, lowering sequence coverage—a worse problem for integral membrane proteins, which already suffer poor sequence coverage in MS (Eichacker et al., 2004). Two strategies can ameliorate these problems. With DSP-related cross-linkers, cross-linked proteins can be fragmented using enzymes (e.g., chymotrypsin or V8 protease) or chemicals (CNBr) that do not require primary amines (Helbig, Heck, & Slijper, 2010). Promising new strategies involve the use of modified or novel cross-linkers (Kluger & Alagic, 2004; Sinz, 2006) that are compatible with, or ideally improve, the identification of cross-linked proteins by MS, e.g., as reported by Muller, Dreiocker, Ihling, Schafer, and Sinz (2010), although this strategy has not yet been used to identify protein interactions in complex systems.

1.4 Protein–Protein Interactions at the Nuclear Envelope

Most characterized NE proteins have a variety of known binding partners (Table 1), including A- or B-type lamins, consistent with their functional diversity and their ability to interact with tissue-specific partners. However, this field is challenged by the fact that NE proteins are notoriously difficult to solubilize (Dwyer & Blobel, 1976; Hallberg, Wozniak, & Blobel, 1993; Radu, Blobel, & Wozniak, 1993; Snow, Senior, & Gerace, 1987; Wozniak, Bartnik, & Blobel, 1989). Classical coprecipitation is therefore an unattractive strategy for identifying novel partners for lamins and NE membrane proteins. Biochemical pull-down studies using recombinant purified domains of various NE proteins have been used successfully to identify potential protein–protein interactions (Crisp et al., 2006; Haque et al., 2006). However, other studies are always required to determine whether such interactions actually take place in living cells.

Samp1 (also known as NET5 or TMEM201) illustrates the challenges faced by researchers who seek to identify functional partners for INM proteins. Samp1 is a highly conserved integral INM protein (Buch et al., 2009). Its N-terminal half includes four conserved zinc finger (CXXC) motifs and is located in the nucleoplasm, and its C-terminal half consists of four transmembrane domains (Gudise et al., 2011). Samp1 is partially extracted by 1% Triton X-100 with 250 m*M* NaCl (Buch et al., 2009), but these same conditions do not solubilize two known partners: Sun1 and lamin B1. This situation is common for NE proteins, emphasizing the need for new or alternative methods to unravel the dynamic NE interactome.

1.5 Overview of the MCLIP Method

A membrane permeable reversible *in vivo* cross-linker, *d*ithiobis-*s*uccinimidyl *p*ropionate (DSP) was previously used to identify labile protein complexes in the nucleus (Percipalle et al., 2002). We adapted DSP cross-linking as a method to covalently link INM proteins to their partners prior to extraction under denaturing conditions. Before immunoprecipitation, the extracts were diluted to urea concentrations that can be tolerated by the antibodies as described by Jafferali et al. (2014).

Specific antibodies are only available for a few NE proteins. Thus for broader applicability of the MCLIP method, NE proteins of interest can be expressed as epitope-tagged fusion proteins (e.g., GFP or Myc), and immunoprecipitated using commercially available high-affinity antibodies, including single-chain "nanobodies" (e.g., GFP-Trap or Myc-Trap; ChromTeK). Immunoprecipitating from isolated nuclei, as described in Section 3, helps enrich samples and also provides evidence that the identified interaction takes place in the nucleus. Using CoIP methods, Borrego-Pinto et al. (2012) showed that Samp1 interacts with Sun2 and lamins A/C in NIH 3T3 cells (from mouse fibroblast). Using a different splice variant of Samp1 as bait, and studying a different cell type, Jafferali et al. (2014) used MCLIP to show that Samp1 interacts with Sun1, lamin B1, Emerin, and Ran in living U2OS cells (derived from human bone osteosarcoma). Thus, MCIP is a promising method that may expand the number of identified partners in the NE proteomes of specific cell types and tissues (Korfali et al., 2012).

2. MCLIP METHOD

Presented here is a detailed protocol for MCLIP analysis of tissue culture cells that stably express the protein of interest fused to an epitope tag. This method can be used to study endogenous INM proteins for which immunoprecipitating antibodies are available, and to study protein interactions during specific short timespans, such as mitosis (Jafferali et al., 2014).

1. Culture cells that stably express the (tagged) protein of interest on two 75-cm^2 flasks until they are confluent. As controls, grow parallel cultures of the parental cell line, or cells that stably express the tag alone, and cells expressing a control protein with tag.
2. Cross-link proteins by incubating cells (15 min, room temperature) in standard culture media containing 1 m*M* DSP, made from a fresh stock solution of 100 m*M* DSP dissolved in DMSO. Tip: Consider titrating the concentration of cross-linker, and cross-linking time, in pilot studies that examine a known partner and noninteracting control.
3. Remove medium; add 5 ml of 15 m*M* Tris–Cl (pH 7.4) and incubate 10 min at room temperature to quench cross-linker.
4. Wash cells twice with ice-cold PBS.
5. Scrape cells using a rubber scraper in ice-cold PBS, or trypsinize and collect by centrifugation.
6. Wash the cell pellet twice with ice-cold PBS. Keep on ice.
7. If needed, isolate nuclei as described in Section 3. Keep on ice.
8. To each cell pellet (or isolated nuclei), add 5 volumes of lysis buffer (7 *M* urea, 1% TX-100, and protease inhibitor cocktail) with benzonase (25 U/mL). Incubate on ice 20 min, and homogenize at 10 min intervals by drawing the lysate vigorously up and down (15 times) through a 23-gauge needle to disrupt membranes. Note: prepare the 7 *M* urea stock several days in advance; solubilization is endothermic, hence slow. *Do not* try to speed this reaction by heating. Heat converts urea to a reactive form that modifies (carbamylates) proteins, interfering with MS analysis (Sun, Zhou, Yang, & Zhang, 2014).
9. Save an aliquot of each sample, as the input control. Mix with an equal volume of 2 × SDS sample buffer containing 200 m*M* DTT, boil for 10 min, and store at −70 °C.
10. Dilute each lysate eightfold with ice-cold PBS containing protease inhibitor cocktail to enable subsequent immunoprecipitation.
11. Sonicate the lysate at 50% amplitude (5 pulses of 5 s on & 5 s off for each cycle; VCX 130 Sonics & Materials, Inc.) to shear the DNA. Keep samples on ice during sonication. If benzonase was used during lysis, this sonication step can be skipped.
12. Centrifuge lysate at 800 × *g* (5 min, 4 °C). Recover the supernatant (clarified lysate). Use immediately for immunoprecipitation in Step 13, or snap-freeze in liquid N_2 and store at −70 °C.

13. Preclear each lysate by rotating end-over-end with BSA-coated control beads for 1 h at 4 °C. Pellet the beads by centrifugation at 800 *g* (5 min, 4 °C), and carefully recover the supernatant (precleared lysate).
14. Incubate the precleared lysates with beads coupled to specific antibodies (bait trap) for 2 h at 4 °C with end-over-end rotation. Antibodies can be covalently coupled to beads (e.g., protein G Sepharose) to reduce the level of contaminating immunoglobulin heavy and light chains, which interfere with both immunoblot analysis and MS. These problems can be greatly reduced by using small single-chain commercial llama antibodies (nanobodies) against GFP (Trinkle-Mulcahy et al., 2008) or Myc.
15. Pellet beads. If you wish to recover proteins that did not immunoprecipitate, TCA precipitate this first supernatant as described by Koontz (2014). Wash beads twice with wash buffer (250 m*M* NaCl, 10 m*M* HEPES, pH 7.4, 0.5% TX-100, 0.1% Tween-20).
16. Elute bound proteins and reverse the cross-links in one step, by adding twice the bead volume of 2 × SDS sample buffer containing 200 m*M* DTT and boiling for 10 min. Collect the supernatant after centrifugation at 800 × *g* (5 min, 4 °C).
17. Resolve small aliquots by SDS-PAGE and monitor results by Western blotting.
18. Resolve larger amounts of sample and cut gel bands to submit for MS analysis. Before doing this, consult your MS experts and follow their advice.

3. METHOD FOR ISOLATING NUCLEI FROM TISSUE CULTURE CELLS

Here, we describe a simple method to isolate nuclei from tissue culture cells (Jafferali, Beckman, Kihlmark, & Hallberg, 2015).

1. Collect cells from two 75-cm^2 flasks (cross-linked or not), and wash with ice-cold PBS.
2. Suspend each cell pellet in 1 mL ice-cold hypotonic buffer (10 m*M* HEPES pH 7.4, 5 m*M* $MgCl_2$, 10 m*M* NaCl, protease inhibitor cocktail) and incubate on ice for 30 min. Transfer to a chilled Dounce homogenizer.
3. Rupture the plasma membrane by 50–100 gentle strokes. Number of strokes required can differ between cell types. Monitor cell breakage and release of free nuclei by phase contrast microscopy in combination with Trypan blue staining.

4. To the homogenate, add 30 μL of 5 *M* NaCl while constantly swirling the tube, to restore physiological salt concentrations. Add fresh 100 × protease inhibitor cocktail.
5. Transfer each homogenate (~1 mL volume) to a 4-mL capacity SW60Ti tube (11 × 60 mm; #328874; Beckman Instruments, Inc.). Carefully underlay with 500 μL precooled isotonic buffer containing 2.3 *M* sucrose, to create a "cushion." Disrupt the interface by carefully stirring five circular turns with a Pasteur pipette at the interface.
6. Fill each SW60Ti tube completely with ice-cold isotonic buffer (10 m*M* HEPES, pH 7.4, 5 m*M* $MgCl_2$, 160 m*M* NaCl, 1 × protease inhibitors) and balance. Centrifuge 1 h at 105,000 × *g* in a Beckman ultracentrifuge swing-out bucket rotor at 4 °C.
7. To avoid disturbing the pellet, pour out the supernatant by inverting the tubes. Cut the tubes just above the pellet. If the nuclear pellet is not obvious, cut above the bottom mark on the tubes.
8. Suspend the pellet in 100 μL ice-cold isotonic buffer. A small aliquot of isolated nuclei can be inspected by phase contrast microscopy in combination with Trypan blue staining.
9. Use the purified nuclei immediately for interaction studies in Section 2 (Step 8), or snap-freeze in liquid nitrogen and store at −70 °C.

4. CONCLUSIONS

Functional analysis of new and emerging NE and INM proteins, and their interactions with lamins, will be critical to understand nuclear structure, genome biology and human laminopathy diseases. Emerging techniques such as BioID and MCLIP in combination with MS holds great promise for defining the NE interactome in different tissues.

ACKNOWLEDGMENTS

This work was supported by grants from the Swedish Research Council #621-2010-448, Cancerfonden #110590, and the Foundation Olle Engkvists minne.

REFERENCES

Al-Haboubi, T., Shumaker, D. K., Koser, J., Wehnert, M., & Fahrenkrog, B. (2011). Distinct association of the nuclear pore protein Nup153 with A- and B-type lamins. *Nucleus*, *2*, 500–509.

Berggard, T., Linse, S., & James, P. (2007). Methods for the detection and analysis of protein-protein interactions. *Proteomics*, 7, 2833–2842.

Borrego-Pinto, J., Jegou, T., Osorio, D. S., Aurade, F., Gorjanacz, M., Koch, B., et al. (2012). Samp1 is a component of TAN lines and is required for nuclear movement. *Journal of Cell Science*, *125*, 1099–1105.

Buch, C., Lindberg, R., Figueroa, R., Gudise, S., Onischenko, E., & Hallberg, E. (2009). An integral protein of the inner nuclear membrane localizes to the mitotic spindle in mammalian cells. *Journal of Cell Science, 122*, 2100–2107.

Chang, W., Folker, E. S., Worman, H. J., & Gundersen, G. G. (2013). Emerin organizes actin flow for nuclear movement and centrosome orientation in migrating fibroblasts. *Molecular Biology of the Cell, 24*, 3869–3880.

Crisp, M., Liu, Q., Roux, K., Rattner, J. B., Shanahan, C., Burke, B., et al. (2006). Coupling of the nucleus and cytoplasm: Role of the LINC complex. *The Journal of Cell Biology, 172*, 41–53.

Dechat, T., Korbei, B., Vaughan, O. A., Vlcek, S., Hutchison, C. J., & Foisner, R. (2000). Lamina-associated polypeptide 2alpha binds intranuclear A-type lamins. *Journal of Cell Science, 113*, 3473–3484.

Dwyer, N., & Blobel, G. (1976). A modified procedure for the isolation of a pore complex-lamina fraction from rat liver nuclei. *The Journal of Cell Biology, 70*, 581–591.

Eichacker, L. A., Granvogl, B., Mirus, O., Muller, B. C., Miess, C., & Schleiff, E. (2004). Hiding behind hydrophobicity. Transmembrane segments in mass spectrometry. *The Journal of Biological Chemistry, 279*, 50915–50922.

Foisner, R., & Gerace, L. (1993). Integral membrane proteins of the nuclear envelope interact with lamins and chromosomes, and binding is modulated by mitotic phosphorylation. *Cell, 73*, 1267–1279.

Georgatos, S. D., Stournaras, C., & Blobel, G. (1988). Heterotypic and homotypic associations between the nuclear lamins: Site-specificity and control by phosphorylation. *Proceedings of the National Academy of Sciences of the United States of America, 85*, 4325–4329.

Gudise, S., Figueroa, R. A., Lindberg, R., Larsson, V., & Hallberg, E. (2011). Samp1 is functionally associated with the LINC complex and A-type lamina networks. *Journal of Cell Science, 124*, 2077–2085.

Hallberg, E., Wozniak, R. W., & Blobel, G. (1993). An integral membrane protein of the pore membrane domain of the nuclear envelope contains a nucleoporin-like region. *The Journal of Cell Biology, 122*(3), 513–521.

Haque, F., Lloyd, D. J., Smallwood, D. T., Dent, C. L., Shanahan, C. M., Fry, A. M., et al. (2006). SUN1 interacts with nuclear lamin A and cytoplasmic nesprins to provide a physical connection between the nuclear lamina and the cytoskeleton. *Molecular and Cellular Biology, 26*, 3738–3751.

Haque, F., Mazzeo, D., Patel, J. T., Smallwood, D. T., Ellis, J. A., Shanahan, C. M., et al. (2010). Mammalian SUN protein interaction networks at the inner nuclear membrane and their role in laminopathy disease processes. *The Journal of Biological Chemistry, 285*, 3487–3498.

Harlow, E., Whyte, P., Franza, B. R., Jr., & Schley, C. (1986). Association of adenovirus early-region 1A proteins with cellular polypeptides. *Molecular and Cellular Biology, 6*, 1579–1589.

Helbig, A. O., Heck, A. J., & Slijper, M. (2010). Exploring the membrane proteome—Challenges and analytical strategies. *Journal of Proteomics, 73*, 868–878.

Hirano, Y., Hizume, K., Kimura, H., Takeyasu, K., Haraguchi, T., & Hiraoka, Y. (2012). Lamin B receptor recognizes specific modifications of histone H4 in heterochromatin formation. *The Journal of Biological Chemistry, 287*, 42654–42663.

Jafferali, M. H., Beckman, M., Kihlmark, M., & Hallberg, E. (2015). *Nucleus and nuclear envelope: Methods for preparation. eLS.* Hoboken, NJ: John Wiley & Sons, pp. 1–4.

Jafferali, M. H., Vijayaraghavan, B., Figueroa, R. A., Crafoord, E., Gudise, S., Larsson, V. J., et al. (2014). MCLIP, an effective method to detect interactions of transmembrane proteins of the nuclear envelope in live cells. *Biochimica et Biophysica Acta, 1838*, 2399–2403.

Ketema, M., Wilhelmsen, K., Kuikman, I., Janssen, H., Hodzic, D., & Sonnenberg, A. (2007). Requirements for the localization of nesprin-3 at the nuclear envelope and its interaction with plectin. *Journal of Cell Science, 120*, 3384–3394.

Kluger, R., & Alagic, A. (2004). Chemical cross-linking and protein-protein interactions-a review with illustrative protocols. *Bioorganic Chemistry*, *32*, 451–472.

Koontz, L. (2014). TCA precipitation. *Methods in Enzymology*, *541*, 3–10.

Korfali, N., Wilkie, G. S., Swanson, S. K., Srsen, V., de Las Heras, J., Batrakou, D. G., et al. (2012). The nuclear envelope proteome differs notably between tissues. *Nucleus*, *3*, 552–564.

Lane, D. P., & Crawford, L. V. (1979). T antigen is bound to a host protein in SV40-transformed cells. *Nature*, *278*, 261–263.

Lomant, A. J., & Fairbanks, G. (1976). Chemical probes of extended biological structures: Synthesis and properties of the cleavable protein cross-linking reagent [35S]dithiobis(succinimidyl propionate). *Journal of Molecular Biology*, *104*, 243–261.

Mansharamani, M., & Wilson, K. L. (2005). Direct binding of nuclear membrane protein MAN1 to emerin in vitro and two modes of binding to barrier-to-autointegration factor. *The Journal of Biological Chemistry*, *280*, 13863–13870.

Masters, S. C. (2004). Co-immunoprecipitation from transfected cells. *Methods in Molecular Biology*, *261*, 337–350.

Mitchell, J. M., Mansfeld, J., Capitanio, J., Kutay, U., & Wozniak, R. W. (2010). Pom121 links two essential subcomplexes of the nuclear pore complex core to the membrane. *The Journal of Cell Biology*, *191*, 505–521.

Muller, M. Q., Dreiocker, F., Ihling, C. H., Schafer, M., & Sinz, A. (2010). Fragmentation behavior of a thiourea-based reagent for protein structure analysis by collision-induced dissociative chemical cross-linking. *Journal of Mass Spectrometry*, *45*, 880–891.

Padmakumar, V. C., Libotte, T., Lu, W., Zaim, H., Abraham, S., Noegel, A. A., et al. (2005). The inner nuclear membrane protein Sun1 mediates the anchorage of Nesprin-2 to the nuclear envelope. *Journal of Cell Science*, *118*, 3419–3430.

Percipalle, P., Jonsson, A., Nashchekin, D., Karlsson, C., Bergman, T., Guialis, A., et al. (2002). Nuclear actin is associated with a specific subset of hnRNP A/B-type proteins. *Nucleic Acids Research*, *30*, 1725–1734.

Polioudaki, H., Kourmouli, N., Drosou, V., Bakou, A., Theodoropoulos, P. A., Singh, P. B., et al. (2001). Histones H3/H4 form a tight complex with the inner nuclear membrane protein LBR and heterochromatin protein 1. *EMBO Reports*, *2*, 920–925.

Radu, A., Blobel, G., & Wozniak, R. W. (1993). Nup155 is a novel nuclear pore complex protein that contains neither repetitive sequence motifs nor reacts with WGA. *The Journal of Cell Biology*, *121*, 1–9.

Roux, K. J., Kim, D. I., Raida, M., & Burke, B. (2012). A promiscuous biotin ligase fusion protein identifies proximal and interacting proteins in mammalian cells. *The Journal of Cell Biology*, *196*, 801–810.

Sakaki, M., Koike, H., Takahashi, N., Sasagawa, N., Tomioka, S., Arahata, K., et al. (2001). Interaction between emerin and nuclear lamins. *Journal of Biochemistry*, *129*, 321–327.

Salpingidou, G., Smertenko, A., Hausmanowa-Petrucewicz, I., Hussey, P. J., & Hutchison, C. J. (2007). A novel role for the nuclear membrane protein emerin in association of the centrosome to the outer nuclear membrane. *The Journal of Cell Biology*, *178*, 897–904.

Schirmer, E. C., Florens, L., Guan, T., Yates, J. R., 3rd., & Gerace, L. (2003). Nuclear membrane proteins with potential disease links found by subtractive proteomics. *Science*, *301*, 1380–1382.

Shimi, T., Koujin, T., Segura-Totten, M., Wilson, K. L., Haraguchi, T., & Hiraoka, Y. (2004). Dynamic interaction between BAF and emerin revealed by FRAP, FLIP, and FRET analyses in living HeLa cells. *Journal of Structural Biology*, *147*, 31–41.

Sinz, A. (2006). Chemical cross-linking and mass spectrometry to map three-dimensional protein structures and protein-protein interactions. *Mass Spectrometry Reviews*, *25*, 663–682.

Smith, A. L., Friedman, D. B., Yu, H., Carnahan, R. H., & Reynolds, A. B. (2011). ReCLIP (reversible cross-link immuno-precipitation): An efficient method for interrogation of labile protein complexes. *PLoS One*, *6*, e16206.

Snow, C. M., Senior, A., & Gerace, L. (1987). Monoclonal antibodies identify a group of nuclear pore complex glycoproteins. *The Journal of Cell Biology*, *104*, 1143–1156.

Sun, S., Zhou, J. Y., Yang, W., & Zhang, H. (2014). Inhibition of protein carbamylation in urea solution using ammonium-containing buffers. *Analytical Biochemistry*, *446*, 76–81.

Trinkle-Mulcahy, L., Boulon, S., Lam, Y. W., Urcia, R., Boisvert, F. M., Vandermoere, F., et al. (2008). Identifying specific protein interaction partners using quantitative mass spectrometry and bead proteomes. *The Journal of Cell Biology*, *183*, 223–239.

Vaughan, A., Alvarez-Reyes, M., Bridger, J. M., Broers, J. L., Ramaekers, F. C., Wehnert, M., et al. (2001). Both emerin and lamin C depend on lamin A for localization at the nuclear envelope. *Journal of Cell Science*, *114*, 2577–2590.

Wilkie, G. S., Korfali, N., Swanson, S. K., Malik, P., Srsen, V., Batrakou, D. G., et al. (2011). Several novel nuclear envelope transmembrane proteins identified in skeletal muscle have cytoskeletal associations. *Molecular & Cellular Proteomics*, *10,* M110 003129.

Worman, H. J., & Schirmer, E. C. (2015). Nuclear membrane diversity: Underlying tissue-specific pathologies in disease? *Current Opinion in Cell Biology*, *34*, 101–112.

Worman, H. J., Yuan, J., Blobel, G., & Georgatos, S. D. (1988). A lamin B receptor in the nuclear envelope. *Proceedings of the National Academy of Sciences of the United States of America*, *85*, 8531–8534.

Wozniak, R. W., Bartnik, E., & Blobel, G. (1989). Primary structure analysis of an integral membrane glycoprotein of the nuclear pore. *The Journal of Cell Biology*, *108*, 2083–2092.

Ye, Q., & Worman, H. J. (1996). Interaction between an integral protein of the nuclear envelope inner membrane and human chromodomain proteins homologous to Drosophila HP1. *The Journal of Biological Chemistry*, *271*, 14653–14656.

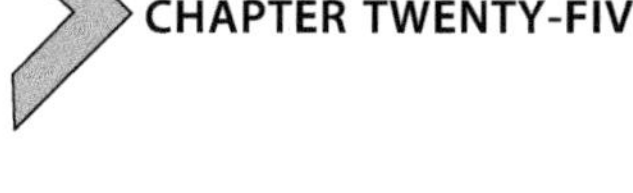

CHAPTER TWENTY-FIVE

Preparation of the Human Cytomegalovirus Nuclear Egress Complex and Associated Proteins

Mayuri Sharma*[,1], Jeremy P. Kamil*[,†], Donald M. Coen*[,1]

*Department of Biological Chemistry and Molecular Pharmacology, Harvard Medical School, Boston, Massachusetts, USA

†Department of Microbiology and Immunology, Louisiana State University Health Sciences Center, Shreveport, Louisiana, USA

[1]Corresponding authors: e-mail address: mayuri_sharma@hms.harvard.edu; don_coen@hms.harvard.edu

Contents

Abstract

Herpesviruses, like most DNA viruses, replicate their genomes in the host cell nucleus. Their DNA is then packaged and assembled into viral nucleocapsids, which, in most cases, are too large to pass through the nuclear pore complex. Instead, herpesviruses use a complex multistep pathway, termed nuclear egress, to exit the nucleus. Key players in this process include two conserved viral proteins that form the nuclear egress complex (NEC). In human cytomegalovirus, these NEC proteins are UL50, embedded in the inner nuclear membrane, and its nucleoplasmic partner UL53. Both are essential for viral nuclear egress. However, other viral components as well as host nuclear envelope proteins may also participate in nuclear egress. Identifying these viral and cellular factors may provide important insight into the herpesvirus lifecycle and its relationship to the underlying, yet still-mysterious, host nuclear egress pathway. We developed an immunoprecipitation-based protocol, described herein, to identify protein–protein interactions involving the NEC from the nuclear fraction of infected cells that express an epitope-tagged version of NEC subunit UL53.

Methods in Enzymology, Volume 569
ISSN 0076-6879
http://dx.doi.org/10.1016/bs.mie.2015.08.020

1. INTRODUCTION

Human cytomegalovirus (HCMV) is a beta-herpesvirus that can cause life-threatening complications in immunocompromised individuals and is the leading infectious cause of congenital defects in newborns (Britt, 2008; Mercorelli, Lembo, Palu, & Loregian, 2011; Pultoo, Jankee, Meetoo, Pyndiah, & Khittoo, 2000). Current therapies to control HCMV infection are limited in terms of their efficacy, potency, selectivity, and pharmacokinetic properties (Coen & Schaffer, 2003; Hakki & Chou, 2011). Knowledge of the viral life cycle can provide insight into pathways that might be targeted for development of new therapeutics. The process of nuclear egress during herpesvirus replication offers such a target. Nuclear egress is a multistep pathway during which herpesvirus nucleocapsids translocate from the nucleoplasm to the cytoplasm, without going through nuclear pore complexes (Hennig & O'Hare, 2015; Johnson & Baines, 2011; Mettenleiter, 2002). Instead, following virus-promoted disruption of nuclear lamina structure, nucleocapsids access the inner nuclear membrane (INM) and are then enveloped by the INM as they bud into the nuclear envelope lumen, a step known as "primary envelopment." These enveloped nucleocapsids then fuse with the outer nuclear membrane and are released into the cytoplasm. Nuclear egress is orchestrated by two viral proteins that form the nuclear egress complex (NEC). In HCMV, these viral subunits consist of UL50 (an integral INM protein) and its nucleoplasmic partner UL53. Both are required for nuclear egress and viral replication; interestingly, they also recruit other proteins to facilitate specific steps during nuclear egress (Sharma, Kamil, Coughlin, Reim, & Coen, 2014).

Herpesvirus nuclear egress may be related to a recently discovered host pathway by which large ribonucleoproteins exit the nucleus (Razafsky & Hodzic, 2015; Speese et al., 2012). However, there appear to be key differences in the players involved in the two pathways, suggesting their molecular details may differ. Notably, a virus-encoded kinase is critical for HCMV nuclear egress, whereas a cellular kinase is important for the host pathway (Sharma & Coen, 2014; Sharma et al., 2014; Speese et al., 2012). To understand these pathways and to identify differences that might be exploited as antiviral drug targets, we first need to identify key viral and cellular proteins involved in nuclear egress. The protocols below were designed to facilitate the identification of NEC-associated proteins in HCMV-infected cells.

2. SAMPLE PREPARATION FOR ISOLATION OF THE NEC

2.1 Cells

Human cells infected with HCMV serve as the source for NECs. HCMV is propagated in primary human foreskin fibroblasts (HFFs) (e.g., strain Hs27, ATCC CRL-1684; American Type Culture Collection; Manassas, VA). The cells are passaged in Dulbecco's modified Eagle's medium (DMEM) (ATCC 30-2002), supplemented with 10% fetal bovine serum (FBS), 100 U/ml of penicillin, and 100 μg/ml of streptomycin, at 37 °C and 5% CO_2 in ~95% humidity. After reaching confluence, cells are maintained in DMEM containing 5% FBS (maintenance medium).

2.2 Viruses

We use a virus expressing a version of UL53 bearing a FLAG-epitope at the C-terminus, designated UL53-FLAG AD169rv (referred to as "53-F") (Sharma et al., 2014). 53-F was derived from a bacterial artificial chromosome (BAC) clone of HCMV strain AD169, named AD169rv, with the BAC-derived wild-type virus as the control. High titer stocks for each virus are generated by infecting HFFs at low multiplicity of infection (MOI) (typically 0.001), with a previously titrated stock, and harvesting the supernatant after 80–90% of the cells show cytopathic effects. The supernatant containing the virus is centrifuged at $1000 \times g$ for 10 min at room temperature to pellet cell debris. The supernatant is dispensed in 1 ml aliquots and frozen in liquid nitrogen until further use. One aliquot of each virus is used for titration by serial dilution as described (Kamil & Coen, 2007).

2.3 Infection

Seed HFFs ($\sim 3 \times 10^6$ cells per 100-mm dish) and incubate until they reach ~90% confluence (typically ~3 days). Count cells from a representative dish, and then infect two 100-mm dishes of HFFs ($\sim 1 \times 10^7$ cells per dish) at an MOI of 1 with either UL53-FLAG AD169rv or, as a specificity control, with wild-type AD169rv virus. To infect cells, replace medium with 7 ml maintenance medium containing diluted virus; incubate 2 h at 37 °C and then replace with fresh maintenance medium lacking virus. To determine the optimal time postinfection at which to isolate NECs, we measured UL50 and UL53 expression levels by Western blotting (Fig. 1A). In cells infected at an MOI of 1, we found NEC expression peaks at 72 h postinfection (hpi).

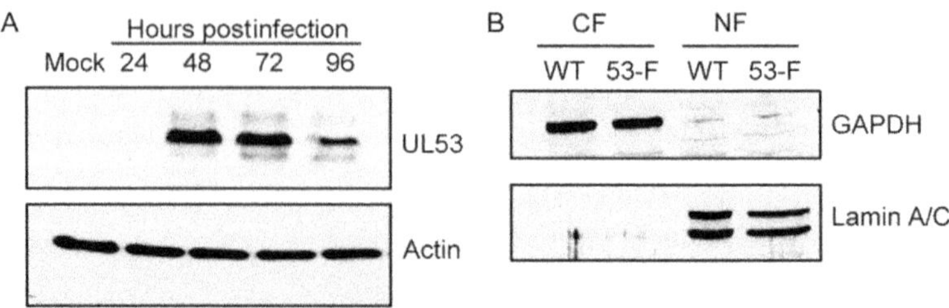

Figure 1 Western blots of UL53 expression time course and nuclear fractionation. (A) Human foreskin fibroblasts (HFFs; 1×10^5 cells per well of a 24-well plate) were infected at an MOI of 1 with UL53-FLAG AD169rv. At each indicated time, cells were washed with 1×DPBS, harvested into 100 µl Laemmli sample buffer, heated, resolved by SDS-PAGE, and immunoblotted with antibodies specific for UL53 or actin as the loading control. (B) At 72 hpi, cells infected with either WT or UL53-FLAG (53-F) AD16rv were fractionated to separate nuclei (nuclear fraction, "NF") from the cytoplasmic fraction (CF), and immunoblotted with antibodies against GAPDH or A-type lamins (lamin A/C) as cytoplasmic and nuclear markers, respectively.

2.4 Prepare Nuclear Fractions from Infected Cells

Prechill all equipment and glassware on ice, and follow the steps outlined in Fig. 2 and (Sharma et al., 2014).

1. At 72 hpi, remove medium from infected cells and wash cells once with 5 ml ice-cold 1 × Dulbecco's phosphate-buffered saline (DPBS).
2. Resuspend cells by scraping into 3 ml hypotonic buffer (10 m*M* HEPES, pH 7.4, 1 m*M* dithiothreitol (DTT), 1 m*M* $MgCl_2$, 10 m*M* KCl, and protease inhibitor cocktail (Roche Applied Sciences, Indianapolis, IN)).
3. Incubate resuspended cells on ice for 15 min. This incubation in hypotonic buffer allows cells to swell.
4. Transfer cells into a 5-ml Dounce homogenizer (Wheaton, USA) and homogenize cells using 40 strokes on ice.
5. Centrifuge the homogenate (1000 × *g*, 4 °C, 10 min) to pellet the nuclei. Remove the supernatant (cytoplasmic fraction) and retain a small aliquot (mix with Laemmli sample buffer) for subsequent Western blotting analyses, and store remainder at −80 °C if needed. Resuspend the nuclear pellet in 3 ml lysis buffer (50 m*M* HEPES, pH 7.4, 1 m*M* DTT, 200 m*M* NaCl, 5 m*M* EDTA, 50 m*M* arginine, 0.5% Triton X-100, EDTA-free protease inhibitor cocktail (Roche Applied Sciences, Indianapolis, IN)). Arginine reduces nonspecific protein binding to resins.
6. Rotate the nuclear fraction end-over-end at 4 °C for 15 min.
7. Homogenize the nuclear suspension with 40 strokes, on ice, to disrupt nuclear membranes.

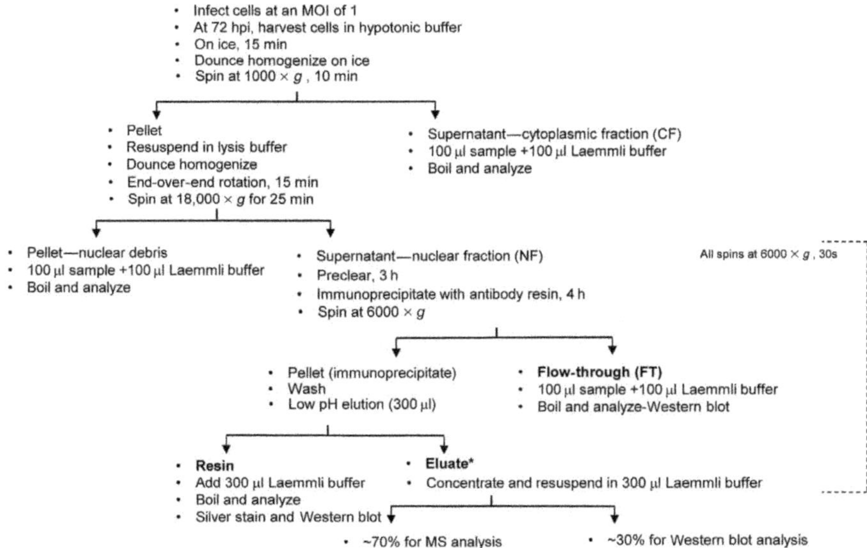

Figure 2 Overview of NEC immunopurification protocol from HCMV-infected cells. Fractions destined for silver staining or immunoblot analysis are in bold. The sample destined for gel-separation and mass spectrometry (MS) is indicated in bold and with an asterisk.

8. Centrifuge 25 min at 18,000 × *g*, 4 °C (Eppendorf centrifuge), and recover the supernatant (nuclear lysate) for further steps.
9. Monitor enrichment in each fraction by SDS-PAGE and immunoblotting with antibodies specific for cytoplasmic and nuclear markers. We use glyceraldehyde 3-phosphate dehydrogenase (GAPDH) as the cytoplasmic marker (Abcam ab8245; dilution 1:5000) and lamin A/C as the nuclear marker (Santa Cruz, N-18; dilution 1:1000). This protocol greatly enriches for nuclear proteins as shown in Fig. 1B.

2.5 Immunoprecipitation

Preclear lysate. To reduce nonspecifically associated proteins, "preclear" the nuclear lysate from Step 8 by incubating with resin bearing an isotype-matched control of the antibody to be used for immunoprecipitation. We prepare IgG agarose (Sigma) resin as suggested by the manufacturer. Briefly, centrifuge 600 µl of the resin slurry (handled with wide-bore tips) for 30 s (6000 × *g*, 4 °C), and remove the supernatant. Wash the settled resin three times (3 ml DPBS each), centrifuging for 30 s between washes. Mix the nuclear lysate from Step 8 with 300 µl of washed/settled anti-IgG agarose resin. Rotate 3 h end-over-end at 4 °C. Centrifuge (30 s, 6000 × *g*, 4 °C) and recover the cleared nuclear supernatant for use in Step 10, below.

Immunoprecipitate with anti-FLAG antibody. We use the EZ-View anti-FLAG M2 monoclonal antibody-conjugated agarose (Sigma) for immunoprecipitation. Prepare the resin for use as described above.

10. Add the entire precleared nuclear supernatant from Step 9, to 300 μl of settled, EZ-View anti-FLAG M2 affinity resin. Rotate 4 h at 4 °C.

11. Centrifuge (6000 × *g*, 4 °C for 30 s) to pellet the beads. Remove the supernatant (flow-through, FT; unbound proteins); mix 100 μl of this supernatant with 100 μl Laemmli sample buffer, boil and retain for subsequent Western blotting analyses. Store the rest at −80 °C.

12. Wash resin (bearing the immunoprecipitated proteins) four times by rotating 15 min in 3 ml lysis buffer at 4 °C, and repelleting (6000 × *g*, 30 s).

13. Elute bound proteins by low pH: add 270 μl of 0.1 *M* glycine, pH 2.5, and incubate 5 min at room temperature. Centrifuge at 6000 × *g* for 10 s, and collect the supernatant. Neutralize the supernatant by adding 30 μl neutralization buffer (0.5 *M* Tris, pH 7.4, 1.5 *M* NaCl) and glycerol (final concentration 10%) to a final volume of 300 μl. Split this eluate into three 100-μl aliquots. Two of the 100-μl aliquots are stored at −80 °C until further processing for mass spectrometry (MS) analysis (Section 2.6). The last 100-μl aliquot is reserved for Western blot validation of candidates identified by MS; add 100 μl Laemmli sample buffer, boil and freeze until needed.

14. Add 300 μl Laemmli sample buffer to the postelution resin, heat denature at 85–90 °C for 5 min, then centrifuge (6000 × *g*, 30 s), and recover the supernatant ("resin"; proteins that resisted low-pH elution). This "resin" sample has a protein composition similar to the eluate (unpublished results) and is used for qualitative assessment by silver staining and Western blotting in Step 15 (shown in Fig. 3), prior to MS.

15. To monitor results, resolve aliquots of boiled "resin" samples by SDS-PAGE and silver staining (Fig. 3A). As expected for conventional antibodies, immunoglobulin heavy and light chains are abundant contaminants (Fig. 3A, IgHC and IgLC, respectively). Bands corresponding to UL50 and UL53 should be highly enriched in the 53-F lane, compared to WT-infected samples (Fig. 3A), as confirmed by MS analysis of this region of the gel (bracket, Fig. 3A) and immunoblotting (Sharma et al., 2014). Importantly, at least 14 additional bands representing candidate NEC-associated proteins also specifically coprecipitated from 53-F-infected samples (Fig. 3A).

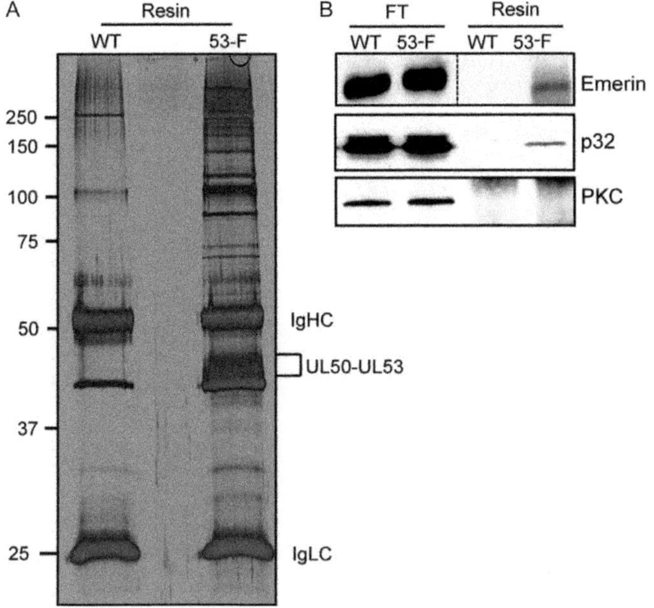

Figure 3 Qualitative assessment of "resin" and "FT" samples prior to MS. (A) Silver-stained visualization of proteins immunoprecipitated from cells infected with WT or FLAG-tagged UL53 (53-F) virus. These samples comprise the "resin" fractions from Fig. 2, which have a protein composition very similar to (or in some cases more concentrated than) the eluate. The bracketed region contains the UL50 and UL53 bands. Immunoglobulin heavy and light chains are indicated by IgHC and IgLC, respectively. (B) Immunoblot analysis of the post-immunoprecipitation flow-through (FT; unbound proteins), and the same "resin" fraction shown in (A), using antibodies specific for each indicated protein. Images are from the same samples, resolved and blotted separately.

2.6 Sample Processing for MS

Please consult your MS experts before proceeding. They may suggest useful alternatives. Based on the advice of our MS facility, we concentrate the samples, resolve proteins by SDS-PAGE, and excise proteins from specific regions of each lane for MS identification. Concentrate each low-pH eluate (100 μl) to a final volume of 30 μl using the Wessel Flügge protocol (Wessel & Flugge, 1984), as follows:

1. Add 400 μl methanol and vortex, add 100 μl chloroform and vortex, add 300 μl water and vortex, then centrifuge (14,000 × *g*) for 1 min.
2. Remove the top (aqueous) layer carefully and discard. Add 400 μl methanol to the organic phase, vortex, and centrifuge (14,000 × *g*, 2 min).
3. Remove the supernatant without disturbing the pellet.

4. Air-dry the pellet. Resuspend the pellet in 30 μl Laemmli buffer, and resolve proteins by gradient (4–20%) SDS-PAGE.
5. Visualize proteins using Colloidal blue (Life Technologies), according to manufacturer's instructions.
6. The samples and gels are handled with gloves. Excise gel slices containing protein bands in the molecular weight range of 10–25, 25–37, 37–50, 50–75, 75–100, 100–150, and 150–250 kDa using separate razor blades for each gel slice. Avoid the immunoglobulin heavy and light chain bands that can be observed at 50 and 25 kDa, respectively. Place each gel slice in its own microcentrifuge tube, containing water to cover the gel slice, and submit for MS analysis.

3. CONCLUDING REMARKS

MS provides a starting point by identifying candidates that might interact with the UL53 subunit of NEC egress complexes. The untagged WT AD169rv (WT) control sample helps identify proteins that bind the FLAG-antibody resin nonspecifically. Indeed, we found no viral proteins in eluates from this sample. Proteins present at similar levels in WT versus 53-F eluates, such as antibody contaminants, are excluded from further analysis. Proteins unique to 53-F eluates, with three or more unique peptide matches, are considered confidently identified, and strong candidates for further investigation. One such candidate is the nuclear lamina-associated protein, emerin (Fig. 3B). Emerin was also identified in a recent proteomics study that immunoprecipitated HCMV NECs from whole cell lysates (Milbradt et al., 2014). Another strong candidate is the host protein, p32 (Fig. 3B; Milbradt et al., 2014), previously reported to colocalize with NEC subunit UL50, or with the viral kinase UL97, in cotransfection experiments (Marschall et al., 2005; Milbradt, Auerochs, Sticht, & Marschall, 2009). However, as p32 associates with a variety of proteins (Muta, Kang, Kitajima, Fujiwara, & Hamasaki, 1997; van Leeuwen & O'Hare, 2001), the functional relevance of its association with the HCMV NEC remains unclear.

Proteins with one or two peptide matches should also be considered. This group can include rare (low-abundance) proteins, weakly interacting proteins, small proteins, and proteins for which peptide sequence coverage is low. For example, we recovered two peptides corresponding to the HCMV protein kinase, UL97, which is important for disruption of the nuclear lamina during nuclear egress (Krosky, Baek, & Coen, 2003;

Reim et al., 2013; Sharma & Coen, 2014). Further studies showed that UL97 indeed associates with the NEC, and this association is dependent on each NEC subunit (Sharma et al., 2014).

These protocols can be extended to other epitope-tagged components of the HCMV NEC, and likely other epitope-tagged host and viral proteins involved in nuclear events during infection. Key features of this protocol include the control wild-type virus expressing a native (untagged) NEC (to exclude nonspecifically associating proteins), and the use of nuclear fractions rather than whole cell lysates. Our experience is that nuclear lysates permit identification of associated proteins that are not observed when using whole cell extracts and help exclude proteins that are likely to be irrelevant to NEC functions during nuclear egress. We emphasize that the protocols described here are merely first steps, and identified candidates must be tested rigorously to determine their biological validity for viral replication and potential roles in nuclear egress.

ACKNOWLEDGMENTS

We thank Brian Bender for technical assistance. We are grateful to our colleagues at the Taplin Biological Mass Spectrometry Facility, Harvard Medical School, for acquisition and analysis of mass spectrometry data. This work was supported by NIH grant R01 AI026077 to D.M.C.

REFERENCES

Britt, W. (2008). Manifestations of human cytomegalovirus infection: Proposed mechanisms of acute and chronic disease. *Current Topics in Microbiology and Immunology*, *325*, 417–470.

Coen, D. M., & Schaffer, P. A. (2003). Antiherpesvirus drugs: A promising spectrum of new drugs and drug targets. *Nature Reviews. Drug Discovery*, *2*, 278–288.

Hakki, M., & Chou, S. (2011). The biology of cytomegalovirus drug resistance. *Current Opinion in Infectious Diseases*, *24*, 605–611.

Hennig, T., & O'Hare, P. (2015). Viruses and the nuclear envelope. *Current Opinion in Cell Biology*, *34*, 113–121.

Johnson, D. C., & Baines, J. D. (2011). Herpesviruses remodel host membranes for virus egress. *Nature Reviews. Microbiology*, *9*, 382–394.

Kamil, J. P., & Coen, D. M. (2007). Human cytomegalovirus protein kinase UL97 forms a complex with the tegument phosphoprotein pp 65. *Journal of Virology*, *81*, 10659–10668.

Krosky, P. M., Baek, M. C., & Coen, D. M. (2003). The human cytomegalovirus UL97 protein kinase, an antiviral drug target, is required at the stage of nuclear egress. *Journal of Virology*, *77*, 905–914.

Marschall, M., Marzi, A., aus dem Siepen, P., Jochmann, R., Kalmer, M., Auerochs, S., et al. (2005). Cellular p32 recruits cytomegalovirus kinase pUL97 to redistribute the nuclear lamina. *The Journal of Biological Chemistry*, *280*, 33357–33367.

Mercorelli, B., Lembo, D., Palu, G., & Loregian, A. (2011). Early inhibitors of human cytomegalovirus: State-of-art and therapeutic perspectives. *Pharmacology & Therapeutics*, *131*, 309–329.

Mettenleiter, T. C. (2002). Herpesvirus assembly and egress. *Journal of Virology*, *76*, 1537–1547.

Milbradt, J., Auerochs, S., Sticht, H., & Marschall, M. (2009). Cytomegaloviral proteins that associate with the nuclear lamina: Components of a postulated nuclear egress complex. *The Journal of General Virology*, *90*, 579–590.

Milbradt, J., Kraut, A., Hutterer, C., Sonntag, E., Schmeiser, C., Ferro, M., et al. (2014). Proteomic analysis of the multimeric nuclear egress complex of human cytomegalovirus. *Molecular and Cellular Proteomics*, *13*, 2132–2146.

Muta, T., Kang, D., Kitajima, S., Fujiwara, T., & Hamasaki, N. (1997). p32 protein, a splicing factor 2-associated protein, is localized in mitochondrial matrix and is functionally important in maintaining oxidative phosphorylation. *The Journal of Biological Chemistry*, *272*, 24363–24370.

Pultoo, A., Jankee, H., Meetoo, G., Pyndiah, M. N., & Khittoo, G. (2000). Detection of cytomegalovirus in urine of hearing-impaired and mentally retarded children by PCR and cell culture. *The Journal of Communicable Diseases*, *32*, 101–108.

Razafsky, D., & Hodzic, D. (2015). Nuclear envelope: Positioning nuclei and organizing synapses. *Current Opinion in Cell Biology*, *34*, 84–93.

Reim, N. I., Kamil, J. P., Wang, D., Lin, A., Sharma, M., Ericsson, M., et al. (2013). Inactivation of retinoblastoma protein does not overcome the requirement for human cytomegalovirus UL97 in lamina disruption and nuclear egress. *Journal of Virology*, *87*, 5019–5027.

Sharma, M., & Coen, D. M. (2014). Comparison of effects of inhibitors of viral and cellular protein kinases on human cytomegalovirus disruption of nuclear lamina and nuclear egress. *Journal of Virology*, *88*, 10982–10985.

Sharma, M., Kamil, J. P., Coughlin, M., Reim, N. I., & Coen, D. M. (2014). Human cytomegalovirus UL50 and UL53 recruit viral protein kinase UL97, not protein kinase C, for disruption of nuclear lamina and nuclear egress in infected cells. *Journal of Virology*, *88*, 249–262.

Speese, S. D., Ashley, J., Jokhi, V., Nunnari, J., Barria, R., Li, Y., et al. (2012). Nuclear envelope budding enables large ribonucleoprotein particle export during synaptic Wnt signaling. *Cell*, *149*, 832–846.

van Leeuwen, H. C., & O'Hare, P. (2001). Retargeting of the mitochondrial protein p32/gC1Qr to a cytoplasmic compartment and the cell surface. *Journal of Cell Science*, *114*, 2115–2123.

Wessel, D., & Flugge, U. I. (1984). A method for the quantitative recovery of protein in dilute solution in the presence of detergents and lipids. *Analytical Biochemistry*, *138*, 141–143.

AUTHOR INDEX

Note: Page numbers followed by "*f*" indicate figures and "*t*" indicate tables.

A

B

C

D

E

F

G

H

I

J

L

M

P

Q

R

S

T

U

V

W

X

Y

SUBJECT INDEX

Note: Page numbers followed by "*f*" indicate figures and "*t*" indicate tables.

E

F

N

R

S

T

U

V

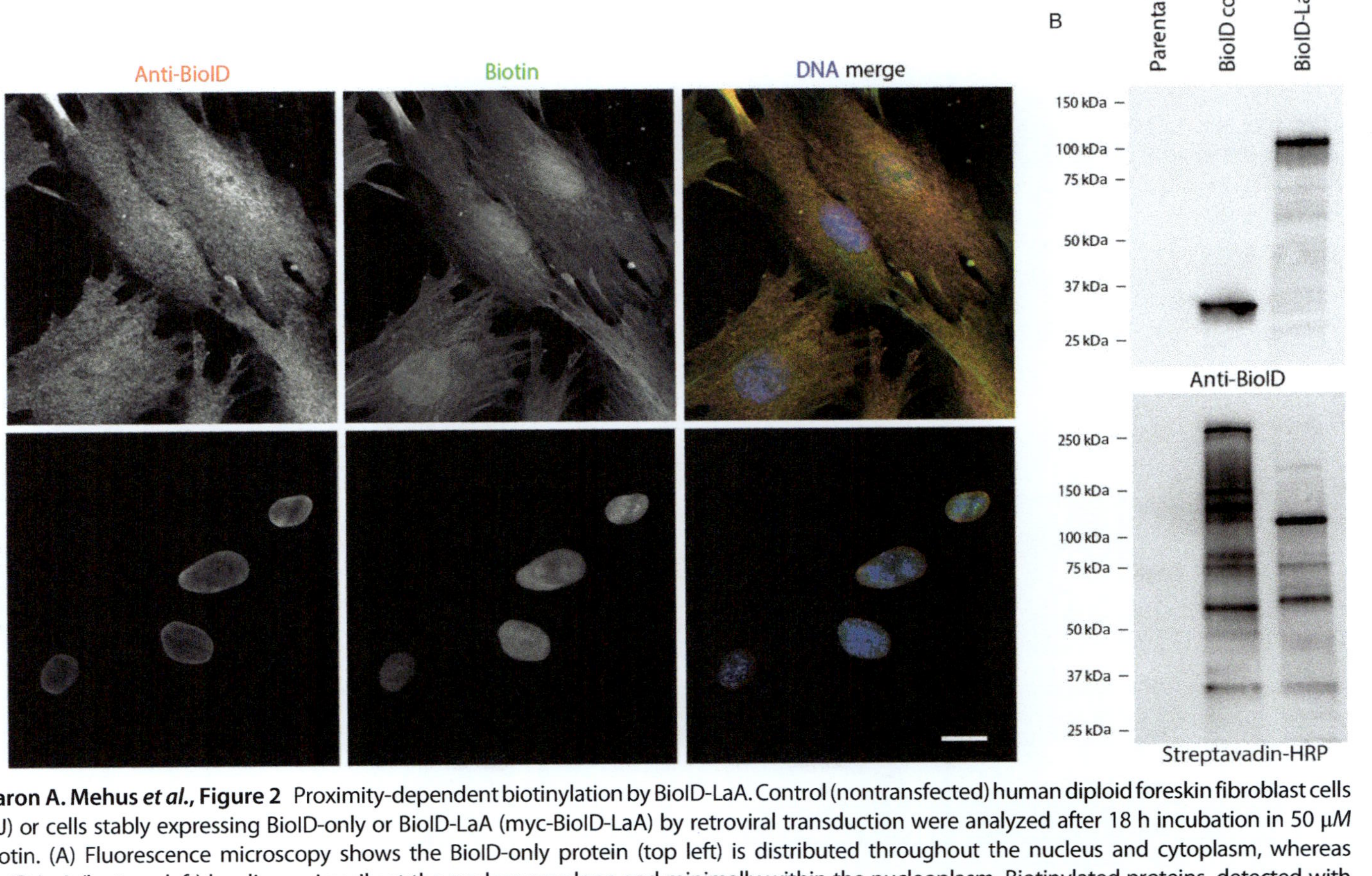

Aaron A. Mehus *et al.*, Figure 2 Proximity-dependent biotinylation by BioID-LaA. Control (nontransfected) human diploid foreskin fibroblast cells (BJ) or cells stably expressing BioID-only or BioID-LaA (myc-BioID-LaA) by retroviral transduction were analyzed after 18 h incubation in 50 μ*M* biotin. (A) Fluorescence microscopy shows the BioID-only protein (top left) is distributed throughout the nucleus and cytoplasm, whereas BioID-LaA (bottom left) localizes primarily at the nuclear envelope and minimally within the nucleoplasm. Biotinylated proteins, detected with labeled streptavidin (top or bottom middle), colocalize with the fusion proteins (top or bottom left). DNA is labeled with Hoechst (top or bottom right). (B) Immunoblot probed with antibodies specific for the fusion protein (anti-BirA); biotinylated endogenous proteins were detected using streptavidin-conjugated HRP.

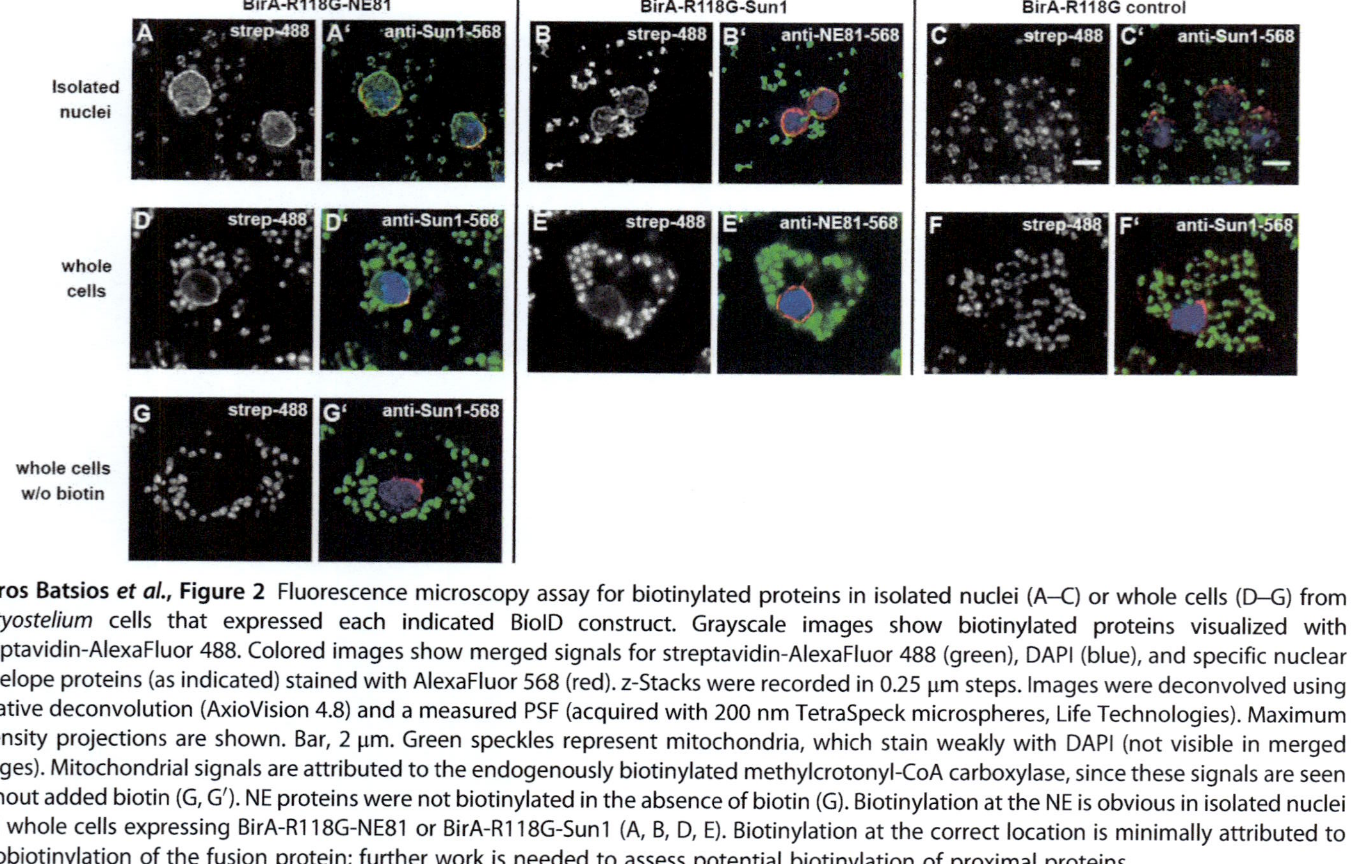

Petros Batsios *et al.*, Figure 2 Fluorescence microscopy assay for biotinylated proteins in isolated nuclei (A–C) or whole cells (D–G) from *Dictyostelium* cells that expressed each indicated BioID construct. Grayscale images show biotinylated proteins visualized with streptavidin-AlexaFluor 488. Colored images show merged signals for streptavidin-AlexaFluor 488 (green), DAPI (blue), and specific nuclear envelope proteins (as indicated) stained with AlexaFluor 568 (red). z-Stacks were recorded in 0.25 μm steps. Images were deconvolved using iterative deconvolution (AxioVision 4.8) and a measured PSF (acquired with 200 nm TetraSpeck microspheres, Life Technologies). Maximum intensity projections are shown. Bar, 2 μm. Green speckles represent mitochondria, which stain weakly with DAPI (not visible in merged images). Mitochondrial signals are attributed to the endogenously biotinylated methylcrotonyl-CoA carboxylase, since these signals are seen without added biotin (G, G′). NE proteins were not biotinylated in the absence of biotin (G). Biotinylation at the NE is obvious in isolated nuclei and whole cells expressing BirA-R118G-NE81 or BirA-R118G-Sun1 (A, B, D, E). Biotinylation at the correct location is minimally attributed to autobiotinylation of the fusion protein; further work is needed to assess potential biotinylation of proximal proteins.

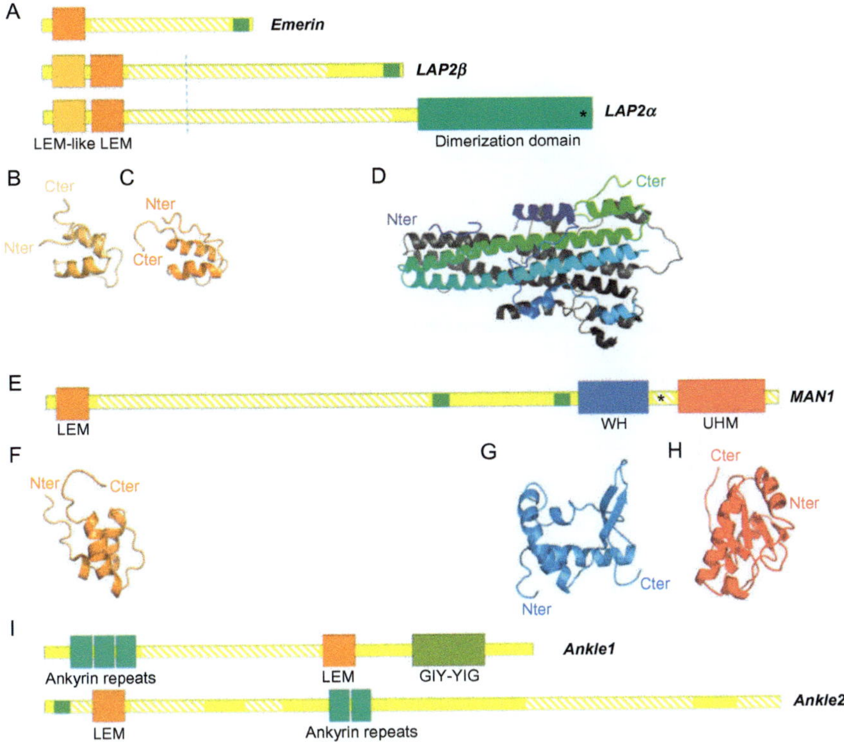

Isaline Herrada *et al.*, Figure 1 Schematic of LEM-domain protein architecture. Three families of LEM-domain proteins, illustrated by six human proteins. Hatched yellow regions are predicted as intrinsically disordered by Disopred3. Small green boxes indicate transmembrane domains. Blue dotted line indicates that LAP2 isoforms share residues 1–186 but have different C-terminal regions (after the line). (A) LEM-domain proteins with one transmembrane domain (emerin, LAP2β) localize at the INM; alternatively spliced isoform LAP2α has no transmembrane domain. The star in LAP2α indicates the Arg690Cys mutation linked to dilated cardiomyopathy. (B)–(D) 3D structures of the LEM-like domain in light orange (PDB code 1H9E), the LEM domain in dark orange (PDB code 1H9F), and LAP2α-specific dimerization domain in green (PDB code 2V0X). (E) INM-localized LEM-domain proteins with two transmembrane segments are Lem2 (not shown) and MAN1. The star in MAN1 indicates the UHM ligand motif Lys/Arg$_{763}$-X-Trp$_{765}$-Gln$_{766}$-X-X-Ala$_{769}$-Phe$_{770}$. (F)–(H) 3D models of MAN1 globular domains, calculated by homology for the LEM domain in orange, from NMR data for the WH domain in blue (PDB code 2CH0; Caputo et al., 2006), and from NMR chemical shift data and molecular modeling for the UHM domain in red (Konde et al., 2010). (I) LEM-domain proteins with ankyrin repeats: Ankle1 is soluble; Ankle2 is ER-localized.

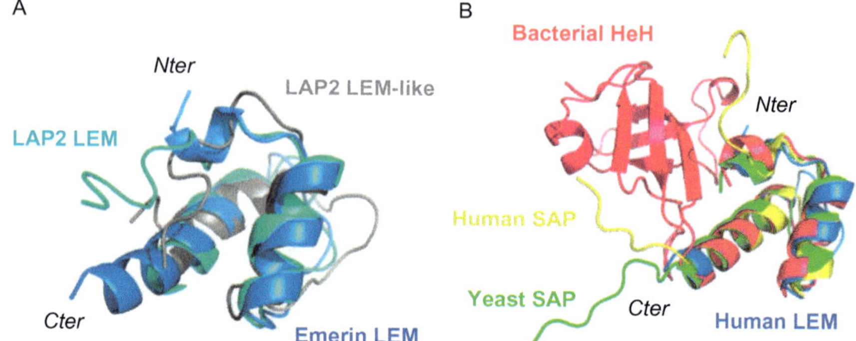

Isaline Herrada *et al.*, Figure 2 The LEM-domain fold. (A) Superimposition of the LEM structure of emerin (PDB 2ODC in marine blue) with the LEM (PDB 1H9F in cyan; DALI rmsd 2.2 Å) and LEM-like (PDB 1H9E in gray; DALI rmsd 3.0 Å) structures of LAP2. (B) Superimposition of the LEM fold with the bacterial HeH structure from the RNA-binding domain of the transcription termination factor rho (PDB 1A62 in magenta; DALI rmsd 1.7 Å), the yeast SAP structure of the RNA-binding protein Tho1 (PDB 4UZW; DALI rmsd 1.4 Å) and the human SAP structure of the ribonucleoprotein Hcc-1 (PDB 2DO1; DALI rmsd 2.2 Å).

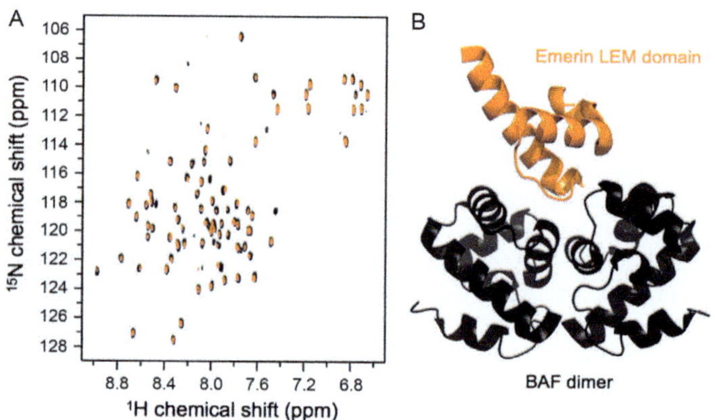

Isaline Herrada *et al.*, Figure 3 The emerin–BAF interaction. (A) Superimposition of two NMR ^{1}H–^{15}N HSQC spectra revealing the interaction between emerin 1–187 and BAF: the red and black spectra were recorded on ^{15}N-labeled BAF alone and in complex with emerin, respectively (in 40 m*M* phosphate buffer, pH 6.7, 150 m*M* NaCl, 1 m*M* tris(2-carboxyethyl)phosphine, 1 m*M* EDTA at 20 °C on a 600 MHz spectrometer). Disappearance of ^{1}H–^{15}N signals indicates binding to the unlabeled partner. (B) 3D structure of the LEM–BAF complex: the emerin LEM-domain is blue, the BAF dimer is red (PDB code 2ODG).

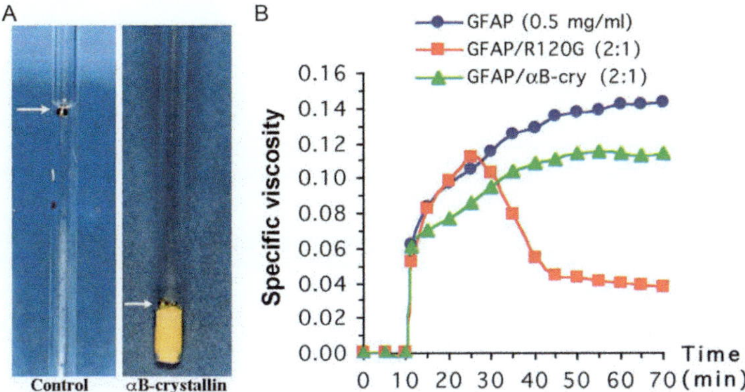

Ming-Der Perng *et al.*, Figure 3 Viscometric analyses of GFAP assembly in the absence or presence of αB-crystallin. (A) Falling ball viscosity assay was performed by assembling of GFAP (0.5 mg/ml) in a glass tube at 37 °C. Prior to this assay, a measurement of the solution viscosity was made. After assembly of GFAP alone in the control tube, a gel had formed preventing the ball from falling. The arrow indicates the ball supported at the top of the filament solution in the control tube (Control, arrow). In contrast, in the presence of αB-crystallin the ball was able to drop to the bottom of the tube (αB-crystallin, arrow), suggesting that αB-crystallin prevents GFAP filaments from forming a gel. (B) Ostwald-type viscometer assay. GFAP (0.5 mg/ml) was assembled either alone or coassembled with either WT or R120G αB-crystallin in a molar ratio of 1:2 at 37 °C. Assembly was initiated at the 10th minute time point by the addition of a 20-fold concentrated binding buffer. Specific viscosity was measured 1 min after assembly and then at 5 min intervals for 60 min. The profile obtained for GFAP reflects the normal increase of specific viscosity after initiation of filament assembly. In the presence of WT αB-crystallin, GFAP exhibited a slightly decreased specific viscosity over a period of 40–45 min before reaching a plateau. In contrast, for GFAP assembled in the presence of R120G αB-crystallin, the specific viscosity increased normally only during the first 25 min. Afterward it decreased dramatically, indicative of GFAP filament aggregation.

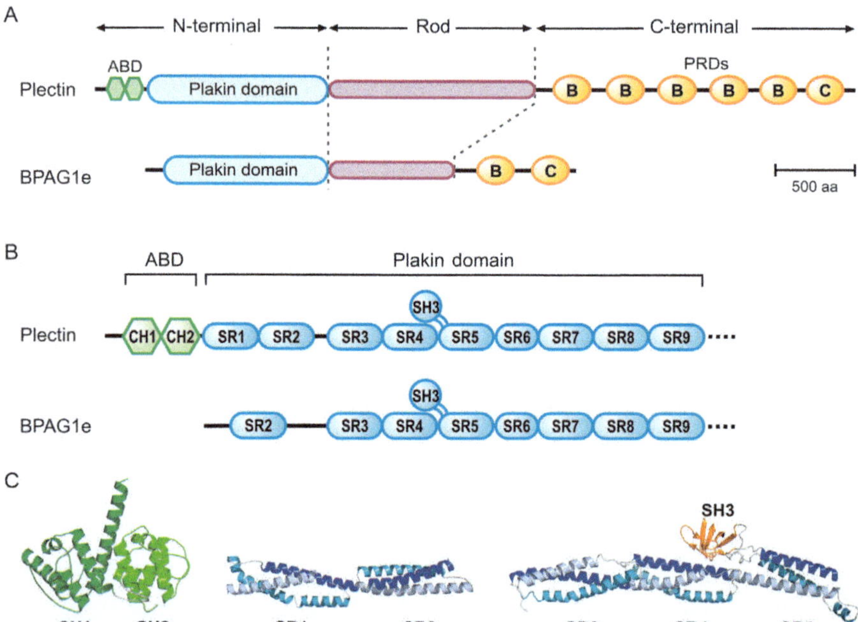

José A. Manso *et al.*, Figure 1 Structure of plectin and BPAG1e. (A) Overall domain organization. (B) Detailed domain structure of the N-terminal regions of plectin and BPAG1e. (C) Structures of the ABD (PDB entry 1MB8) and SR1–SR2 (2ODU) of plectin, and composite 3D model of the SR3–SR5 region constructed using the crystal structures of the SR3–SR4 (3PDY) and SR4–SR5–SH3 (3PE0).

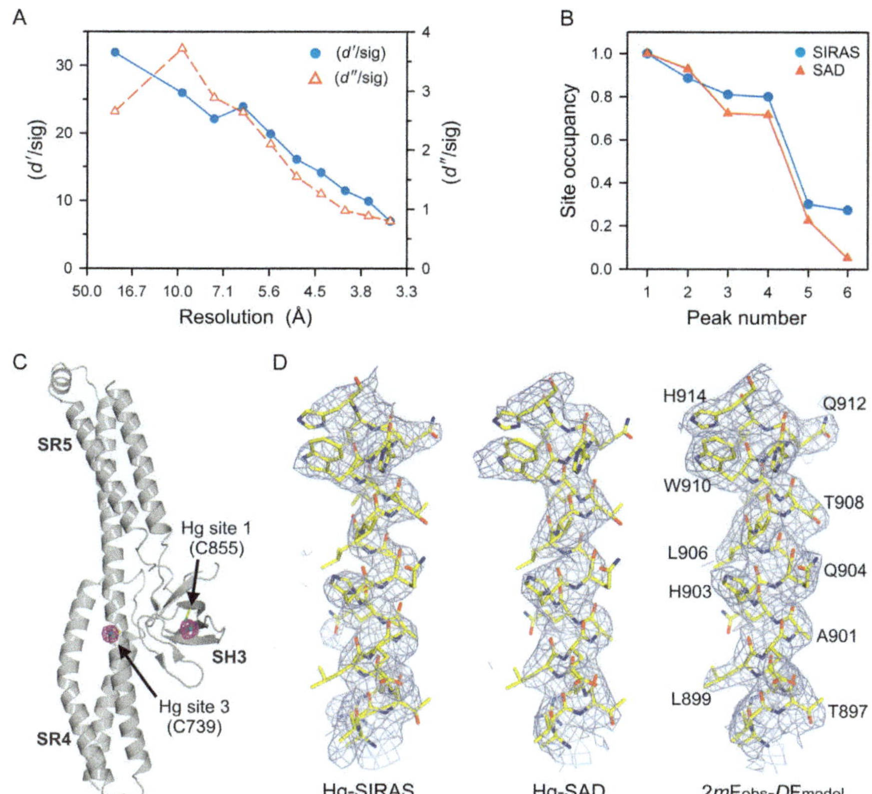

José A. Manso *et al.*, Figure 2 Phasing of the SR4–SR5–SH3 of plectin by Hg-SIRAS and Hg-SAD. (A) Statistics of the isomorphous differences between the EMTS and native datasets (circles) and anomalous signal of the EMTS dataset (triangles) calculated with SHELXC. (B) Occupancies of the Hg sites identified by SIRAS or SAD with SHELXD. The Hg-substructure was solved including data to a maximum resolution of 3.7 Å and 3.9 Å for SIRAS and SAD, respectively. (C) Structure of a plectin molecule (shown as ribbon) and position of the two Hg atoms (sites 1 and 3 in A) attached to cysteines C855 and C739, respectively. The asymmetric unit contains a second plectin molecule with two Hg attached to equivalent positions. A difference anomalous density map calculated using the EMTS-derivative dataset is shown contoured at 5σ. (D) Representative region of the electron density maps calculated after density modification with SHELXE using SIRAS (left) or SAD (center). A $2mF_{obs}\text{-}DF_{model}$ map calculated using phases of the refined structure is show for comparison (right). Maps are contoured at 1σ. The refined structure is superimposed in all maps.

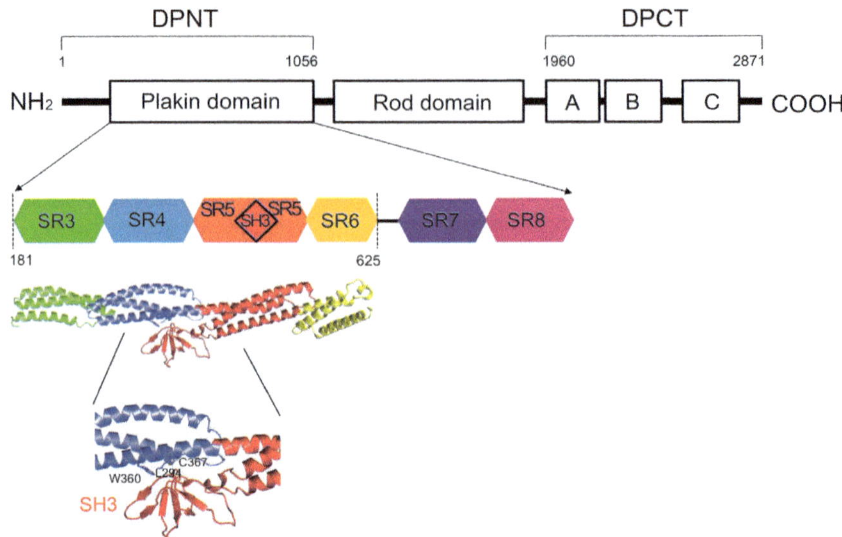

Hee-Jung Choi and William I. Weis, Figure 1 Structure of human desmoplakin I. *Top*, primary structure. The amino acid boundaries of DPNT and DPCT are indicated. Small rectangles labeled as A, B, and C in DPCT represent PRD-A, PRD-B, and PRD-C, respectively. The plakin domain in DPNT, containing six spectrin repeats (SRs) is shown as a close-up view. *Bottom*, crystal structure of DP(175–630). A close-up view of the SH3 domain and the interacting hydrophobic residues from SR4 is also shown. Top: *Adapted with permission from Choi and Weis (2011).*

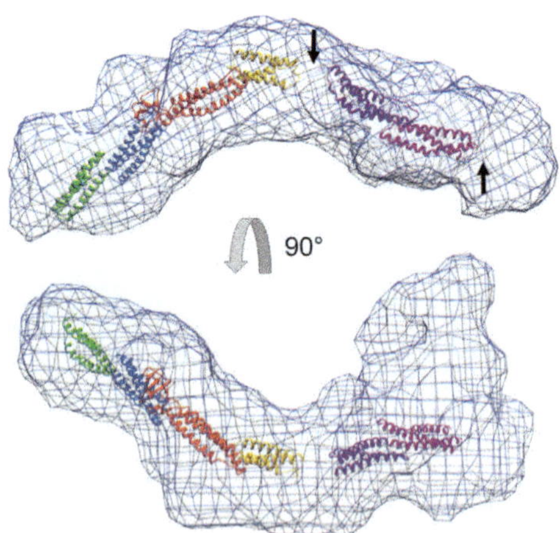

Hee-Jung Choi and William I. Weis, Figure 2 Molecular envelope of DP(180–1022) obtained by SAXS data analysis. The best fit for the structure of SR3-6 and the SR7-8 model is shown. Each SR is colored as Fig. 1. The flexible hinge region between SR6 and SR7 is not modeled and is indicated by a black arrow. The unmodeled envelope density at the right (black arrow) corresponds to non-SR sequence that links to the rod domain. *Adapted with permission from Al-Jassar et al. (2011).*

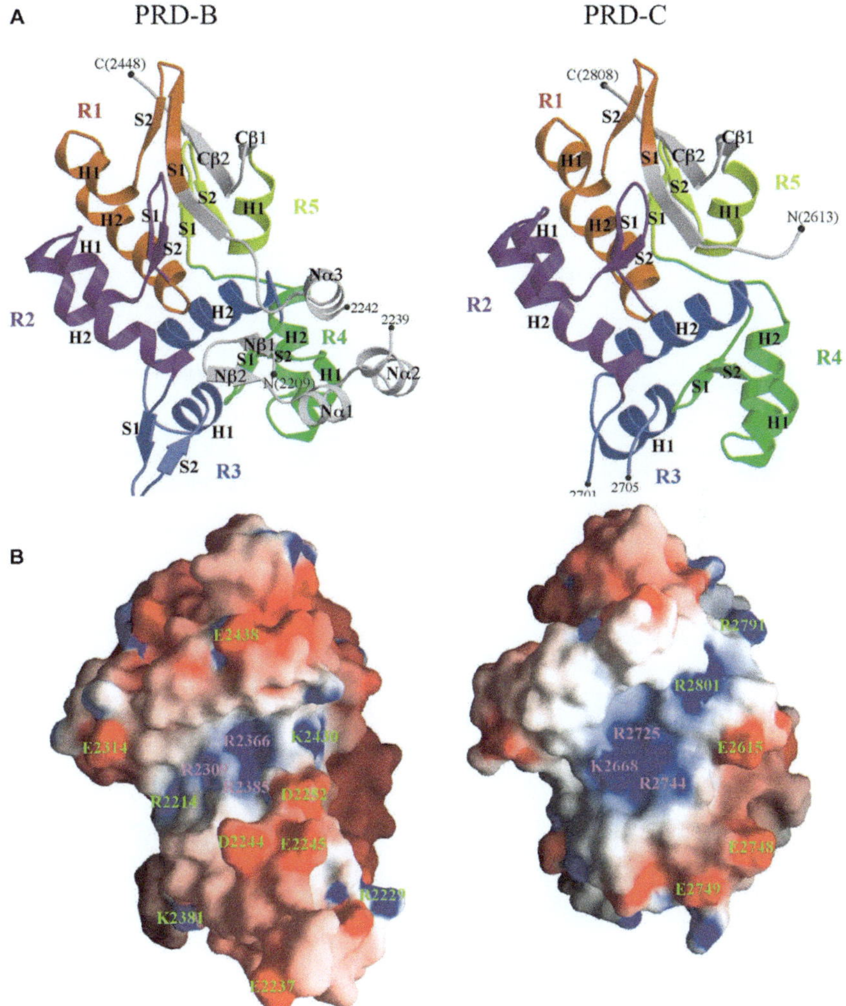

Hee-Jung Choi and William I. Weis, Figure 3 Crystal structures of PRD-B (left) and PRD-C (right). (A) Ribbon representation of the PRDs. Plakin repeats (PRs) 1–5 are colored in orange, magenta, blue, green, and yellow green. The N-terminal and C-terminal regions that flank the core PRD are shown in gray. The secondary structure elements of each PR are labeled as S1, S2, H1, and H2. Secondary structures in the N-terminal PR-like motif are labeled as Nβ1, Nβ2, Nα1, Nα2, and Nα3. The N- and C-termini, and the ends of breaks in the structure, are indicated with small black spheres. (B) Electrostatic surface representations of PRD-B and PRD-C. Negatively charged and positively charged regions are colored red and blue, respectively. Contoured at $\pm 10\ k_BT/e$. Conserved basic residues, which are labeled in yellow constitute positively charged groove in PRD. *Adapted from Choi et al. (2002).*

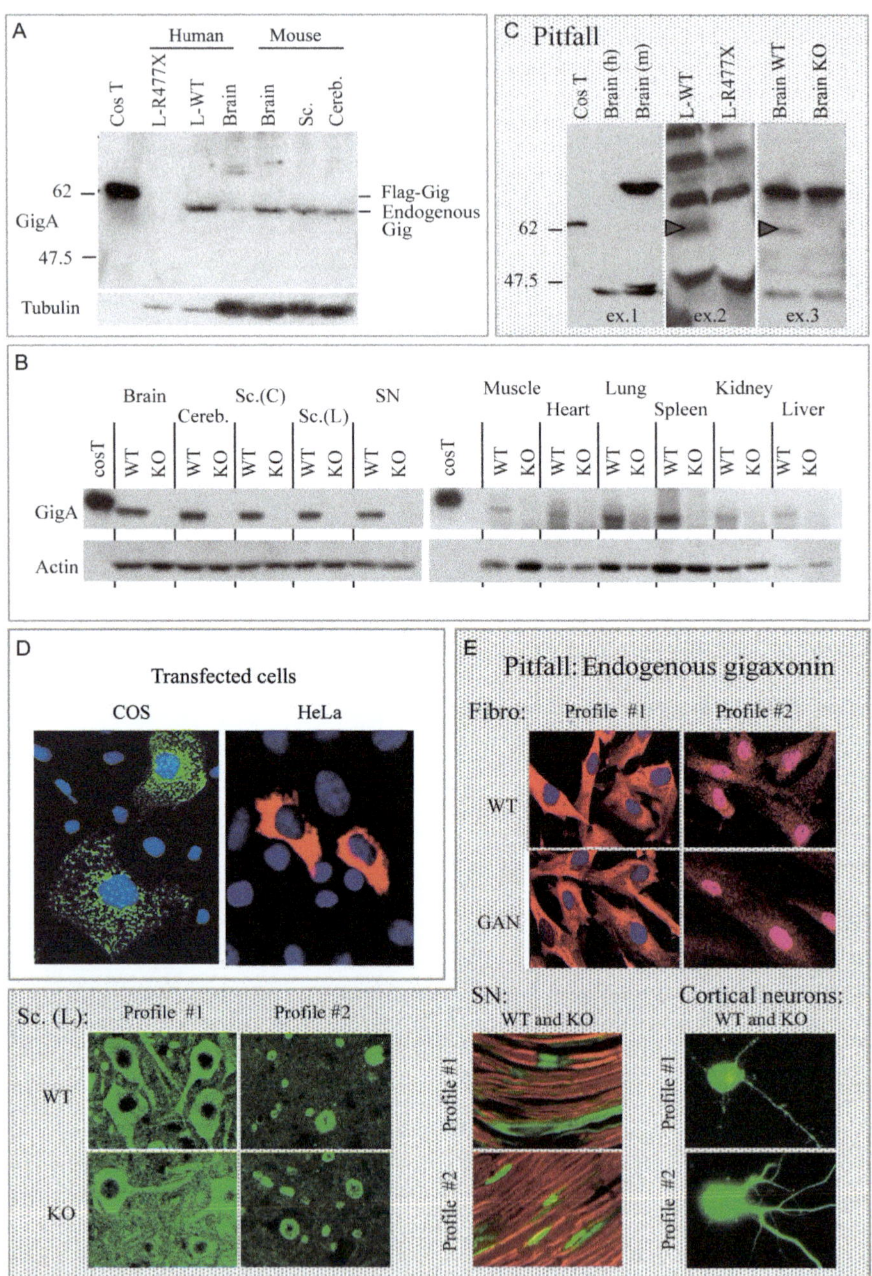

Pascale Bomont, Figure 1 See legend on opposite page.

Pascale Bomont, Figure 1 Broad and low abundance of gigaxonin. (A) The monoclonal GigA antibody detects endogenous gigaxonin at a molecular weight lower than expected (a 65 kDa band with a Flag-GAN construction transfected in Cos cells: Cos-T), in human samples ("L"-human lymphoblast cell line, "Brain") and mouse tissues ("Brain," spinal cord "Sc.," and cerebellum "Cereb."). GigA antibody is described in Cleveland et al. (2009). Specificity of detection is verified in a lymphoblast cell line derived from a patient bearing a homozygous nonsense mutation at amino acid position 477 (L-R477X). (B) Ubiquitous but preferential expression of gigaxonin in mouse neuronal tissues. Control for specificity is demonstrated by the corresponding tissues from the GAN knockout model "KO.," "Sc. (C)," "Sc. (L)," and "SN" stand for cervical, lumbar sections of the spinal cord, and sciatic nerve, respectively. (C) Examples of aspecific detection using other polyclonal antibodies (ex. 1–3 correspond, respectively, to Ding et al. (2002), Cullen et al. (2004), and Sigma #SAB4200104). "Brain (h) and (m)" correspond to human and mouse brain samples, respectively. The arrowhead points to the endogenous gigaxonin. Cellular distribution of gigaxonin. (D) A fine or large granular distribution in the cytoplasm of transfected HeLa and Cos cells defines ectopic expression of gigaxonin. (E) Absence of specific detection of the endogenous gigaxonin with available gigaxonin antibodies. Examples presented include primary fibroblasts "Fibro," lumbar sections of the spinal cord "Sc. (L)" and longitudinal sections of sciatic nerves "SN" of mouse, as well as the derived E15.5 cortical neurons. As demonstrated by the internal controls (GAN fibroblasts and GAN KO tissues/cells), the two distinct profiles revealed by the screening of all 32 gigaxonin antibodies are not specific. *Panel (A) adapted from Cleveland et al. (2009) by permission of Oxford University Press; Panel (B) adapted from Ganay, Boizot, Burrer, Chauvin, and Bomont (2011). Copyright (2011) with permission from Biomed Central.*

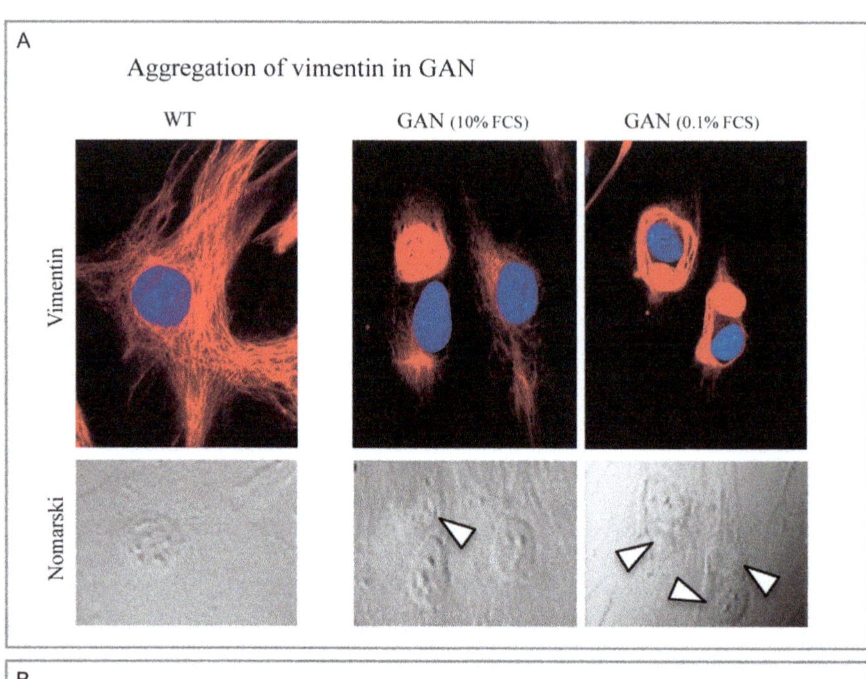

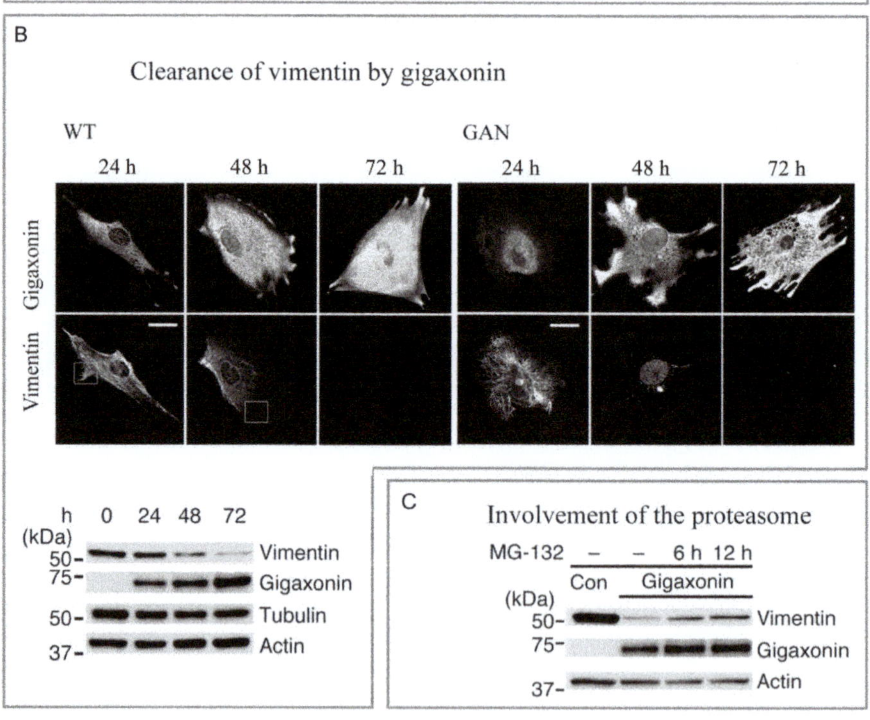

Pascale Bomont, Figure 3 See legend on opposite page.

Pascale Bomont, Figure 3 Regulation of vimentin by gigaxonin in health and disease. (A) Aggregation of vimentin in GAN fibroblasts is partial, conditional, and highly dynamic. Vimentin forms dense perinuclear ovoid bundles in GAN fibroblasts, which are visible by Nomarski contrast (arrowheads) and co-exist with a well-organized vimentin network within the same cell. The phenotype is partial as only a subset of cells contains such aggregates but depends on the conditions, as it is exacerbated upon serum deprivation (0.1% FCS). Under this condition, not only the proportion of cells which aggregate increases drastically, but the aggregates are also more pronounced and form dense perinuclear rings. (B) Clearance of vimentin upon gigaxonin over-expression. The time course of vimentin degradation in human fibroblasts upon viral-mediated delivery of Flag-tagged gigaxonin. The progressive destruction of vimentin, assessed both by Western blot on total cell extracts and immunofluorescence, begins 48 h after gigaxonin surexpression and is effective on vimentin aggregates at 72 h posttransduction in GAN fibroblasts. (C) Involvement of the proteasome in the gigaxonin-mediated clearance of vimentin. Human fibroblasts deprived of vimentin by the expression of gigaxonin for 72 h are treated with the proteasome inhibitor MG-132 for 6–12 h. A slight increase in the amount of vimentin is obtained upon treatment, indicating a partial blocking of vimentin degradation by the proteasome in primary fibroblasts. *Panel (A) adapted from Bomont and Koenig (2003) by permission of Oxford University Press; Panels (B) and (C) reproduced from Mahammad et al. (2013) with permission of J Clin Invest.*

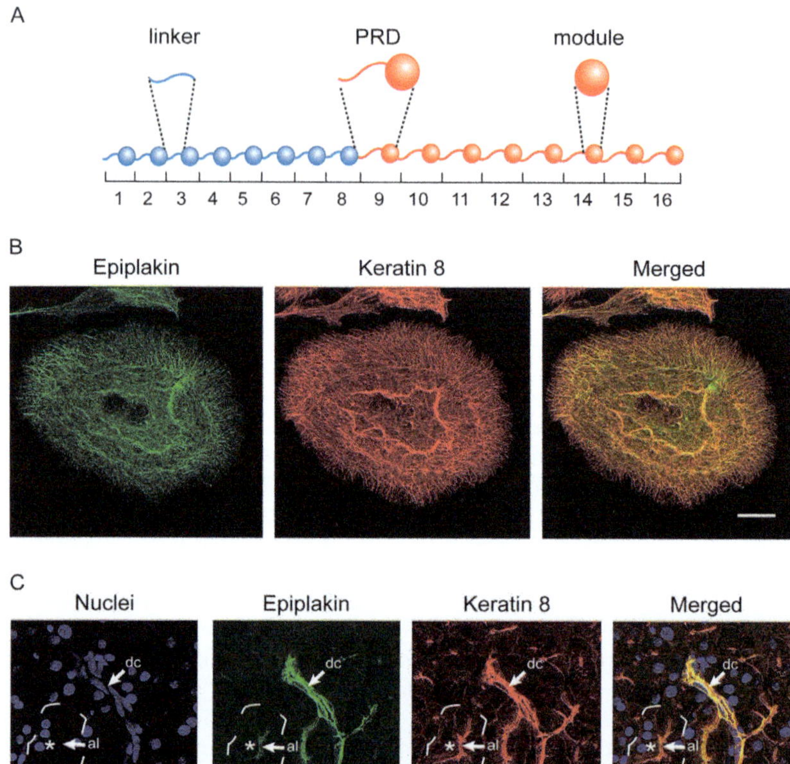

Sandra Szabo *et al.*, Figure 1 *Schematic illustration of murine epiplakin and immunofluorescence microscopy of epiplakin and keratin 8 in murine cells and tissue.* (A) Domain structure of mouse epiplakin showing 16 PRDs consisting of linkers (curved lines) and modules (circles). Modules that are virtually identical in amino acid sequence are shown in red (modules 9 to 16), those with lower homology in blue (modules 1 to 8). (B) Immunofluorescence microscopy of epiplakin and keratin 8 in primary hepatocytes. Cells were fixed with methanol and subjected to immunofluorescence microscopy using antibodies to epiplakin and K8. Epiplakin exhibits a dotted staining pattern lining keratin filaments. Scale bars, 20 μm. (C) Paraffin sections, prepared from pancreata of adult wild-type mice, were subjected to immunofluorescence microscopy using antibodies to epiplakin and K8. Epiplakin is found primarily in ductal cells (dc) and the apicolateral compartment (al) of acinar cells, where it colocalizes with K8-positive filaments. K8-positive filaments are also seen in perinuclear areas of acinar cells. Nuclear staining is shown for easier distinction of ductal and acinar cells. Whole acini are outlined by dashed lines for orientation. Asterisks indicate cytosolic/perinuclear areas. Scale bar, 10 μm. *Panel (A) illustration was originally published in* Journal of Hepatology; *Szabo et al. (2015). Panel (C) image was originally published in* PLoS One; *Wögenstein et al. (2014).*

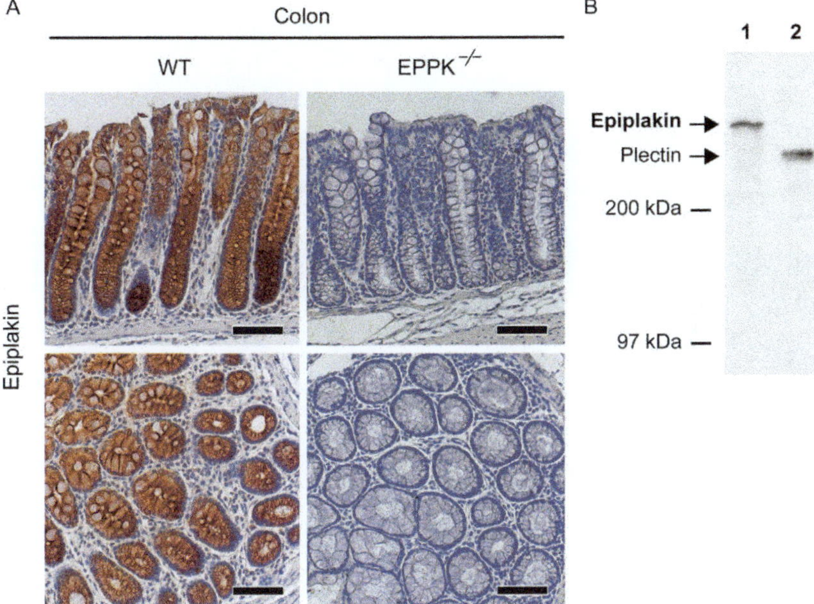

Sandra Szabo *et al.*, Figure 2 *Immunohistochemistry staining and immunoblots of epiplakin.* (A) Representative immunohistochemistry staining depicting the expression of epiplakin in murine colons from WT and epiplakin-deficient mice. Longitudinal (upper panels) and cross sections (lower panels) of colon crypts are shown. Epiplakin is expressed in epithelial cells of WT colon whereas the protein is completely absent in epiplakin-deficient colon, demonstrating the specificity of the antibody used. Scale bars, 50 μm. (B) Immunoblotting of epiplakin and plectin. A protein extract from newborn mouse was separated via SDS-PAGE, transferred to a nitrocellulose membrane and analyzed using antibodies to epiplakin (*lane 1*) and to plectin (*lane 2*). The positions of additional size markers are indicated. A 3.75% polyacrylamide gel was used. *Panel (B) image was originally published in* The Journal of Biological Chemistry*; Spazierer et al. (2003). © the American Society for Biochemistry and Molecular Biology.*

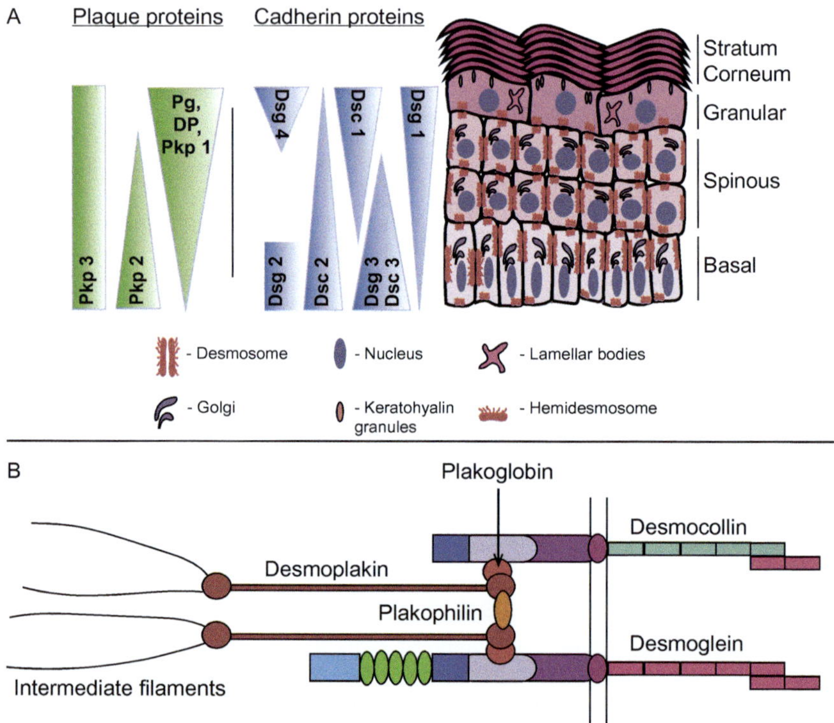

Christopher Arnette *et al.*, Figure 1 Organization and expression of desmosomal components within the skin. (A) There are four layers within the stratified epidermis: basal, spinous, granular, and cornified. Each layer contains a complement of differentiation-specific junctional proteins critical to support epidermal cytoarchitecture and drive tissue morphogenesis. (B) Desmosomes are specialized, calcium-dependent adhesive structures that are composed of three families of proteins: desmosomal cadherins, armadillo proteins, and plakin proteins. These structures link to the intermediate filament cytoskeleton to maintain tissue integrity.

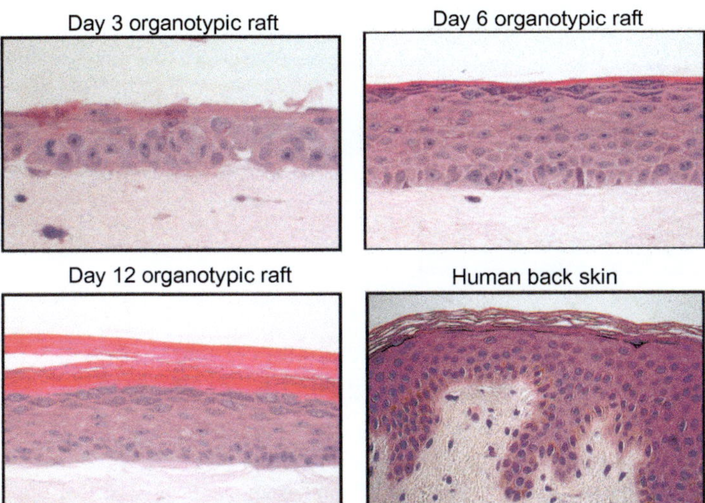

Christopher Arnette *et al.*, Figure 2 Comparison to human epidermis and time course of 3D raft development. Keratinocytes were seeded on collagen plugs and were grown submerged conditions for 2 days. The cultures were then lifted to the air–liquid interface. Organotypic rafts were harvested 3, 6, or 12 days after being lifted to air. *Representative images of keratinocyte stratification in 3D raft cultures (provided by Paul Hoover of the Northwestern University Skin Disease Research Core).*

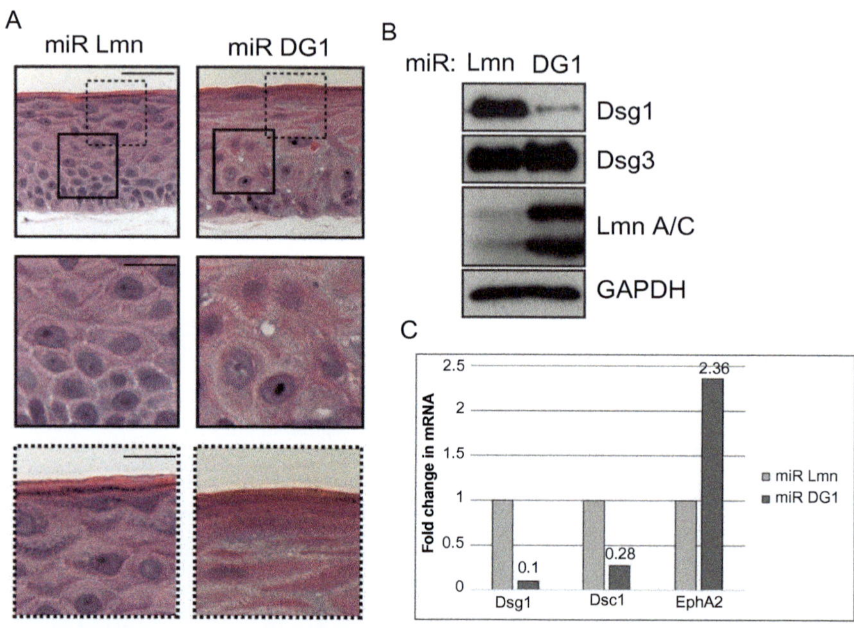

Christopher Arnette *et al.*, Figure 4 Manipulation and analysis of Dsg1 in 3D organotypic raft cultures. (A) H&E-stained sections of 6-day-old rafts expressing either miR Lmn or miR DG1. Silencing of DG1 results in altered suprabasal morphology (insets with continuous lines) and impaired differentiation (insets with dashed lines) compared to control. (B) Western blot analysis of desmosomal proteins, desmoglein 1 and 3, lamin A/C (Lmn A/C), and GAPDH (glyceraldehyde-3-phosphate dehydrogenase) in 6-day-old 3D raft cultures. Either Dsg1 or Lamin expression was silenced in raft cultures using retroviruses engineered to express microRNA (miRNA)-like sequences specific for each protein. KD reveals ~90% reduction of protein levels under each condition. (C) Real-time PCR analysis show decreased levels of desmocollin 1 mRNA levels following Dsg1 knockdown in 3D raft cultures, while a gene target of EGFR, EphA2, mRNA levels are increased. Note: *Figure originally published in Getsios et al. (2009) (© 2009* The Journal of Cell Biology. *doi: 10.1083/jcb.200809044); used with permission.*

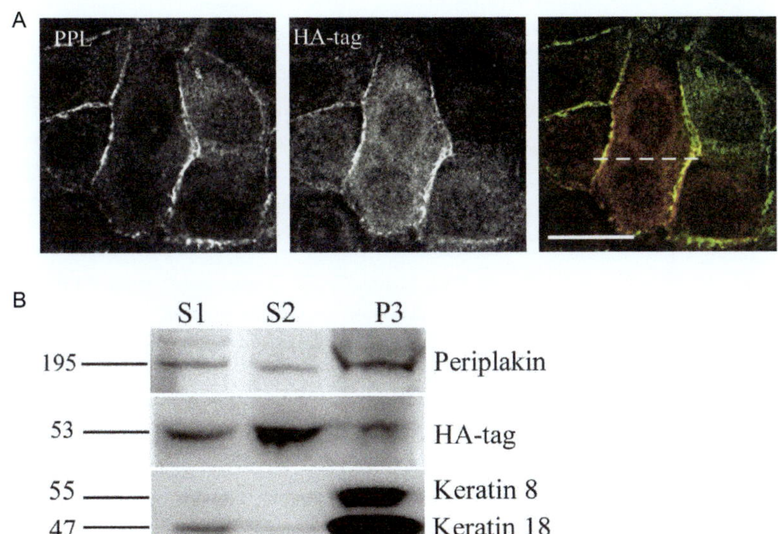

Veronika Boczonadi and Arto Määttä, Figure 2 (A) Stably transfected MCF-7 cells expressing p1/2-HA periplakin construct. The N-terminal construct shows targeting to cell membrane (anti-HA tag immunofluorescence mAb Y-11, 1:200 dilution) and colocalization with endogenous periplakin (PPL panel, immunofluorescence with TD2, 1:100 dilution). Merged panels at right (TD2 green (labeled PPL), HA-tag red (note localization at cell borders)). Scale bar = 20 μm. (B) Subcellular fractionation of stably transfected MCF-7 cells expressing periplakin N-terminus. S1 is the saponin-soluble fraction and S2 denotes the Triton X-100-soluble fraction. Immunoblots of both endogenous periplakin (using TD2 antibody) and transfected 1/2 N-terminus (HA-tag) are shown. Keratin 8/18 immunoblot is a marker for the Triton-insoluble (P3) pellet fraction. *Reproduced with permission from Boczonadi et al. (2007).*

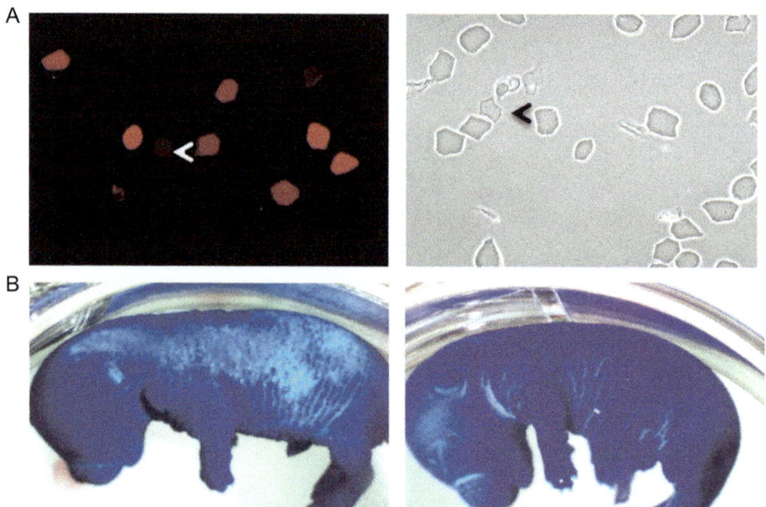

Veronika Boczonadi and Arto Määttä, Figure 3 (A) Corneocytes isolated from mouse skin and stained with nile red (left-hand panel) or unstained and visualized by phase-contrast light microscopy (right-hand panel). Intensely stained hexagonal envelopes are mature whereas faintly stained, irregularly shaped envelopes (arrowheads) represent earlier stages of the terminal differentiation. (B) Toluidine blue assay for barrier development. E16.5 mouse embryos of heterozygous (left) or envoplakin −/− homozygous (right).

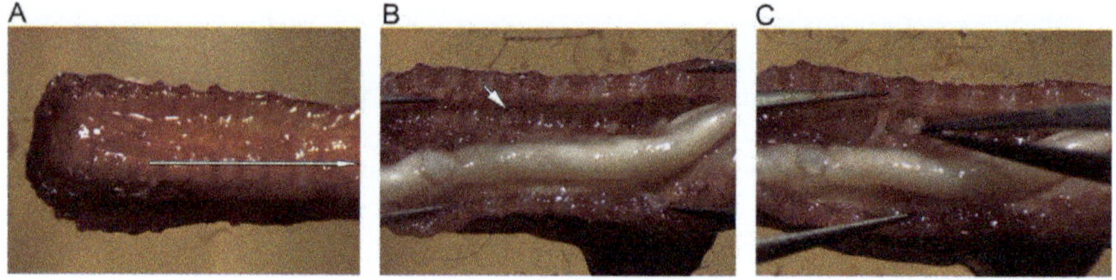

Anisha Lynch-Godrei and Rashmi Kothary, Figure 3 Dissection of dorsal root ganglia (DRG) in mice. (A) Spinal column, arrow indicates the location (midline) and direction (rostral to caudal) of incision. (B) Opened spinal column with the spinal cord pushed to the right side (bottom of the image) to expose the DRG of left side. Arrow points to a single DRG. (C) Removal of a single dorsal root ganglion. To preserve the shape of the DRG and keep cells intact, remove the DRG from behind by holding onto the roots as shown.

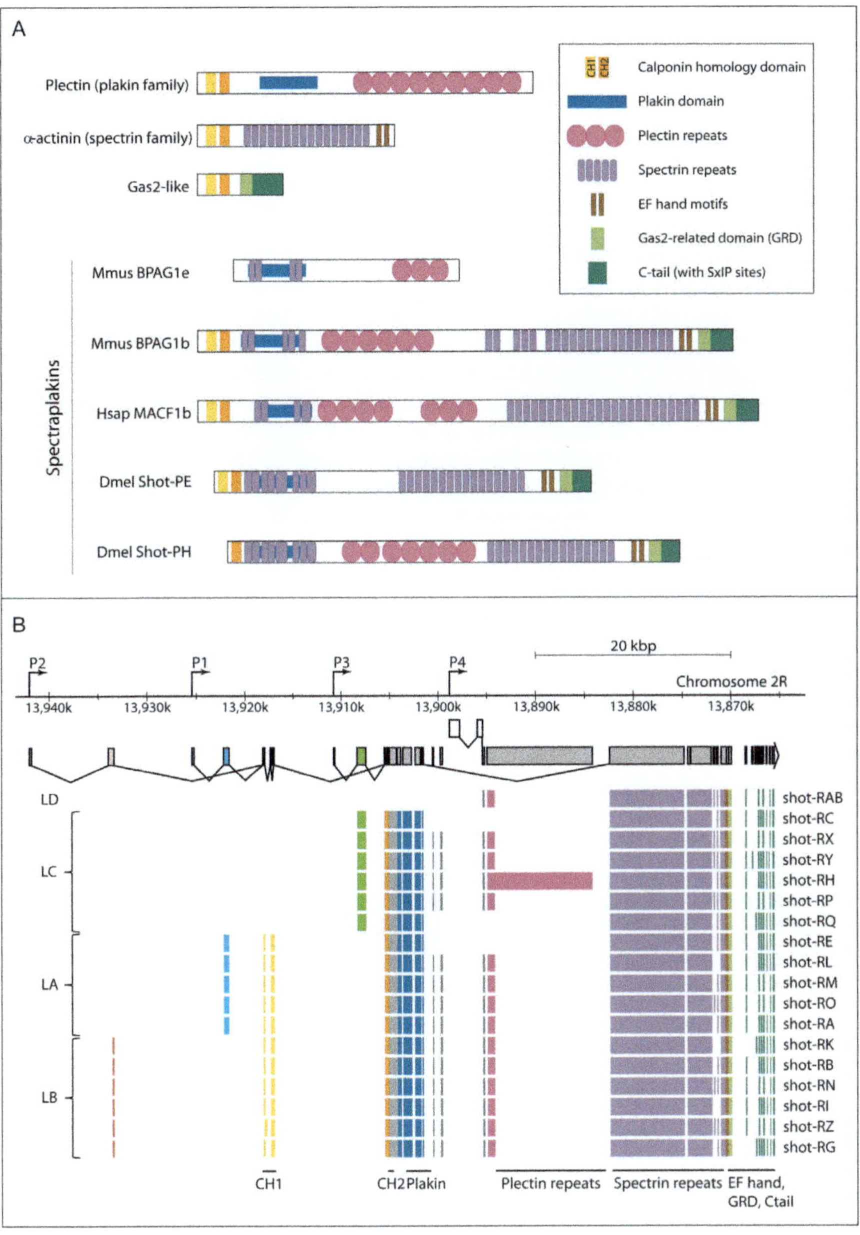

Ines Hahn *et al.*, Figure 1 See legend on next page.

Ines Hahn *et al.*, Figure 1 Family associations, homologues, functional domains, and isoforms of Shot. (A) Images show three typical representatives of the plakin, spectrin, and Gas2-like protein families, compared to specific isoforms of spectraplakins. Here shown are the shortest e- and longest b-isoform of mouse BPAG1 (Young, Pool, & Kothary, 2003), the longest b-isoform of human MACF1, and the long Shot-RE and Shot-RH isoforms (compare B). The key (emboxed) shows the different functional domains: The N-terminus of Dystonin-a2 promotes localization to the nuclear envelope (Young & Kothary, 2007; Young et al., 2003). Two calponin homology domains (CH1 and CH2) form a classical actin-binding domain (ABD) essential for neuronal morphogenesis (Alves-Silva et al., 2012; Jefferson, Leung, & Liem, 2004; Kodama, Lechler, & Fuchs, 2004; Lee & Kolodziej, 2002b; Leung, Sun, Zheng, Knowles, & Liem, 1999). The plakin-homology domain (six to nine spectrin-like repeats and an SH3 domain embedded in the 5th repeat) links to transmembrane collagens and β4 integrin in BPAG1e (Aumailley, Has, Tunggal, & Bruckner-Tuderman, 2006; Koster, Geerts, Favre, Borradori, & Sonnenberg, 2003) and mediates compartmentalization of transmembrane proteins in fly neurons (Bottenberg et al., 2009). The spectrin/dystrophin repeat rod domain likely acts as spacer or in dimerization (Bottenberg et al., 2009; Leung et al., 1999; Machnicka et al., 2014; Röper, Gregory, & Brown, 2002). Plectin-repeat domains (PRD) typically bind intermediate filaments (Aumailley et al., 2006; Jefferson et al., 2004). The two helix–loop–helix EF-hand motifs bind calcium to regulate MT association (Kapur et al., 2012), the putative translational regulator Krasavietz/eIF5C (Kra) important for actin regulation during axonal pathfinding (Lee et al., 2007; Sánchez-Soriano et al., 2009), and intramolecularly to the CH domains to autoinhibit Shot functions (Applewhite, Grode, Duncan, & Rogers, 2013). The Gas2-related domain (GRD) binds MTs and protects them against destabilization. The flexible, positively charged Ctail region associates with MTs and contains one or two SxIP sites required for binding to end-binding proteins (EB1) which is essential for axon growth (Alves-Silva et al., 2012; Applewhite et al., 2010; Honnappa et al., 2009; Sun, Leung, & Liem, 2001). The C-terminus can be phosphorylated to regulate MT association (Wu et al., 2011) and is the region specifically deleted in human BPAG1/dystonin mutations linked to neurodegeneration (Edvardson et al., 2012). (B) According to FlyBase, the *shot* locus spans ~78 kb (genomic position 13,864,237–13,942,110) in cytogenetic map position 50C6–11 of chromosome 2R. The different isoforms Shot-R/P A–Z (R refers to transcript and P to protein isoforms) are generated from 4 transcriptional start sites (P1–4) giving rise to 4 N-terminal isoforms (termed Shot-LA-D as indicated on the left) and from the differential splicing of 39 different coding exons and 4 noncoding exons. *Modified from Röper and Brown (2003), Röper et al. (2002), and FlyBase (McQuilton, St Pierre, & Thurmond, 2012).*

UAS construct	Length [deleted region]	Isoform	Scheme	References
UAS-Shot-L(A)-GFP	5201 aa	shot-RE		1
UAS-Shot-L(A)-N′6myc	5201 aa	shot-RE		2
UAS-Shot-L(A)-N′6myc-C′GFP	5201 aa	shot-RE		2
UAS-Shot-L(C)-GFP	5160 aa	shot-RC		1
UAS-Shot-L(B)-GFP	5390 aa	shot-RB		1
UAS-Shot-L(A)-ΔPlakin-GFP	4364 aa [488–1325]	shot-RE		2
UAS-Shot-L(A)-Δrod1-GFP	1805 aa [1204–4600]	shot-RE		1
UAS-Shot-L(A)-Δrod2-GFP	4560 aa [1023–1664]	shot-RE		1
UAS-Shot-L(A)-Δrod3-GFP	4191aa [3589–4599]	shot-RE		1
UAS-Shot-L(A)-ΔEFhand-GFP	5063 aa [4711–4849]	shot-RE		1
UAS-Shot-L(A)-ΔGRD-GFP	5155 aa [4859–4905]	shot-RE		1
UAS-Shot-L(A)-ΔEB1-GFP	5165 aa [5154–5191]	shot-RE		2
UAS-Shot-L(C)-ΔGRD-GFP	4859–4905 aa of RE	shot-RC		2
UAS-Shot-RE-3MTLS*-GFP	5201 aa [4966–9, 5077–80, and 5141–4 replaced by AAAA]	shot-RE		4
UAS-Shot-RE-ΔCtail-GFP	4918 aa [4919–5201]	shot-RE		4
UAS-APT	1–1202 aa	shot-RE		5
UAS-PT	379–1202 aa	shot-RE		5
UAS-EF-GRD-Ctail-GFP	4780–5201 aa	shot-RE		5
UAS-GRD-GFP	4841–4923 aa	shot-RE		1
UAS-EB1-2HA	5140–5383 aa	shot-RE		2
UAS-Ctail	4918–5201 aa	shot-RE		4
UAS-Ctail-3MTLS*-GFP	4966-9, 5077–80, and 5141–4 replaced by AAAA	shot-RE		4

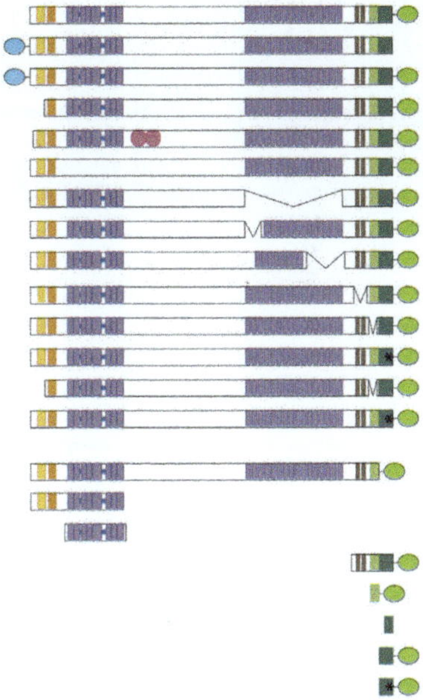

Ines Hahn *et al.*, Figure 2 Published transgenic fly lines carrying UAS-linked *shot* constructs. Published name of the construct (1st column), length in amino acids (2nd column), the respective mother construct (3rd column), and schematic representations of the constructs (4th column). The domain key is given in Fig. 1; asterisks indicate mutated sequences; blue circles, c-myc; green/red circles, GFP/RFP. References: (1) Lee and Kolodziej (2002a), (2) Bottenberg (2006), (3) Bottenberg et al. (2009), (4) Alves-Silva et al. (2012), (5) Subramanian et al. (2003).

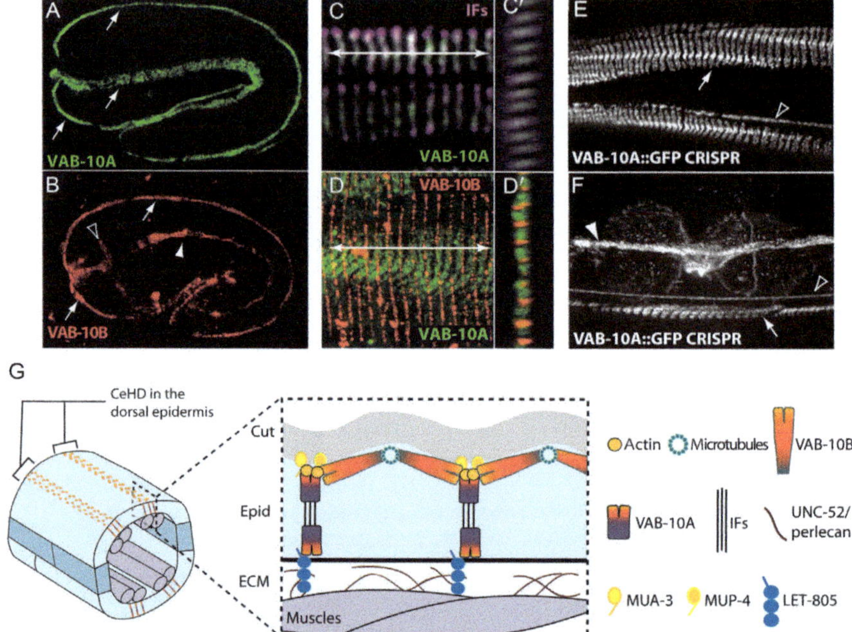

Christelle Gally *et al.*, Figure 2 Subcellular localizations of VAB-10A and VAB-10B. (A) VAB-10A (MH5) and (B) VAB-10B (K22) immunostainings on wild-type elongating embryos. Arrows, visible hemidesmosomes; arrowhead, apical side of the intestine; open arrowhead, nerve ring. Anterior is to the left and dorsal is to the top. (C) Coimmunostaining of VAB-10A (4F2) and IFs (MH4) on L1 larvae showing that they perfectly colocalize. (D) Coimmunostaining of VAB-10A (MH5) and VAB-10B (K22) on L1 larvae showing that both proteins are intercalated. Panels (C′ and D′) show an optical section through the apico–basal axis along the area marked with a double arrow in C and D, respectively. (E and F) Confocal images of the VAB-10A::GFP CRISPR knock-in in adults. Hemidesmosomes (arrow) and mechanosensory neurons (open arrowhead) are clearly visible (E and F) as well as the utse (arrowhead in F). (G) Schematic representation of CeHD during embryonic elongation (left) and model for the distributions of VAB-10 isoforms with respect to the different cytoskeletal elements in the epidermis (right). Note that the lateral epidermal cells (dark blue in the left panel) are not connected to muscles and therefore do not present CeHDs. The VAB-10B localization is speculative and deduced from the respective distribution of VAB-10A, VAB-10B, actin filaments, microtubules, and intermediate filaments.

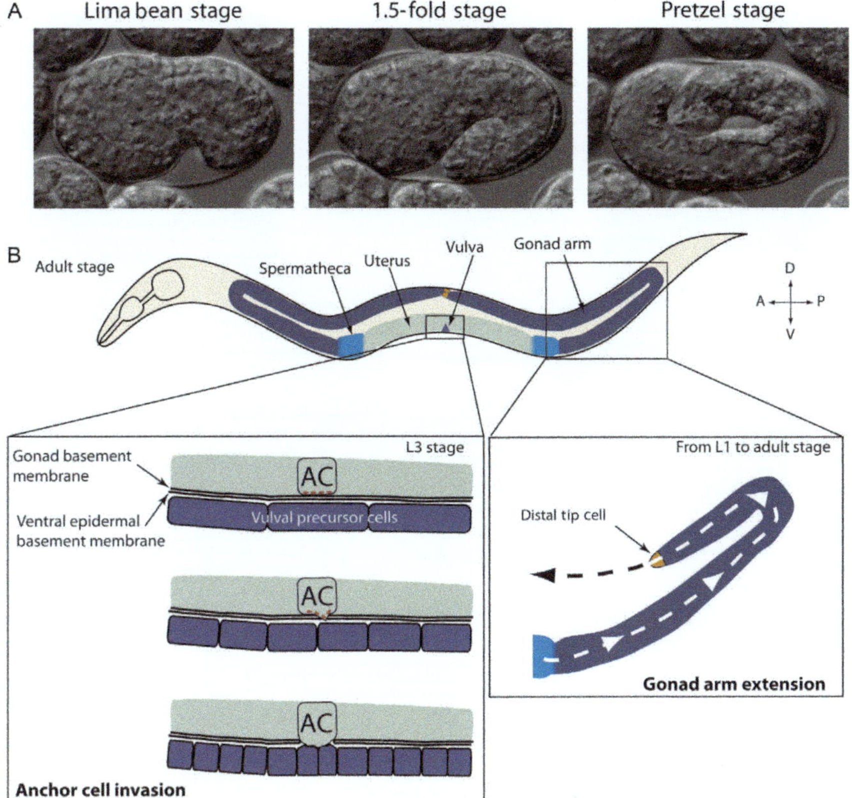

Christelle Gally *et al.*, Figure 3 Cellular processes affected by *vab-10* mutants. (A) Embryonic elongation is a process that transforms a ball of cells into worm-shaped larvae ready to hatch. Embryonic elongation stages are described based on embryo shape from lima bean (1.2-fold stage, where the embryo length is 1.2 × that of the egg-shell) to pretzel (or 4-fold stage, the final length of the embryo). (B) Simplified adult representation showing the reproductive system affected in *vab-10* mutants (upper panel). During the third larval stage, the anchor cell (AC) breaches the two layers of the basement membranes separating the uterine from the vulval precursors (left panel). This process requires the presence of the VAB-10A protein at the basal side of the AC (red dashed line). The AC invasion subsequently allows the formation of a passageway for egg laying between the ventral uterine tissue and the vulva, after the fusion of the AC with the uterine-seam cell (utse). From the L1 to the adult stages, the left and right arms of the somatic gonad extend laterally and reflex after migrating dorsally to form a U-shaped extension (right panel). This process is mainly driven by one cell at the tip of each arm, the distal tip cell (DTC). The DTCs undergo a rearrangement of their microtubule network that is dependent of the VAB-10B gonad isoform when they undergo dorsal migration.

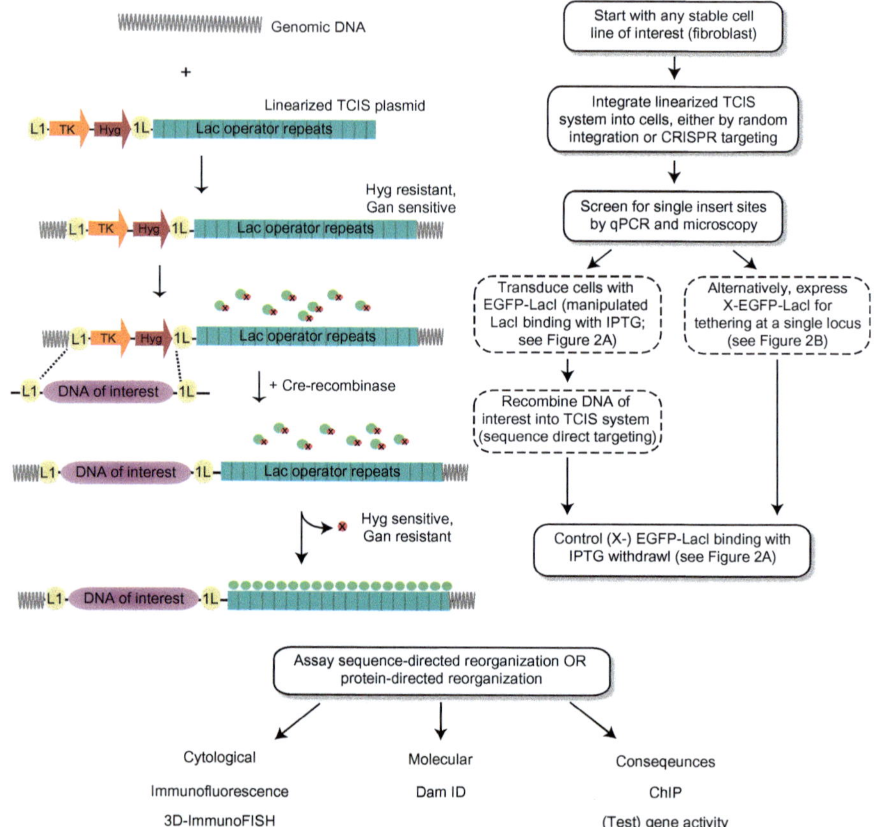

Jennifer C. Harr and Karen L. Reddy, Figure 1 The TCIS system and a flow chart showing the main steps for creating TCIS cell lines and integrating a DNA fragment of interest. The TCIS system is designed to allow for site-specific integration of any "DNA of interest" into the genome. This is accomplished by integrating the TCIS construct into the genome of a cell line of choice (e.g., murine fibroblasts). The linearized TCIS construct (derived from two previous technologies, the *lacO*/LacI system and recombination-mediated cassette exchange [RMCE]) is integrated into the cell line. The TCIS construct includes a thymidine kinase (TK)/hygromycin gene, and successful integration renders cells hygromycin-resistant. Isolated individual clones are screened by qPCR and microscopy (visualized via EGFP-LacI protein accumulation at *lacO* arrays in the TCIS site) to identify those with single integrations. These cell lines can be used to study recruitment of specific proteins to the *lacO* array (protein of interest fused to EGFP-LacI) or to study the effects of specific DNA fragments inserted into the TCIS site. Binding of EGFP-LacI proteins is controlled by adding or removing IPTG. Without IPTG, EGFP-LacI proteins will bind the *lacO* arrays in the TCIS system.

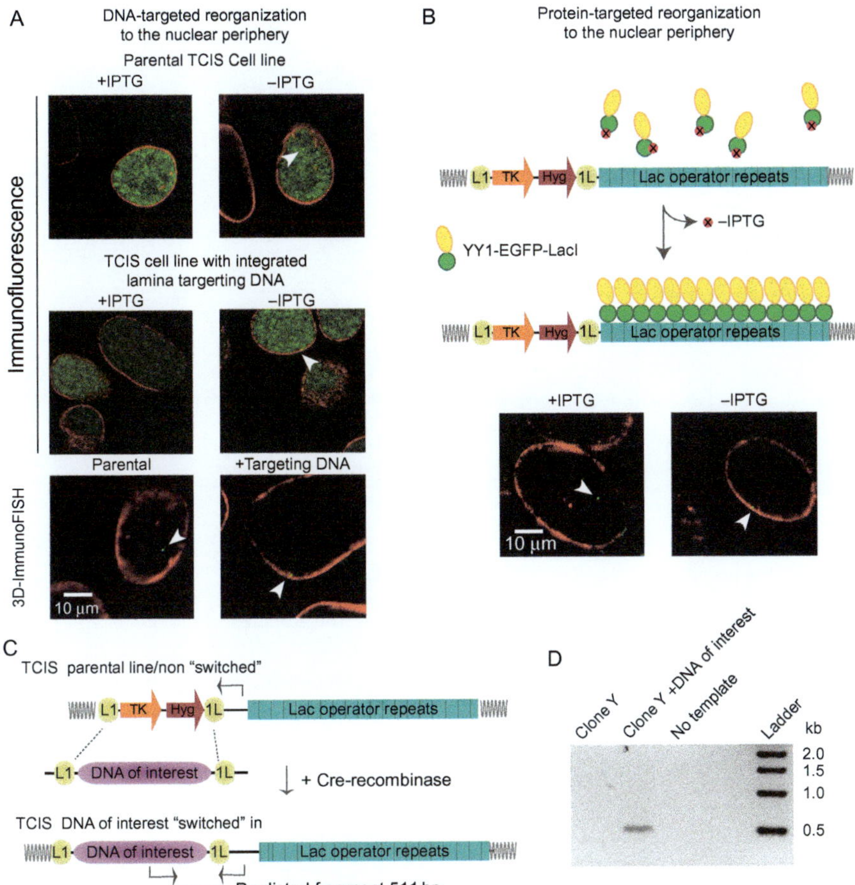

Jennifer C. Harr and Karen L. Reddy, Figure 2 TCIS system and recruitment. (A) The TCIS system has been used to investigate the mechanisms by which chromatin regions relocalize to the nuclear periphery. Analysis by different methods—live cell imaging, immunofluorescence (IF), and 3D-immunoFISH—yields similar results. Interaction with a subcompartment of interest (e.g., the nuclear periphery) is determined by overlap of IF signals (e.g., EGFP-LacI and anti-Lamin B1). (B) The TCIS system is used to study protein-targeted reorganization of chromatin to a subcompartment of interest (e.g., the nuclear periphery) by tethering a protein of interest (POI) to the *lacO* arrays through EGFP-LacI (POI-EGFP-LacI). Top: schematic showing that IPTG inhibition of LacI binding to *lacO* is reversed by removal of IPTG. Bottom, YY1-EGFP-LacI tethering to the TCIS locus causes this locus to target the nuclear periphery. (C) The TCIS system contains a RMCE region that harbors a thymidine kinase hygromycin gene. This allows for negative selection against cells that did not recombine the DNA of interest. PCR can be used to verify the integration of your DNA of interest, using primers that flank the integrated DNA. Successful integration yields a gancyclovir-resistant culture and a PCR band of predicted size (D).

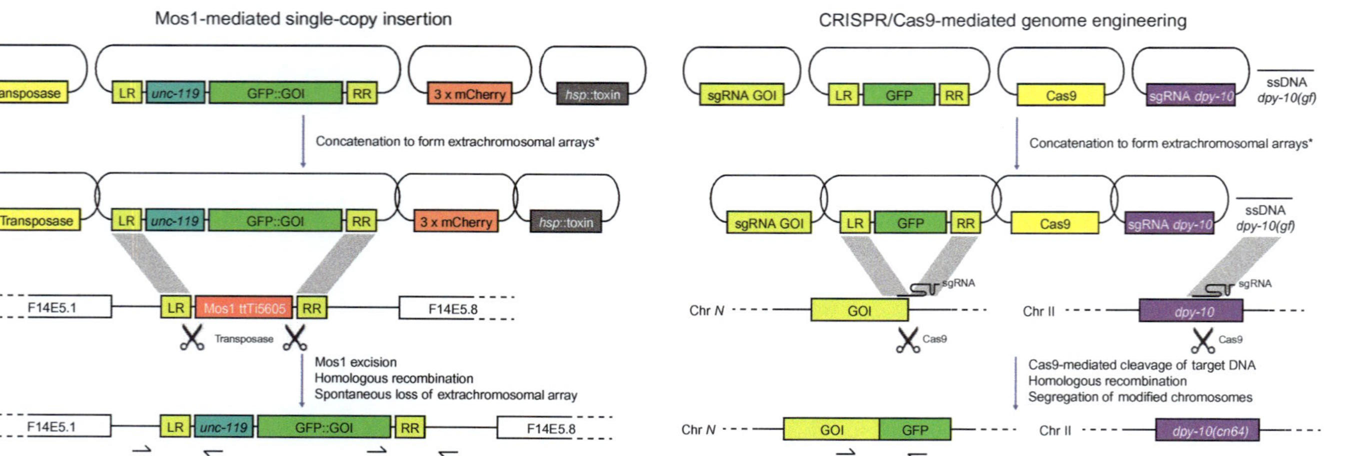

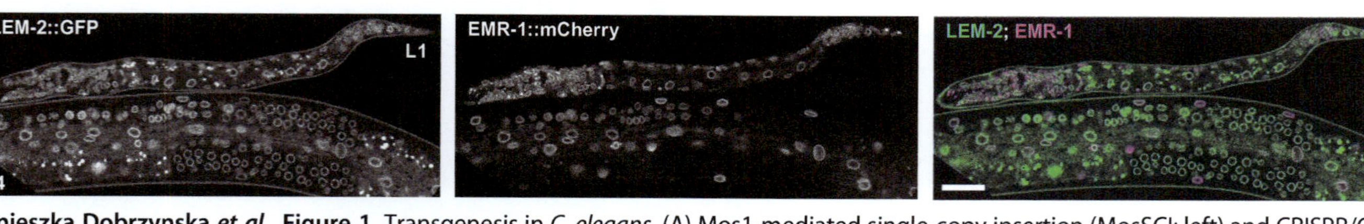

Agnieszka Dobrzynska *et al.*, Figure 1 Transgenesis in *C. elegans*. (A) Mos1-mediated single-copy insertion (MosSCI; left) and CRISPR/Cas9-mediated genome engineering (right) are based on site-directed generation of double-stranded DNA breaks in the genome followed by repair via template-mediated homologous recombination. MosSCI is typically used to insert entire transgenes (e.g., gene of interest [GOI] fused to GFP) including transgenesis markers (e.g., *unc-119*), whereas CRISPR/Cas9 is mainly used to introduce protein tags into endogenous loci. However, both methods are versatile and offer many other options. The genomic position of an Mos1 transposon in the host strain determines the integration site in MosSCI, whereas sgRNA molecules are designed to guide the Cas9 nuclease to a genomic target site. Co-conversion to generate an easy tractable allele (e.g., *dpy-10*(*cn64*)) may increase CRISPR/Cas9-mediated transgenesis. LR and RR denote homologous left and right recombination sequences, respectively. Asterisks indicate that genome modification may take place in the germ line of the injected animals or their progeny, in which case the injected DNA molecules presumably form extrachromosomal arrays. See text for details. (B) Confocal micrographs of an L1 larva and central part of an L4 larva expressing EMR-1::mCherry and LEM-2::GFP from single-copy transgenes (Morales-Martínez

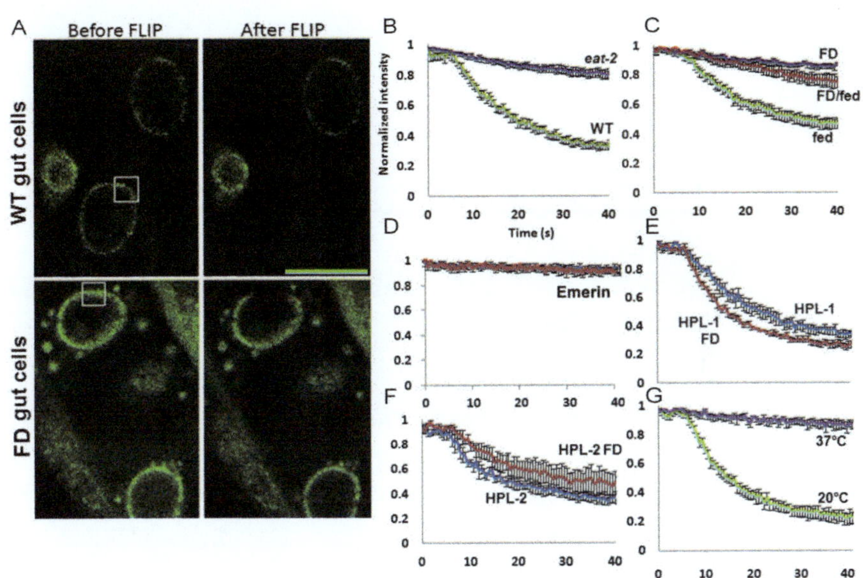

Agnieszka Dobrzynska *et al.*, Figure 2 FLIP analysis of intestine cells in L1 larvae reveals reduced GFP::BAF-1 mobility in response to dietary restriction, food deprivation, or 1 h heat shock. (A) FLIP analysis of GFP::BAF-1 in intestine cells of L1 larvae in wild-type (upper panel) or *eat-2* animals (lower panel). Scale bar, 5 μm. (B) Mobility plot of the relative intensity of the GFP::BAF-1 in wild-type (green line) and *eat-2* animals (purple). (C) Mobility plot of the relative intensity of the GFP::BAF-1 in *C. elegans* L1 larvae that were either well fed (blue line), animals that were food deprived (FD) overnight (red line), or following 2 h recovery from FD (green line). FLIP analysis of emerin (D), HPL-1 (E), and HPL-2 (F) fused to GFP did not reveal any difference in mobility after food deprivation. (G) Mobility plot of the relative intensity of GFP::BAF-1 in wild-type animals with time following 1 h heat shock at 37 °C. Error bars indicate SEM. For each experiment in (B), $n = 7$; in (C–G), $n = 6$. *X*-axis: time. *Y*-axis: normalized fluorescence intensity. Bleaching area: 2 μm^2. *Figure taken from Bar et al. (2014).*

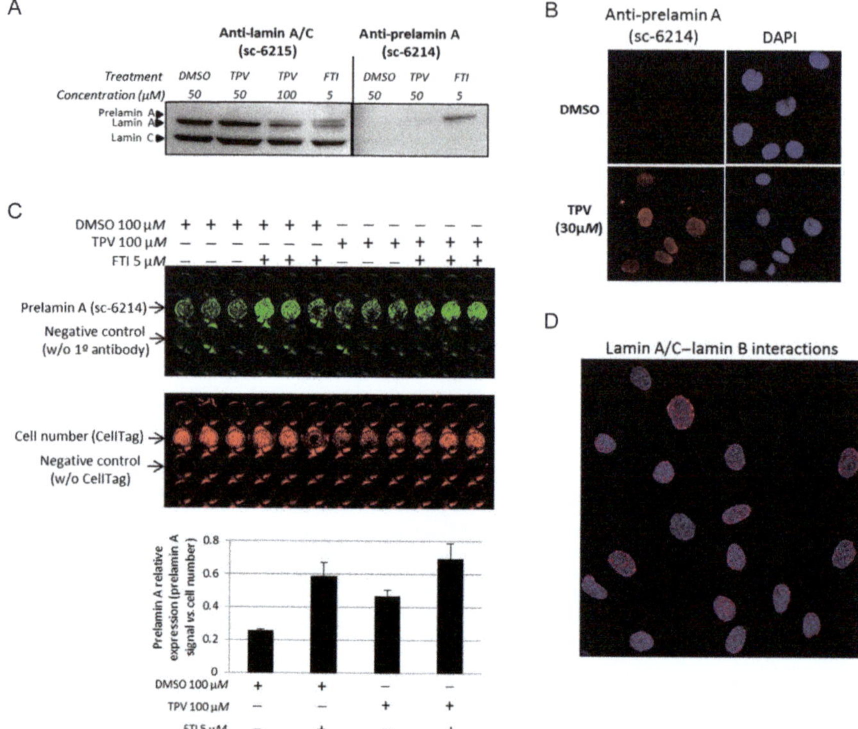

Arantza Infante and Clara I. Rodríguez, Figure 2 (A) Western blot showing prelamin A accumulation triggered by high concentrations of TPV, probed with antibodies that recognize either all A-type lamins (sc-6215), or only prelamin A (sc-6214). Negative control cells were treated with vehicle alone (DMSO). As the positive control for prelamin A accumulation, hMSCs were treated with a farnesyl transferase inhibitor (FTI). (B) Confocal immunofluorescence images showing prelamin A accumulation in hMSCs (red) treated with a low concentration of TPV. Cells were stained with the prelamin A-specific antibody (sc-6214). Nuclei are stained with DAPI (blue). Scale bar: 20 μm. (C) In-cell Western detection of prelamin A expression in treated cells in triplicate. This image shows a 96-well two-color in-cell Western with the 800 and 700 nm channels detecting prelamin A (sc-6214) and total cell numbers, respectively. Results are quantified in the graph below (mean ± standard deviation of triplicates). (D) PLA detection of lamin A/C interactions with lamin B in hMSCs. Each red dot reveals a physical protein–protein interaction. Nuclei are stained with DAPI (blue).

CPI Antony Rowe
Chippenham, UK
2016-01-20 17:21